International Yearbook of Soil Law and Policy

Volume 2020

The *International Yearbook of Soil Law and Policy* is a book series that discusses the central questions of law and policy with regard to the protection and sustainable management of soil and land. The Yearbook series analyzes developments in international law and new approaches at the regional level as well as in a wide range of national jurisdictions. In addition, it addresses cross-disciplinary issues concerning the protection and sustainable management of soil, including tenure rights, compliance, food security, human rights, poverty eradication and migration. Each volume contains articles and studies based on specific overarching topics and combines perspectives from both lawyers and natural scientists to ensure an interdisciplinary discourse.

The *International Yearbook of Soil Law and Policy* offers a valuable resource for lawyers, legislators, scholars and policymakers dealing with soil and land issues from a regulatory perspective. Further, it provides an essential platform for the discussion of new conceptual approaches at the international, national and regional level.

Harald Ginzky • Elizabeth Dooley •
Irene L. Heuser • Patricia Kameri-Mbote •
Robert Kibugi • Till Markus • Oliver C. Ruppel
Editors

International Yearbook of Soil Law and Policy 2020/2021

Springer

Editors
Harald Ginzky
German Environment Agency
Dessau, Germany

Elizabeth Dooley
Agriculture and Food Development Authority
Teagasc
Fermoy, Ireland

Irene L. Heuser
IUCN World Commission on
Environmental Law
Kleinmachnow, Germany

Patricia Kameri-Mbote
University of Nairobi
Nairobi, Kenya

Robert Kibugi
Faculty of Law
University of Nairobi
Nairobi, Kenya

Till Markus
UFZ
Helmholtz Centre for Environmental Research
Leipzig, Germany

Oliver C. Ruppel
Faculty of Law, Stellenbosch University,
Stellenbosch, South Africa; Research
Center for Climate Law
University of Graz
Graz, Austria

ISSN 2520-1271 ISSN 2520-128X (electronic)
International Yearbook of Soil Law and Policy
ISBN 978-3-030-96346-0 ISBN 978-3-030-96347-7 (eBook)
https://doi.org/10.1007/978-3-030-96347-7

This Springer imprint is published by the registered company Springer Nature Switzerland AG
The registered company address is: Gewerbestrasse 11, 6330 Cham, Switzerland

Foreword by Keriako Tobiko, Cabinet Secretary in the Kenyan Ministry of Environment and Forestry

Land Tenure Rights and Sustainable Soil Management in Kenya

In recent decades, issues of soil and land have been raised on international platforms. Humankind is facing challenges the world over as the global population continues to grow, with cities expanding and diets changing. Land users are increasingly struggling with soil degradation, erosion, and drought. With priority placed on food security, sustainable soil management has never been more important.

Indeed, it is not surprising therefore that the United Nations adopted 17 Sustainable Development Goals with the aim of protecting the planet and ensuring prosperity for all while leaving no one behind. 9 SDGs affect land and soil management and in extension agricultural productivity and environmental resilience and sustainability. More specifically, SDG 15 Life on Land, particularly target 15.3, focuses on land by demanding action against land degradation and efforts to achieve a land degradation-neutral world. This will impact positively on food security, climate change, and several other interconnected SDGs like No Poverty, Good Health and Well-being, Clean Water and Sanitation, Affordable and Clean Energy, and Responsible Consumption and Production.

The full potential of soil must be unlocked to support food production, store and supply clean water, maintain biodiversity, sequester carbon, and increase resilience in changing climate; this requires universal implementation of sustainable soil management. Soils are the foundation of food production and many essential ecosystem services. It has been shown that sustainable soil management is linked to land tenure rights.

Whilst the World Charter contains the key principles and guidance towards sustainable soil management, it is important that it is complemented by tools, books, writings, and journals elaborating land tenure principles and practices for incorporation into policies and decision-making especially in Africa. In Kenya for example, the economy is agriculture-based meaning that the contribution of agriculture to the GDP, employment generation, food security, and foreign exchange earnings remains unrivalled. However, for the last two or so decades the contribution of agriculture to the national GDP has continued to decline rapidly due to population explosion, rapid urbanisation, and shortage of arable land.

Like in most other sub-Saharan African countries, soil erosion and land degradation have become a major environmental concern and a formidable threat to food security and sustainable agricultural production. Livelihoods of several households have over the years been sustained by poor farming methods and pastoral systems. These are indeed linked to tenure security in the adoption of soil management practices. Decisions to invest in soil and water conservation structures will be made by farmers more secure about their land ownership and land tenure rights.

Sustainable development is specifically domiciled in Article 10 of the Kenya's 2010 Constitution. In addition to several articles on sustainable land and ecological management, one key sustainable soil management practice is mentioned in Article 69 of the CoK2010, stating that Kenya shall have a minimum tree cover of 10% of the land area in Kenya. It is worth noting that Kenya has the Agriculture (Farm Forestry) Rules, 2009. These Rules apply for the purpose of promoting and maintaining farm forest cover of at least 10 per cent of every agricultural land holding and to preserve and sustain the environment not only in combating climate change and global warming but also sustainable soil management, with the overall result being the realisation of a clean and healthy environment in line with Article 42 of the COK 2010.

I therefore expect that this timely fifth volume of IYSLP on land tenure and sustainable soil management could not have come at a better time with its critical look at the clash of modern and traditional tenure concepts, illegal or illegitimate land acquisitions in developing countries, FAO voluntary guidelines on tenure rights, and UNCCD, FAO, and African Union tools to assist effective soil governance in African states coupled with comparative studies on both soil and tenure rights law from EU, Germany, New Zealand, Iran, and India amongst other cross-cutting issues and experiences from Cameroon and South Africa. Our Land Tenure and Soil Management discourse in Africa will be enriched. Practitioners and policymakers and the general public will get the necessary guidance to work towards increasing the land area under sustainable soil management worldwide. Efforts of the editorial team that enabled this volume are commendable, and Africa will not be left behind in implementing effective policies on land tenure and soil management.

Ministry of Environment and Forestry,
Nairobi, Kenya

Keriako Tobiko

Preface

This fifth volume of the *International Yearbook of Soil Law and Management* is a remarkable one. First, it is—so to say—the "pandemic volume". It was certainly conceptualised before the pandemic, but the entire drafting, reviewing, and revising process took place during the COVID-19 pandemic, which kept the world—yes, we all know the whole world—breathless. The focus was, amongst other things, on numbers of infected persons, incidents, scary developments, new variants of the virus, and appropriate measures to deal with the unprecedented challenges. In later stages of the pandemic, the general effects on society at large and lessons learnt for other global threats, such as the climate and biodiversity crises, have been intensively discussed. This leads to the second point—why this volume is inherently special. It is the first volume in the decade during which many say that humanity as a whole, but also individual societies, must realise a socio-ecological transformation towards climate neutrality and sustainability.

The year 2021 has seen many landmark court decisions concerning climate change. The German Constitutional Court ruled that it is a constitutional responsibility of the German parliament to put in place an effective planning mechanism to achieve climate neutrality by 2045, including sector-specific targets as well as monitoring and control measures (including specified sanctions). The core argument of the Court was that it would constitute a breach of the constitutional obligations if later generations face the risk of having to bear stringent and thus disproportional restrictions on their freedoms.

The structure of the volume conforms to all previous volumes—four main parts provide relevant and recent information on soil governance topics for academics, legislators, and policymakers:

- Part I: The Theme
- Part II: Recent Developments of Soil Regulation at International Level
- Part III: National and Regional Soil Legislation
- Part IV: Cross-Cutting Topics

The theme of the volume was chosen to address how concrete forms of tenure rights can either enable or hinder sustainable soil management. By addressing this theme, we align the discussion with the requirement set by the German Constitutional Court as appropriate land tenure is a kind of precondition to the implementation of a comprehensive planning mechanism which was seen as essential by the German Constitutional Court. Moreover, we contribute to the debate around how the socio-economic transformation could be implemented on the ground. The chapters contributing to this theme address very different aspects of tenure rights, such as clash of legal systems, three countries' perspectives, illegal land acquisition, land take in general, and management options to strengthen land rights.

Part II—Recent international developments entails two chapters: one on the outcome of the last Conference of the Parties (COP) of the UNCCD, and the second provides a critical analysis of support from the international regime to achieve sustainable soil management in Africa.

Part III—Regional and national reports provide insights on soil protection governance in Africa (Kasimbazi/Yahyah), the European Union through the new Green Deal (Heuser/Itey), Iran (Faryadi), and South Africa (Ruppel/Knutton/Marivate).

Part IV—Cross-cutting issues includes contributions on many diverse topics. Firstly, the effects of the COVID-19 pandemic are analysed, particularly with regard to countries in the Global South (Sambo). A further chapter by Vanheusden/Jacobs discusses the pros and cons of the concept of soil and land stewardship. Stubenrauch addresses the interlinkages of soil health and phosphorus extraction and use. The newly established FAO SoiLEX, a database providing a fantastic overview of national soil legislation, is explained by Bhorris. Finally, the chapter by Ruppel explains the nexus of soil protection, food security, and the international regulations on climate change and trade.

At this juncture, we would like to cordially thank two colleagues, namely Prof. Emmanuel Kasimbazi, Makerere University, and Prof. Tianbao Qin, University of Wuhan, who have left our team of editors. Both of them have wholeheartedly supported the first four volumes of the *International Yearbook of Soil Law and Policy* with their outstanding expertise. We would also like to welcome our new editor, Prof. Patricia Kameri-Mbote, University of Nairobi.

Finally, we would like to express our deepest gratitude to all authors in this volume for their engagement, commitment, contributions, and, not to be forgotten, patience—unfortunately necessary due to delays caused by the pandemic. Moreover, we thank the members of the Advisory Board for providing their important insights during the review process, particularly Marc Shepard, Pradeep Singh, and Mercy Teko for conducting language reviews of about half of the chapters. Last but definitely not least, to the publishing house SPRINGER, and Laura Hofmann in particular, for their ongoing operational support and technical assistance.

Dessau, Germany — Harald Ginzky
Fermoy, Ireland — Elizabeth Dooley
Kleinmachnow, Germany — Irene L. Heuser
Nairobi, Kenya — Patricia Kameri-Mbote
Nairobi, Kenya — Robert Kibugi
Leipzig, Germany — Till Markus
Graz, Austria — Oliver C. Ruppel
July 2021

Contents

Part I
Land Tenure Rights and Sustainable Soil Management

Clash of Modern and Traditional Tenure Concepts: An Overview

Patricia Kameri-Mbote

Abstract Modern and traditional tenure concepts are closely aligned to property and share property's complexity and dynamism. The term property defies definition, has different meanings to different people and establishes entitlements through recognition and protection. Property thus connotes different things, has a broad meaning and requires different institutions and mechanisms to actualize the property castle. At the core of property is the relationship between an individual and the community with regard to the use and exploitation of resources and is dependent on enforcement mechanisms of the state. This chapter looks at differences between modern and traditional tenure concepts arguing that property conceptions are contextual and geographically situated. Indeed while the right to exclude is viewed by many as a defining feature of property and exists in modern property systems, it does not exist in traditional property systems that allow multiple rights over property.

1 Introduction

Modern and traditional tenure concepts are closely aligned to property, and therefore, share property's complexity and dynamism. Indeed, commentators opine that the term property defies definition[1] and means different things to different people, operating 'as both an idea and an institution'.[2] According to Thomas Grey, the common notion of property is the ownership of things. He assigns the difficulty in defining property to the divergent conceptions of property from the point of view of 'specialists', such as lawyers and lay people.[3] However defined, property establishes

[1] Waldron (1988), p. 26.

[2] Alexander et al. (2009), p. 743.

[3] Alexander and Penalver (2012).

P. Kameri-Mbote (✉)
University of Nairobi, Nairobi, Kenya
e-mail: Pkmbote@uonbi.ac.me

H. Ginzky et al. (eds.), *International Yearbook of Soil Law and Policy 2020/2021*, International Yearbook of Soil Law and Policy 2020,
https://doi.org/10.1007/978-3-030-96347-7_1

entitlements through recognition and protection.[4] Honore's[5] incidents of property define the range of entitlements that a property owner has over their property. Change in the range of justified claims of competing public interest threatens property.[6] In the case of land, increasing concerns for sustainable development, relating largely to resources on land, has eaten into the range of entitlements of landowners.

While to the lay person property is a thing represented in the physical *res*,[7] to lawyers property comprises the 'collection of individual rights people have against one another with respect to owned things'.[8] Land can be categorized as a thing but land as property is a concept.[9] This is the meaning that law ascribes to property—a conception of the mind. Property in this sense is nothing but a basis of expectation of deriving certain advantages from a thing that we are said to possess; in consequence of the relation in which we stand towards it. Premised on this view, only through the protection of law is one able, for instance, to enclose a field as property.[10]

Property is also viewed as a "bundle of sticks"[11] granted to property owners. In this rendition, property connotes different things, has a broad meaning and requires different institutions and mechanisms to actualize.[12] The definition of property from its attributes opens the door for other conceptions: property as a place of refuge;[13] as entitlement;[14] as expression;[15] as creativity and innovation;[16] ideas;[17] to name a few. Inherent in all these meanings are the centrality of the individual; the role of contract in dealing with property and the permissible boundaries of incursions of other citizens into the property castle.[18] Boundaries are a critical factor in property and may have informed the old adage that 'good fences make good neighbours'.[19]

At the core of property are relationships between the thing and the holder of the thing; between different holders of things; between holders of things and non-holders; and between holders and non-holders of things and the agency that grants and guarantees the rights, usually the government. The right to exclude

[4] Underkluffer (2016), p. 2.
[5] Honore (1961), p. 107.
[6] Underkluffer (2016), p. 2.
[7] Munzer (1990).
[8] Alexander and Penalver (2012).
[9] Bentham (1853).
[10] Ibid.
[11] Honore (1961), p. 107.
[12] Alexander and Penalver (2012).
[13] Alexander (2018).
[14] Honore (1961), p. 107 on the bundle of sticks that a property holder has.
[15] Cornish (2019).
[16] Ibid.
[17] Ibid.
[18] Alexander and Penalver (2010).
[19] Frost (undated).

interventions is viewed by many as a *sine qua non* feature of property, yet, there are property systems that allow multiple rights over the thing and allow 'non-owners' rights to roam.[20] In a nutshell, property is the relationship between an individual and the community with regard to the use and exploitation of resources and is dependent on enforcement mechanisms of the state.

This chapter looks at the clash between modern and traditional tenure concepts. It is divided into five sections. Section 1 is the introduction while Sect. 2 provides a background to tenure. Sections 3–5 discuss traditional and modern tenure concepts respectively. Section 6 concludes and highlights the clash between traditional and modern tenure concepts.

2 Background to Tenure

Tenure is derived from the Latin word *tenere*, which means "to hold", connoting the nature of the relationship that exists between individuals in relation to a specified thing.[21] Tenure denotes the methods by which individuals or groups acquire, hold, transfer or transmit property rights in land.[22] It is a system used to determine who can use land, the period for such use and under what terms and conditions. Tenure is based on official laws and policies and even informal customs.[23] In essence, land tenure means a system that outlines how land is held by an individual or the actual user of the land.[24] It stipulates the rights and responsibilities that owners enjoy with regard to the connection with their holding.[25]

Ownership of land historically constitutes one of the main categories of property rights holding, conveying an array of rights upon the owner.[26] Property rights in land exist against other people with regard to the land, not against other parcels of land.

> Ownership of property is a creation of law whereby a bundle of sticks/entitlements are sanctioned by law against many persons. Property is that *bundle of rights and expectations in a tangible or intangible thing that are enforceable against 3rd parties including the government.* These include entitlements to possess; to use; to exclude; allow others to use; sell; give away; dispose of by will; recover from thief and compensation for damage.[27]

---

20 Alexander and Penalver (2010), p. 4. Example from Sweden on pastoralists also accord one another reciprocal grazing rights to cope with droughts.

21 Field (2005), pp. 279–290.

22 Okoth-Ogendo (1991).

23 Coldham (1979), pp. 615–627.

24 Ogolla and Mugabe (1996), p. 85.

25 Ondiege (1996).

26 It confers the right to extract minerals from the land, to use and abuse and dispose of as the property holder wills. Megarry (1984).

27 Blocher (2006), pp. 166 and 177.

A search for the tenure system operative in a particular society is an attempt to answer the tripartite question as to *who* holds *what interest* in *what land*.[28] Land tenure is complicated in both traditional and modern societies and systems of law because the term land has a wide connotation. For instance, the English Law of Property Act, 1925 defines land to include land of any tenure, mines, minerals whether or not held apart from the surface, buildings, or parts of buildings. From this definition, it is clear that the surface of the soil and the things on the soil are enjoyed as part of the land such as the air, water and growing trees or artificial attachments such as houses, buildings and other structures. Land also encompasses interests or rights to collect things or hunt on the land. Land tenure systems have thus wide-ranging implications.

Consequently, the notion of tenure has very wide-ranging implications and may have distorted as much as it has clarified.[29] Land tenure systems vary from community to community and are influenced by the unique historical development of each political grouping and consequent variation of legal and institutional structures.[30] Land tenure represents the relations of people in society with respect to the essential and often scarce land. It also refers to possession or holding of the rights associated with each parcel of land and ordinarily has at least three dimensions, namely, people, time and space. In so far as people are concerned, it is the interaction between different persons that determines the exact limits of the rights any one person has over a given parcel of land. These rights are not absolute since there are rules that govern the manner in which the person with tenure is to utilise their rights. While the time aspect of tenure determines the duration of one's rights over the land, spatial dimensions limit the physical area over which the rights can be exercised. The spatial dimension of tenure may be difficult to delineate in exclusive terms since different persons may exercise different rights over the same space at different times.[31]

Tenure systems are culture-specific and dynamic, changing as the social, economic and political situations of groups change.[32] They are shaped by economic, political, social and legal parameters. Under both African and Western systems of land holding, for instance, ownership can be sub-divided and lesser interests can be (and are frequently) held by different persons simultaneously. While questions have arisen as to whether the notion of legal rights as a cluster of claims, powers and immunities[33] have a place in primitive or pre-capitalist societies, it is clear that landholders in these societies have entitlements that are respected by all among whom they live.

For instance, while most African customary laws recognised a measure of individual control over the broad interests that were hosted by land, paramount or

[28] Okoth-Ogendo (1991).

[29] Bohanan (1963).

[30] Crocombe (1968) and Ojwang (1992).

[31] Fortmann and Riddell (1985).

[32] Lawry and Bruce (1987).

[33] Hohfeld (1922).

allodial title was perceived as vested in society and whatever rights any one person had to the land were subordinate to the entire community's rights.[34] Ghanaian Chief Nana Ofori opined as follows:

> I conceive that land belongs to a vast majority of whom many are dead, a few are living and countless host are still unborn.[35]

This statement captures the intra-generational and inter-generational aspects of landholding that is common to many African communities. Colonialism, however, had profound effects on African tenure systems by introducing the notions of individual and state ownership of land in a bid to promote economic development.[36] The push to reform land relations to unlock the economic potential of dead capital that land has have remained for a long time.[37] In some instances, the Torrens title system (based on statutory registration and ownership of individually demarcated plots) was introduced to replace pre-existing customary notions of land ownership.[38] The latter have, however, persisted and been informed in practice by the introduced system. Thus, Bentsi-Enchill notes that the defects of African systems of land tenure have arisen from the fact that these systems have been left to informally adapt to changed circumstances.[39]

Different societies have different ways of holding things that they value. Until the advent of technology and the knowledge associated with it, land was an unrivalled genre of property and colonial acquisitions in the 1800s had the acquisition of land for expansion of territory as the main aim. The mode of land holding among people in the colonised and the colonising spaces differed markedly. In most cases, colonisers introduced modern tenure predicated on individual holding of land with public land as the other category. Many colonised people, however, had land held by communities and not individuals. This leads to a clash between the contemporary and the traditional concepts of tenure.

Holders of land under both traditional and modern tenure fall broadly into three categories, namely, public, private and community. Public tenure is assigned to land that is held by the state or other local authorities and reserved and used for broad societal purposes such as roads, railways and so on. It also includes land abutting watercourses, ocean and national parks. In most countries, it is not amenable to alienation to private actors and is managed by public agencies. Private land tenure is assigned to land held by individuals and corporate entities, while community tenure is assigned to groups who share land in common. Commons are a genus of private land held by groups united by kinship or similar characteristics.

[34] Maini (1967).

[35] Ollennu (1962).

[36] Fortmann and Riddell (1985).

[37] de Soto (2000).

[38] Okoth-Ogendo (1989).

[39] Bentsi-Enchill (1966).

Utilitarianism/instrumentalism has been used to justify private property rights, which include private tenure. The argument is that property institutions should be shaped to maximize net utility,[40] which over time has been reduced to 'everything that an individual might value'. This is based on individual preferences and discounts the broader moral frameworks within which property rights exist. This approach is supported by the law and economics approach, which employs a single metric value to property defined in terms of individual experience; employment of economic tools such as rational choice, and game theory to explain why property rights are granted. Economists opine that problems exist when resource allocations are inefficient or expected to leave future generations worse-off. Inefficiency results from non-transferability in the market or absence of incentives to sustainably manage resources. In their view, the person with the strongest incentives should be assigned property rights to minimise transaction costs and maximise social returns. The expectation is that the market will balance competing uses and force participants to use property in the most efficient way.[41]

Private property is seen as the standard to aim for in stemming the so-called 'tragedy of the commons'. Private property rights proponents argue that market solutions prevent the tragedy of the commons that too often results when incentives to preserve common pool resources do not exist.[42] Such arguments rely on the notion that property held in common encourages a rush by all having access to it to appropriate as much of it as possible while it lasts.[43] The desire to take as much as possible, the argument continues, is fanned by the fact that the negative effects of over-exploitation of the resource are not felt proportionately by any of the takers, and consequently, none of them feels personally compelled to stem the over-exploitation. Hence, what is everybody's property is perceived as nobody's and becomes valued at a rate proportionate to its utility only after it has been individually appropriated.[44] In this sense, private property rights provide incentives to manage resources, reduce uncertainty and ensure predictability.[45]

The major thrust of this argument is that when property rights are assigned in these situations, the market acts to properly balance competing uses and force the participants to use such property in the most efficient way. Private property rights in resources evolve only when demand for those resources makes the extra effort of defining and enforcing those rights worthwhile. They constitute the underlying basis for the operation of any economic system. The rights-holders are able to acquire rights to property and benefit from economic returns from investment in their property.[46]

[40] Bentham (1789).

[41] Hardin (1977), p. 16.

[42] Ibid.

[43] Ostrom (1977), p. 173.

[44] Ostrom (1990), p. 3.

[45] Baden and Stroup (1977), p. 229.

[46] Walden (1995), p. 176.

The assumption here is that all values ascribed to the property can be transacted in the market. With regard to land, it is critical to ask the question whether soil, a burial site for a community's ancestors, or the sentimental value associated with ancestral land, can be transacted in the market. Moreover, environmental goods such as ecosystem services, which are indirectly related to land, are for the most part consumed directly and never marketed, thus resulting in gross undervaluation of these services that are largely consumed as a public good. For instance, it is difficult to allocate a market value to soil for its role in carbon storage, which helps to reduce greenhouse gases in the atmosphere and ultimately in addressing climate change.

Traditional communitarian rights to land have utility and value to the holders and only differ from those in modern property rights' systems because their value and utility is for groups and not individuals. Some of their aspects are also difficult to transact in the market as many merge the personal aspects with the shared aspects.[47]

Deontological theories of property[48] that do not treat property as a means to an end provide a more appropriate lens for looking at traditional rights. These theories reflect underlying moral entitlements in the property that are not necessarily associated with the results they generate. For instance people who labour on an unowned piece of land and use their labour have a moral claim to the land and philosophers like Aristotle would justify their rights to property as necessary to forestall quarrels. Such property rights are instrumentally embedded within the labour and are not strictly utilitarian, according to Thomas Aquinas and David Hume. Property deals with value enhancing relationships regarding assets as the legal enforcement of property rights enhances the owner's probability of retaining possession. It could mesh with assets that are not capable of being commodified[49] as property, such as, belonging to a group for a bereaved widow, kinship and other familial ties. Indeed, the value of property increases with each additional subscriber and the utility of property draws from the network of subscribers who can keep away the free-riders. The state provides the mechanism for public enforcement of property as a public good, ensuring that the law standardizes forms of property and reduces the costs of investigating

It is the duty of law, as the expression of the will of the people expressed through the sovereign, to provide mechanisms for the protection of property in the interest of all its citizens. This duty extends to both modern and traditional tenure, bringing to the fore the conception of property as a social relationship even whilst it has an individual side. The Ghanaian Constitution captures this in the following parlance

[47] Radin (1993), p. 11.

[48] Alexander and Moore (2016).

[49] Radin (1993), p. 11 explaining the distinction between fungible (not unique and not linked to persona) that is easily amenable to transaction in the market and non-fungible property (unique and personal as part of owner's personality; sentimental, emotional link) whose value to the owner is beyond the market.

> The State shall recognise that ownership and possession of land carry a social obligation to serve the larger community and, in particular, the State shall recognise that the managers of public, stool, skin and family lands are fiduciaries charged with the obligation to discharge their functions for the benefit respectively of the people of Ghana, of the stool, skin, or family concerned and are accountable as fiduciaries in this regard.[50]

This supports the social utility theory view that law should promote the maximum fulfillment of human needs and aspirations and that legal protection of rights should ensure that.[51] The social utility theory places emphasis on the individual as the locus for the grant of rights. Within countries with plural legal entities, such as Kenya and Ghana, it is clear that there is need to include other actors (such as communities and families). Article 61 of the Kenyan Constitution provides that all land in Kenya belong to the people of Kenya collectively as a nation, as communities and as individuals, and goes ahead to delineate what land is public, community and private.[52]

3 Traditional Tenure

A lot of land in Africa is held under customary law. The exact amount of land under traditional tenure is not known but is estimated at roughly 75% of the Continent or over two billion hectares.[53]

Traditional tenure comprises a complex system of customary rights of access and use regulated by intricate rules, usages and practices. These are often based on communal solidarity, such as clan or other lineal heritage, and are unwritten. Communities and the people were governed by rules passed down from one generation to another through various forms of communication.[54] In some cases, traditional leaders (such as chiefs) allocated rights of access to and use of land to persons under their authority.[55] For many communities, land was not owned by individuals and individuals in most communities did not enjoy the right to dispose of land.[56] Land was not perceived as a tradable good as it belonged to a group and not to the individual.[57]

Scholars[58] have summarized the hallmarks of African customary land tenure as:

[50] Government of Ghana, Constitution of Ghana (1992 as amended in 1996).

[51] Kameri-Mbote (2009).

[52] Government of Kenya, Constitution (2010) Articles 62–64 of the Constitution.

[53] Wily (2017), p. 108.

[54] Lambert (1949).

[55] Abdulai (2006).

[56] Kuria (2018).

[57] Kameri-Mbote (2016).

[58] Okoth-Ogendo (1991).

- Distinction between rights of access to land and control of those rights;
- Power of control vested in recognized political authority/entity within a specific community;
- Political entity exercised these powers to allocate rights of access to individuals depending on needs and status;
- Rights of access guaranteed by political authority/entity on the basis of reciprocal duties performed by the rights' holders (even in family) to the community;
- Rights to land determined on a continuum of flexibility always adjusting and changing as circumstances demanded
- No element of exclusivity to land

As noted above, property held by communities under traditional tenure is a kind of commons (*res communis*) and not *res nullius*, representing private property for the group that controls it and whose members have access to it.[59] Sara Berry in her work among the Asante, notes that land is owned and administered as a social process and not in relation to a set of rules and enforcement mechanisms.[60] Liz Alden Wily perceives community as connoting both social and spatial sphere and in this regard notes that:

> key to the community in the customary sphere is its existence as a definable community land area, territory or domain, the limits of which are accepted by neighbouring communities ....[61]

In that process, the core elements of culture, kinship, and other social relations are recalled, redefined, and reinforced as they are asserted.[62] This differs markedly from modern tenure that has fixed rights and responsibilities for the rights holder. The state in administering traditional tenure rights must allow for the negotiation of the rights whose exercise is tied to kinship and responsibilities. Effective enforcement of traditional tenure rights requires mechanisms outside the modern administrative and judicial systems.[63]

4 Modern Tenure

Modern tenure refers to the conventional, formal and contemporary mode of land ownership, exported by European countries to their colonies. It formally recognizes two potential holders of land: individuals/legal persons and the sovereign/public. Modern tenure is influenced by the view that property is necessarily exclusive and is informed by William Blackstone's full liberal theory of property as "the sole and

[59] Okoth-Ogendo (2002), pp. 17–29.
[60] Berry (2001).
[61] Wily (2017), p. 106.
[62] Ibid.
[63] Kameri-Mbote et al. (2013).

despotic dominion over land to the total exclusion of all others".[64] It is centered on land registration and formalisation of title. Whoever is registered on the deed of title as the owner of land is the recognized proprietor of that land. Registered property owners and can fence off their property and exclude the whole world. Freehold tenure is the largest quantum of land rights an individual can hold under modern tenure.[65] Modern tenure guarantees the owner absolute rights subject only to permissible regulatory controls for planning and environmental sustainability.[66] Rights held under modern tenure can also be terminated through compulsory acquisition for a public purpose or in public interest.

The main purpose of registration of title to land is to enhance certainty in land or security of tenure and achieve simplicity.[67] This rationale for land registration proceeds from an economic postulate which holds that registration of title promotes confidence among title holders and other third parties that deal in the land that is subject of registration, thereby enhancing the value of such land and giving comfort to dealers in such land. This comfort mainly draws from the fact that registration of title usually means that the state, which is the custodian of registration, indemnifies persons who suffer loss from acting or relying on the strength of the title documents that it produces and also enforces ownership rights whenever infringed or threatened.[68] Registration has the effect of conferring upon the person(s) whose name is on the register an indefeasible title to such land, thereby dispensing with the need by third parties dealing in such land to inquire into the ownership and other interests that may lie in respect of that land. Usually, all the interests that are rightfully on a particular piece of land are to be found on the encumbrance section of the title document, and if not so found, then no right may legitimately lie (but of course, subject to the overriding interests such as customary trusts).[69]

While the certainty of title or security of tenure appears to be the chief purpose of registration, there are other interrelated and important aims of registration. They include: minimization of land disputes and easier administration of the loan system by financial institutions;[70] reduction of transaction costs in conveyancing by clarifying ownership and extinguishing competing claims;[71] enhanced access to credit by land owners since registered land may be used as collateral;[72] and encouragement of

---

[64]Demsetz (1967), p. 347.

[65]Ann (1966), p. 1071.

[66]Greiner (2017), p. 78.

[67]Gray and Gray (2001), p. 976.

[68]Miceli et al. (2001), p. 275.

[69]The various overriding interests are outlined under section 28 of the Land Registration Act, No. 3 of 2012.

[70]Onalo (2008), p. 178.

[71]Atwood (1990), p. 65.

[72]Besley (1995), p. 103.

investments in land due to an assurance to investors that they will have a return on their investment.[73]

Modern tenure proponents see titling and registration of land as a panacea to the perceived poverty problem among community land holders. They are of the view that the process would 'bring dead capital to life' in de Soto's words.[74] This argument has, however, been debunked by many scholars.[75]

5 Clash Between Modern and Traditional Tenure

Tenure is both dynamic and culture-specific and both modern and traditional tenure are greatly influenced by the contexts within which they have developed. In plural legal societies where both co-exist, they influence and affect each other significantly. Modern legal systems have gradually recognized customary law and other community-based norms in formal law, however, formal law tends to take precedence in the hierarchical structure in most legal systems. It follows that traditional tenure is relegated to the lower rungs in the hierarchy of laws despite the fact that it governs many property relationships in many post-colonial societies where modern tenure was superimposed on traditional tenure. This creates the stage for clashes between claimants of rights under the two systems. Colonial subjugation of traditional tenure was geared towards extinguishing the claims of prior holders. The colonisers negated the pre-existing traditional tenure rights using laws and policies granting settler communities modern tenure rights that were accorded higher status than the traditional ones. Law was used as a sword to wrest colonised communities of rights to their land.[76] The colonized communities did not comprehend the import of the modern rights, and in most cases, had to be forcefully removed to make way for the new 'owners'. In fact, a colony became the coloniser's land.[77] This marked the beginning of aggressive conversion of traditional tenure to modern tenure.[78] The introduction process was characterized by the forceful acquisition of land and the displacement of local populations.[79] Members of local communities who were unable to find places to settle were deemed squatters and sometimes enslaved to work without pay.[80] This have been the case in other colonies that were subject to the scramble and partition of Africa.[81]

---

[73] Barber (1970), p. 6.

[74] de Soto (2000).

[75] Cousins (2017), pp. 93–94.

[76] Fimbo (2017), p. 59.

[77] Wakoko (2014).

[78] Elias (1956).

[79] Wanyonyi et al. (2015).

[80] Government of Kenya (2013).

[81] Khamisi (2018).

The multiple layers of entitlements to land, familiar to traditional societies, were subsumed under the titled landowner's rights in total disregard of the fact that for many traditional societies, different rights could be claimed over the same land. The designation of public and private land was not part of the nomenclature of traditional communities since rights to land were assigned for a purpose, which could be for the individual, family or community, but for the overall good of the community. Okoth Ogendo notes that centuries of subjugation of African customary tenure through law and force failed to oust the force of customary norms over land occupied and used by Africans. The ubiquity of the traditional rights threatened the rights of the settlers to the land they obtained.

6 Conclusion

Both traditional and modern tenure are important for sustainable management of land. The co-existence of these tenure types is a reality. However, the clash between the two arises where one (usually modern tenure) is hoisted over the other (customary tenure) with no attempt to understand how the latter works. On the one hand, customary tenure rules are part of the body politic of the community and are accessible. They are also dynamic and responsive to changing circumstances on the ground. Modern tenure, on the other hand, is removed from the communities and has rigidities that communities are not accustomed to dealing with land that is very dear to them and from the basis of many of their activities. The assertion that the grant of title seals the fate of ownership of land is difficult to appreciate among communities for who land is inalienable.[82] For the two tenure types to contribute optimally to sustainable land management, there is need to carefully study the way they work and how they can be best applied. Considering that most land is predominantly held under customary law in Africa, for instance, ignoring customary tenure makes any applicable tenure rules ineffective as they leave out most land users.

References

Abdulai R (2006) Is land title registration the answer to insecure and uncertain property rights in sub-Saharan Africa? Royal Institution of Chartered Surveyors (RICS) Research paper series, 6(6)

Alexander GS (2018) Property and human flourishing. Oxford University Press

Alexander L, Moore M (2016) Deontological ethics. In: Zalta EN (ed) The Stanford encyclopedia of philosophy. https://plato.stanford.edu/archives/win2016/entries/ethics-deontological/

Alexander GS, Penalver EM (2010) Property and community. Oxford University Press

Alexander GS, Penalver EM (2012) An introduction to property theory. Cambridge University Press, New York

[82] Kameri-Mbote (2015), p. 40.

Alexander GS, Penalver EM, Singer JW, Underkuffler LS (2009) A statement of progressive property. Cornell Law Rev 94(4):743
Ann PM (1966) Land law in Kenya. Wis Law Rev, 1071
Atwood D (1990) Land registration in Africa: the impact on agricultural production. World Dev 18: 65
Baden J, Stroup R (1977) Property rights, environmental quality, and the management of national forests. In: Hardin G, Baden J (eds) Managing the commons, p 229
Barber W (1970) Land reform and economic change among African farmers in Kenya. Econ Dev Cult Exch 19:6
Bentham J (1789) An introduction to the principles of morals and legislation
Bentham J (1853) Principles of the Civil Code
Bentsi-Enchill K (1966) Do African systems of land tenure require a special terminology? J Afr Law 9:114–139
Berry S (2001) Chiefs know their boundaries: essays on property, power, and the past in Asante, 1896–1996. Boydell & Brewer, Limited
Besley T (1995) Property rights and investment incentives: theory and evidence from Ghana. J Polit Econ 103:903
Blocher J (2006) Building on custom: land tenure policy and economic development in Ghana. Yale Hum Rights Dev Law J 9:166
Bohanan PJ (1963) 'Land', 'Tenure', and 'Land Tenure'. In: Biebuyck D (ed) African Agrarian systems, p 101
Coldham SFR (1979) Land-tenure reform in Kenya: the limits of law. J Mod Afr Stud 17(4): 615–627. Cambridge University Press. https://www.jstor.org/stable/160742. Accessed 28 May 2020
Cornish W (2019) Intellectual property: patents, copyrights, trademarks & allied rights. Sweet & Maxwell
Cousins B (2017) Beyond private ownership: alternative paradigms for rural and urban tenure reform in post-apartheid South Africa. In: Kameri-Mbote P, Odote C (eds) The Gallant Academic: essays in honour of HWOO Okoth Ogendo. University of Nairobi School of Law, Nairobi, p 79
Crocombe RG (1968) Improving land tenure
de Soto H (2000) The mystery of capital: why capitalism triumphs in the west and fails everywhere else. Basic Books
Demsetz H (1967) Toward a theory of property rights. Am Econ Rev 57(2). Papers and Proceedings of the Seventy-ninth Annual Meeting of the American Economic Association, p 347
Elias TO (1956) The nature of African customary law. Manchester University Press
Field E (2005) Property rights and investment in urban slums. J Eur Econ Assoc 3(2–3):279–290. https://doi.org/10.1162/jeea.2005.3.2-3.279
Fimbo GM (2017) In search of thematic unity in land law: tenure security. In: Kameri-Mbote P, Odote C (eds) The Gallant Academic: essays in honour of HWOO Okoth Ogendo. University of Nairobi School of Law, Nairobi, p 59
Fortmann L, Riddell J (1985) Trees and tenure: an annotated bibliography
Frost R (undated) Mending wall. Poetry Foundation
Government of Ghana, Constitution of Ghana (1992 as amended in 1996)
Government of Kenya (2013) Truth Justice and Reconciliation Commission, The Final Report of the Truth Justice and Reconciliation Commission of Kenya (Volume IV). https://www.jfjustice.net/downloads/1460970274.pdf. Accessed 25 May 2020
Government of Kenya, Constitution (2010) Articles 62–64 of the Constitution
Gray K, Gray SF (2001) Elements of land law. Butterworths, London, p 976
Greiner C (2017) Pastoralism and land-tenure change in Kenya: the failure of customary institutions. Dev Change 48(1):78
Hardin G (1977) The tragedy of the commons. In: Hardin G, Baden J (eds) Managing the commons, p 16

Hohfeld WN (1922) In: Cook WW (ed) Fundamental legal conceptions as applied in judicial reasoning and other essays
Honore AM (1961) Ownership. In: Guest AG (ed) Oxford essays in jurisprudence, p 107
Kameri-Mbote P (2009) The land question in Kenya: legal and ethical dimensions. In: Governance: institutions and the human condition. Strathmore University and Law Africa
Kameri-Mbote P (2015) The land question and voting patterns in Kenya. In: Njogu K, Wekesa PW (eds) Kenya's 2013 General Election: stakes, practices and outcomes. Twaweza Publications, pp 34–47
Kameri-Mbote P (2016) Kenya Land Governance Assessment Report. World Bank Group
Kameri-Mbote P et al (2013) Giving meaning to community rights to land and related resources in Kenya. Strathmore University Legal Press
Khamisi J (2018) Kenya: looters and grabbers: 54 years of corruption and plunder by the elite, 1963–2017. Jodey Book Publishers, Plano
Kuria D (2018) Land reforms. Paper presented at the Institution of Surveyors of Kenya (ISK) regional Conference 2018, Nairobi
Lambert HE (1949) Systems of land tenure in the Kikuyu Land Unit. University of Cape Town
Lawry S, Bruce J (1987) Resource tenure and the management of natural resources in Africa
Maini KM (1967) Land law in East Africa
Megarry RE (1984) The law of real property, 5th edn
Miceli TJ, Sirmans CF, Kieyah J (2001) The demand for land title registration: theory with evidence from Kenya. Am Law Econ Rev 3(2):275
Munzer SR (1990) A theory of property
Ogolla DB, Mugabe J (1996) Land tenure systems. In: Juma C, Ojwang JB (eds) In land we trust: environment, private property and constitutional change. Initiative Publishers, Nairobi, p 85
Ojwang JB (1992) Laying a basis for rights: towards a jurisprudence of development
Okoth-Ogendo HWO (1989) Some issues of theory in the study of tenure relations in African agriculture. Africa 59:6
Okoth-Ogendo HWO (1991) Tenants of the crown: evolution of agrarian law and institutions in Kenya.
Okoth-Ogendo HWO (2002) The tragic African commons: a century of expropriation, suppression and subversion. Keynote Address to African Public Interest Law and Community-Based Property Rights Workshop, USA River-Arusha, Tanzania, published in CIEL/LEAT/WRI/IASCP, Amplifying Local Voices for Environmental Justice: Proceedings of the African Public Interests Law and Community-Based Property Rights Workshop (USA, CIEL, 2002), pp 17–29
Ollennu NA (1962) Customary land law in Ghana
Onalo PL (2008) Land law and conveyancing in Kenya. Law Africa, Nairobi, p 178
Ondiege P (1996) Land tenure and soil conservation. In: Juma C, Ojwang JB (eds) In land we trust: environment, private property and constitutional change. Initiative Publishers, Nairobi
Ostrom E (1977) Collective action and the tragedy of the commons. In: Hardin G, Baden J (eds) Managing the commons, p 173
Ostrom E (1990) Governing the commons: the evolution of institutions for collective action, p 3
Radin MJ (1993) Reinterpreting property. University of Chicago Press, p 11
Underkluffer LS (2016) Property and change: the constitutional conundrum. Tex Law Rev 91(2015):2
Wakoko V (2014) The evolution of land law in Kenya. Nairobi. https://www.academia.edu/8972722/the_evolution_of_land_law_in_kenya. Accessed 17 May 2020
Walden I (1995) Preserving biodiversity: the role of property rights. In: Swanson T (ed) Intellectual property rights and biodiversity conservation, p 176
Waldron J (1988) The right to private property, p 26
Wanyonyi A, Nyadimo E, Kariuki J (2015) Land tenure security in Kenya. In: Kirk M (ed) Report 3. UNON, Publishing Services Section, Nairobi, UN Nairobi
Wily LA (2017) The fate of Res Communis in Africa: unfinished business. In: Kameri-Mbote P, Odote C (eds) The Gallant Academic: essays in honour of HWOO Okoth Ogendo. University of Nairobi School of Law, Nairobi, p 103

Mutually-Reinforcing Transgressions of Justice in Large Scale Land Acquisitions in the 'Public Interest'

Jennifer Clare Mohamed-Katerere

Abstract (In)justice describes land governance in the global south: For centuries, accumulation of wealth has occurred through land dispossession, particularly of indigenous people, peasants and pastoralists, increasing vulnerabilities and decreasing capabilities. The wave of acquisitions from the mid-2000s though focused on the public interests of 'development,' 'human security' and 'conservation' hollows out rights, violating the fundamentals of justice. These acquisitions are for food, fuel and feed, mining, logging, carbon and coercive conservation. Using a multidimensional analysis, the chapter explores the injustice of these acquisitions, identifying layers of mutually reinforcing transgressions. It considers how power and dominant development and conservation cultures creates unjust land governance. At the heart of this is the promotion of markets through extractivism over and above the pursuit of social-ecological justice. Using a justice framework, embedded in redistributive justice, it demonstrates how these acquisitions impact on recognition, inclusion, engagement, distribution of costs and benefits, and structural opportunities of rural citizens. Transitions in the control of production, increased land inequality, and discriminatory distribution of public resources fostered by these interactions impacts on development futures as rural political- and economic-scapes are redefined.

1 Introduction

The mid-2000s marked an unprecedented spike in large-scale land acquisitions of millions of hectares (mha) of land in the global south for food, fuel and feed crops—which is well documented—as well as for mining, coercive conservation, climate

J. C. Mohamed-Katerere (✉)
IUCN Commission on Environment, Economics and Social Policy (CEESP), Johannesburg, South Africa

World Commission on Environmental Law (WCEL), Johannesburg, South Africa

H. Ginzky et al. (eds.), *International Yearbook of Soil Law and Policy 2020/2021*, International Yearbook of Soil Law and Policy 2020,
https://doi.org/10.1007/978-3-030-96347-7_2

mitigation, ecotourism and real estate from rural communities, including indigenous people.[1]

These acquisitions are different from earlier waves in that contemporary interest is, more often than not, about extracting value from land for biofuels, conservation, carbon sequestration and accessing water achieved through new financial mechanisms, commodification and marketization rather than occupation or ownership.[2] This is in contrast to earlier agricultural models that focused on integration of small producers into national economies.[3] By building on colonial legacies and narratives of modernity, swaths of territory are effectively moved from the control of smallholders (and also nations) to global capital[4]—amounting to a foreignization of space.[5] Transnational investors at inter- and intra-regional levels are key acquirers, although in some places national investors are also engaged.[6] For example, by the beginning of 2019, European Union (EU) based companies have been involved in 616 land deals encompassing 23 Mha in the global south.[7] Agrawal et al. estimate that more than 40 Mha of agricultural land in over 35 lower and middle income countries were grabbed.[8]

This trend is significant because the loss of control of land productivity by rural citizens has implications for futures in the global south including for economies, rural livelihoods, vulnerability to climate change, citizen engagement and accountability. These approaches are likely to escalate as markets in environmental goods mature, new global conservation and climate goals are set, and other neoliberal development approaches that encourage the increasing commodification of nature, like the green economy, take hold.[9] Current indications are that large conservation organizations and states, as part of the Convention on Biological Diversity's Post-2020 framework, will demand land enclosures to protect 30% of the world's land, water and oceans by 2030, in an effort to roll back the rate of species extinction which is now 100–1000 times faster than the historical background rate[10] and restore so called 'human–non-human justice'.[11] The *High Ambition Coalition for Nature and People* embracing more than 85 states, mainly from the global north and Latin America are key in driving this approach.[12] While some argue that this will convert

[1] Agrawal et al. (2019) and Borras Jr et al. (2020).

[2] Sikor et al. (2013) and Benjaminsen and Bryceson (2012).

[3] Moyo and Yeros (2005).

[4] Kumar (2020) and McKay (2017).

[5] Zoomers (2010).

[6] Cotula et al. (2014).

[7] Borras Jr et al. (2020).

[8] Agrawal et al. (2019).

[9] Larson et al. (2013) and Fairhead et al. (2012).

[10] Bhola et al. (2021).

[11] Kopnina (2018).

[12] https://www.hacfornatureandpeople.org.

private resources into public goods with intergenerational benefits,[13] others are concerned that because the success of this ambition is dependent on community and indigenous peoples lands, already vulnerable people will be further marginalized as resources are privatized or co-managed and become available for tourism and other commercial use.[14] Others claim that by using diverse land based conservation mechanisms the rights of indigenous people to land will be recognized.[15] Given these diverse views it is important to evaluate these acquisitions through a justice lens and consider what this means for self-determination, a fundamental right in international law.

The main thesis of this chapter is that these land acquisitions although legal are inherently unjust. The term, land and green grabs (or grabs) is used to denote this capture of control over land and natural resources through processes that fall short of the fundamentals of justice or violate human rights.[16] Following this Introduction, Sect. 2 identifies the main forms of acquisition and their consequences of rural political- and eco-scapes. Section 3 explores how these acquisitions are shaped by vast power differentials in governance and unequal access to institutions—law, policies, treaties and contracts—from global to local planes. Section 4 employs a justice framework to elaborate on how well-established dimensions of justice find expression in the processes and outcomes of grabs, including for livelihoods, wellbeing, resilience of rural citizens. Section 5 concludes by summarising the main findings.

2 Acquisition and Their Consequences for Rural Land Economies

Land acquisitions are unevenly distributed among and within regions and countries, with agro-ecological qualities, infrastructure, water abundance and political influence being important factors.[17] In highly-targeted countries the scale of acquisition is staggering—in Tanzania, for example, a third of land is included in the *Southern Agricultural Growth Corridor of Tanzania* and as is typical this combines conservation areas, intensive agriculture and infrastructure development.[18] The extent of recent grabs for mining and conservation is poorly documented and it is seldom accounted for in aggregate studies. But evidence suggests that investments in mining

[13] Kopnina (2018) and Bhola et al. (2021).

[14] Garnett (2018).

[15] Weeber (2016).

[16] International Land Coalition (2011), Borras Jr and Franco (2012), Franco and Borras Jr (2019), Fairhead et al. (2012), and Holmes (2014).

[17] Borras Jr et al. (2011), Borras Jr and Franco (2012), and Zoomers (2010).

[18] Busscher et al. (2018).

have rocketed as global economic volatility increases the demand for secure investments, such as gold.[19] In Africa, a fivefold increase in foreign direct investment between 2000 and 2012 is largely attributable to extractive industries.[20] In response to climate change a market for carbon, which serves to 'offset' greenhouse gas emissions from industrialized countries by averting deforestation or reforesting in countries of the global south in return for payment has developed—and this contributes to grabbing.[21]

The vast majority of grabs are legal—lawful acquisition of rights in land occur through purchase, investments, lease, concessions, rights to conserve and manage biodiversity, carbon sequestration schemes and re-designations of state land including forests, protected areas and communal lands under common property or individual use rights.[22] For example, in Cameroon, the government de-gazetted forest land to make way for palm oil production.[23] In Argentina, multi-millionaire philanthropists acquired land in the protected area of Iberá National Park.[24] Although legal, inequality in power and in negotiations underpin these[25]—potentially compromising the fundamentals of free and fair consent that underpins contract law.[26]

Illegality extends to the use of coercion and duress, whether veiled, subtle or violent, to obtain consent to acquisitions that would otherwise be legal as this violates the requirement for consent to be free and fair.[27] A striking example of veiled coercion is the touting of private investment as the only support available to communities thus inducing them to cooperate.[28] Investors routinely target regions with recent histories of violent conflict or resettlement as this triggers land or asset enclosure, as has been the experience in Cambodia and Laos.[29] In Nicaragua landholders were willing to forfeit land to settle debts.[30] Unlawful acquisition also occurs through forced reclassification of protected areas, that have been deliberately and illegally deforesting, as agricultural land.[31] In Brazil, the legislature diluted constitutional protection of forests after the national administration had signalled impunity for illegal deforestation by land investors.[32] When land tenure is insecure government authorities and traditional leaders are able to assert control over lands

[19] Sikor et al. (2013) and German et al. (2013).

[20] Kinda and Thiombiano (2020).

[21] Franco and Borras Jr (2019).

[22] Hall (2011) and Holmes (2014).

[23] Green Development Advocates (2021).

[24] Busscher et al. (2018).

[25] Oberlack et al. (2016).

[26] Cochrane (2016).

[27] Cochrane (2016).

[28] Li (2014).

[29] Dwyer (2014).

[30] Oberlack et al. (2016) and Gómez et al. (2015).

[31] BBC (2021), Oldekop et al. (2020), and Busscher et al. (2018).

[32] Carrero et al. (2020).

beyond their legal authority and this contributes to illegal acquisition.[33] Political disregard for the law, especially by governments, public authorities and national elites in recipient countries underpins grabs, creates impunity and clouds transparency and accountability.[34] The risk of being ousted politically or convicted along with potential benefits to the individual or economy influences the extent of this kind of corruption.[35]

Rural citizens respond to these new acquisitions in diverse ways—both challenging and accommodating them including by negotiating benefits and roles for themselves.[36] The intrusion by investors into areas where rural citizens have rights can moreover contribute to conflict, illegality and injustice. Most of the land targeted belongs to or is used by local communities, including indigenous people, small farmers and pastoralists, although this is less so in Latin America where frontier forest land is most targeted.[37] Globally, customary land tenure accounts for 50% of all land and supports 2.5 billion people, however this is poorly recognized—in 2015 national laws protected only 10% of land as belonging to communities[38] notwithstanding the recognition of local communities rights in international law.[39] Indigenous Peoples have tenure rights over at least 38 Mkm^2 in 87 countries—this is more than a quarter of the worlds surface.[40] The exclusion of local communities from investor-state agreements exposes their lands to investment and grabbing.[41] Dell'Angelo et al. in a multi-study analysis across 27 countries find that 55% of acquisitions happened on common property, 13% on land subject to more than one property regime, 11% on state land and only 5% on private property.[42]

The underpinning narratives justifying acquisitions on the grounds that large swaths of land are marginal, underutilized or remote belies the reality: some 35% of land deals are on densely populated and easily accessible cropland, while 26% is on populated, moderately accessible grasslands.[43] Only 34% of grabs occur in remote forestlands with lower settlement.[44] Dell'Angelo et al. suggest that grabs impact smallholders in 61% of cases, indigenous people in 20% of cases and pastoralists in 16% of cases.[45]

[33] Schoneveld (2017), Cotula et al. (2014), and German et al. (2013).

[34] Kolstad and Søreide (2009).

[35] Ranjan (2018) and Hitcher (2010).

[36] Green and Adams (2015) and Hall et al. (2015).

[37] Borras Jr et al. (2011) and Dell'Angelo (2017).

[38] Notess et al. (2020).

[39] Anonymous (2016).

[40] Garnett (2018).

[41] Perrone (2019), Sachs (2020), and Cotula (2020).

[42] Dell'Angelo (2017).

[43] Messerli et al. (2014).

[44] Messerli et al. (2014).

[45] Dell'Angelo et al. (2017).

Land set aside for the protection for biodiversity is also impacted: Messereli et al find that 34% of the deals overlap with protected areas, while 10% have more than half of their buffer area within protected areas.[46] The trend to converting forests and protected areas is global and especially evident in Latin America where many investments occur in these areas including by well-established philanthropists and investment companies.[47] Globally there is an increase in protected area downgrades for mining. In Mexico, by 2010, mining concessions were given in 63 of the nation's 169 protected areas.[48] In 2015, the Democratic Republic of Congo adopted the *Hydrocarbon Code* which systematically downgraded all protected areas in the country, including the Virunga and Salonga National Parks to enable oil exploration.[49]

Conversely, grabs may also occur to bring new land under globally-recognized conservation regimes.[50] The transfer of control from *de facto* or *de jure* land holders to private actors, including conservation organizations and business entities, also occurs through other forms of expropriation in which title is not acquired, such as joint ventures and co-management.[51] This enclosure of community and public lands may occur for different kinds of protected areas, including private protected areas, avoided deforestation and climate change mitigation in exchange for some form of financial benefit.[52] There are also examples of large scale mining adopting such approaches with tripartite arrangements among artisanal miners, villagers and the state.[53]

Many of these so-called win-win arrangements are premised on the understanding that the State is the landowner, hence communities are engaged as co-beneficiaries or in management rather than as owners.[54] Frequently a trade-off is made between being granted 'fair and equitable' benefits while recognition and self-determination are denied.[55] In Indonesia, a community whose consent was not obtained for a new national park rejected an offer of benefits from authorities as the community feared that this would be construed as support for the park and acceptance of their non-recognition.[56] Elsewhere local communities may acquiesce or struggle to define approaches from which they are able to glean benefits, including a cash income and

---

[46]Messerli et al. (2014).

[47]Borras Jr and Franco (2012), Ojeda (2012), Green Development Advocates (2021), and Busscher et al. (2018).

[48]Napoletano and Clark (2020) and Armendáriz-Villegas et al. (2015).

[49]Qin et al. (2019).

[50]Fairhead et al. (2012), Benjaminsen and Bryceson (2012), and Napoletano and Clark (2020).

[51]Hall (2011), Murombedzi (2010), Benjaminsen and Bryceson (2012), and Snyman and Spenceley (2019).

[52]Busscher et al. (2018) and Hall (2011).

[53]de Jong and Sauerwein (2021).

[54]Myers and Mumu (2015), Creado (2008), and Chiaravalloti (2019).

[55]Benjaminsen and Bryceson (2012).

[56]Myers and Mumu (2015).

better terms of engagement.[57] In southern Africa, and elsewhere, community based natural resource management (CBNRM) and other co-management approaches connecting environmental management to economic activities, such as eco-tourism, were motivated as an approach to strengthen local participation and empowerment through decentralization and to reverse adverse consequences of coercive conservation while protecting biodiversity.[58] However, the failure to create institutions that recognize communities as owners, has meant that in practice co-management and CBNRM frequently replace local *de facto* control with governance by local 'representatives,' donors and government departments.[59] The role of the private sector as the most 'efficient' actor to run new business activities is key to this dispossession. These approaches are widespread including in Nepal, the Philippines and Madagascar.[60] There are exceptions, such as Namibia, where the territorial interests of communities have been recognized as the institutional basis for more equitable conservation[61] and in Nicaragua where in Bosawas the recognition of local land rights and community visions, rather than conservation, was the starting point.[62]

Approaches that disenfranchise local communities can create conflicts over how land and nature is managed, valued and utilized.[63] This includes limiting or prohibiting traditional practice thus forcing communities to abandon local livelihood and conservation practices in favour of externally-sanctioned (coercive) conservation practices.[64] In 2009, Agrawal estimated that displacement from conservation, although lower than that from development, has effected at least tens of millions of people.[65] The case study literature suggests that this number has continued to grow as local communities are displaced to make way for preservationist, state or private management.[66] A further quandary is that the link between conservation and economic activities may motivate additional grabs, particularly when protected areas create additional value. In the Philippines, the largest REDD+ project of 200,000 ha is also a hot spot for grabbing for real estate and tourism.[67]

Despite variations regionally and nationally, the discussion has illustrated that three broadly differentiated forms of land and green grabs are evident—these are

---

[57] Green and Adams (2015) and Hall et al. (2015).

[58] Dressler et al. (2010) and Green and Adams (2015).

[59] Murombedzi (2010), Nelson (2010), and Benjaminsen and Bryceson (2012).

[60] Dressler et al. (2010).

[61] Nelson (2010).

[62] Dressler et al. (2010).

[63] Mohamed-Katerere (2014).

[64] Busscher et al. (2018), Franco and Borras Jr (2019), and Myers and Mumu (2015).

[65] Agrawal and Redford (2009).

[66] Kashwan (2021), Chiaravalloti (2019), Lunstrum et al. (2016), Brockington and Wilkie (2015), Myers and Mumu (2015), Martin et al. (2015), Lunstrum (2015), and Creado (2008).

[67] Franco and Borras Jr (2019).

Table 1 Forms of land acquisition and the prevalence in different land use groups

	Enclaves	Commercial ventures incorporating smallholders	Expropriation without dispossession (Co-Management)
Acquisitions for food-feed-fuel	X	X	
Expropriations for mining and logging	X		X
Biodiversity Conservation regimes	X		X
Carbon sequestration initiatives	x	X	X

based on new patterns of control that emerge from these lawful processes of acquisition of land rights, that cut across different land use regimes (Table 1):[68]

1. *Land and resource enclaves*, that result in dispossession and displacement of communities. These enclaves (or enclosures) are associated with large agriculture estates, plantations, forests, protected areas and mines. Agribusiness, extractive industries, large conservation organizations, and development agencies are key actors.
2. *Large commercial ventures*, that engage smallholders in the value chain for example as contract farmers are associated with large agricultural estates, small farms and also forests. Agribusiness, conservation and development organizations key actors.
3. *Expropriations-without-dispossession* extinguish, restrict or limit *de facto* or *de jure* entitlements of local communities, including indigenous people, in exchange for financial benefits and engagement in co-management. Large conservation organizations, development agencies, traditional, national or local government, the private sector and local organizations pursuing externally-driven conservation goals, are key actors.

The literature reveals that as a result of these grabs, three critical transitions have occurred that redefine rural economies, political-scapes and wellbeing.

First, the control over land and natural resources has shifted from smallholders to external actors. Investors and corporates have emerged as 'shadow sovereigns' while the role of large conservation organizations, development agencies and foreign governments play in shaping labour laws, finance, food and agriculture policy, conservation, taxes and trade, and dispute resolution is amplified.[69] The appropriation of control festers unequal distribution of benefits and costs resulting in new proletarianization and new rural 'strong men,' reshaping political and economic

[68] Hall (2011) and Hall et al. (2015).

[69] George (2015).

relations.[70] Increasing economic differentiation further marginalizes women, youth, pastoralists and other vulnerable people and brings the newly dispossessed into a place of heightened vulnerability.[71]

Second, changes in land mosaics create high land inequality that rolls back earlier gains made by smallholders in redistributive national law reforms.[72] Concentrations of land holdings by a small number of powerful companies and actors, particularly in Latin America and South East Asia is evident,[73] where about 10% of landholders control 75% of agricultural land while the poorest 50% control only 2% of land.[74] Land concentration exacerbates landlessness, elite capture, food insecurity, gender inequality and restricts access to credit for small farmers.[75] Declines in rural population as a percentage of total population are most significant in south east Asia.[76] In all regions, urbanization has adverse livelihood impacts including unemployment and slum dwelling, that compounds inequality, vulnerability and poverty traps.[77] Consistent with historical trends, concentration also contributes to deforestation, environmental and soil degradation and, as discussed, the conversion of protected areas,[78] and greenhouse gas emissions.[79] Land clearance, for example, increases carbon emissions compared to historical deforestation rates and soil degradation,[80] which impacts immediately, but also has long-term climate change consequences. These social and environmental insecurities feed the occurrence, risk and magnitude of conflict and violence and reduce opportunities.[81]

Third, a significant reallocation of private and public funding away from community-focused agricultural extension and research to strengthen the smallholder sector, to agribusiness research and development, infrastructure development and new conservation forms is taking place.[82] For example, the largest private funder of agriculture, the Bill and Melinda Gates Foundation, reallocated funding in favour of four large agriculture networks engaged in research, technology and implementation of pro-agribusiness policy, which receive approximately half of their grants for agriculture, and another 40% went to organizations in the US, UK and Europe.[83] Fiscal policies are being (re)designed to favour investment rather than the security of

---

[70] Chimhowu (2019).

[71] Oberlack et al. (2016) and Park (2018).

[72] Hall (2011) and Kay (2015).

[73] Kay (2015) and Borras Jr (2011).

[74] Bauluz et al. (2020).

[75] Guerena and Wegerif (2019), Carrero et al. (2020), and Mechiche-Alami et al. (2021).

[76] Bauluz et al. (2020).

[77] Kay (2015) and Nolte and Ostermeier (2017).

[78] Schoneveld (2017) and de Toledo et al. (2017).

[79] Liao et al. (2021).

[80] Liao et al. (2021).

[81] Guerena and Wegerif (2019), Kay (2015), and Mohamed-Katerere (2014).

[82] Moyo and Yeros (2005).

[83] Grain (2014).

poor and marginalized communities.[84] In Brazil, state sponsored support for agribusiness, including through credit, led to 3.7-fold increase in agricultural productivity between 1975 and 2010, leaving only 10% of farms responsible for 85% of the gross value of production.[85] This reduction in support for smallholders reduces the opportunities available to smallholders.

3 The Injustice of Land Governance

Power asymmetries and coercion provide the canvas for grabbing in all regions, more so where land is subject to multiple claims[86] and governance alienates rural citizens.[87] This makes it possible for powerful external actors, as discussed below, to influence development narratives and to use institutions to their advantage.[88] Rural citizens are not simply passive in these processes but actively engage—resisting or identifying opportunities to earn some benefit from the neo-liberalization of rural landscapes.[89] While experiences vary among and within regions, as this part shows, these realities contribute to entrenching business and economic interests over other considerations of justice.[90]

The alignment of government interests with those of foreign and national investors, corporates, global financial and development organizations, global donors and philanthropists, and national and rural elites encourages investment in land.[91] Governments reportedly supported investments in about 89% of cases by lowering barriers to investment and trade, identifying available land, deregulating land markets, and making land available through reclassification, re-designation and reallocation.[92] Potential contributions to development and foreign exchange earnings are key motivations for this support.[93] Global actors play a key role in shaping national land and acquisition policies: the multi-million-dollar World Bank led programme, *Enabling the Business of Agriculture,*[94] for example, drives policy and legal reform to facilitate agribusiness. National laws and policies—such Ethiopia's *Growth and Transformation Plan* and the Zambia *Lands Act* 1995 which convert customary land to leasehold land are national tools facilitating the

[84] Fairhead et al. (2012).

[85] de Toledo et al. (2017).

[86] Dell'Angelo et al. (2017), Cochrane (2016), and Schoneveld (2017).

[87] Ece et al. (2017), Schoneveld (2017), Notess et al. (2020), and Dell'Angelo et al. (2017).

[88] Scoones (2016), Benjaminsen and Bryceson (2012), and Holmes (2014).

[89] Hall et al. (2015) and Green and Adams (2015).

[90] Brent (2015).

[91] Dell'Angelo et al. (2017), Schoneveld (2017), Wolford et al. (2013), and Busscher et al. (2018).

[92] Borras Jr et al. (2020), Wily (2014), Dell'Angelo et al. (2021), and Busscher et al. (2018).

[93] Cotula et al. (2014) and Schoneveld (2017).

[94] World Bank (2013).

'business of agriculture' by opening up land to foreign investment for export markets.[95]

Policy convergences between investor and host countries are important[96]—such as the multilateral and bilateral agreements for climate change mitigation.[97] The EU biofuels policies are a major component of many land deals and are motivated by a desire to achieve energy security in Europe given an existing deficit and growing consumption, while reducing CO_2 emissions[98] as agreed in the *Paris Agreement*. In Zambia, between 2003 and 2016, companies from the EU expressed interest in 370,000 ha of land to produce crops, that included biofuels.[99] EU-based corporations engagement in land grabs is facilitated and legitimized by finance capital actively seeking investment in foreign resources, public-private partnerships that support investment in foreign resources and development financing institutions that back and facilitate grabs[100] and narratives of scarcity.[101] In Cambodia, Myanmar, Laos and Indonesia, where government climate policies focus on producing biofuels logging companies are able to secure concessions for biofuels production in hardwood forests.[102] Many companies abandon these lands once clearing operations are completed without planting for biofuels to reap quick gains and this results in massive deforestation—in Cambodia, nearly half of the area granted in concessions between 2000 and 2012 was forested in 2000.[103] In these contexts developed countries benefit disproportionately and at the expense of the global south.[104]

Acquisitions occur in the context of the enduring legacy of colonial land histories in which land ownership and governance is contested and inequalities are linked to structural imbalances in opportunity and wealth.[105] In southern Africa, legal pluralism and associated tenure limits entitlements to use of local communities, while the state remains the owner or trustee.[106] In Nepal, a similar state-as-landlord relationship means the power of government to grant, confiscate land or designate its use is absolute.[107] Habitually contemporary tenure regimes fail to incorporate layered tenure rights such as the seasonal or migratory rights of pastoralist that exist alongside rights of smallholders.[108] Land reform that formalized individual title in

[95] Nolte and Ostermeier (2017) and Chilombo et al. (2019).

[96] Dell'Angelo et al. (2021).

[97] Büscher and Fletcher (2019) and Borras Jr and Franco (2018).

[98] Carroccio et al. (2016).

[99] Chilombo et al. (2019) and Cotula et al. (2014).

[100] Borras Jr et al. (2020).

[101] Scoones et al. (2019).

[102] Franco and Borras Jr (2019) and McAllister (2015).

[103] Franco and Borras Jr (2019).

[104] Kotzé (2019).

[105] Cochrane (2016) and German et al. (2013).

[106] Cotula et al. (2014), Schoneveld (2017), and German et al. (2013).

[107] Lai and Saumik (2014).

[108] Davies et al. (2016).

communal areas increased market influence over land and investor penetration of rural areas.[109] In sub-Saharan Africa between 1990 and 2017, 32 new land laws reformed rights on customary land inadvertently increasing local vulnerability to land and green grabbing.[110]

Worldwide legal regimes for mining vest title of the subsoil in the state or define mineral exploitation as a public utility that takes precedence over other uses. In these circumstances, even secure community rights are at risk: in Chhattisgarh, India, the district administration rescinded approved community forest title to allow for coal mining.[111] The recognition of community rights to free, prior informed consent (FPIC) is beginning to gain ground in regional and national. In South Africa, for example, a high court decision related to the protracted claims of the Xolobeni community against the government and an Australian mining company ruled, in line with the principle of FPIC, that the community has a right to say no to the projects.[112]

Many global land governance tools focus on ensuring 'responsible' investments, including certification, standards and safeguards against harm.[113] Safeguard approaches are not effective where recognition, enforcement, and the protection of economic and social rights are weak and political and economic power is highly uneven.[114] The core rights of indigenous peoples and local communities to recognition and self-determination are often inadequately protected in responsible investment instruments; many global agreements focus on ensuring benefits accrue to communities but do not recognise local land rights[115] and frequently there is no appropriate international mechanism to include rural people in decision making exists.[116] Certification schemes are often criticized for prioritizing economic benefits over land rights.[117] The Forest Stewardship Council, for example, controversially certified industrial tree plantations on high value conservation land the Harvard Management Company acquired in Iberá, Argentina that resulted in displacement.[118] A further challenge with investment agreements is that communities are not party to them and thus not able to secure justice through the law.[119] Strengthening the

[109] Chimhowu (2019) and Hall (2011).
[110] Chimhowu (2019).
[111] Kashwan (2021).
[112] Huizenga (2019).
[113] Lynggaard and Ravborg (2020).
[114] Ranjan (2017).
[115] Lynggaard and Ravborg (2020).
[116] Ranjan (2017).
[117] Fortin and Richardson (2013).
[118] Busscher et al. (2018).
[119] Cotula (2020) and Perrone (2019).

regulatory role of the state related to transnational corporations and other business enterprises is needed to protected the land rights of rural citizens.[120]

4 Justice Depraved

Justice and law should not be conflated. Justice is concerned with what is right and wrong and provides a normative framing for evaluating land acquisitions and determining whether they are illicit or not, irrespective of their legality. Approaches to justice are not universal, thus the adoption of any justice framework privileges one world view over another. For example, the approach taken here is that the achievement of justice requires existing patterns of social privilege—based on class, gender, race and ethnicity—and their intersection with the allocation of power and resources, including the rules and policies that reproduces inequality and dominance, to be dismantled. In contrast, the assumption in neoliberal approaches is that justice is attainable through distributive benefits in wealth and food security from trickle-down economics.[121]

Drawing on environmental justice and political ecology,[122] the justice framework employed here (Fig. 1) includes the well-established dimensions of recognition, procedural justice, distributional justice, and capabilities.[123] In addition, as distributive justice is commonly interpreted as restricted to the allocation of benefits and costs, a fifth dimension of redistributive justice is added. Redistributive justice focuses on economic and material inequality in property, food, education, work and multiple other aspects of society and addresses how groups experience domination, alienation and social inequality—and it has at core the deconstruction of the underlying structure.[124]

The failure to fulfil any of these dimensions compromises the ability to fulfil others, however, the way in which different remedies are implemented can have adverse outcomes for transformative justice. Despite tension between the different dimensions of justice, this should not detract from the ways in which the proposed dimensions can be mutually reinforcing.[125] For example, the struggle of indigenous people for recognition is closely aligned with struggles to improve material wellbeing while delinking this from inequitable extractivist productive systems and to pursue alternative values and pathways.

In each of these dimensions, human rights can make a significant contribution as they increasingly used to address the rights of collectives in public

---

[120] Kashwan (2021).

[121] Kashwan (2021).

[122] Suiseeya (2017), Sikor (2013), and Martin et al. (2016).

[123] Dirth et al. (2020).

[124] Fraser (1995).

[125] Young (undated).

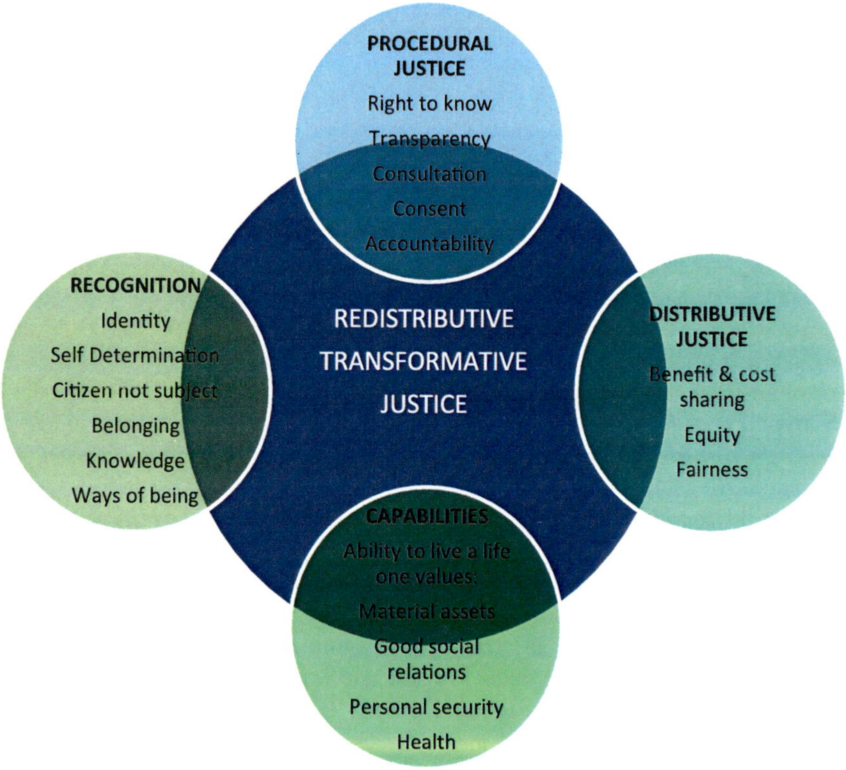

Fig. 1 Justice Framework

decision-making, although they have had limited value in addressing structural causes of inequality.[126] Human Rights applications have also failed to provide a sufficient methodology for addressing and curtailing abuses by corporates.[127] This is partly because they are not universally recognized as creating positive obligations and limitations on power of corporates and non-state organizations engaged in public decision-making, policy and related activities.[128] A further challenge is that many local communities and indigenous people are put in the precarious position of being pressurized to accept the solutions because they participated in certain processes.[129]

The following sections elaborate on these dimensions, with examples of their expression in the law, policy or norms that frame land and green grabs. The discussion illustrates that these acquisitions, even when legal, are fraught with

[126] Myon (2014) and Kotzé (2019).
[127] Myon (2014).
[128] Cotula (2020) and Huizenga (2019).
[129] Hendry and Tatum (2016).

political tension among diverse actors of completing interests and worldviews. It is also apparent that these acquisitions place the burden of addressing food insecurity, conservation, climate change on rural people who are already politically marginalized, vulnerable and frequently poor. [130]

4.1 Redistributive Justice

Redistributive justice provides the foundation for transformation and creating the context in which citizens can achieve lives and futures they value. It requires a just distribution of resources and power that challenges the political-economic structure of society including associated patterns of inequality and inequity. Without addressing the economic marginalization and deprivation experienced by people as a group and the related norms of allocation, discrimination and privilege,[131] the opportunities available to local communities to change their realities remain insufficient. Recognition provides the foundation for redistributive justice. [132]

Land and green grabs counter redistributive justice by increasing land inequality and, in particular, seizing the land of local communities and indigenous people. While redistributive is not limited to restitution, this is an important remedy increasingly ordered by courts in land rights cases of indigenous people. The landmark African Commission on Human Rights decision on the rights of the Endorois people triggered negotiations towards land restitution.[133] One challenge is that restitution can be purely symbolic if it is not accompanied by control of land, this occurs for example when communities are pressured into co-management arrangements or into leasing their land back to the state.[134]

In the context of land, redistributive justice encompasses the twin principles of guaranteeing a minimum land access and imposing a maximum land size to establish fairness for historically-disadvantaged groups—both diametrically opposed to land and green grabs.[135] The first principle relates to sufficiency and the capabilities dimensions of justice, that is the idea that all people should be able to live meaningful lives that they value. The second principle focuses on power and is transformative by preventing those with financial resources or political connections from endless accumulation that forces others off the land, as is the case with land and green grabs. Land and green grabs have led to land concentration and increasing power disparities between governments and corporates, and local communities and indigenous people.[136] Globally about 16% of farms control 88% of land, the

[130] Kashwan (2021).

[131] Fraser (2000).

[132] Fraser (1995) and Young (undated).

[133] Open Society Foundation (2017).

[134] Koot and Büscher (2019).

[135] Borras Jr and Franco (2018).

[136] de Toledo et al. (2017) and Hall (2011).

remaining 12% of land is shared among 84% of farms; in the most unequal countries less than 1% of owners control 50% or more of agricultural land.[137]

Achieving redistributive justice requires transformative remedies that aim to reduce inequalities and redistribute power in institutions, relationships between people, and discourses and norms by addressing underlying causes, such as cultural domination and exclusion. In this sense, it is distinct from other elements of the framework, which can also be implemented through affirmative remedies that address inequitable outcomes without disturbing the underlying political-economic system that generates them.[138]

4.2 Recognition

Recognition requires acknowledging the identity—knowledge, values, culture and histories—and legal rights of groups and respecting these in decision-making.[139] The lack of recognition of local communities' and indigenous peoples' underpins land grabs by marginalizing them politically and not recognizing their cultural land values.[140] A run-on consequence of this exclusion is deprivation—indigenous people are three times more likely to live in poverty compared to others and have a high risk of vulnerability.[141]

The hegemony of global discourses that link ideas of modernity and progress to portrayals of local communities as inefficient, backward and incapable of making needed technical interventions to manage scarce resources lead to political and cultural dominance, making it difficult for rural citizens to prevent land grabs.[142] Many governments value large-scale interventions, whether for agriculture or conservation, over small-scale approaches as a means to achieve food security, as is evident since the food price crisis of 2006–2008.[143] Frequently, interventions adopted, including those focused on the commodification of nature or the replacement of traditional agriculture with monocultures, run counter to the widely held indigenous knowledge systems that incorporate responsibility to nature in their world view and see identity of people and land as interdependent.[144] Repeatedly, the outcome of these kinds of impositions is displacement, imposed management practices, the criminalization of local use and political exclusion.[145] In this context

[137] Guerena and Wegerif (2019).

[138] Fraser (1995).

[139] Martin et al. (2015, 2016).

[140] Fladvad et al. (2020).

[141] Open Society Foundation (2017).

[142] Borras Jr et al. (2011) and Scoones et al. (2019).

[143] Busscher et al. (2018) and German et al. (2013).

[144] Koot and Büscher (2019), Hendry and Tatum (2016), and Open Society Foundation (2017).

[145] Martin et al. (2016), Hulme and Murphree (2001), and Schoneveld (2017).

indigenous people and local communities face an onerous legal burden to protect their land.[146]

Institutions at the global level provide a framework for communities to demand recognition, protect their land rights and claim self-determination. The *United Nations Declaration of the Rights of Indigenous People* (UNDRIP), which codifies and builds on international customary law and other conventions, establishes FPIC and self-determination of Indigenous People regarding lands, territories and other resources as rights. The *United Nations Declaration on the Rights of Peasants and other People living in rural areas* recognizes the rights to land, both individually and collectively, to sustainably use and manage resources, including land, water and pastures, needed for an adequate standard of living, to live in peace and with security and dignity, and to culture development. Regional instruments and decisions of regional and national courts are important, but limited mechanisms, to claim and protect rights.[147] In *Mayagna (Sumo) Awas Tingi community v Nicaragua,*[148] the Court ordered the government to delimit, demarcate and title indigenous property according to indigenous law.[149] A key challenge is the gap between rhetorical recognition and actual fulfilment, including the failure to implement court decisions, however, even when not implemented these legal victories have helped strengthen the confidence and agency of communities and foster new partnerships as more people become aware of the injustice inherent in land grabs.[150]

Even where institutions exist that provide for recognition, internalized cultural hierarchies and informal institutions of prejudice may limit meaningful enjoyment of these rights.[151] For example, the use of criteria for recognition that differ from those used by indigenous people or local communities in administrative process and development and conservation projects can prevent them from enjoying their rights.[152] Static understandings of indigeneity that imagine indigenous people as primordial, for example, by requiring that the community maintains traditional culture, property regimes, and social and leadership structures and demonstrates genealogical descent, fail to acknowledge processes of change.[153] The imposition of such requirements may deny indigenous status to communities whose ancestors arrived generations ago, but later than the ancestors of other people. In Brazil, national policy recognizes self-recognition as the formal way to identify traditional communities but in the Pantanal, conservation organizations justified the displacement of local fisher folk because they are genealogical dissimilar to other indigenous

[146] Open Society Foundation (2017) and Anonymous (2016).

[147] Open Society Foundation (2017).

[148] Inter-American Court no 79, 31 August 2001.

[149] Anonymous (2016).

[150] Open Society Foundation (2017).

[151] Martin et al. (2016).

[152] Gil and Linares (2019).

[153] Creado et al. (2008) and Koot et al. (2019).

inhabitants, even though they have a long history in the region, use sustainable customary practices and self-identify as indigenous.[154]

4.3 *Procedural Justice*

Although states are responsible for development, it is now widely accepted that citizens should consent to activities that impact on their wellbeing and livelihoods,[155] even where these are purported to be in the public interest. Procedural justice relates to how citizens are engaged in decision-making, who is involved and the influence they yield. Representation, accountability, participation, consultation, FPIC, access to information, transparency and redress are procedural mechanisms recognized to various extents in national and international law and can support local bargaining and more equitable outcomes.

However, for rural citizen these procedural rights are regularly hollowed out because political systems marginalize rural people by casting them as subjects rather than citizens.[156] Frequently the implementation mechanisms for procedural rights in administrative or project standards fall short of global requirements even when community land rights are formally protected.[157] In Peru, for example, rights to FPIC are negated by denying recognition to communities in the mining law[158] and this led to a steep rise in mining concessions as a percentage of land from 7.3% in 2009 to 20.23% in 2014.[159] A further challenge is that many countries limit the right of FPIC to displacement and dispossession and to indigenous people.[160] Further, to be effective FPIC must include the right to say no, rather than simply to collaboratively frame the investment, to do otherwise is inherently undemocratic.[161]

In general, communities have limited opportunities to challenge investing companies in national courts as already mentioned they are not parties to investment agreements and have no rights to demand review by national courts.[162] However, communities can and do challenge state actions in national and regional courts. For example, in Indonesia, the courts have ruled that forest designation without FPIC is arbitrary and illegitimate.[163] Communities are also able to defend their rights in

[154] Chiaravalloti (2019).
[155] Ece et al. (2017).
[156] Mamdani (1997) and German et al. (2013).
[157] Schoneveld (2017), Gil and Linares (2019), Ece et al. (2017), and Nuesiri (2017).
[158] Gil and Linares (2019).
[159] Borras Jr and Franco (2012).
[160] Anonymous (2016).
[161] Ece et al. (2017).
[162] Perrone (2019) and Sachs (2020).
[163] Myers and Mumu (2015).

regional courts, the *Inter-American Court of Human Rights* has ruled in favour of communities in disputes over mining, logging and agribusiness.[164]

Opportunities for rent capture by traditional leadership and local government may serve as a disincentive for genuine representation and consent when deciding inland investments.[165] In Mozambique, the 1997 land law requires investors to consult communities to identify land that is not occupied or to negotiate use with them, but investors tend to 'sell the project' to a few representatives on the promise of jobs, food security or development.[166] The selection of representatives by developers or governments rather than by the impacted communities further distorts consultation.[167] Ece et al, in a study of 13 African countries, find a "systematic disjuncture between donor and government claims of participation, representation in policy-making, and project development and implementation."[168]

The lack of legitimacy and ability to lawfully contest grabs frequently leads to conflict between the state and adversely impacted people, and between investors and communities, as well as within communities.[169] Communities may engage in direct protests or 'everyday' protest such as pilfering, arson, and preventing tourists from entering protected areas.[170] In the absence of acceptance of oppositions, those contesting investments may experience exclusion and social conflict that contributes to increased vulnerability.[171] Human rights and environmental defenders, especially in the extractive and agrarian sectors are often targeted and face criminalization, violence and assassinations.[172] In 2020, 69% of the 331 defenders killed land and indigenous peoples rights and the environment.[173] This disregard continues to take place despite a 2019 United Nations Human Rights Council resolution requiring states to create a safe and enabling environment for defenders and calling upon corporates to respect human rights.

In a meta-analysis Oberlack et al. identify four main archetypes of vulnerability that are directly linked to the governance of land and green grabs, these are: asset enclosure, elite capture, selective marginalization in which one group experiences a reduction in assets, and the polarization of development discourses among citizens leading to conflict.[174] Contributing governance injustices include asymmetries in negotiation, selective exclusion and weak accountability and consultation that privileges private capital interests over local rights and interests in national law and

---

164 Anonymous (2016).

165 Schoneveld (2017).

166 Shilomboleni and Yildiz (2017).

167 Baruah (2017).

168 Ece et al. (2017), p. 358.

169 Borras Jr and Franco (2012), Hall et al. (2015), and Mohamed-Katerere (2014).

170 Holmes (2007), McAllister (2015), Fletcher et al. (2015), Springer (2015), and Kay (2015).

171 Springer (2015) and Oberlack et al. (2016).

172 Scheidel and Gingrich (2020) and Open Society Foundation (2017).

173 Frontline Defenders (2021).

174 Oberlack et al. (2016).

contractual agreements.[175] Vulnerability may also be linked to failure of states and corporate sector to fulfil their human rights obligations to impacted peoples.[176] It is plausible that less democratic and more repressive state responses will become the norm as states respond to discontentment, as evidenced in structural adjustments, land and green grabs, and in response to the CoVID-19 Pandemic.[177]

4.4 Distributive Justice

Distributive justice is concerned with the distribution of benefits and costs of development and conservation interventions and is widely recognized as key to fairness. Empirical research finds that people, across different cultures, commonly consider sufficiency as an important benchmark for assessing justice.[178] However, the implementation of distributive justice is commonly restricted to the share of direct costs and benefits and does not address the structural and systemic distribution of resources.

Land and green grabs, in general, magnify the skewed distribution of costs and benefits nationally and globally by continuing inequitable development and conservation traditions[179] of placing the adverse impacts of displacement, elite capture, environmental damage and violent outcomes on poor people and rural citizens.[180] In some approaches, cost-benefit analysis considers benefits at a global or regional level (such as carbon sequestration) from grabs as being in the 'public interest' and offsetting costs or harm suffered locally.[181]

Displacement is a major cost of grabs—it magnifies vulnerability when social support networks are disrupted, affected people unable to gain access to alternative lands of similar extent, or receive adequate compensation.[182] Even where compensation is awarded, it is normally limited to individually held land and not benefits attained from common property resources, such as forest resources.[183] In Nepal, given the focus on present entitlements and not overall historical and cultural transformations, displaced Rana families without land registration received around 11% of their actual land, compared to an average compensation rate of 56% for those with registration.[184] The losses occurred in grabs are generally not offset by employment,

[175] Nel (2015), Ogwang et al. (2019), and Ece et al. (2017).

[176] Kok et al. (2010).

[177] Abouharb and Cingranelli (2006), Franklin (1997), Poe et al. (1999), Cotula (2021), and Frontline Defenders (2021).

[178] Martin et al. (2015).

[179] Hufe and Heuermann (2017) and Hulme and Murphree (2001).

[180] Kotzé (2019).

[181] Sikor et al. (2013).

[182] Oberlack et al. (2016).

[183] Schoneveld (2017).

[184] Lai and Saumik (2014).

which is the main benefit.[185] This cost-benefit approach effectively aggregates different, incommensurate, components of life and reduce these to a simple metric or a net balance to determine the validity of any action or decision.[186] This approach runs counter to a human rights approach which is based on the idea that an intolerable harm cannot be offset or justified by another good. Indeed, the popularity of human rights indicates the limited appropriateness this approach.[187]

4.5 *Capabilities, Sufficiency and Equality*

Capabilities are concerned with what citizens are able to do and be, and include the achievement of wellbeing and human dignity.[188] The *Millennium Ecosystem Assessment* proposes that the freedom to achieve a life one values is based on sufficient access to material assets, health, good social relations and personal security,[189] and this understanding is incorporated into the *Millennium Development Goals*.

The unequal opportunity structure of land and green grabs fails to ensure sufficiency and constrains the capabilities and wellbeing of impacted people by reducing access to required material assets from healthy soils to education, health, good social relations and physical security—as land inequality is increased, public and private support for small holders is diminished and control over land is lost. Importantly these impacts cumulatively reduces the freedom of choice of citizens.

The risk of the sufficiency failures is that the vulnerability of marginal groups including newly displaced or (partially) dispossessed people to future shocks such as climate change, extreme weather events, conflict, epidemics, environmental degradation, market volatility and so on will increase thus contributing to chronic poverty.[190] Vulnerability to food insecurity is a glaring outcome of land grabs particularly in Africa and Asia.[191] In a study of 38 Africa countries, Mechiche-Alami et al. find that only 20% of these countries are likely to improve food security, although they also face the risk of increase land pressure, conflict or deforestation which could adversely impact on food security.[192] In 50% of the countries studied, agricultural produce was primarily non-food or intended for export, while only 15.7% of produce was targeted for the local market.[193] Importantly land grabbing coincides with water grabs, including both rainwater and irrigation, affecting health

[185] Kay (2015), Schoneveld (2017), and Hufe and Heuermann (2017).

[186] Nussbaum (2007) in Martin (2016).

[187] Greiber et al. (2013).

[188] Martin et al. (2016).

[189] MEA (2005).

[190] Kok et al. (2010) and Sobhan (2014).

[191] Borras Jr et al. (2011).

[192] Mechiche-Alami et al. (2021).

[193] Mechiche-Alami et al. (2021).

and livelihoods in surrounding and downstream lands.[194] Pastoralists, given their cyclic use of land resources and their lack of continuous physical control, lose entitlements through land and green grabs.[195]

The tensions in social relationships that arise in the context of land and green grabs impact on agency and freedom. Many civil society initiatives face considerable social and political risk including being labelled as anti-development by powerful coalitions that support grabs.[196] In general, the judicial system has not held national and local government and traditional leaders responsible for multiple justice failings and adverse social outcomes.[197] National, local and traditional leaders are often embroiled in these conflicts as gate keepers and deal makers.[198] The proposed law in Brazil to allow title to be granted for illegal land claims on the basis of self-declaration is an extreme example of this.[199] Further, land and green grabbing has strengthened the presence and power of the state in rural areas, although this tends to be uneven and concentrated in particular zones, including transport or development corridors where special policy and legal regimes may apply including new tax regimes, separate administrative and judicial institutions to settle disputes, and a shift in regulatory authority from the legislature to the executive.[200] In Brazil dense infrastructure networks to meet the demands of the energy, agribusiness and mining place new pressures on territories of IPLC[201] and a similar direction is evident in eastern and southern Africa development corridors.

The conflict from grabs compromises the collaborative social relationships that could drive peaceful change processes to more equitable societies.[202] However, the emergence of networks and alliances in response to grabs is evident, including at a transnational level, and this strengthens local agency.[203] Many have been able to use formal institutions to deconstruct power and drive institutional change through diverse strategies including human rights approaches and extending rights claiming.[204] For example, REDD+ has been strongly influenced by the assertion of cultural ties to land by indigenous people that focus on the value of land as a place of belonging, for the exercise of sovereignty and revenue sharing,[205] even though practice has lagged behind the rhetoric.[206] Strategic litigation has also been used to

194 Rulli et al. (2013).

195 Davies et al. (2016).

196 Wolford et al. (2013).

197 Anonymous (2016).

198 Hall (2011).

199 Carrero et al. (2020).

200 Cotula (2020).

201 de Toledo et al. (2017).

202 Martin et al. (2016) and Mohamed-Katerere (2014).

203 Avellino and Wittmayer (2016).

204 Park (2018) and Fladvad et al. (2020).

205 Sikor et al. (2013).

206 Suiseeya (2017).

create awareness around rights of local communities and indigenous people. Some governments have responded by transferring forest tenure to excluded groups to address historical injustice.[207] Cotula suggests that concerns around 'grabs' has led to resource nationalism and that the world might be entering a phase of economic disintegration.[208] There is no certainty as to what the contemporary political-scape will mean for capabilities and the ability of rural citizen to shape their own futures, much depends on 'unruly politics'[209] and the agency of those marginalized through land and green grabs.

5 Conclusion

This chapter has analysed multiple interlinked justice failings in land and green grabs. It has demonstrated that the current governance and institutional framework privileges powerful coalitions of global and national elites over and above citizens living with the targeted resources. It is evident that the emergence of these new neoliberal models of accumulation take advantage of the colonial legacies of weak tenure and reducing rural citizens to subjects. This is linked to imbalances in governance and the very direct ways in which procedural rights are hollowed out. Human rights approaches are only partially implemented and this as well as the lack of clear limitations on the actions of corporates lies at the heart of this institution failure. Additionally, the lack of recognition of local communities, including their land rights and cultural and political values compounds the governance injustices.

Social justice and human rights do not have the same level of political commitment or legal mandate as market based development for multiple reasons,[210] instead there is an assumption that justice will be achieved through material improvements in wellbeing from trickle down mechanisms.[211] Measurable adverse impacts on wellbeing and vulnerability have occurred, particular as the opportunities created, such as employment, do not outweigh the costs imposed. Many of the benefits associated with grabs are experienced globally or at a national level, such as reductions in carbon emissions or increases in foreign exchange earnings respectively. Justice will require more serious consideration of equity in distributive and redistributive mechanisms.

Importantly, as demonstrated, governance and livelihood injustice combine to hook rural communities into these neoliberal approaches and reduce the choices available to them. This is underpinned by the active denial and locking out of other options and paths to prosperity. It is clear that political and institutional systems,

---

[207] Sikor et al. (2014).

[208] Cotula (2020).

[209] Scoones (2016).

[210] Mechiche-Alami (2021), Wolford et al. (2013), Fletcher et al. (2015), and Koot et al. (2019).

[211] Kashwan et al. (2021).

including at a global level, need to develop mechanism for listening to and respecting the views and aspirations of those that live in rural communities. As long as these citizens are imposed upon, development and conservation policies will remain coercive and unjust.

References

Abouharb MR, Cingranelli DL (2006) The human rights effects of World Bank Structural Adjustment, 1981–2000. Int Stud Q 50(2):233–262

Agrawal A, Redford K (2009) Conservation and displacement: an overview. Conserv Soc 7:1–10

Agrawal A, Brown DGJ, Sullivan A (2019) Are global land grabs ticking socio-environmental bombs or just inefficient investments? One Earth 1(2):159–162

Anonymous (2016) The double life of international law: indigenous people and extractive industries. Harv Law Rev 129:1755

Armendáriz-Villegas EJ, de la Ángeles Covarrubias-García E, Troyo-Diéguez E, Lagunes E, Arreola-Lizárraga A et al (2015) Metal mining and natural protected areas in Mexico. Environ Sci Policy 48:9–19

Avellino F, Wittmayer JM (2016) Shifting power relations in sustainability transitions: a multi-actor perspective. J Environ Policy Plann 18(5):628–649

Baruah M (2017) Facipulation and elite formation: community resource management in Southwestern Ghana. Conserv Soc 15:371–383

Bauluz L, Govind Y, Novokmet F (2020) Global inequality. World inequality database 10/2020

BBC (2021) Amazon rainforest plots sold via Facebook market place. 5 March 2021. https://www.bbc.com/news/technology-56168844

Benjaminsen TA, Bryceson I (2012) Conservation, green/blue grabbing and accumulation by dispossession in Tanzania. J Peasant Stud 39(2):335–355

Bhola N, Klimmek H, Kingston N, Burgess ND, van Soesbergen A, Corrigan C, Harrison J, Kok MTJ (2021) Perspectives on area-based conservation and its meaning for future biodiversity policy. Conserv Biol 35(1):168–178

Borras SM Jr, Franco JC (2012) Global land grabbing and trajectories of agrarian change: a preliminary analysis. J Agrarian Chang 12(1):34–59

Borras SM Jr, Franco JC (2018) The challenge of locating land-based climate change mitigation and adaptation politics within a social justice perspective: towards an idea of agrarian climate justice. Third World Q 39(7):1308–1325

Borras SM Jr, Franco J, Gómez S, Kay C, Spoor M (2011) Land grabbing in Latin America and the Caribbean. FAO

Borras SM Jr, Mills EN, Seufert P, Backes S, Fyfe D, Herre R, Michéle L (2020) Transnational land investment web: land grabs, TNC and the challenge of global governance. Globalizations 17(4): 608–628

Brent Z (2015) Territorial restructuring and resistance in Argentina. J Peasant Stud 42(3-4): 671–694

Brockington D, Wilkie D (2015) Protected areas and poverty. Philos Trans R Soc Lond Ser B Biol Sci 370(1681):20140271

Büscher B, Fletcher R (2019) Towards convivial conservation. Conserv Soc 17:283–296

Busscher N, Parra C, Vanclay F (2018) Land grabbing within a protected area: the experience of local communities with conservation and forestry activities in Los Esteros del Iberá, Argentina. Land Use 78:572–582

Carrero GC, Fearnside PM, do Valle DR et al (2020) Deforestation trajectories on a development frontier in the Brazilian Amazon: 35 years of settlement colonization, policy and economic shifts, and land accumulation. Environ Manag 66:966–984

Carroccio A, Crescimanno M, Galati A, Tulone A (2016) The land grabbing in the international scenario: the role of the EU in land grabbing
Chiaravalloti RM (2019) The displacement of insufficiently 'Traditional' communities: local fisheries in the Pantanal. Conserv Soc 17:173–183
Chilombo A, Fisher JA, van Der Horst D (2019) A conceptual framework for improving the understanding of large scale land acquisitions. Land Use Policy 88:104184
Chimhowu A (2019) The 'new' African customary land tenure. Characteristic, features and policy implications of a new paradigm. Land Use Policy 81:897–903
Cochrane L (2016) Land grabbing. In: Kaplan DM, Thompson PB (eds) Encyclopedia of food and agricultural ethics. Springer
Cotula L (2020) (Dis)integration in global resource governance: extractivism, human rights, and investment treaties. J Int Econ Law 23(2):431–454
Cotula L (2021) Towards a political economy of the COVID-19 crisis: reflections on an agenda for research and action. World Dev 138:105235
Cotula L, Oya C, Codjoe EA, Eid A, Kakraba-Ampeh M, Keeley J, Rizzo M (2014) Testing claims about large land deals in Africa: findings from a multi-country study. J Dev Stud 507:903–925
Creado ESJ, Mendes ABV, da Costa Ferreira L, de Campos SV (2008) Between "Traditional" and "Moderns": negotiations of rights in two protected areas of the Brazilian Amazon. http://socialsciences.scielo.org/pdf/s_asoc/v4nse/scs_a04.pdf
Davies J, Herrera P, Ruiz-Mirazo J, Batello C, Hannam I, Mohamed-Katerere J (2016) Improving governance of pastoral lands: implementing the voluntary guidelines on the responsible governance of tenure of land, fisheries and forests in the context of National Food Security. FAO Governance of Tenure Technical Guides 6
De Jong T, Sauerwein T (2021) State-owned minerals, village-owned land: how a shared property rights framework helped formalize artisanal diamond miners in Cote d'Ivoire between 1986 and 2016. Resourc Policy 70:101920
de Toledo PM, Dalla-Nora E, Vieira ICG, Aguiar APD, Araújo R (2017) Development paradigms contributing to the transformation of the Brazilian Amazon: do people matter? Curr Opin Environ Sustain 26–27:77–83
Dell'Angelo J, D'Odorico P, Rulli MC, Marchand P (2017) The tragedy of the grabbed commons: coercion and dispossession in the global land rush. World Dev 92:1–12
Dell'Angelo J, Navas G, Witteman M, D'Alisa G, Scheidel A, Temper L (2021) Commons grabbing and agribusiness: violence, resistance and social mobilization. Ecol Econ 184:107004
Dirth E, Biermann F, Kalfagianni A (2020) What do researchers mean when talking about justice? An empirical review of justice narratives in global change research. Earth Syst Gov 6:100042
Dressler W, Büscher B, Schoon M, Brockington D, Hayes T, Kull CA, McCarthy J, Shrestha K (2010) From hope to crisis and back again? A critical history of the global CBNRM narrative. Environ Conserv 37(1):5–15
Dwyer MB (2014) Micro-geopolitics: capitalising security in Laos's golden quadrangle. Geopolitics 19(2):377–405
Ece M, Mutombedzi J, Ribot J (2017) Disempowering democracy: local representation in community and carbon forestry in Africa. Conserv Soc 15:357–370
Fairhead J, Leach M, Scoones I (2012) Green grabbing: a new appropriation of nature? J Peasant Stud 39(2):237–261
Fladvad B, Klepp S, Dünckmann F (2020) Struggling against land loss: environmental (in)justice and the geography of emerging rights. Geoforum 117:80–89
Fletcher, R, Dressler W, Büscher B (2015) NatureTM Inc: nature as neoliberal capitalist imaginary. In: International handbook of political ecology
Fortin E, Richardson B (2013) Certification schemes and the governance of land: enforcing standards or enabling scrutiny? Globalizations 10(1):141–159
Franco JC, Borras S Jr (2019) Grey areas in green grabbing: subtle and indirect interconnections between climate change politics and land grabs and their implications for research. Land Use Policy 84:192–199

Franklin J (1997) IMF conditionality, threat perception, and political repression: a cross-national analysis. Comp Polit Stud 30:576–606
Fraser N (1995) From redistribution to recognition? Dilemmas of justice in a post-socialist age. New Left Review 1/212
Fraser N (2000) Rethinking recognition. New Left Rev 3(3):107–118
Frontline Defenders (2021) Global Analysis 2020. https://www.frontlinedefenders.org/sites/default/files/fld_global_analysis_2020.pdf
Garnett ST, Burgess ND, Fa JE et al (2018) A spatial overview of the global importance of indigenous lands for conservation. Nat Sustain 1:369–374
George S (2015) Shadow sovereigns. How global corporations are seizing power
German LA, Schoneveld GC, Mwangi E (2013) Contemporary processes of large-scale land acquisition in sub-Saharan Africa: legal failures, or elite capture of the rule of law? World Dev 48:1–18
Gil AG, Linares CC (2019) Mineralizing the right to prior consultation: from recognition to disregard of indigenous and peasant rights in Peru. Glob Jurists 20(1):20190019
Gómez CJ, Sánchez-Ayala L, Vargas GA (2015) Armed conflict, land grabs and primitive accumulation in Colombia: micro processes, macro trends and the puzzles in between. J Peasant Stud 42(2):255–274
GRAIN (2014) How does the Gates Foundation spend its money to feed the world. https://www.grain.org/article/entries/5064-how-does-the-gates-foundation-spend-its-money-to-feed-the-world
Green KE, Adams WA (2015) Green grabbing and the dynamics of local-level engagement with neo-liberalization in Tanzania's wildlife management areas. J Peasant Stud 42(1):97–117
Green Development Advocates (2021) De-gazetting and granting of 60 000 hectares of prime forest for oil palm production in CAMPO and Nyete sub divisions – Ocean division south Region of Cameroon. https://www.farmlandgrab.org/post/view/30289-degazetting-and-granting-of-60-000-ha-of-prime-forest-for-oil-palm-production-project-in-ocean-division-south-region-of-cameroon
Greiber T, Janki M, Orellana M, Savaresi A, Shelton DL (2013) Conservation with justice: a rights-based approach. International Union for Conservation of Nature, Environmental Law & Policy Paper No. 71
Guerena A, Wegerif M (2019) Land inequality framing document. International Land Coalition
Hall R (2011) The many faces of the investor rush in Southern Africa: towards a typology of commercial land deals. ICAS Rev Paper 2:28–29
Hall R, Edelman M, Borras S Jr, Scoones I, White B, Wolford W (2015) Resistance, acquiescence or incorporation? An introduction to land grabbing and political reactions 'from below'. J Peasant Stud 42(3-4):467–488
Hendry J, Tatum ML (2016) Human rights, indigenous peoples, and the pursuit of justice. Yale Law Policy Rev 34:351–386
Hitcher SL (2010) Heart of Borneo as a 'Jalan Tikus': exploring the links between indigenous rights, extractive and exploitative industries and Conservation at the World Conservation Congress 2008. Conserv Soc 8:320–330
Holmes G (2007) Protection, politics and protest: understanding resistance to conservation. Conserv Soc 5:184–201
Holmes G (2014) What is a land grab? Exploring green grabs, conservation, and private protected areas in southern Chile. J Peasant Stud 41(4):547–567
Hufe P, Heuermann DF (2017) The local impacts of large–scale land acquisitions: a review of case study evidence from sub-Saharan Africa. J Contemp Afr Stud 35(2):168–189
Huizenga D (2019) Governing territory in conditions of legal pluralism: living law and free, prior, and informed consent (FPIC) in Xolobeni, South Africa. Extractive Ind Soc 6(3):711–721
Hulme D, Murphree M (eds) (2001) African wildlife and livelihoods: the promise and performance of community conservation. James Currey Ltd., Oxford
International Land Coalition (2011) Global assembly 2011, Tirana, Albania

Kashwan P, Biermann F, Gupta A, Okereke C (2021) Planetary justice: prioritizing the poor in earth system governance. Earth Syst Gov 6:100075

Kay C (2015) The Agrarian question and the neoliberal rural transformation in Latin America. Eur Rev Latin Am Caribb Stud 100(100):73

Kinda H, Thiombiano N (2020) Effects of extractive industries in developing countries on deforestation. https://www.researchgate.net/project/Tax-Policies-and-Reforms-in-Developing-Countries-Big-Data-Analysis-Approaches

Kok MTJ, Jäger J, Karlsson SI, Lüdeke MB, Mohamed-Katerere JC, Thomalla F (2010) Vulnerability of people and the environment - challenges and opportunities. Background studies. Netherlands Environment Agency and UNEP

Kolstad I, Søreide T (2009) Corruption in natural resource management: implications for policy makers. Resourc Policy 34:214–226

Koot S, Büscher B (2019) Giving land (back)? The meaning of land in the indigenous politics of the South Kalahari Bushman Land Claims, South Africa. J South Afr Stud 42(19):357–374

Koot S, Hitchcock R, Gressier C (2019) Belonging, indigeneity, land and nature in Southern Africa under neoliberal capitalism: an overview. J South Afr Stud 45(2):341–355

Kopnina H, Washington H, Gray J, Taylor B (2018) The 'future of conservation' debate. Biol Conserv 217:140–148

Kotzé LJ (2019) A global environmental constitution for the anthropocene? Transnatl Environ Law 8(1):11–33

Kumar P (2020) "Modernization" and Agrarian Development in India, 1912–52. J Asian Stud 79(3):633–658

Lai L, Saumik P (2014) Disputed land rights and conservation-led displacement: a double whammy on the poor. Conserv Soc 12(1):65–76

Larson AM, Brockhaus M et al (2013) Land tenure and REDD+: the good, the bad and the ugly. Glob Environ Chang 23(3):678–689

Li T (2014) Land's end: capitalist relations on an indigenous frontier. Duke University Press, Durham

Liao C, Nolte K, Sullivan JA et al (2021) Carbon emissions from the global land rush and potential mitigation. Nat Food 2:15–18

Lunstrum E (2015) Green grabs, land grabs and the spatiality of displacement: eviction from Mozambique's Limpopo National Park. Area 48(2)

Lunstrum E, Bose P, Zalik A (2016) Environmental displacement: the common ground of climate change, extraction and conservation. Area 48(2):130–133

Lynggaard N, Ravborg HM (2020) Governing large-scale land-based investments: a literature review. Danish Institute for International Studies

Mamdani M (1997) Citizen and subject. Contemporary Africa and the legacy of late colonialism. Princeton University Press, Princeton

Martin A, Akol A, Gross-Camp N (2015) Towards an explicit justice framing of the social impacts of conservation. Conserv Soc 13:166–178

Martin A, Coolsaet B, Corbera E, Dawson NM, Fraser JA, Lehmann I, Rodriguez L (2016) Justice and conservation: the need to incorporate recognition. Biol Conserv 197:254–261

McAllister K (2015) Rubber, rights and resistance: the evolution of local struggles against a Chinese rubber concession in northern Laos. J Peasant Stud 42(3–4):813–837

McKay B (2017) Agrarian extractivism in Bolivia. World Dev 97:199–211

Mechiche-Alami A, Yagoubi J, Nicholas KA (2021) Agricultural land acquisitions unlikely to address the food security needs of African countries. World Dev 141:105384

Messerli P, Giger M, Dwyer B, Breu T, Eckert S (2014) The geography of large-scale land acquisitions: analyzing socio-ecological patterns of target contexts in the global South. Appl Geogr 53:449–459

Mohamed-Katerere JC (2014) Human rights: an opportunity for making adaptation conflict-sensitive. In: Bob U, Bronkhorst S (eds) Climate change and conflict: conflict-sensitive climate change adaptation in Africa. Berlin Science Publisher

Moyo S, Yeros P (2005) The resurgence of rural movements under neoliberalism. In: Moyo S, Yeros P (eds) 2005. Zed Books, Reclaiming the land: the resurgence of rural movements in Africa, Asia and Latin America
Murombedzi J (2010) Agrarian social change and post-colonial natural resource management interventions in Southern Africa's 'Communal Tenure Regimes'. In: Nelson F (ed) Community rights, conservation and contested land. The Politics of Natural Resource Governance in Africa, Earthscan
Myers R, Mumu M (2015) Searching for justice: rights vs benefits in Bukit Baka Bukit Raya National Park, Indonesia. Conserv Soc 13(4):370–381
Myon S (2014) A powerless companion: human rights in the age of neoliberalism. https://scholarship.law.duke.edu/cgi/viewcontent.cgi?article=4711&context=lcp
Napoletano B, Clark B (2020) An ecological-Marxist response to the half-earth project. Conserv Soc 18(1):37–49
Nel A (2015) The choreography of sacrifice: market environmentalism, bio politics and environmental damage. Geoforum 65:246–254
Nelson F (2010) The politics of Natural Resource Governance in Africa. In: Nelson F (ed) Community rights, conservation and contested land. The Politics of Natural Resource Governance in Africa, Earthscan
Nolte K, Ostermeier M (2017) Labour market effects of large-scale agricultural investment: conceptual considerations and estimated employment effects. World Dev 98:430–446
Notess L, Veit P, Monterroso A, Sulle E, Larson A, Gindroz S, Quaedvlieg J, Williams A (2020) Community land formalization and company land acquisition procedures: a review of 33 procedures in 15 countries. Land Use Policy 110:104461
Nuesiri EO (2017) Feigning democracy: performing representation in the UN-REDD funded Nigeria-REDD Programme. Conserv Soc 15:384–399
Nussbaum M (2007) Frontiers of justice: disability, nationality, species membership. Harvard University Press, Cambridge, MA
Oberlack C, Tejada L, Messerli P, Rist S, Giger M (2016) Sustainable livelihoods in the global land rush? Architypes of livelihood vulnerability and sustainability potentials. Glob Environ Chang Part A 41:153–171
Ogwang T, Vanclay F, van den Assem A (2019) Rent-seeking practices, local resource curse, and social conflict in Uganda' s emerging oil economy. Land 8(4):53
Ojeda D (2012) Green pretexts: ecotourism, neoliberal conservation and land grabbing in Tayrona National Natural Park, Colombia. J Peasant Stud 39(2):357–375
Oldekop JA, Rasmussen LV, Agrawal A et al (2020) Forest-linked livelihoods in a globalized world. Nat Plants 6:1400–1407
Open Society Foundation (2017) Strategic litigation impacts. Indigenous peoples' land rights. Technical Report. OSF
Park CMY (2018) "Our lands are our lives": gendered experiences of resistance to land grabbing in rural Cambodia. Fem Econ 25(4):22–44
Perrone N (2019) The 'invisible' local communities: foreign investor obligations, inclusiveness, and the international investment regime. AJIL Unbound 113:16–21
Poe SC, Tate C, Neal CKL (1999) Repression of the human right to physical integrity revisited: a global cross-national study covering the years 1976–1993. Int Stud Q 43:291–313
Qin S, Kroner RG, Cook C, Tesfaw AT, Braybrook R, Rodriquez CM, Poelking C, Mascia M (2019) Protected area downgrading, downsizing, and degazettement as a threat to iconic protected areas. Conserv Biol 33(6):1275–1285
Ranjan R (2017) Unravelling the narratives of Adivasi Dispossession: a case study of land acquisition in Nagri Village, Jharkhand. Development 60(3–4):227–234
Ranjan R (2018) The role of political-industry nexus in promoting illegal extraction of mineral resources and deforestation: a case of iron ore mining in Goa. Resourc Policy 57:122–136
Rulli MC, Saviori A, D'Odorico P (2013) Global land and water grabbing. PNAS 110:892–897

Sachs LE, Johnson L, Merrill E (2020) Environmental injustice: how treaties undermine human rights related to the environment. La Revue des Juristes de Sciences Po 18:90
Scheidel A, Gingrich S (2020) Toward sustainable and just forest recovery: research gaps and potentials for knowledge integration. One Earth 3(6):680–690
Schoneveld GC (2017) Host country governance and the African land rush: 7 reasons why large-scale farmland investments fail to contribute to sustainable development. Geoforum 83:119–132
Scoones I (2016) The politics of sustainability and development. Annu Rev Environ Dev 41(1): 293–319
Scoones I, Smalley R, Hall R, Tsikata D (2019) Narratives of scarcity: framing the global land rush. Geoforum 101:231–241
Shilomboleni H, Yildiz F (reviewing ed) (2017) A sustainability assessment framework for the African green revolution and food sovereignty models in southern Africa. Cogent Food Agric 3: 1
Sikor T, Auld G, Bebbington AJ, Benjaminsen TA, Gentry BS, Hunsberger C, Upton C (2013) Global land governance: from territory to flow? Curr Opin Environ Sustain 5(5):522–527
Sikor T, Martin A, Fisher J, He J (2014) Toward an empirical analysis of justice in ecosystem governance. Conserv Lett 7(6):524–532
Snyman S, Spenceley A (2019) Private sector tourism in conservation areas in Africa. CABI
Sobhan R (2014) Vulnerability traps and their effects on human development. UNDP HDR Occasional paper
Springer S (2015) The violence of neoliberalism. In: The handbook of neoliberalism. Routledge
Suiseeya MK (2017) Contesting justice in global forest governance: the promises and pitfalls of REDD+. Conserv Soc 15(2):189–200
TNI Millennium Ecosystem Assessment (2005) Ecosystems and human well-being: synthesis. Island Press, Washington, DC
Weeber S (2016) Nodes of resistance to green grabbing: a political ecology. Environ Soc Psychol 1(2):116–129
Wily LA (2014) The law and land grabbing: friend or foe? Law Dev Rev 7(2):207–242
Wolford W, Borras S Jr, Hall R, Scoones I, White B (2013) Governing global land deals: the role of the state in the rush for land. Dev Chang 44(2):189–210
World Bank (2013) Growing Africa: unlocking the potential of agribusiness. The World Bank, Washington DC
Young IM (undated) Unruly categories: a critique of Nancy Fraser's dual systems theory. http://bibliopreta.com.br/wp-content/uploads/2018/01/Fraser-Redistribution-Recognition-Dilema-1.pdf
Zoomers A (2010) Globalisation and the foreignization of space: seven processes driving the current global land grab. J Peasant Stud 37(2):429–447

Soil Steering Law in Germany

A Contribution to the Systematics of Soil Law

Christian Busse

Abstract The systematics of soil law is a challenging, but often neglected topic. The study offers an inside look at this topic, using the situation in Germany as an example. Special reference is made to the real estate transaction law within the fields of agriculture and forestry, established in Germany since 1918.

1 Objective of This Contribution

Soil forms the national territory of a state and is thus one of the foundations of any statehood, along with the "state people" and the "state power". The law on landed property, which determines the content, use, and transaction of the soil, is of eminent importance for every society and, therefore, for every state order. Not uncommon in the history of humankind, landed property relationships have been the starting point of revolutions. Hence, it would be expected that the law dealing with the various aspects of soil would in its entirety represent a well-researched legal-scientific matter. However, the opposite seems to be the case. In law, fundamental research and legal dogmatic discourses exist in numerous legal fields on a national and international level. The areas of sales law, marriage law, corporate law, constitutional law, and criminal law, which are classic fields of comparative law endeavours, are just but a few of the examples. Contrarily, the regulations that relate to land in a variety of ways, which will be referred to as soil law in the following, have received little attention as an independent area of law. So far, it has predominantly been

Any views expressed are purely personal.

All non-English quotations were translated by the author himself.

C. Busse (✉)
Faculty of Law and Economics of Rheinische Friedrich-Wilhelms-Universität Bonn, Bonn, Germany
e-mail: Christian.Busse@uni-bonn.de

H. Ginzky et al. (eds.), *International Yearbook of Soil Law and Policy 2020/2021*, International Yearbook of Soil Law and Policy 2020,
https://doi.org/10.1007/978-3-030-96347-7_3

individual aspects of soil law that have formed the subject matter of jurisprudential debates. Examples from the international discourse include land-related environmental law, legal aspects of land grabbing, state registration of land-related property rights, and the drawing of territorial boundaries between states.

The preceding volumes of the present Yearbook bear witness to this state of affairs. They contain a lot of valuable contributions towards the various facets of soil law, without a transnational discourse on the content and dogmatics of soil law having emerged in the Yearbook. In the self-description of the Yearbook, numerous aspects of soil law are addressed with a focus on soil protection and sustainable soil management, but the structure and dogmatics of soil law are not mentioned.[1] However, the fact that soil law should also be considered from such a point of view appears briefly in the preface to the first Yearbook, which was published in 2017, when it alludes "to establish a platform for discussion among legislators, lawyers and policymakers regarding regulatory concepts and approaches on the international, regional and national level."[2]

It is the general fate of legal doctrine to be a stepchild of the law, since—especially for soil law—the first priority is justifiably to describe and deal with current legal issues. However, it seems generally helpful in practical work on the law to have a clear concept of the respective area of law. This is an important instrument to quickly recognise and take into account interactions and to structure and formulate norms clearly enough. If the structure of an area of law remains in the dark, in the worst case, provisions may arise in different sub-areas that contradict each other or even cancel each other out in their effects. Cumulative interventions in fundamental rights may also occur, which only in combination exceed the limit of a constitutionally permissible limitation of fundamental rights. Soil law particularly faces such dangers, as its objectives are very heterogeneous, and numerous different bodies are involved in law-making and subsequent law application.

If the situation in Germany is considered, soil law as an independently conceived field of law is only rudimentary existent, although Germany is traditionally considered as a country with a penchant for jurisprudential system building. An important reason for this situation from the author's perspective seems to be the fact that soil law is divided into civil law, public law, and criminal law provisions, which stand side by side in a predominantly incoherent manner; both within these three areas of law and between these three areas of law. Above all, there is no central law that regulates the fundamentals of soil law. It therefore, remains difficult to form an external and internal system of soil law. Supposing the author's assessment is correct, this situation exists not only in Germany, but also in other states of the world. Against this background, the present contribution intends to provide some thought-provoking impulses for the development of a system inherent in soil law. In particular, the steering function as an important aspect of soil law will be examined more closely.

[1] IYSLP (2017), p. II.

[2] Ginzky et al. (2017), p. V.

An important starting point for the author is his research on the law of the transfer of agricultural land.[3] In Germany, there exists a legal regulation that subjects the transaction of agricultural and forestry landed property governed by civil law to public-law control. The core of this regulation was first created at the central state level shortly before the end of the First World War. In German law, which is based on the freedom of economic action, such a regulation is an exception.

Although the regulation's systematics have remained substantially unchanged to this day, it is currently under discussion in detail, as the ownership structures of agricultural landed properties have been undergoing a gradual, but substantive profound change for about the last fifteen years. In this respect, the agricultural policy asks for reform possibilities and this has therefore triggered a jurisprudential debate. Up until now this debate has missed being embedded in an overall concept of soil law. This point was examined in further detail by the author in 2020 in an expert report for the North Rhine-Westphalian state parliament.[4] The state parliament of North Rhine-Westphalia set up an Enquete-commission on the future of agriculture in North Rhine-Westphalia, in which soil law aspects were of major importance. In preparation of the report, the author confirmed his opinion, which is that there is a lack of proper dovetailing of the individual sub-areas of soil law in Germany. Simultaneously, the author used the report to work out the exemplary character of agricultural soil law for a land-related regulation with a steering effect. Hence, in the following, this regulation of agricultural soil law is to serve as an illustration of a steering mechanism in soil law.

2 Aspects of the Historical Development of German Soil Law

To understand the situation in Germany, a brief review of juridical history is useful. At the beginning of the nineteenth century, a state-controlled change in the ownership of agricultural land occurred in Germany. Since by far the largest proportion of land used by humans during that time catered to agricultural purposes, significant parts of land ownership were affected by this reform. The land reform's core was that the land belonging to aristocratic and ecclesiastical huge landowners became the property of those farmers who had previously cultivated the land for the same landowners. The feudal property relations that dated back to the Middle Ages and the partially still existing serfdom were thereby abolished. This process is generally known as 'peasant liberation'. It took more than a hundred years for the land reform to be implemented in all the single German states. The land reform paved the way for today's agricultural structure, which consists of a large number of farms, some of

[3] Busse (2019a, b). Over the course of two study visits, the author was able to gain insights into Japanese and Chinese real estate transaction law.

[4] Busse (2020a).

which are very small. At the same time, more and more technical devices found their way into the agricultural industry. This agricultural revolution went hand in hand with the industrial revolution. More people could now be fed, which therefore led to a considerable increase in population. These people were available for work in the emerging industries.

Ultimately, this is how today's capitalist economic order came into existence, which, like the socialist social order as its counterpart, is decisively determined by the question of landed property. Given that systemic importance of soil law, it is even more surprising that in Germany, soil law as an independent field of law has been virtually non-existent for a long period of time. That is probably partly due to the fact that the capitalist economic order in Germany has never been fundamentally called into question. One exception was the German Democratic Republic, which first carried out a land reform to implement the socialist idea and then nationalised most of the land ownership. After German reunification, a broad political majority argued that this socialist land rights system, which based on severe violations of property rights, had led to considerable inefficiencies and as such was in conflict with the traditional social structure, which should not be continued. As a result, the previous capitalist land system was restored on the territory of the German Democratic Republic.

German capitalist soil law is anchored in Constitutional law too. Both the Constitution of the German Empire of 1871 and the Weimar Constitution of 1919 are based on the premise that soil law is determined by the principle of private property and the free transferability of land ownership. Currently, the Basic Law for the Federal Republic of Germany (hereinafter: Grundgesetz) of 1949 enshrines the institute of private ownership of land. The free availability of landed property and the underlying principle of unrestricted use of private property are limited by the so-called 'social obligation of property.' This principle means that the private property of a person is not only to be distinguished from the private property of other persons and limited, but is also limited by the general welfare. Therefore, the fact that the individual is a member of a state community means that the individual must take common goods into account when using his/her private property.

In the Weimar Constitution as well as in the Grundgesetz, this aspect of the general welfare is emphasised. Article 153 (1) of the Weimar Constitution states that: "Property shall be guaranteed by the Constitution. Its nature and limits shall be prescribed by law." Paragraph 3 adds: "Property entails obligations. Its use by its owner shall at the same time serve the public good." Article 14 (1) and (2) of the Grundgesetz has taken over these provisions partly verbatim: "(1) Property and the right of inheritance shall be guaranteed. Their content and limits shall be defined by the laws. (2) Property entails obligations. Its use shall also serve the public good."

Furthermore, both Constitutions contain provisions on the so-called 'social economy' (in German: Gemeinwirtschaft), i.e., a form of economy that is not based on private property but on communitised property. For this purpose, private enterprises in a specific economic sector are to be compulsorily obliged by the state to operate in a way that replaces the profit motive of the individual enterprises with the benefit of the general public or at least of a larger group of people. Thus, both Constitutions recognise a form of property that is seemingly a hybrid between private and state property. Specifically, the first sentence of Article 156 (1) of the Weimar Constitution reads: "The Reich may by law, without prejudicing the right of compensation, and with due application of the provisions in force with regard to expropriation, transfer to public ownership private economic enterprises suitable for socialization." That is further elaborated in the associated sentences.

The basic idea is taken up by Article 15, sentence 1 of the Grundgesetz, which states that: "Land, natural resources and means of production may, for the purpose of nationalisation, be transferred to public ownership or other forms of public enterprise by a law that determines the nature and extent of compensation." Whereas at the beginning of the Weimar period there was at least an attempt to realise the idea of a social economy in some branches of the economy,[5] such an approach was never seriously undertaken by the state under the Grundgesetz. After 1949, the principle of the social market economy quickly established itself, with the establishment of the idea that the social economy is not compatible. It may also have played a role that the German Democratic Republic had largely withdrawn both the land and the business enterprises from private ownership and that West Germany wanted to distance itself from this situation. The trade unions then founded social economy enterprises in West Germany voluntarily, e.g., in the banking and housing sector, but some of them failed spectacularly. Overall, it should be noted that, at least in land ownership, no attempt was made by the state under the Constitutions of 1919 and 1949 to organise land ownership or land use based on the public economy and thus to abandon private ownership.

However, the realisation of the principle of free use and transferability of private property in land does not mean that there are no steering measures in the area of usage and transfer of land in Germany. Such measures contribute to realising the above-mentioned social obligation of property by pursuing public interest objectives. To understand where steering measures can be applied and how they work within soil law, it is necessary to have an idea of the systematics of soil law, as such steering measures are regularly linked to classic legal institutions of soil law.

3 The Systematics of Soil Law in German Literature

It is not easy to gain such an idea for Germany. As already emphasised, this conclusion results from the circumstance that soil law does not currently exist in Germany as a fully formed and sufficiently researched field of law. Since this situation has existed for a long time, it has led to the fact that there is no recent textbook on soil law and also no deeper systematics of soil law. Again, it is worth taking a brief look at the history of law.

[5] Cf. in more detail the dissertation by Zacher (2002), p. 102 ff.

The last major study on the entirety of German soil law, as far as the author was able to survey, stems from the German civil law Professor Justus Wilhelm Hedemann, a pioneer in the emergence of commercial law as an independent field of law. Born in 1878, Hedemann received his legal training at the time of the creation of the Civil Code (in German: Bürgerliches Gesetzbuch) and was influenced by the economic liberalism of the German Empire. As a young professor, he conceived the plan to write a history of the "Die Fortschritte des Zivilrechts im XIX. Jahrhundert" (Progress of Civil Law in the XIXth Century), consisting of four parts, which should not only be based on German law but also on the legal systems of Austria, France and Switzerland. In 1910, the first part, "Die Neuordnung des Verkehrswesens" (The Reorganisation of Transport), was published. It dealt with the "transport" (transferral) of private legal positions as manifestation of individual's freedom of action under civil law.[6] The second part was dedicated to the "Entwicklung des Bodenrechts von der französischen Revolution bis zur Gegenwart" (Development of Soil Law from the French Revolution to the Present), which found itself exceeding the time frame set by the main title of the work.

The first half-volume of this second part was published in 1930, and the second half-volume followed in 1935. The first half-volume contained 421 pages on "Materielles Bodenrecht" (Substantive Soil Law),[7] while the second half-volume contained 406 pages on the "Entwicklung des formellen Bodenrechts mit einem Anhang: Bodenrecht und neue Zeit" (Development of Formal Soil Law with an Appendix: Soil Law and the New Era).[8] Hedemann thus chose as his fundamental systematics as division between formal and substantive law and consequently no content-related criteria that were oriented towards the subjects of soil law. Hedemann understood formal soil law as the registration of civil soil law rights in the state-managed land register, as he described numerous individual areas of soil law under the keyword of substantive soil law. This presentation was not based on a legal system, but on the political aspect of—as Hedemann called it—the "fight against land egoism".

For Hedemann, large parts of substantive soil law were characterised by the endeavour to restrict the freedom of property law as achieved in the nineteenth century. Accordingly, he used as keywords; the "fight against land fragmentation" and the "fight against land indebtedness" as well as the "history of expropriation", the "history of ownership restrictions", the "prehistory of socialisation" and the "socialisation movement of the years 1918/19" as sub-headings of the "fight against land egoism". To this, he added a sketch of the "new settlement law". This approach can best be described as a combination of juridical history and ideological history. Hedemann, who died in 1963, did not get around to get into the planned parts 3 and 4, which were intended to deal with labour law and family law. His

[6] Hedemann (1910).

[7] Hedemann (1930).

[8] Hedemann (1935).

comparative analysis of soil law in the overall development of civil law from the nineteenth century onwards thus still remains unfortunately incomplete.

In 1939, Hedemann summarised his findings on commercial law in the first edition of his textbook "Deutsches Wirtschaftsrecht – Ein Grundriss" (German Economic Law—An Outline). In this publication, he succinctly described the situation of soil law in Germany at that time, which until today, has not changed. Amongst other things, he said: "Even if a final classification is not possible, the legal system cannot ignore the soil. It is therefore constantly striving for the 'law of the soil'. Hundreds of provisions are dedicated to it." He remarked: "By far the majority of legal provisions on soil are of an economic nature." This is "not to say that they all fall into the area of 'economic law'. On the contrary: just as private property is mainly laid down in general civil law, so it is with soil law." But "unfortunately, there is no central land law that would bring together the abundance of rules that belong here". It is "also not to be assumed that such a far-reaching consolidation of the material will take place in the future. Not even a central soil management law is to be expected. Therefore, the following presentation is based on a very fragmented and uneven individual legislation."[9]

This prediction of a lack of a legislative centre for soil law, which has proven true to this day, is all the more remarkable because it was written during the National Socialist era. The Nazi regime did not hesitate to reshape entire areas of law according to its ideological guidelines. Examples of this include; civil service law, marriage law, the will and testament law, stock corporation law, labour law and juvenile law. Hedemann, who was committed to a Nazi-influenced law and had a fundamentally positive attitude towards National Socialism, played a leading role in the project of the Nazi-oriented "Volksgesetzbuch" (Peoples' Law Code), which was put in place to replace the Civil Code. However, soil law proved to be too unwieldy for a speedy new regulation. On the one hand, there was the principle of "Blut und Boden" (blood and soil), which gave rise to a new peasant inheritance law. Stemming from this, land ownership was brought into a mystical connection with the peasant family. On the other hand, the Nazi economic system was largely based on capitalist mechanisms, which stood in the way of full state regulation of landed property. Consequently, soil law remained fragmented also during the Nazi era.

This historical description contributes to the thesis that the design of national soil law is less of a question of the respective form of government—monarchy, democracy or dictatorship—but rather seems to depend on the economic system pursued in each particular case. While the change of the German form of government from the Empire to the Weimar Republic and the Third Reich to the Federal Republic of Germany hardly brought about any fundamental changes in the orientation of soil law towards the free transferability of land and the fragmented structure of soil law, in the German Democratic Republic there was a change of principles and, as a result, a different orientation of soil law. Another proficient example of this is the Chinese property right to land. Just as the Chinese economic system consists of a mixture of

[9] Hedemann (1939), p. 214 f.

socialism and capitalism, land ownership there is also characterised by a state of limbo between state ownership and private rights. The land belongs to the state, while the rights to use it are similar to private rights and are transferable.

Interestingly, the last classic textbook on soil law in Germany was written during the Nazi era. It was published in 1938 by Franz Wieacker, known above all as a specialist in Roman law, under the title "Bodenrecht" (Soil Law) and reflected on the described problems of soil law through the Nazi regime.[10] Wieacker excluded agricultural soil law from his description. As already mentioned, agricultural soil law was part of a newly conceived peasant inheritance law. As an expression of this ideological connection between soil law and peasant law, the series "Beiträge zum Bauern- und Bodenrecht" (Contributions to Peasant and Soil Law), published from 1935 to 1941 within the framework of the Reichsnährstand (central organisation of agriculture in Nazi era), may be mentioned. The publisher of Wieacker's textbook intended to deal with agricultural soil law in a corresponding textbook on peasant law, which, however, was not finished until the end of the Third Reich.[11]

In contrast to Hedemann, Wieacker did not pursue a legal-historical approach guided by the opposition of freedom and boundedness of landed property, but rather an approach oriented towards the subject areas of soil law in the sense of a textbook suitable for teaching law. To this end, he first presented the embedding of soil law in regional planning as a superordinate point of view. He then turned to the base structures of ownership of land. In this regard, Wieacker dealt with landed property in private and public hands, the limits of landed property through individual and public interests, and expropriation. That was followed by private-law soil law with, on the one hand, the regulations of the Civil Code on the exact content of landed property and its transfer and, on the other hand, the provisions on land register law and the various forms of landed property. Wieacker went on with tenancy law, hereditary building law, homestead law and the law of small settlements. His textbook concluded with mining law. However, Wieacker did not present a concrete system of soil law that embeds the individual subject matters. The impression remained as that of a collection of soil laws. Although Wieacker was also a supporter of a reform of the law in the sense of Nazi ideology, he did not develop a legal dogmatic structure of soil law any more than Hedemann.

In the Federal Republic of Germany, only one work has been published to date that has attempted to comprehend soil law as a whole, i.e., both civil law and public soil law. The latter has been considerably more extensive in terms of the number of norms. It dates from 1962, having been written by the surveyor Wilhelm Matthes and bears the title "Deutsches Bodenrecht" (German Soil Law).[12] In contrast to Hedemann's predominantly legal-historical study and Wieacker's textbook,

[10]Wieacker (1938).

[11]Cf. on this issue Busse (2018).

[12]Matthes (1962); later from an educational perspective Bernhard and Görgmaier (1980), as a publication of the Bayerische Landeszentrale für Politische Bildungsarbeit.

Matthes' work was primarily intended to serve legal practice and, therefore, essentially presented a merely briefly annotated collection of legal texts.

Different to Wieacker's textbook, the work included both urban and agricultural soil law. Its structure was divided purely formally into historical texts, Constitutional Law, Civil Law, Land Registry Law and Administrative laws. As a result of this, an internal system of soil law could therefore not be derived from it. If the literature in the German Democratic Republic is also considered, several books with the title "Soil Law" were also published. However, since they are based on the concept of socialist land ownership, which was abolished with reunification in 1990, they are only referred to here.[13]

4 Attempt at a Systematics of Soil Law

For the sake of clarification, it should to be emphasised that Germany does not lack soil law literature. Several publications such as; textbooks, commentaries, and scientific studies exist on the main areas of soil law. Collected works on Administrative Law usually contain a description of Building Law, compilations on Environmental Law offering an overview of soil protection law and compilations on Agricultural Law contributions to agricultural real estate transactions. The Civil Law side of soil law is dealt with in the Civil Code commentaries. However, there is no overall view that brings the various sub-areas of soil law into a contextual basis and, e.g., reveals its developing inconsistencies. Nor is there a jurisprudential institute that deals comprehensively and exclusively with aspects of soil law. The answer to—in terms of legal policy—the exciting question of the extent of the state's guiding measures in soil law in Germany and the concrete range of objectives pursued in this context is currently lacking.

Against this background, the author would like to attempt to present general systematics of soil law. In doing so, he can draw on his intensive efforts to develop systematics and dogmatics of agricultural law.[14] Agricultural law is even more fragmented than soil law. At the same time, aspects of soil law belong to agricultural law.

Soil law could be divided into three levels. The first level forms a kind of basis for soil law. It contains the legal definition of land, the rules for determining land ownership and the general provisions on the transfer of landed properties. In Germany, this first level is contained in the Civil Code and the supplementary land

[13] Deutsche Akademie für Staats- und Rechtswissenschaft (1961); Arlt and Rohde (1967); Rohde et al. (1969); Rohde (1976, 1979, 1989); Zänker and Boenicke (1979); see also Ulbrich (1987) as a collection of texts and Hoffmann and Schinke (1973) and Gallien (2007) for analyses. Horst (1995), among others, deals with the special regulations for the transition of the GDR soil use law to the Federal German soil law.

[14] Cf. as a sum of the author's previous publications on the contours of agricultural law as an independent field of law Busse (2016) and most recently Busse (2020b).

register law. It is substantially Civil Law in the form of Property Law and the Law of Sale. Property Law defines landed property and its associated components in more detail, while the law of sale regulates its transfer. The supplementary land register regulations stipulate that—in simple terms—all owners of German land are to be entered in the land register. The land register is created and administered by the state. It carries the presumption of correctness. If it does not correspond to the substantive legal situation, it is subject to correction, whereby the correction must in principle be applied for by the specific persons concerned.

The second level of soil law shapes the ownership of property. The rights and obligations associated with the landed property are regulated. The starting point is the well-known § 903, sentence 1 of the Civil Code: "The owner of a thing may, to the extent that a statute or third-party rights do not conflict with this, deal with the thing at his discretion and exclude others from every influence." The rights of other persons or the general public to private landed property result from contractual or legal regulations. All restrictions on the property that go beyond contractually agreed restrictions can be considered as being legal barriers. Such restrictions on the property are locatable in large parts of the German legal system. E.g., building law restricts the possibilities of building on a plot of land at will. Residential property law defines the ownership of flats in more detail. Tenancy law and lease law for buildings and land standardise the fundamental rights and obligations of both contracting parties.

Neighbouring property law regulates the co-existence of adjoining properties. Immission protection law limits the extent to which a property can become the source of immissions. Environmental protection law also establishes many limitations on the use of land. These include water law, pesticide and fertiliser law, as well as the Federal Soil Protection Act (FSPA), which includes soil in its title and whose primary purpose, according to § 1, sentence 1 is to "to protect or restore the functions of the soil on a permanent sustainable basis". This therefore refers to harmful interventions in the substance or natural functioning of the soil. Contrastingly, the law on the protection of monuments affects the use of land. These examples may suffice to illustrate the second level of soil law.

The soil steering law forms the third level. Through this, the state aims to influence the transfer of landed property. This goal is primarily achieved, on the one hand, by erecting barriers or at least obstacles to the acquisition of landed property for certain groups of people; and, on the other hand, vice versa, by making it possible or at least easier for certain groups of people to acquire landed properties. That can be accomplished directly or indirectly. The direct way interferes with the acquisition process, whereas the indirect way mainly consists of making the acquisition either more attractive or less attractive through restrictions on use of the land. Building law is a proficient example of the second type.[15] To this extent, it generally

[15]Cf. from the extensive German literature on building law as publications with a reference to soil law in the title, listed in chronological order: Froemer (1974); Deutscher Verein für Vermessungswesen (1974); Ernst and Hoppe (1978, 1981); Rutkoswky and Schulz (1986).

prohibits the construction of new residential buildings outside of housing estates, making such plots unattractive to the prospective group of people who would like to build residential buildings. Land-use plans that determine the type of use—for commercial purposes, for residential development or other uses—also indirectly control land acquisition by only allowing a certain group of people to realise their building project. Since this restricts the rights arising from landed property, such regulations are at the same time part of the second and third level of soil law; whereby it is debatable as to what intensity the threshold to soil steering law and thus to the third level is crossed.

Tax law, e.g., also has a governing effect by setting the tax on acquisition at different levels depending on the type of use of the land or the type of acquirer, therefore making the acquisition more financially attractive to certain groups of people. Employing the right of expropriation, the state can even compulsorily seize landed property in pursuit of public purposes—e.g., for the construction of public roads—although it must pay compensation with effect to this. That is the strongest interference on landed property, in that the state actively transfers the land to another legal entity.

This brief outline also confirms the observation made at the beginning that soil law in Germany is regulated in a very scattered manner and, above all, that a soil law code that summarises soil law provisions and which regulates general soil law principles does not exist. Additionally, soil law provisions are not only existent at the federal level. Also, at the level of the German sixteen federal states ("Länder") and at the municipal level, there are provisions. Furthermore, there is no system put in place to which every land acquisition beyond the mere registration of the new owner has to be notified and permitted in advance.

This therefore, also results in the absence of an overarching idea of steering soil property that might impact the entire soil law. A state supervisory system for all land transactions with the associated rights of intervention has been alien to Germany on a federal level since the foundation of the German Empire in 1871. This also applies to the period of the Third Reich. That is an expression of the liberal economic constitution, which bases on the freedom of property. Traditionally, landed property in Germany has been widely dispersed. Residential houses, flats, business premises, gardens, agricultural farms, forests, and other land belonging to millions of individual owners. The largest owner being the state, followed by the Christian churches. However, neither of these two dominate the land market. The state's land ownership is predominantly limited to those properties needed for functioning of the state or purposes of general interest. Even if these areas—such as administrative buildings, military properties, and transport infrastructure—account for a larger share of

---

Furthermore, a loose-leaf collection "Bau- und Bodenrecht" (Building and Soil Law, 1950–1966) was published. It is particularly noteworthy that the work by Ernst/Hoppe does not present a systematic approach to soil law, although it is a legal textbook. Only spatial planning law and urban soil law are presented.

German land, they are only available for disposal and thus for real estate transactions to a limited extent.

Hence, an important basic structure of German soil law may be seen as the combination of a broad distribution of ownership and liberal transfer rules. This basic structure permeates the first soil law level and affects the second level. At the same time, the effects of the social obligation begin at the second level and shape the third level. This reflects the antagonism depicted by Hedemann. The "fight against land egoism" has today donned the garb of the social obligation of landed property.

5 The Special Case of the Agricultural and Forestry Real Estate Transaction Act

However, there are two important areas of soil in which the state exercises control over the transaction of soil. These are the agricultural and forestry areas, which together, account for about eighty percent of Germany's total area. The corresponding control system was introduced shortly before the end of the First World War and has survived to this day.[16] It is not a particularly strict system, as it also has a considerable regulatory gap. The system should be briefly presented with its historical development and the reform discussion that has been intensifying for about a decade.

The system came into being in 1918 primarily to prevent commercial enterprises from buying agricultural land for the sole purpose of securing its value. The commercial industry had made good money during the First World War and now wanted to invest their profits in a crisis-proof manner because of the economically uncertain period. At the same time, many farms were impoverished and ready to sell their land. The imminent fear that was present was that a large amount of land could be taken away from agriculture and that this would result in a weakening of agriculture and thus of the food supply. This then led to the introduction of a permission requirement for the acquisition of agricultural and forestry land.

This permission system, which is now more than one hundred years old, has progressed in its basic structure up to the present day and is currently in force in the form of the "Grundstückverkehrsgesetz" (Real Estate Transaction Act—RETA). It is based on the principle that any acquisition of agricultural or forestry land requires official permission. However, this permission must be granted if there is no reason for the refusal. The reasons for a refusal are limited. Consequently, the state has no general discretion as to whether or not to approve the acquisition. It is often concluded from this circumstance that it is not a real steering instrument for the agricultural and forestry land market.

[16]Cf. in detail on the following Busse (2019a) and with a focus on Union law the dissertation by Hoffmeister (2018).

In 1969, e.g., the German Federal Constitutional Court emphasised in an early decision on the RETA by stating that: "The approval procedure does not serve the purpose of steering the real estate transactions on agricultural land, but rather to avert threats to the agricultural structure."[17] All subsequent case law and the prevailing view in the literature on the RETA have followed this trendsetting statement.[18] It is true that at first glance the governmental explanatory memorandum to the draft RETA of 1958 can be referred to, as it stated that "the authorities entrusted with the supervision of real estate transactions would be overburdened if they had to pursue a steering soil policy". However, this statement only referred to the shaping of the right of sale and thus not to the RETA as a whole. On the contrary, the government's explanatory memorandum stated that it was not possible to "refrain from intervening more than before in the land market in a guiding and regulating manner".[19]

In the author's view, the concept of steering should not be understood too narrowly. Steering already exists if the legal provision intends to achieve a specific soil-related goal. In general terms, the RETA aims towards the continuance of agricultural and forestry land in the ownership of farmers and thus within the agricultural and forestry sector. This objective bases on the idea of a functioning agricultural and forestry structure as a matter of public interest.

The permission system is, on the one hand, secured by the circumstance that a transaction requiring permission but not being notified for permission, is null and void. During the permission procedure, the transaction is provisionally ineffective. On the other hand, when registering the change of ownership of agricultural and forestry properties, the land register offices must check ex officio whether the transaction requires permission. If the answer is affirmative, the registration of the change of ownership without permission is not allowed. If the land register office enters the transaction in the land register despite the lack of permission, the parties concerned are entitled to file an objection or demand a correction of the land register. If no objection is lodged and no adjustment requested, the permission is deemed to have been granted—for reasons of legal certainty—by the state after one year.

The regulation is relatively complex in its details and cannot be presented here in its full depth. In this respect, it is referred to the pertinent commentaries.[20] In simplified terms, the regulation initially allows the transaction without restriction in certain constellations. These are transactions between spouses and close relatives or in the context of anticipated succession or the sale of an entire agricultural or forestry business. In these cases, it is assumed that the land in question will remain in agriculture or forestry. In addition to a few other exceptional circumstances, permission to transfer can be refused in all other cases if, firstly, "the sale means an unhealthy distribution of the land" or, secondly, the purchase price is "grossly

---

[17] German Federal Constitutional Court (1969), p. 223.

[18] See Martinez (2016), p. 473; different view Faßbender (1964), p. 22.

[19] Printed in: Busse (2019a), p. 168.

[20] See especially the commentaries by Martinez (2016) and Netz (2018).

disproportionate to the value of the land" or, thirdly, contiguous areas would be "uneconomically reduced or divided".

It is immediately obvious that in the third case the land structure is damaged. In the second case, it should be prevented that purchase bids from outside the agricultural or forestry sector outdo those from within the agricultural or forestry sector only because of their disproportionate amount. That expresses a fundamentally lower level of capital in agriculture and forestry. When exactly such a "gross disproportion" exists is controversially discussed, and has also triggered questions from state aid law.[21] However, the first group of cases is the most difficult to handle. The RETA stipulates that the sale of land in question must not contradict "measures to improve the agricultural structure." This refers to the agricultural and forestry structural measures undertaken by the state in Germany. However, these change over time and cannot always be defined in a precise manner.

At this point, therefore, it could be assumed that the authorities have discretionary powers. However, jurisprudence countered this, based on the Constitutional protection of property, at an early stage and has concretised the criterion of "measures to improve the agricultural structure" through case law.[22] In simple terms, an interested agricultural buyer always enjoys priority over an interested party from outside agriculture, provided the farmer is prepared to pay the purchase price demanded by the seller. If there is no interested agricultural party, the purchase must be permitted. The latter is not stated in the RETA. There are, therefore, good reasons to doubt whether the legislator had intended such a consequence. However, the jurisdiction was and is not willing to approve the massive state intervention into property rights, which would otherwise result, meaning the property being unsaleable at the desired price.

In this way, the agricultural real estate transaction authorities are deprived of an essential steering option. If the sale to a non-agricultural interested party could not be permitted solely because of the fear that detrimental effects on the agricultural or forestry structure may result, in many cases, the seller would subsequently reduce the purchase price. That would lead to the effect that farmers could still be found who wished to purchase the land but could not afford the original purchase price. If two farmers alone compete for a plot of land—according to the jurisdiction—the RETA does not allow any intervention, because the proportions of the agricultural holdings are in principle, irrelevant. It is, therefore, not possible to counteract the concentration of agricultural land in the hands of a few farms through the approval procedure. According to the jurisdiction of the German Federal Court of Justice, it is only examined whether—especially in the case of a very large distance between the land

[21] See European Court of Justice (2015) and subsequently German Federal Court of Justice (2016). The case concerned the constellation whether the refusal to permit a purchase contract for a plot of land on the grounds of a purchase price that is not customary on the market, constitutes indirect state aid because it may create the possibility for another buyer to acquire the land at a lower price.

[22] See in more detail for an overview of the extensive and differentiated case law Netz (2018), p. 804 ff.

in question and the actual farm of the farmer—it can be assumed that the farmer is capable of cultivating the newly acquired land.

6 The Connection of Agricultural Real Estate Transaction Law to Settlement Law

The agricultural and forestry real estate transaction law (in the following, for the sake of simplification, referred to as agricultural real estate transaction law) also contains a connection to settlement law. The "Reichssiedlungsgesetz" (Reich Settlement Act) was passed in 1919, one year after the first regulation on agricultural real estate transaction law was entered into force. This Act still remains in force until now. It stipulates that "non-profit settlement companies" can be established throughout Germany.[23] Particularly, those that are supported by the German "Länder". Their task is the creation of agricultural holdings and the strengthening of pre-existing ones. To this end, the settlement companies, which currently exist mostly in the form of settlement or land companies and are united in a federal association of non-profit land companies, have to acquire agricultural land or obtain it from state ownership so as to sell this land to farmers.

In order to facilitate the acquisition of agricultural land by the non-profit settlement societies, they have a pre-sale right within the framework of the RETA. Hence, if a settlement company has an interest in agricultural land, it is treated in the same manner as that of a farmer in the approval procedure. The authorising authority is, therefore, obliged to inform the locally competent settlement company of the approval procedure before deciding to permit the acquisition by a non-farmer. Thus, in cases where no farmer—willing and able to pay the purchase price demanded by the seller—is found, it can be prevented that the agriculture loses the corresponding land.

However, the financial power of the settlement companies is limited. They are primarily fed by a foundation granted by the state and the income from their sales. Consequently, the settlement company has to choose exactly which areas of land it will exercise a pre-sale right on. It also happens that the authorising authorities do not inform the settlement company on time about the approval procedure so that the settlement company does not have the opportunity to examine and assert its pre-sale right.

[23] See on the Reich Settlement Act and the settlement companies Lück (2017) and with numerous examples from practice Bundesverband der gemeinnützigen Landgesellschaften (2019).

7 The Reform Discussion on Agricultural Real Estate Transaction Law

This description already shows that a reform of the RETA could be taken into consideration.[24] Before dealing with individual points, however, it is crucial to point out a general problem that plays an important role in the reform discussion. This problem is the Constitutional basis of competence for agricultural real estate transaction law, which has changed over time and should, therefore, be outlined briefly.

In 1919, the Weimar Constitution included an explicit competence of the federal level for regulation of soil law. Article 10 No. 4 stated that the Reich "may by law prescribe principles with respect to ... 4. land titles, land distribution, land colonization and homesteads, entail, housing, and distribution of the population". That was therefore not a full competence, but merely the competence to regulate "principles". Since it was controversial as to whether the 1918 regulation on the transaction of the land went beyond such principles. A transfer of the provision issued in 1918 by ordinance into a parliamentary law did not come about during the Weimar Republic. In the Third Reich, the regulation continued substantially unchanged. The most important change was the integration of the previous grounds for refusal into a newly created general clause, which allowed for refusal based on a "substantial public interest". After 1945, however, this general clause ceased again.

In 1949, Article 74 No. 18 (old version) of the Grundgesetz regulated the relevant legal matters as a concurrent full-fledged federal competence: "Concurrent legislative power shall extend to the following matters: ... 18. real estate transactions, land law (except for laws regarding development fees) and agricultural leases, housing, housing estates and homesteads". In 2006, the competence for agricultural real estate transactions as well as leasing, settlement, housing and homesteads was transferred to the exclusive legislative power of the "Länder". The RETA continues to apply in the individual "Länder" until the respective "Land" has adopted its regulations on agricultural real estate transactions. However, only Baden-Württemberg, and thus just one of the "Länder" has so far adopted its own regulations so that the RETA is still valid in almost the whole of Germany.

As far as the current reform discussion is concerned, a regulatory gap in the RETA that has existed from the onset has become steadily graver over the past decade. That is because the structural development in agriculture has resulted in an increasing number of agricultural enterprises run in the form of associations, predominantly as a civil law association and as a Gesellschaft mit beschränkter Haftung (GmbH—"company with limited liability"). However, the RETA does not cover the transaction of shares in a company that owns agricultural or forestry land. Consequently, such a transfer does not require permission under the RETA.

[24] Cf. on the different aspects of the current reform discussion Busse (2019a), p. 56 ff., with further references.

This loophole was already identified shortly after the first regulation was issued in 1918, using the example of the takeover of all shares in the inheritance of co-heirs by a third person, and the associated danger of bypassing the law was pointed out.[25] For decades, however, the loophole played no role in practice. Now, however, more and more agricultural and forestry land is de facto transferred within the framework of associations by changing the shareholders. But closing the gap is associated with specific difficulties.

If in the event a joint-stock company holds an agricultural estate, it would have to be clarified whether every acquisition of shares triggers the permission requirement. In this respect, there is widespread agreement to set a minimum percentage of shares, in the sense of a threshold value. In the case of listed companies, the effects of the suspended validity of the original sale of shares on the resale of these shares would have to be examined. Similar questions arise in the transaction of cooperative shares. A restriction to certain types of companies as a solution for such questions leads to the problem of unequal treatment. Attention should also be paid to the difficulty of control, as there is no general register for company shares in Germany. Finally, there is controversy as to whether the inclusion of the acquisition of company shares does not at the same time fall within the legislative powers for company law, which is not held by the "Länder" but by the federal level. However, the control of real estate transactions is not a matter of regulating individual forms of company, but merely a requirement under public law for the acquisition of company shares.

In addition to the matter of including the transaction of company shares, another aspect of reform is the introduction of an additional reason for refusing the acquisition of land. The goal is to counteract a concentration of land in the hands of a few owners if the concentration arises or increases as a result of the acquisition. The first question to be answered here is which area—"Land", county, municipality, local district—is to be targeted to prevent a concentration of land. It must be taken into consideration that even within the German "Länder" the respective areas differ considerably in size. Furthermore, a concentration threshold must be defined, which may not be surpassed. In this respect, it would also have to be clarified whether already existing larger concentrations should enjoy grandfathering. Besides, the issue arises of how to deal with owners who are made up of shareholders—e.g., in the form of a cooperative—and have contributed their individual areas to farm them jointly. Henceforth, a problem of competence arises in the form of as to whether the prevention of a concentration of ownership in the hands of agricultural or forestry enterprises does not at the same time fall within the federal competence for cartel law. However, not every regulation under public law that affects market conditions can automatically be assigned to cartel law. There is some evidence to suggest that the legislative power for a sector-specific regulation is connected to the general competence of the sector concerned.

At present, several "Länder" are working on their own agricultural real estate transaction law acts to replace the RETA in whole or in part. They are supported in

[25] Cf. Busse (2019a), p. 31.

this by a working group of experts from the federal government and the "Länder".[26] Following the failure of reform projects at the "Länder" level at various stages in recent years,[27] sample building blocks are now to be prepared to tackle problem areas,[28] which can serve as a guideline for the new law of the "Länder". It will be interesting to see whether a majority for the respective draft will be found in the subsequent legislative process at the "Länder" level. The political endeavour to reconcile the partly very heterogeneous interests of the different groups operating in the rural area will probably not be easy to realise. In addition to this, the regulation found must be designed to be constitutionally sound, since a judicial review of newly introduced restrictions is to be expected.

A crucial recent development is that environmental protection aspects are beginning to play a more important role. In a ruling of May 2020, the German Federal Court of Justice confirmed that "the non-agricultural purpose pursued by a recognised nature conservation association can override the reason for the restriction based on a farmer's interest in acquiring the land".[29] According to the agricultural reports of the federal government and the "Länder", the "measures to improve the agricultural structure" also include the interests of nature conservation. Agriculture is dependent, inter alia, on "the long-term functioning and usability of the natural balance".

If a nature conservation association competes with a farmer, neither can claim priority. It then depends on who the owner of the real estate wants to sell to. In most cases, the amount of the purchase price offered is likely to be the decisive factor. In such a case, the agricultural real estate transaction authorities are prevented from making their own preferential decision, as the Federal Court of Justice states: "Since the approval procedure under the Real Estate Transaction Act does not serve to positively govern agricultural real estate transactions, it is not incumbent on the authorising authority and the agricultural courts to establish a ranking for several competing agricultural structure improvement measures by means of an agricultural policy assessment."

It is now up to the "Länder" to decide within the framework of their new provisions whether they wish to follow this broad interpretation of the concept of agricultural structure. In the meantime, the problem enlarges. For instance, there exists a so called "Gemeinschaftsaufgabe 'Agrarstruktur'" (Common Task

---

[26] Cf. fundamentally: Working Group of the Federal Government and the "Länder" (Bund-Länder-Arbeitsgruppe) on land market policy (2015). More information on current developments can be found on the website of the German Federal Ministry of Food and Agriculture.

[27] Documented up to and including 2017 in Busse (2020a), p. 279 ff. The alone federal state law replacement regulation in Baden-Württemberg to date has only taken up some reform aspects discussed.

[28] Cf. also German Federal Ministry of Food and Agriculture (2019) on the Agriculture Ministers' Conference in Mainz, at which the issue of amending the law on the transaction of real estate was discussed.

[29] German Federal Court of Justice (2020). In this judgement, further decisions from the previous case law are proven.

"Agricultural Structures"), which is used by the federal government and the "Länder" to co-finance EU projects for developing rural areas. More and more of these projects—such as the creation and renewal of village, traffic-related and technical infrastructure—has only an indirect connection with agriculture. To label those projects as measures to improve agricultural structures and, consequently, take them into account within the RETA, undermines the original intention to provide for farmers a preferential access to agricultural land.

8 Urban Soil Law, Fundamental Freedoms, and Investment Aspects

This overview of the state of affairs and reform aspects in agricultural and forestry land transaction law should not obscure the fact that other areas of soil law are also currently discussed in detail in Germany. An example of this is that an intensive debate is taking place about the considerable increase in rents in German conurbations. As the influx from rural areas into larger cities continues, and the German population has been growing continuously since 2011 due to immigration, the demand for housing in the cities is rising. Because it cannot be met by construction measures for various reasons, rents are inevitably increasing in a market-based housing market. Since the financial crisis, this has been fuelled by abundant liquidity in search of capital investments. Individual cities are trying to mitigate this development with so-called rent control measures, but those have been partially stopped by the judiciary. That is because tenancy law is a competence fundamentally exercised by the federal government. Where exactly the competence of the "Länder" for housing ends has not yet been clarified.[30] This has led to the situation that the "Länder" and even more so, individual cities encountering difficulties in enacting court-proof regulations on steering the latitude of rents.

The problem of rents, however, is not primarily a land-use management problem, but an aspect of structural planning. If jobs are concentrated more and more in the urban agglomerations, and at the same time, for environmental and climate reasons, individual transport is to be restricted, then the movement of population into the cities will automatically increase. The design of building law is also related to this. The discussion about the structure of urban development goes back a long way in history. In Germany, after 1945, it was conducted for a while under the heading of soil law. E.g., the title of a collection of building policy documents published in 1968 was "Material zur Reform des Bodenrechts" (Material on the Reform of Soil Law).[31]

[30] Cf. from the current literature with further references in each case: Farahat (2020); Kingreen (2020); Schmid (2020).

[31] Institut für kommunalpolitische Bildung und Forschung der Politischen Akademie Eichholz (1968); see also in chronological selection: Gesellschaft für Wohnungs- und Siedlungswesen (1969); Evangelische Akademie Bad Boll (1970); Storck (1970); Werner (1970); Conradi et al.

In 1971, the federal Parliament passed an Urban Development Promotion Act, which is still in force today. At the same time, social housing was systematically developed temporarily.

As early as 1965, the federal Parliament enacted a "Raumordnungsgesetz" (Regional Planning Act), which established guidelines for the spatial organisation of Germany and set a framework for the Regional Planning Acts of the "Länder". Such a Regional Planning Act still exists today. If it is considered from the perspective of soil steering, the results are very meagre. The word "soil" appears in only four places, which are not related to the aspect of soil steering. The property dimension is not mentioned at all. That is consistent with the finding that in Germany the principle of free transactions with the land is assumed, and therefore landed property, including its transaction, is not supposed to be an aspect of land planning. At most, the Regional Planning Act indirectly touches on this point by stipulating in the fourth sentence of § 2 (2) No. 2 that, "settlement activity should be spatially concentrated" and "primarily directed towards existing settlements with sufficient infrastructure and towards central places". This is ultimately a requirement primarily for urban land use planning.

All in all, the law on agricultural and forestry land transfer gives the impression of a solitaire, which lacks a counterpart in the urban area and the embedding in an overall soil law concept. After all, § 2 (2) No. 4, sentence 7 of the Regional Planning Act states: "The spatial conditions for agriculture and forestry in their importance for the production of food and raw materials are to be maintained or created." This can be understood to mean that the necessary production areas for agriculture and forestry must be available, to which real estate transaction law makes its contribution.

It should be remembered that any restriction on the movement of real estate must be measured against the "Four Freedoms" of the European Single Market.[32] The German legislator is, therefore, not only limited by German Constitutional law but also by Union law in shaping this matter. If considering the corresponding practice of the European Commission and the jurisdiction of the European Court of Justice, it promptly becomes evident that other member states have a much more restrictive law on transactions of real estate than others. It often orients against the acquisition of landed property by foreigners. On several occasions, the European Court of Justice has ruled against the so-called "native" regulations, according to which either an acquisition is only possible for nationals or local residents or the acquirer must take up residence on the property. In principle, such restrictions are violations of the free movement of capital and the freedom of establishment. In Germany, such regulations occur to a limited extent locally, e.g., when the sale of municipal properties for building purposes is linked to requirements that can primarily only

(1972); Henrich (1972); Knirsch (1972); Deutscher Städte- und Gemeindebund (1975); Dietrichs (1976).

[32] Cf. in more detail Busse (2019a), p. 47 ff., and Hoffmeister (2018), p. 112 ff.

be fulfilled by locals. Such regulations need to be closely examined based on the criteria established by the European Court of Justice.

The argument of non-local buyers also plays a considerable role in the discussion on the further development of German agricultural real estate transaction law. The fear exists that—as it was the case when the first regulation on agricultural real estate transactions was enacted in 1918—agricultural land will be acquired by commercial investors only in order to participate in the increase in the value of such land and the further rise in rental fees for agricultural land. Such participants in the agricultural real estate market—e.g., an international investment fund—are regularly able to pay much higher purchase prices than local farmers, especially since the financial crisis with its continuing low-interest phase high amounts of capital are available for such investments.[33] If only a plot of land is acquired within the framework of a traditional purchase contract, there is definitive protection for agriculture that is ensured by the RETA. However, if the acquisition is made by taking over shares in a company, there is no protection whatsoever, as described above. As early as 1918, it was proposed to argue that the legal regulation could be circumvented and that such acquisition transactions could be classified as null and void based on § 134 of the Civil Code. The author does not know whether the agricultural real estate transaction authorities of the "Länder" have attempted such a course. Moreover, it may only be applied in cases by which a company was founded specifically for the purpose of acquiring the land.

Meanwhile, the question of principle could be raised as to whether the transfer of competence for agricultural and forestry real estate transaction law to the "Länder" was the right way to go. For one thing, the "Länder" have not managed to deal with the problems outlined for more than a decade. Second, if the "Länder" were to one day replace the still valid federal RETA with their own regulations, there would be a risk of a patchwork rug of provisions. As of now, the 8th edition of Netz's leading commentary on the agricultural real estate transaction law, published in 2018, has grown to be more than 1600 pages. If a large number of new full regulations on the level of the "Länder" had to be taken on board, the commentary would no longer be sufficient in one volume, thus being a symbol of an accumulating complexity.

Moreover, differences in the strictness of the provisions on permissions could lead to a concentration of investors on those "Länder" with the less stringent hurdles to land acquisition. Viewed from a national wide perspective, it might take place a competition between investor-hostile and investor-friendly real estate transaction laws, depending on the agricultural structural orientation of the individual regulation of the "Länder". The potential consequences for farm structures must, therefore, be determined and weighed up. This applies particularly to the possible setting of permissible upper limits for the land ownership of single farms, as such limits can affect the competitiveness of farms in the medium term.

The fact that the agricultural structure in the "Länder" differs was already considered from the beginning in the agricultural real estate transaction law, e.g.,

[33] Cf. from research: Forstner et al. (2011); Forstner and Tietz (2013); Tietz (2017).

by authorising the “Länder” to exempt areas below a certain hectare size from the requirement to obtain a permit. However, hitherto no “Land” has demanded that the competence for the control of agricultural real estate transactions be transferred back to the federal level. A new amendment to the Grundgesetz on this point, therefore, seems unrealistic, as it would require a two-thirds majority of the “Länder” in the German Bundesrat. In addition, the aforementioned development of sample building blocks counteracts the danger of a contradictory multiplicity of state regulations.

At the same time, there seems to be an agreement in agricultural policy at both the federal and “Länder”-level that the instrument of agricultural and forestry real estate transaction law makes sense and should therefore be retained. It is an illustrative example of cautious direct steering in soil law that achieves a balance between a liberal land market and the public interests in a functioning agricultural and forestry structure. It would be conceivable to use this instrument in urban land law as well, by giving priority to the acquisition of land, houses, and flats by private individuals for their own use over the acquisition by investors.[34] Also, a corresponding permission system involving the land register offices could be set up there.

9 Relevance and Change of Seven Guiding Principles of Agricultural Law from 1950

As this article and its historical links illustrate, soil law objectives are subject to constant change. At the same time, however, it is possible to identify some constants. To illustrate this fact, the essay “Landwirtschaftsrecht in 7 Leitsätzen zur Einführung” (Agricultural Law in Seven Guiding Principles as an Introduction), published seventy years ago by the lawyer Friese in the now-forgotten legal training journal “Der Jurist” (The Lawyer), published in Wiesbaden from 1949 to 1955, was based on the legal situation at that time.[35] All seven principles dealt with agricultural land, thereby connecting agricultural law strongly to land law. The first principle read: “As far as possible, new land is to be made available for agricultural purposes, i.e., land that has not been used for agricultural purposes up to now.” Instead of gaining new land for agricultural purposes, the areas used for agriculture in Germany has been steadily decreasing in recent decades. Previously unused land that could be converted into agricultural land is basically no longer available. That is because land in its natural state should be preserved for environmental reasons.

---

[34] Up to now, the contrasts between urban and rural soil law have mostly been emphasised, cf. e.g., Döpfner and Dietzfelbinger (1973), p. 7: “Among the authors on both sides, there was agreement from the outset in the conviction that the problems of building land law and those of agricultural land readjustment are of a completely different nature, indeed are subject to opposing developments.”

[35] Friese (1950).

The second guiding principle followed from the first guiding principle stated: "Land which has hitherto been used for agricultural purposes may not be alienated from this purpose." On the contrary, the rededication of agricultural land to other uses—especially for construction purposes—has been taking place in Germany on a daily basis for a long time and contributes significantly to soil sealing. Compensatory measures for soil sealing are also mostly at the expense of agricultural land. The third guiding principle is to be seen against the background of the food shortage that still existed at the time: "Agricultural land must be managed in the best possible way to achieve the maximum amount of produce". At present, the focus is no longer on maximum yield, but on sustainable management.

The fourth guideline stated that "agricultural land should not come into the hands of persons who are not suitable for its cultivation". In this respect, the permission for the transaction of land was more restrictive in that it included the personal suitability for cultivating agricultural land. In current agricultural real estate transaction law, this aspect only plays a role when a decision has to be made between two interested buyers. If only one farmer asserts his priority, he must show that he has a functioning agricultural business into which the land in question fits in a meaningful way. The fifth guiding principle addressed the agricultural right to debt relief, which no longer exists today: "The indebtedness of agricultural land must be prevented."

The sixth guiding principle took up a reason for refusal that still is valid today in agricultural real estate transaction law, which asserts that: "Agricultural land should not be broken up in an uneconomical manner." The seventh guiding principle, on the other hand, has been superseded by the development of agricultural structures since then, which has led to the fact that only a fraction of the agricultural holdings that existed at that time still survived: "The economic independence of existing agricultural holdings should not, in principle, be abolished by combining them with another holding." It was explained in more detail that the aim was to prevent "the emergence of large agricultural enterprises through the absorption of small farms, thereby reducing the number of independent agricultural livelihoods. That is an agricultural policy and at the same time a socio-political consideration, which also underlies the Land Reform Act." However, there is a certain similarity to the presented idea of reforming the agricultural real estate transaction law by introducing a locally fixed maximum limit for the accumulation of agricultural land.

The legal situation at that time, which was essentially based on Allied occupation law, was concisely described in the "Provisional Guidelines for the Cooperation of the Governments and District Administrative Authorities in Farmers' Court Proceedings for the Approval of Legal Transactions in Agricultural Real Estate Transactions" that the Bavarian State Ministry of Food, Agriculture and Forestry issued on 14 September 1949 and which were printed in the journal "Recht der Landwirtschaft" (Law of Agriculture) in 1950.[36] Here it was stated, among other things that: "The acquisition of land used for agriculture and forestry for recognisable reasons of capital investment, speculation, self-sufficiency and

[36] Bayerisches Staatsministerium (1949).

hobbyism is to be strictly rejected . . .” Nowadays, farming for “hobby” is no longer undesirable. E.g., a number of hobby farms contributes to the preservation of old types of fruit and rare breeds of livestock.

In the same journal, Henrici, who was a higher civil servant in the field of agriculture in the British occupation zone, also wrote in 1950 under the subheading “Lenkung des Bodens in die Hand tüchtiger Berufslandwirte” (Steering the Land into the Hands of Capable Professional Farmers): “Real estate transactions should [. . .] not be left unrestricted to the free or blind play of forces. The new agricultural law, therefore, contains provisions which guarantee a migration or rather a steering of agricultural land into the hands of competent land managers.”[37] That makes it clear that the agricultural real estate transaction law was, without a doubt, understood as a measure of soil steering.

10 Concluding Remarks

The German agricultural real estate transaction law shows on a small scale how controversial interventions in the liberal land market are. The German economic system orients towards a dualism between capitalist property and the social obligation of property. It is the task of the state to ensure that there is an appropriate balance between the two faces of property. The current COVID-19-pandemic has abruptly brought forth the high importance of a nationally functioning economic order with its simultaneous integration into the European Single Market into the general consciousness. That also includes the guarantee of food supply as a so-called critical infrastructure. As was the case in 1918 with the creation of the first agricultural real estate transaction regulation, a sound agricultural structure is considered as an essential element of such a guarantee. That means that a significant proportion of agricultural land must remain in the hands of farmers. For the urban land market, it is also unfavourable if a large part of the land is possessed by persons who do not use the land and the buildings on it for their own, but only want to make the highest possible rate of return from it.

If the landed property becomes the subject of financial speculation, this tends to lead to instability, so that the state should keep a particularly vigilant eye on this aspect. Unlike inter alia in Africa, land grabbing—i.e., large-scale land purchases and leases by foreign powers such as others states and global corporations[38]—is not yet a serious issue in Europe.[39] There undoubtedly exists a considerable need for

---

[37] Henrici (1950).

[38] The ownership structure of land in large parts of Africa is closely linked to the colonial past. Jahnel (2009) offers an instructive legal-historical insight into soil law in the former German colonies, using the example of German Southwest Africa.

[39] The situation is different with “technology grabbing”. §§ 55 et seq. of the Foreign Trade and Payments Regulation contain licensing requirements for the acquisition of shareholdings in a domestic company by a non-EU citizen. Regulation (EU) 2019/452 has for the first time provided

investment in the housing and commercial sectors as well as in the agricultural sector. In the housing and commercial sectors, the existing buildings must be adapted to the ecological challenges, as well as to the changing world of work. Concerning agriculture, the prerequisites for sustainable land management and, thus, agriculture must be created in terms of business premises, operating materials, and business management. However, in both sectors, the demand for the necessary investment resources should not be met by transferring urban and agricultural land—e.g., through sale-and-lease-back transactions and similar constructions—to investors who are not themselves users of the land. That poses the risk of destroying existing property and social structures.

Already in 2010, the German Federal Court of Justice emphasised this limit in a land acquisition case and established relevant criteria.[40] A farmer had split his farm into a holding and an operating company. The holding company owned the agricultural land and made it available to the operating company on a contractual basis. Subsequently, the farmer wanted to expand his farm and therein acquire additional agricultural land for this purpose through the holding company. Since a settlement company was also interested in this land, the agricultural real estate transaction authority refused the acquisition, stating that the holding company could not be regarded as a farmer.

The Thuringian Oberlandesgericht (Higher Regional Court) did not follow this argumentation, because it saw the necessity to make the "widespread forms of entrepreneurial action" existing in the "rest of the medium-sized economy" also usable for agriculture. However, to prevent abuse, it demanded "a factual and personal link between the holding company and the operating company" in such a way that "the handover of the land to the agricultural enterprise is ensured" and "the persons behind the enterprises also have the uniform will to practise agriculture". The German Federal Court of Justice approved this line of reasoning. Admittedly, the question arises as how to ensure compliance with the aforementioned conditions. However, the RETA provides for the permission to be subject to the requirement that the purchaser of the land must lease the land to a farmer for a certain period.

Jurisprudence is thus attempting the balancing act of not preventing modern economic constructions and at the same time inhibiting the migration of landed property from the hands of famers. That shows once again that the RETA acts as a framework that has been decisively filled out by case law. Since the courts are civil courts, the case law has been guided by the general property law principles of the Civil Code, in which property law freedom is paramount. Whether this enhances justice to the intention of achieving a specific soil law objective through an

this with a Union legal framework, to be followed by the Member States from 11 October 2020 on. Article 4 (1) lit. c identifies "food security" as an aspect of the "supply of critical inputs", the protection of which may be pursued by the Member States. In connection with the protection of "critical infrastructure", Article 4 (1) lit. a mentions investments in "land and real estate crucial for the use of such infrastructure". It remains to be seen to what extent Germany will use this framework.

[40] German Federal Court of Justice (2010).

administrative approval procedure can be questioned. The legislator has neither codified nor modified the direction taken by case law for decades. Triggered above all by the rapid structural development in the agricultural sector, a considerable need for reform has arisen in the meantime, which must be overcome to achieve the originally pursued objective again. In this respect, the agricultural real estate transaction law continues to face challenging times. It remains an open discussion as to whether it can serve as a model for steering measures in other areas of soil law.

Altogether, it is worthwhile to have a clear overall picture of a state's soil law. The three levels of soil law presented here act like a pyramid. If the ownership and the process of transacting are not regulated in a legally secure manner, there is no basis for shaping the other two levels in a meaningful way. For, if it is not clear who owns what land, no rights and obligations can be effectively attached to the landed property. Then, the implementation of a steering mechanism in the area of the transaction of landed property is all the more impossible. The first level is undoubtedly sufficiently regulated in Germany, and the land register system, in particular, has served as a model for other states. But in Germany, it currently lacks a systematic, comprehensive order and at the same time complete overview of the regulations on the second and third levels. As a result, rights and obligations from different areas of soil law may not be coordinated with each other or may even contradict each other. Steering effects pursued with one measure can be counteracted by another.

A strong example is the municipalities' drive for more residential and commercial space that goes substantially at the expense of the agricultural landed property surrounding the municipalities. Simultaneously, a healthy agricultural structure should be preserved, and the land consumption resulting from building development ought to be considerably reduced throughout Germany. It is apparent that these objectives are contradictory, and reconciliation is needed.[41] Though, municipal building law and agricultural real estate transaction law are regulated in separate legal acts that lack closer interlocking.

Combining all soil law regulations in one soil law codex seems to be an ideal vision from a legal system perspective, but it would probably neither be possible to realise in practice nor make sense. This is because many provisions form—as the example of the law on the purchase of real property in the Civil Code shows—an integral part of other legal areas and can only be separated from them with great difficulty and not without legal system losses. Therefore, it might be worth considering a basic act on soil law that defines only the objectives pursued under soil law. In this way, existing conflicts of interest and inconsistencies would be revealed and could be resolved. In the end, each state in the world must choose the path appropriate for its historical, social, and economic context. Regardless of this, jurisprudential exchange on the legal-structural and legal-dogmatic foundations of soil law appears to be useful: to sharpen the view of one's own soil law, and to facilitate the comparative legal work on soil law, which has to look at its details.

[41] Cf. in more detail Busse (2020a), p. 2 ff.

References

Arlt R, Rohde G (1967) Bodenrecht. Berlin

Bau- und Bodenrecht (1950 bis 1966) Lose-leaf collection. Hamburg

Bayerisches Staatsministerium (1949) Vorläufige Richtlinien für die Mitwirkung der Regierungen und Kreisverwaltungsbehörden im bauerngerichtlichen Verfahren zur Genehmigung von Rechtsgeschäften des landwirtschaftlichen Grundstücksverkehrs vom 14. September 1949. Recht der Landwirtschaft 1950: 54

Bernhard O, Görgmaier D (1980) Bodenrecht. Munich

Bundesverband der gemeinnützigen Landgesellschaften (2019) Herausgeber. Aufgaben der gemeinnützigen Landgesellschaften im Wandel. Landentwicklung aktuell. Berlin

Bund-Länder-Arbeitsgruppe zur Bodenmarktpolitik (2015) Landwirtschaftliche Bodenmarktpolitik: Allgemeine Situation und Handlungsoptionen. Berlin (available at the website of the Federal Ministry for Food and Agriculture)

Busse C (2016) Sonderrechtlicher und funktionalrechtlicher Ansatz im agrarrechtlichen Widerstreit – Ein Beitrag zur agrarrechtlichen Systembildung. Hagen

Busse C (2018) Agrarrecht und Agrarrechtswissenschaft im Dritten Reich einschließlich des Fortwirkens in der Nachkriegszeit. Teile A und B. Manuskriptdruck als Arbeitsunterlage für die Unabhängige Historikerkommission beim Bundesministerium für Ernährung und Landwirtschaft. Bonn

Busse C (2019a) Ein Jahrhundert landwirtschaftliches Grundstücksverkehrsrecht in Deutschland – Rechtsgeschichtlicher Überblick, aktueller Diskussionsstand und Materialien. Baden-Baden

Busse C (2019b) Grundstückverkehrsrecht in Deutschland – Rechtsgeschichtlicher Überblick und aktueller Diskussionsstand. Waseda Law Rev 94:225 ff

Busse C (2020a) Sachverständige Stellungnahme im Rahmen der zweiten Anhörung „Rechtliche Herausforderungen in der Landwirtschaft" der Enquetekommission V des Landtages Nordrhein-Westfalen. Drucksache 17/3227

Busse C (2020b) Von der agrarrechtlichen Theoriendiskussion zur agrarrechtlichen Strukturdiskussion – Überlegungen zur Viertauflage von Holzers "Agrarrecht". Agrarrecht – Jahrbuch 2020:312 ff

Conradi P, Dieterich H, Hauff V (1972) Für ein soziales Bodenrecht – Notwendigkeiten und Möglichkeiten, 1st edn. Frankfurt am Main

Deutsche Akademie für Staats- und Rechtswissenschaft (1961) Herausgeber. Bodenrecht – Eine Sammlung von Beiträgen. Berlin

Deutscher Städte- und Gemeindebund (1975) Herausgeber. Bodenrecht und Bodenrechtspolitik. Göttingen

Deutscher Verein für Vermessungswesen (1974) Herausgaber. Bodenrecht, Bauleitplanung und Bodenordnung. Stuttgart

Dietrichs H-E (1976) Herausgeber. Diskussion um ein neues Bodenrecht – eine Literaturdokumentation 1963-1975. Cologne

Döpfner J, Dietzfelbinger, H D (1973) Herausgeber. Soziale Ordnung des Baubodenrechts. Gütersloh – Trier

Ernst W, Hoppe W (1978) Das öffentliche Bau- und Bodenrecht, Raumplanungsrecht. 1st edn. Munich

Ernst W, Hoppe W (1981) Das öffentliche Bau- und Bodenrecht, Raumplanungsrecht. 2nd edn. Munich

European Court of Justice (2015) Judgement of 16th July 2015. C-39/14 (available at the website of the European Court of Justice)

Evangelische Akademie Bad Boll (1970) Herausgeber. Städtebau und Bodenrecht. Bad Boll

Farahat A (2020) Eigentum verpflichtet: die Sozialbindung des Eigentums am Beispiel des Berliner Mietendeckels. Juristenzeitung: 602 ff.

Faßbender HJ (1964) Die Privilegierungen und die Versagungsgründe des landwirtschaftlichen Grundstückverkehrsrechts. Dissertation. Cologne

Federal Ministry for Food and Agriculture (2019) Beispiele für Regulierungslücken im landwirtschaftlichen Bodenrecht. Anlage zur Pressemitteilung Nr. 196/2019 vom 27. September 2019 zur Agrarministerkonferenz in Mainz. Berlin (available at the website of the Federal Ministry for Food and Agriculture)

Forstner B, Tietz A (2013) Kapitalbeteiligung nichtlandwirtschaftlicher und überregional ausgerichteter Investoren an landwirtschaftlichen Unternehmen in Deutschland. Thünen Report 5

Forstner B, Tietz A, Klare K, Kleinhanss W, Weingarten P (2011) Aktivitäten von nichtlandwirtschaftlichen und überregional ausgerichteten Investoren auf dem landwirtschaftlichen Bodenmarkt in Deutschland. Endbericht, 2. aktualisierte Fassung. Landbauforschung Sonderheft 352

Friese H (1950) Landwirtschaftsrecht in 7 Leitsätzen zur Einführung. Der Jurist 1950:67 ff

Froemer F (1974) Herausgeber. Neues Bodenrecht – 52 Modelle. Meinungen und Entwürfe mit einer ausführlichen Einführung und dem Gesetzentwurf zur Änderung des Bundesbaugesetzes. Opladen

Gallien K (2007) Bodenrecht und -bewertung in der DDR. Munich

German Federal Constitutional Court (1969) Decision of 19th June 1969. File reference 1 BvR 363/67. Entscheidungen des Bundesverfassungsgerichts Band 26(1970):215 ff

German Federal Court of Justice (2010) Judgement of 26th November 2010. File reference BLw 14/09 (available at the website of the German Federal Court of Justice)

German Federal Court of Justice (2016) Decision of 29th April 2016. File reference BLw 2/12 (available at the website of the German Federal Court of Justice)

German Federal Court of Justice (2020) Judgement of 8th May 2020. File reference BLw 2/18 (available at the website of the German Federal Court of Justice)

Gesellschaft für Wohnungs- und Siedlungswesen (1969) Herausgeber. Verfassung, Städtebau, Bodenrecht. Hamburg

Ginzky H u.a. (2017) Foreword. IYSLP 2016: V f

Hedemann JW (1910) Die Fortschritte des Zivilrechts im XIX. Jahrhundert – Ein Überblick über die Entfaltung des Privatrechts in Deutschland, Österreich, Frankreich und der Schweiz. 1. Teil: Die Neuordnung des Verkehrslebens. Berlin

Hedemann JW (1930) Die Fortschritte des Zivilrechts im XIX. Jahrhundert – Ein Überblick über die Entfaltung des Privatrechts in Deutschland, Österreich, Frankreich und der Schweiz. 2. Teil: Die Entwicklung des Bodenrechts von der französischen Revolution bis zur Gegenwart. 1. Hälfte: Das materielle Bodenrecht. Berlin

Hedemann JW (1935) Die Fortschritte des Zivilrechts im XIX. Jahrhundert – Ein Überblick über die Entfaltung des Privatrechts in Deutschland, Österreich, Frankreich und der Schweiz. 2. Teil: Die Entwicklung des Bodenrechts von der französischen Revolution bis zur Gegenwart. 2. Hälfte: Die Entwicklung des formellen Bodenrechts mit einem Anhang: Bodenrecht und neue Zeit. Berlin

Hedemann JW (1939) Deutsches Wirtschaftsrecht – Ein Grundriss, 1st edn. Berlin

Henrich F (1972) Herausgeber. Eigentum und Bodenrecht, Munich

Henrici (1950) Allgemeine Grundgedanken des neuen Landwirtschaftsrechts der britischen Zone. Recht der Landwirtschaft 1950:104

Hoffmann M, Schinke E (1973) Bodenrecht und Bodennutzung in der Landwirtschaft der DDR. Cologne

Hoffmeister F (2018) Steuerung des landwirtschaftlichen Grundstücksverkehrs – Bestand und Weiterentwicklung des Grundstückverkehrsrechts unter besonderer Berücksichtigung unionsrechtlicher Grenzen. Baden-Baden

Horst HR (1995) Bodennutzungsrecht – Sachenrechtsbereinigung und Schuldrechtsanpassung in den neuen Bundesländern. Düsseldorf

Institut für kommunalpolitische Bildung und Forschung der Politischen Akademie Eichholz (1968) Herausgeber. Material zur Reform des Bodenrechts, Eichholz

IYSLP (2017)

Jahnel MJ (2009) Das Bodenrecht in „Neudeutschland über See“ – Erwerb, Vergabe und Nutzung von Land in der Kolonie Deutsch-Südwestafrika 1884-1915. Frankfurt am Main – Berlin – Bern – Brussels – New York – Oxford – Vienna
Kingreen T (2020) Die Verteilung der Gesetzgebungskompetenzen für das Mietpreisrecht bei Wohnraum, Neue Zeitschrift für Verwaltungsrecht: 737 ff.
Knirsch H (1972) Herausgeber. Bodenrecht – Beiträge zur Reform des Grundeigentums. Berlin – New York
Lück H (2017) Zur Entwicklung des landwirtschaftlichen Siedlungs- und Grundstücksrechts seit dem späten 19. Jahrhundert – Eine rechtshistorische Skizze. Halle an der Saale
Martinez J (2016) Grundstückverkehrsgesetz. In: Düsing/Martinez (Herausgeber), Agrarrecht: 473 ff
Matthes W (1962) Deutsches Bodenrecht. Hamburg
Netz J (2018) Grundstückverkehrsgesetz – Praxiskommentar. 8. edn. Butjaden-Stollhamm
Rohde G (1976) Herausgeber. Bodenrecht: Lehrbuch. 1st edn. Berlin
Rohde G (1979) Herausgeber. Bodenrecht: Ein Lehrbuch. 2nd edn. Berlin
Rohde G (1989) Herausgeber. Bodenrecht, Berlin
Rohde G, Schietsch E, Zänker S (1969) Bodenrecht der DDR. Berlin
Rutkoswky H, Schulz WR (1986) Handbuch Bau- und Bodenrecht – Eine Einführung für die tägliche Praxis. Wiesbaden – Berlin
Schmid CU (2020) Falsche Fronten im Mietendeckelstreit – Zur Bedeutung eines effektiven und gerechten Gesetzesvollzugs, Juristenzeitung: 610 ff
Storck L (1970) Stadterneuerung und Bodenrecht in der BRD. Bonn
Tietz A (2017) Überregional aktive Kapitaleigentümer in ostdeutschen Agrarunternehmen: Entwicklungen bis 2017. Thünen Report 52
Ulbrich M (1987) Herausgeber. Sammlung von Rechtsvorschriften zum Bodenrecht der DDR. Teile 1 und 2. Dresden
Werner E (1970) Bodenrecht – Fragen zur Neuordnung. Cologne-Mülheim
Wieacker F (1938) Bodenrecht. Hamburg
Zacher C (2002) Die Entstehung des Wirtschaftsrechts in Deutschland – Wirtschaftsrecht, Wirtschaftsverwaltungsrecht und Wirtschaftsverfassung in der Rechtswissenschaft der Weimarer Republik. Berlin
Zänker S, Boenicke F (1979) Bodenrecht. Berlin

Quantitative Targets, Tradable Planning Permits and Infrastructure Cost Calculators: Examples of Instruments Addressing Land Take in Europe

Jana Bovet and Elisabeth Marquard

Abstract Excessive "land take" is unsustainable because it causes detrimental and partly irreversible changes to ecosystems as well as adverse social and economic impacts. Here, we showcase three policy instruments for reducing land take, namely *quantitative targets*, *tradable planning permits* and *infrastructure cost calculators* and give an overview of their current application in selected European countries. We conclude that proven and field-tested instruments are available and that it is important to tap their full potential for further reducing land take in Europe.

1 Introduction

Throughout Europe, agricultural and semi-natural land is lost due to expanding settlements or infrastructure developments. This phenomenon is referred to as "land take" by the European Environment Agency (EEA).[1] Excessive land take signals an inefficient use of natural resources and contravenes ecologic and social sustainability targets. It is therefore important to identify promising approaches for reducing land take, e.g. by comparing land policies and related instruments across different states.[2] Here, we portray selected policy instruments for reducing land take that are currently being applied or discussed in different EU Member States and Switzerland, drawing mainly on results of the SURFACE project, which was funded

[1] European Environment Agency (2019b).

[2] Gerber et al. (2018); Naumann et al. (2019); Ronchi et al. (2019).

J. Bovet (✉)
UFZ – Helmholtz Centre for Environmental Research, Department of Environmental and Planning Law, Leipzig, Germany
e-mail: jana.bovet@ufz.de

E. Marquard
UFZ – Helmholtz Centre for Environmental Research, Department of Economics, Leipzig, Germany
e-mail: lisa.marquard@ufz.de

H. Ginzky et al. (eds.), *International Yearbook of Soil Law and Policy 2020/2021*, International Yearbook of Soil Law and Policy 2020,
https://doi.org/10.1007/978-3-030-96347-7_4

by the German Environment Agency from 2017 to 2022. We focus on *quantitative targets* as one way of fostering political commitment and accountability with respect to reducing land take, and on *tradable planning permits* as a tested mechanism for operationalizing quantitative targets at the sub-national level. Furthermore, we discuss the potential of *infrastructure cost calculators* as an example of a decision support tool for the municipal level.

1.1 Current Land Take Trends in Europe and Their Impacts

The total land take of the 28 EU Member States is over 500 km^2 per year (data basis: Corine Land Cover 2012–2018).[3] The amount and distribution of land take is very heterogeneous among countries and influenced by demographic and economic factors, the planning system, and taxes and subsidies, which in turn affect decisions on public infrastructure, and private consumption and mobility patterns.[4] Even within countries there are significant regional differences, driven, e.g., by geography (in mountainous regions, less land is available for development which is all the more under pressure of being built-up) or the political set-up (in federal/decentralized countries, sub-national administrative entities may have far-reaching powers). A commonality across European countries is that land take occurs primarily at the expense of agricultural land;[5] in fact, during the period 2000–2018, about 50% resulted in the loss of arable land and permanent crops and another almost 30% in the loss of pastures and mosaic farmlands[6]—i.e., land take affects mostly soils of high quality.

Widespread land take has manifold impacts. Soil-sealing and landscape fragmentation destroy or impair species' habitats and ecosystem functions, and may affect the local water and climate regime. Additionally, the conversion of land into built-up areas reduces the amount of land available for the production of renewable energy or the provision of other ecosystem services. With a focus on public spending, land take is undesirable for municipalities when it leads to scattered settlements with low population densities (i.e., urban sprawl) because they are then faced with high per-capita costs for the installation and maintenance of technical infrastructures (s. also Sect. 2.3). Furthermore, inefficient land use may have adverse social implications, e.g., where it results in high numbers of commuters and in areas under-supplied with local services, such as medical care and retail stores. Reduced accessibility generally increases the dependency on cars and lowers the prospect of expanding alternative means of transportation.

[3] European Environment Agency (2019b).

[4] Colsaet et al. (2018).

[5] European Environment Agency (2019a), p. 118; Hennig et al. (2015); Naumann et al. (2019).

[6] European Environment Agency (2019a), p. 118.

All in all, land take is nevertheless generally accepted as a by-product of settlement and infrastructure expansion, as developments are mostly associated with growing prosperity. With respect to housing, a trend toward a greater number of smaller households and an increase in living space per person can be observed in many European countries, further fuelling the demand for new buildings. At the same time, the number and size of roads and other traffic routes grows continuously in the EU, triggered by an increase in individual use of cars as well as by the infrastructure demands of particular branches of industry—in particular, tourism, logistics and production sites.[7]

1.2 International and European Policies for Reducing Land Take

In 2012, the Member States of the United Nations (UN) declared at the Conference on Sustainable Development in Rio de Janeiro (Rio+20), that they "will strive to achieve a land-degradation neutral world".[8] In 2015, the UN General Assembly adopted the 2030 Agenda for Sustainable Development including the 17 Sustainable Development Goals (SDGs).[9] Each of the goals has a list of time-bound targets which are partially quantitative, and two of them (SDG 11 and 15) implicitly request the reduction of land take. The EU and its Member States have repeatedly declared their commitment toward the 2030 Agenda,[10] and although the 2030 Agenda is non-binding and its implementation at the national level is voluntary, it "carries a strong moral obligation or social contract on the part of governments".[11] Accordingly, the call for "resilient and sustainable cities" (SDG 11—requiring, inter alia, a reduction of the ratio of the land consumption rate to the population growth rate) and for "a land-degradation neutral world" (SDG 15) urge the EU and its Member States to develop and implement policies for confining land take.

The EU has no general competence for spatial planning in its Member States but does impact spatial planning indirectly through instruments designed primarily for other purposes, such as safeguarding nature and the environment (European environmental legislation), fostering cohesion (e.g. European Grouping of Territorial Cooperation, EGTC) or supporting research (e.g. European Spatial Planning Observation Network, ESPON). Furthermore, there are some specific infrastructure planning instruments at the European level (e.g. Trans-European Networks) and areas of shared competence (e.g. Marine Spatial Planning). However, for the main part, EU Member States rely on their national regulations with regard to land-related

[7] Naumann et al. (2019); Prokop et al. (2011); Ronchi et al. (2019).

[8] United Nations (2012).

[9] https://sdgs.un.org/goals.

[10] European Commission (2016).

[11] TAP Network (2018).

planning. As a result, much variation exists in this context across Europe, but actors in all Member States face a common challenge when carrying out spatial planning: reconciling (urban) growth with environmental concerns. Among the latter, land degradation is of major importance and a very serious problem.[12]

When discussing strategies for combatting land degradation, a differentiation is made between approaches addressing qualitative aspects (such as pollution from noxious substances and mechanical stresses) and quantitative aspects (such as the loss of natural and semi-natural areas). The call for combatting *land take* clearly emphasizes the quantitative perspective and respective political objectives aim to reduce or control land consumption by means of spatial planning.

In 2011, the EU Commission proposed that: "By 2020, EU policies take into account their direct and indirect impact on land use in the EU and globally, and the rate of land take is on track with an aim to achieve no net land take by 2050 [. . .]".[13] Building on this initiative, the goal of achieving "no net land take by 2050" was incorporated into the 7th Environment Action programme[14] of the EU which was adopted by the European Parliament and the Council of the EU in 2013. Similar to the SDGs, the "no-land-take by 2050" target is only a non-binding political declaration, intended as an orientation for the Member States.

In 2014, land take was given greater consideration by the EU when the Directive on environmental impact assessments (EIAs) was altered (adoption of Directive 2014/52/EU amending Directive 2011/92/EU). EIAs are carried out for certain public and private projects to ensure that environmental considerations are explicitly addressed and incorporated into the related decision making process. In 2014, the EU included 'land' in the list of factors that need to be considered during EIAs (Art. 3), and 'land take' was mentioned in recital 9 as an example of a relevant impact.

Despite these initiatives, policies targeted to the prevention of land and soil degradation are fragmented throughout the EU, since the planning systems themselves have remained to this day mostly in the competence of the Member States and an initiative for creating a legally binding European policy for soil protection has failed: after years of debate, a proposal by the European Commission for a Soil Framework Directive was withdrawn in 2014. Instead of a Directive, the non-binding "Soil Thematic Strategy" of 2006 has remained in place. As a result, there is an enormous inhomogeneity with regard to land and soil protection strategies and instruments among the Member States,[15] even though they face similar challenges as they strive for reconciling (urban) growth on the one hand with environmental concerns on the other. Comparing instruments and approaches of different Member States may therefore help identify promising starting points to curtail land take. This comparison could allow countries to reflect on their own instruments against the background of other approaches, and the EU could benefit from such an

[12]Montanarella et al. (2016); IPBES (2018), pp. XX–XXV.

[13]European Commission (2011).

[14]European Parliament and the Council (2013).

[15]Flasbarth (2017); Frelih-Larsen et al. (2016).

analysis as differences become apparent, particularly in terms of terminology, definitions and indicators. Such differences are, on the one hand, legitimate and even necessary for ensuring that instruments and approaches are fit for purpose and address the specific needs of a country, region, or municipality. On the other hand, however, terminological and methodological variation may also hamper the effective exchange of relevant knowledge. In particular with regard to monitoring and reporting in the EU- and SDG-context, conceptual and methodological clarifications would much improve the communication among relevant actors, as well as the comparability of data across different countries.[16]

1.3 Research Design of the Project SURFACE

Within the project "Standards and Strategies for Reducing Land Consumption" (SURFACE), funded 2017–2022 by the German Federal Environment Agency (*Umweltbundesamt*), we investigated whether and how the objective of reducing land take has found its way into political and legal governance in various countries, with a particular focus on the EU. To this end, a survey was conducted among 25 experts from 16 EU Member States, another three members of the OECD, and two BRICS[17] countries (for a total of 21 countries). To supplement the obtained data, a workshop was held in April 2019 with experts from twelve EU-countries who had a background in science or administration. At this workshop, commonalities and differences with respect to country-specific objectives and instruments for the reduction of land take were discussed.[18]

2 Policy Instruments Applied or Discussed in European States

Here, we present three instruments that experts who participated in our survey and/or workshop (s. above Sect. 1.3) mentioned as being promising for reducing land take. These are: *quantitative targets for reducing land take* as a strategic policy tool (Sect. 2.1), *tradable planning permits* as an economic instrument (Sect. 2.2) and the *infrastructure cost calculator* as a decision support tool for planning (Sect. 2.3). All three instruments presented in this chapter are adaptable to various contexts and can therefore presumably be incorporated into all national or sub-national planning frameworks.

[16] Nicolau et al. (2019); Marquard et al. (2020).

[17] BRICS = acronym for Brazil, Russia, India, China, South Africa.

[18] Bovet et al. (2019); Marquard et al. (2020).

2.1 Quantitative Targets for the Reduction of Land Take

Quantitative environmental targets define a level of ecological quality that sustainably balances social and economic interests. Hence they can be means for expressing and implementing a societal consensus about the kind and level of environmental quality that we aspire to preserve for ourselves and also for the sake of future generations.[19] In accordance with this intention, *quantitative land take targets* specify a maximum amount or a rate of land take that is permitted during a defined period or per unit of time. Particularly in combination with a monitoring framework, such quantitative targets may prove effective because they provide a reference value against which land use changes are evaluated and to which developments that are not on track should be re-adjusted.

Some of the European states currently have quantitative targets for the reduction of land take in place (e.g. Germany, Switzerland and Belgium/Flanders) or have adopted such targets in the past (e.g. Luxemburg, Austria).[20] All of these quantitative targets are or have been legally non-binding.

As early as 2002, **Germany** was one of the first countries in the EU to address land take using a quantitative target. Facing at that time a land take of about 110 ha/d,[21] the German federal government set the target to reduce land take by 2020 to 30 ha/d. This so called "30 ha target" was included in the National Sustainability Strategy.[22] It did not differentiate between types (or locations) of land development to be permitted or prevented but defined an overall reference value to aim at, and thus has had a guiding function. In 2016, the National Sustainability Strategy was revised and the German government decided to postpone the target by ten years and to adjust it to "less than 30 ha/d".[23] This can be interpreted as a political reaction to the realization that the 30 ha target was not going to be achieved by 2020, since, in 2016, the daily land take in Germany was about 50 ha/d.[24] The reformulation of the target, however, gives no indication about interim steps or, generally, about the shape of the intended reduction curve. A plausible alternative to promoting "less than 30 ha/d by 2030" would have been to adhere to the 30 ha target with a less drastically adjusted timeline and in combination with a more ambitious, yet specific, target for 2030. In this way, the 30 ha target would have been kept as an interim target on a time line until 2030, thereby articulating more explicitly the intention of continuously reducing land-take.

---

[19] Bovet et al. (2018); Reese and Bedtke (2015).

[20] See also Naumann et al. (2019).

[21] https://www.destatis.de/DE/Themen/Branchen-Unternehmen/Landwirtschaft-Forstwirtschaft-Fischerei/Flaechennutzung/Publikationen/Downloads-Flaechennutzung/anstieg-suv.pdf?__blob=publicationFile.

[22] German Federal Government (2002), p. 99.

[23] German Federal Government (2016), p. 154.

[24] Ibid.

In **Switzerland**, a quantitative land take target was part of the Swiss National Sustainability Strategy during 2012–2015.[25] The target read "max. 400 m^2 settlement area (*Siedlungsfläche*)/inhabitant", which was close to the actual use of settlement areas at that time.[26] The target in Switzerland may therefore be interpreted as a stabilization target rather than a reduction target. The Swiss Sustainable Development Strategy 2016–2019[27] aimed to confine urban sprawl, and to protect cultural and natural land as far as possible from being built-up, but the former land take target was not updated for the new Strategy. Recent data suggest that land take has increased again and now exceeds the value of the former target (407 m^2 settlement area per inhabitant[28]). In May 2020, the Swiss Federal Council adopted a Soil Strategy[29] that focuses on the sustainable protection of soil as a resource and sets a target of "no-net soil consumption" (*Netto-Null Bodenverbrauch*) from 2050 onwards. According to this Soil Strategy, it is still possible to build, but if soil functions are lost, they must be compensated for. Whether the compensation is sufficient is decided based on qualitative and not on quantitative criteria. The Swiss no-net soil consumption target is explicitly justified by SDG 15.3 and the 7th Environmental Action Programme of the European Union (see above Sect. 1.2).

In **Flanders/Belgium**, a quantitative land take target was announced by the regional government in 2018, calling for a step-wise process to reduce land take in Flanders to 3 ha/d until 2025 and to 0 ha/d by 2040 (Draft of the Strategic Vision of the Spatial Policy Plan of Flanders[30]). However, after the regional elections in 2019, the newly formed Flemish government weakened this goal by not including and confirming it in its coalition agreement. At the same time the political discourse has changed from a 'concrete stop', to a 'construction shift' (i.e. further building in the cores and no more urban sprawl). For the Belgian **Walloon** region, a similar step-wise reduction of land take has been declared in the revised strategic master plan (SDT—Schéma de Développement du Territoire): 6 km^2/a until 2030 and 0 km^2/a until 2050.[31] However, although this SDT was adopted by the regional government of Wallonia in 2019, it has not yet entered into force (as at December 2020) because a restrictive clause was included stipulating that this will only happen at a day to be determined by the government.

In **Luxembourg**, the National Sustainability Plan 2010 formulated the target of reducing land consumption from 1.3 to 1 ha/d by 2020.[32] Measures such as a soil

---

[25] Schweizerischer Bundesrat (2012), p. 30.

[26] Federal Statistical Office of Switzerland (2014).

[27] Swiss Federal Council (2016).

[28] https://www.are.admin.ch/are/de/home/raumentwicklung-und-raumplanung/grundlagen-und-daten/fakten-und-zahlen/siedlungsflaechen.html.

[29] Bundesamt für Umwelt (BAFU) (2020).

[30] Departement Omgeving (2018), p. 26.

[31] Wallonian Government (2019), p. 111.

[32] Commision interdépartemental du développement durable du gouvernement luxembourgeois (2010), p. 35.

sealing tax, a Soil Protection Act and remediation of contaminated sites were to contribute to achieving the target. The "3ème Plan National pour un Développement Durable" of December 2019[33] does not set a new target for the period after 2020, but envisages that, in future, a new framework law will allow a better coordination of all aspects of soil protection and that a national soil protection plan will be established. Meanwhile, this law has been drafted. It contains two main parts: a preventive part focusing on soil protection in the strict sense and a curative part describing the principles of management of polluted or potentially polluted sites.[34]

In **Austria**, a quantitative target for the reduction of land take was set in 2002 as part of the Austrian Strategy for Sustainability. However, this target ("a maximum of one tenth of the current growth rate should be achieved by the year 2010", which was equivalent to 2.5 ha/d)[35] was not met and it was not explicitly renewed when the Sustainability Strategy was replaced by a new version in 2010 (Strategy for Sustainable Development of the Federal Government and Federal Provinces, ÖSTRAT[36]). The governmental programme 2020–2024[37] of the New Austrian People's Party and the Green Party includes a target path ("*Zielpfad*") for reducing the net land take to 2.5 ha/d by 2030, but only as an indicative ambition. In April 2020, an initiative for creating a binding legal framework for reducing land take and for shifting more political competences to the national scale failed in the Austrian National Council.[38] The Council, however, agreed on the development of a national soil protection strategy aiming at the implementation of the "2.5 ha/d net land take target", the establishment of a nationwide monitoring of land take, the promotion of brownfield recycling, and the strengthening of supra-regional spatial planning.[39]

Table 1 gives an overview over the currently and previously applicable, legally non-binding quantitative land take targets in selected European countries:

Caution is advised when comparing and interpreting the land take targets listed in Table 1 because their units of measurement differ. While the given figures for Wallonia can be easily adapted to the unit "ha / d" used by the majority of the countries (see squared brackets in Table 1), the Swiss figures cannot be converted into this unit because they are expressions of settlement density instead of changes in area occurring over time. Furthermore, it seems adequate to evaluate land take not only by its pace but also in relation to meaningful reference values, such as the size of the population (as does the Swiss target), the total area of the country or the total area available for development. These possible reference values differ substantially between the listed countries.

[33] Le Gouvernement du Grand-Duché de Luxembourg (2019).

[34] https://environnement.public.lu/fr/natur/sol/projet-de-loi-sols.html.

[35] Austrian Federal Government (2002), p. 70.

[36] Österreichische Bundesregierung und Landesregierungen (2010).

[37] Die neue Volkspartei/Die Grünen (2020), p. 104.

[38] https://www.parlament.gv.at/PAKT/VHG/XXVII/A/A_00362/imfname_784433.pdf.

[39] https://www.parlament.gv.at/PAKT/VHG/XXVII/E/E_00025/fname_793311.pdf.

Table 1 Quantitative land take targets in selected European countries

	Quantitative target & year when it was set	Current land take[a]	Previous target	Remark
1a) Selection of quantitative targets for land take				
Germany	Less than 30 ha/d until 2030 (2016)	56 ha/d (2018)	30 ha/d until 2020 (2002)	Target relaunched and modified in 2016
Switzerland	No-net soil consumption from 2050 onwards (2020)	407 m^2 settlement area/ inhabitant (2018)	Max. 400 m^2 settlement area/ inhabitant (2012) (stabilization target)	The target max. 400 m^2 settlement area/inhabitant was removed in 2016
Flanders	3 ha/d until 2025 and 0 ha/d until 2040 (2018)	6 ha/d (2018)		No longer included in the current coalition agreement
Wallonia	6 km^2/a [1.6 ha / d] until 2030 and 0 km^2/a until 2050 (2019)	10.7 km^2/a [2.9 ha/d] (between 2015 and 2018)		Included in the strategic master plan 2019 not yet in force
1b) Selection of former quantitative targets				
Austria	2.5 ha/d until 2010 (2002)	11.8 ha/d (between 2017 and 2019 yearly averaged)		Removed in 2010; mentioned in the governmental programme 2020–2024 but without implications
Luxemburg	1 ha/d until 2020 (2010)	0.5 ha/d (2019)		Removed in 2019

[a]References: DE: https://www.destatis.de/DE/Themen/Branchen-Unternehmen/Landwirtschaft-Forstwirtschaft-Fischerei/Flaechennutzung/Publikationen/Downloads-Flaechennutzung/anstieg-suv.pdf?__blob=publicationFile; CH: https://www.bfs.admin.ch/bfs/de/home/statistiken/raum-umwelt/umweltindikatoren.assetdetail.9066372.html; FL: https://publicaties.vlaanderen.be/download-file/28360; WL: https://www.iweps.be/wp-content/uploads/2019/10/CC2019-1.pdf; AT: https://www.umweltbundesamt.at/fileadmin/site/publikationen/REP0684.pdf; LU: https://environnement.public.lu/dam-assets/documents/developpement-durable/PNDD.pdf

As already mentioned in the context of the German 30 ha target above, in addition to setting a "final target" for the land take, it would also be very useful to specify intermediate targets. Firstly, different reduction paths lead to a different total amount of land take (e.g. a linear reduction leads to a lower total amount of land take than one in which land take continues to be high for a considerable period of time, even if it decreases steeply toward the end of the curve). Secondly, (projected) intermediate targets are indispensable, for effective readjustments are only possible when path deviations are recognized in time. Against this background, it seems inadequate that only Flanders and Wallonia have set intermediate targets.

In conclusion, despite their shortcomings, such as their non-binding character, and the fact that some are no longer in force, the European quantitative land take targets of the present and the past have spurred the scientific and political discussion on how to manage land and soils more sustainably. In Germany, some federal states have meanwhile set their own (also non-binding) targets for reducing land take (e.g. Rheinland-Pfalz: <1 ha/d[40]) and in 2019, a referendum was held in the State of Bavaria with the aim of establishing a quantitative land take target.[41] In February 2019, a referendum[42] was held in Switzerland with the aim to amend the Federal Constitution and to introduce a compensation scheme which would couple the designation of building zones (*Bauzonen*) with the obligation to remove building zones that have already been designated but are still unsealed and have at least the same size and a comparable potential agricultural yield value. Both referendums failed, but the reported activities show that the topic of "land take" has become important in the public and political debates in Germany as well as in Switzerland.

2.2 *Tradable Planning Permits*

In Flanders,[43] Switzerland[44] and Germany,[45] tradable planning permits are discussed as a means to reduce land take. First and foremost, these discussions are held among academics and a few practitioners; tradable planning permits are not actually practiced in any of the countries mentioned. However, in Germany, tradable planning permits have been tested in empirical studies,[46] in laboratory experiments,[47] and in a large-scale simulation, in which nearly 100 municipalities took part (2012–2017).[48] Therefore, some experience has been gained for the German context, and findings on the legal conformity of such trading schemes as well as on their feasibility and effectiveness have been published.[49]

The main idea behind tradable permit schemes in general is to artificially tie the use of a resource to a politically defined cap, e.g. tonnes of CO_2 emissions or hectares of natural land for development. Every unit, e.g. tonne or ha, below this cap is turned into a commodity by issuing permits, allowing the holder of a certificate to access or use the resource. Since the total quantity of permits is limited by the cap, a well-functioning tradable permit scheme reliably prevents the use of more of the resource in question than the overarching target defines. By making the permits

[40] https://mueef.rlp.de/de/themen/klima-und-ressourcenschutz/bodenschutz/flaechensparen/.

[41] http://betonflut-eindaemmen.de/.

[42] https://www.zersiedelung-stoppen.ch/zersiedelungsinitiative-offiziell-zustande-gekommen/.

[43] Crabbé and Coppens (2019); Ruimte Vlaanderen (2016).

[44] Menghini et al. (2015).

[45] Bovet et al. (2013); Henger and Bizer (2010); Köck et al. (2018); Siedentop et al. (2013).

[46] Bizer et al. (2012).

[47] Meub et al. (2016).

[48] Grimski (2019).

[49] Henger et al. (2019).

tradable, economists expect that the resources would be directed towards their most profitable use.[50]

In the large scale simulation game for trading planning permits in the context of the German 30 ha target (s. above Sect. 2.1), municipalities were allowed to develop a certain amount of natural land for settlements and traffic infrastructure.[51] This defined quantity was issued in the form of permits, and the municipalities, responsible for urban land use planning in Germany, received certificates for free via a distribution key. If a municipality wanted to develop land by designating areas for settlement, it had—as before—to design the land use plan in compliance with all legal provisions, but additionally, it had to present a corresponding number of permits. If a municipality owned fewer permits than it needed for a plan, it could purchase permits from other municipalities. If it had more permits than it needed for its land use plan(s), it could sell permits to other municipalities or keep them for later projects. Furthermore, special kinds of permits (so-called "white permits") were issued when municipalities withdrew adopted but not realized plans. A municipality could use these white permits itself or it could sell them. By allowing white permits, municipalities could at least partially offset compensation obligations and dissolution costs that potentially arose through the withdrawal of a plan. The results of the simulation showed that municipalities were able to handle this instrument well and that the possibility of white permits created an incentive to withdraw unrealized land use plans and to enhance inner development.[52]

Trade is not in itself a necessary component of a system of land planning permits but the tradability of the permits gives stakeholders higher flexibility of action and discretion since they can decide whether they want to use their permits themselves or sell them.[53] Furthermore, by selling their permits to other municipalities, shrinking municipalities may generate revenue that they can then use for urban consolidation measures.

With regard to the distribution key allocating the capped permits, several different indicators have been proposed: e.g. the population size of a municipality, the future population size or its economic prosperity. In the German simulation, a degressive distribution key based on the number of inhabitants was chosen. Consequently, small municipalities received slightly more certificates per inhabitant than bigger cities, where denser construction is common and where less land is needed per inhabitant. This key was accepted widely by local authorities as fair and transparent.[54]

The results from the large-scale simulation conducted in Germany are very promising. The researchers found out that the existence of such a trading scheme encourages municipalities to actually check the cost-benefit relations of their

---

[50] Nuissl and Schroeter-Schlaack (2009).

[51] Grimski (2019).

[52] Nuissl and Schroeter-Schlaack (2009).

[53] Ibid.

[54] Grimski (2019).

planning much more precisely than they otherwise would. Municipalities also were capable of incorporating this calculation into the framework of urban land use planning. Independently of how the municipalities decided—whether using, selling or purchasing permits—the overarching land take target was always met.[55] Based on these finding of the simulation, there have been initiatives from opposition parties in regional parliaments (e.g. in Bavaria and Saxony) to implement this instrument into practice but they were rejected and so far, no tradable permit system for reducing land take exists in Germany or, to our knowledge, elsewhere in Europe.

2.3 *Infrastructure Cost Calculator*

The realization of residential or commercial buildings leads to technical, social and cultural infrastructure costs which in general have to be paid by the municipalities. The costs are often downplayed by the responsible actors, because of the assumption that these costs are compensated for by tax revenues and key allocations, or that technical infrastructures like roads, drinking water supply, sewage, gas or electricity will be part of an urban development contract ("the investor pays it all"). Nevertheless, not all projects include such contracts, and even if the investor is responsible for installing the technical infrastructure, municipalities have to bear the long-term maintenance and repair costs for it, and often cover the entire costs for necessary social and cultural infrastructure (e.g. kindergartens, primary schools, libraries). Past studies have shown that the settlement structure has an influence on the infrastructure costs a municipality has to bear. Increasing land take and demographic changes in suburban and especially in rural areas generally lead to a decreasing utilisation density and higher infrastructure costs.[56]

Infrastructure cost calculators are useful here. These decision support tools estimate the investment and ongoing infrastructure costs of a project or a plan and thus provides valuable information for the planning process. Such cost-benefit analyses do not replace the ecological and social criteria or the appreciation of values that a municipality has to comply with when it takes its planning decision, but should rather be regarded as supplementary.

A wide variety of information may be included in such analyses; e.g. type of plan (residential, commercial, mixed areas), different costs and benefits which are considered for different periods of time, or comparisons of different location and/or development variants. Here, we present the infrastructure cost calculator NIKK[57]

[55]Henger et al. (2019).

[56]Ecoplan et al. (2017); Ewing and Hamidi (2015); Hesse et al. (2020).

[57]NIKK - Niederösterreichischer Infrastrukturkostenkalkulator: https://www.raumordnung-noe.at/index.php?id=148.

from Austria in some more detail.[58] The Federal State of Lower Austria offers this calculator to its municipalities. The standard application is a free online tool for municipalities and potential stakeholders in local land use planning. Data of six thematic categories can be entered, including (1) settlement types, (2) technical infrastructure, (3) housing type proportions, (4) land tax, (5) social infrastructure, and (6) development costs. Additionally, various time horizons can be selected, and background information is provided by the federal state—such as a local population prognosis—which can be integrated into the application.[59] The quality of the result obviously correlates with the realistic assessment of the situation at hand and with the proper input of the required data. Therefore, municipalities should invest some time in data collection and input in order to obtain a valid result. Both, the findings on the NIKK by Humer et al. (2019) and feedback from municipal representatives in the context of the German large-scale simulation on tradable planning permits[60] recommended that such fiscal impact analyses should be applied more extensively during planning processes in order to make the costs transparent for the municipalities and the decision-makers (e.g. city council). At present, the use of all infrastructure cost calculators is voluntary but it would certainly make sense to include this application in planning law and to make it an obligatory part of the local statutory planning procedure.[61]

3 Discussion

Although the extent and pace of land take varies across the European countries and regions, it has in common that it takes place primarily on high quality soils and that the drivers and the manifold impacts are often comparable. Urban sprawl processes are projected to accelerate worldwide, and in Europe nearly the entire land has already been subject to heavy anthropogenic influence. In this context, wise and effective implementation of policies for reducing land take prove to be extremely important. In the EU, where the competences for spatial planning lie almost exclusively at the national or sub-national level, comparing the different approaches seems a promising way to extract best practice examples and lessons learnt. The given overview of selected instruments (some already in use, others only discussed in a scientific context) illustrates a variety of existing approaches and the potential they may have for enriching discussions on land take reduction strategies at the national or European level.

[58] More infrastructural cost calculators can be found, e.g., at: https://aktion-flaeche.de/folgekosten-rechtzeitig-kalkulieren.

[59] Humer et al. (2019).

[60] Henger et al. (2019).

[61] Humer et al. (2019).

Since instruments are always embedded in specific national policy and legal frameworks, however, it is not possible to transfer them directly from one state to another; they always need to be adapted to the legal and planning context at hand. Nevertheless, it is worth analysing the circumstances and policies of different countries, as this may help to identify promising starting points as well as clarify methodological questions.

Here, we have given an overview on quantitative land take targets which set standards and time lines. Looking at the present European picture, it is apparent that there are quantitative targets in a few countries but not a single one has ever been legally binding. They are all programmatic, and as such subject to politics—they can also be cancelled or readjusted (like in Austria or Germany), depending on changing political constellations.

Of the five countries listed, only two countries (Luxembourg and Switzerland) met their land take target within the given timeline (note, however, that the target in Switzerland was a stabilisation target). These targets have therefore not been reinforced after their expiration, but in both countries the topic of land take and soil consumption is being pursued further politically. In Switzerland, e.g., a new "no-net soil consumption target" (from 2050 onwards) is in place, which widens the focus beyond the rather narrow indicator of "settlement area/inhabitant". Although this indicator is still used for describing soil consumption, the focus of Swiss planning policy has shifted and the indicator of urban sprawl ("*Zersiedlung*") is gaining importance.[62]

In Germany, the land take target set in 2002 was not met in 2020 and was postponed by ten years; in Austria, the target was simply cancelled as it became apparent that it would not be achieved. These undesirable developments could explain why the latter two countries are working on and experimenting with additional innovative instruments to reduce land take (tradable planning permits and infrastructure cost calculator for municipalities).

Furthermore, some other European countries work with quantitative targets, e.g. France, where the law of agricultural and fishery modernization (Law n° 2010-874 of July the 27th 2010) lays down the objective that, by 2020, land take on agricultural land should be halved, and according to the 'Grenelle II law' (Law n° 2010-788 of July the 12th 2010), strategic urban planning documents have to set targets for reducing urban sprawl.[63] The reason we have not presented France together with the other countries above is that the French target does not set a cap for land take but restricts land conversions at the expense of agricultural land. Thus, France does not define the maximum permissible land take for a certain (unit of) time (i.e., a quantitative land take target as understood in this chapter). However, it ought to be mentioned that, prospectively, France wants to include the goal of halting land take, with a target of no net land take by 2025, in a Plan for Biodiversity.[64]

[62] Schwick et al. (2018).

[63] Desrousseaux (2017).

[64] https://www.gouvernement.fr/en/biodiversity-france-s-strategy-for-action.

Furthermore, our description of tradable planning permits shows that such a scheme is a promising regulatory mechanism, especially if it is linked to a quantitative land take target. The simulation which was carried out in Germany demonstrated that such a scheme can be implemented consistently with the existing legal framework and that it may be implemented in a way that is feasible for the participating municipalities as well as effective in securing progress toward the target. Certainly, the German example of tradable planning permits cannot be transferred 1:1 to other European countries, because planning systems differ. But in general, tradable permit schemes for natural land can be incorporated in all land use planning systems—either as a market of land use rights, in which land owners trade the right to use a plot of land for construction purposes, or as a market of land designation rights, in which municipalities trade with each other (like in Germany). A well-functioning tradable planning permit scheme has the advantage that an overarching target value is not exceeded, and that the tradability of the permits gives stakeholders the broadest scope for freedom of action since they can decide if they want to use or to sell the permits. Since the municipalities in Germany have a strong right of self-administration, which is guaranteed by the constitution and grants them planning sovereignty, it can be questioned whether a system with tradable planning permits violates the municipalities' autonomy. Juridical experts have, however, approved the compatibility of land planning permits with the German constitution since the instrument limits the municipalities only quantitatively but not conceptually, and the possibility of selling or buying permits provides them sufficient flexibility.[65]

An infrastructure cost calculator provides important information for local authorities when making planning decisions. Compared with a traditional urban development calculation, the presented NIKK is much more comprehensive and takes into account long-term-costs and the demographic development of the municipality. Based on such cost transparency, better decisions can be made because it becomes obvious that in the long term, inner urban development is often the more cost-effective alternative compared with urban sprawl.

The tradable planning permits as well as the infrastructure cost calculator give economic considerations more weight in planning decisions. In the German simulation, it was intensively discussed with the stakeholders how much increased importance of economic aspects would change the designation behaviour of the municipalities. It became clear that even under a tradable planning permit scheme, planners do not decide solely on fiscal aspects such as opportunity costs, but also on other criteria such as urban planning considerations or political priorities.

In conclusion, the examples presented here show that proven and field-tested instruments are available that can help to reduce further land take in Europe. Quantitative targets are strategically very important but they need to be supplemented by further policies and implementation mechanisms. A scheme of tradable planning permits is one option for such a mechanism. It creates incentives

[65] Köck et al. (2018); Grimski (2019).

for municipalities to question their development plans and to accurately evaluate their needs. Infrastructure cost calculators can support them in their assessments and related decision-making processes. Thus, the three instruments presented here seem indeed promising, as the experts consulted during the SURFACE project had suggested, and they will likely work most effectively in combination. Tapping their full potential is required for arriving at more sustainable outcomes of planning processes and for tackling the still high rates of land take in Europe.

4 Further Research Needs

Here we presented selected results from project SURFACE (2019-2022) focusing on different instruments for reducing land take. With this overview, we aim at facilitating a comparative view on the most effective policies and approaches for reducing land take in the EU. In the course of the project we encountered significant questions which could not be investigated in depth within SURFACE. Among those, the question of how to reconcile efficient land use with other important societal goals, e.g. preserving urban biodiversity and maintaining or improving the physical and mental health of cities' inhabitants, struck us as most urgent. Urban ecosystems, including brownfields, have proven valuable for recreation, for mitigating climate change effects and for providing habitats for plant and animal species.[66] These ecosystem services may be under pressure by prioritizing inner urban development over greenfield development. Bearing this in mind seems particularly relevant in the light of ecologically and aesthetically impoverished agricultural landscapes which harbor sometimes less biodiversity than urban areas.[67] Furthermore, in 2020, the COVID-19 crisis has demonstrated risks of high population densities in a pandemic situation. This has raised further caution against efforts for strengthening inner urban development, as they may create situations in which inhabitants are more likely exposed to transmittable diseases. At the same time, it is important to note in the context of public health that dense urban constructions need not always be the unhealthier alternative to low-density (suburban) settlement patterns, since a healthy urban environment encompasses many aspects, including, e.g., the availability of an efficient (medical) infrastructure. In our view, the densification of cities should be regarded as an important approach for containing land take but its benefits need to be traded off carefully against its potential drawbacks. Studying arguments for and against densification as well creative solutions for arriving at compact and healthy settlements are therefore important avenues for further research. They should include exploring the potentials not only of recycling brownfields but also of other strategies for using urban space more efficiently, e.g., elevating low-story buildings and converting excessive private-car related infrastructure, such as sizeable parking lots.

[66]Mathey et al. (2015).

[67]Kowarik (2011).

Acknowledgement The project "Standards and Strategies for Reducing Land Consumption – SURFACE" was funded by the Federal Environment Agency (funding code: FKZ 3717 18 110 0). We thank all the 'land take experts' across Europe and beyond who we consulted for our research and who provided valuable insights.

References

Austrian Federal Government (2002) The Austrian Strategy for Sustainable Development, p. 182. http://sdgtoolkit.org/wp-content/uploads/2017/02/The-Austrian-Strategy-for-Sustainable-Development_en.pdf (last accessed: 1 September 2021)

Bizer K, Bovet J, Henger R, Jansen N, Klug S, Ostertag K, Schleich J, Siedentop S (2012) FORUM: Handel mit Flächenzertifikaten. Fachliche Vorbereitung eines überregionalen Modellversuchs. In: Umweltbundesamt (ed) UBA-Texte 60/2012, Dessau-Roßlau, p. 252. https://www.iwkoeln.de/fileadmin/publikationen/2012/124947/Handel_mit_Flaechenzertifikaten.pdf (last accessed: 1 September 2021)

Bovet J, Bizer K, Henger R, Ostertag K, Siedentop S (2013) Handelbare Flächenzertifikate – vom akademischen Diskurs über einen Modellversuch in die Planungspraxis? Raumforschung und Raumordnung 71:497–507

Bovet J, Reese M, Kock W (2018) Taming expansive land use dynamics - sustainable land use regulation and urban sprawl in a comparative perspective. Land Use Policy 77:837–845

Bovet J, Marquard E, Schröter-Schlaack C (2019) SURFACE Workshop Report. International Expert Workshop on Land Take. 4-5 April 2019 in Berlin, Germany. Helmholtz Centre for Environmental Research – UFZ, Leipzig, p. 36. https://www.ufz.de/surface (last accessed: 1 September 2021)

Bundesamt für Umwelt (BAFU) (2020) Bodenstrategie Schweiz für einen nachhaltigen Umgang mit dem Boden. Schweizerischer Bundesrat, Bern, p 64. https://www.bafu.admin.ch/dam/bafu/de/dokumente/boden/ud-umwelt-diverses/bodenstrategie-schweiz.pdf.download.pdf/Bodenstrategie_2020-05-01.pdf (last accessed: 1 September 2021)

Colsaet A, Laurans Y, Levrel H (2018) What drives land take and urban land expansion? A systematic review. Land Use Policy 79:339–349

Commision interdépartemental du développement durable du gouvernement luxembourgeois (2010) PNDD (Plan National de Développement Durable) Luxembourg. Un Luxembourg durable pour une meilleure qualité de vie. Ministère du Développement durable et des Infrastructures, Luxembourg, p 86. https://environnement.public.lu/content/dam/environnement/documents/developpement-durable/Un-Luxembourg-plus-durable-pour-une-meilleure-qualite-de-vie-2010.pdf (last accessed: 1 September 2021)

Crabbé A, Coppens T (2019) Swapping development rights in swampy land: strategic instruments to prevent floodplain development in flanders. In: Hartmann T, Slavíková L, McCarthy S (eds) Nature-based flood risk management on private land: disciplinary perspectives on a multidisciplinary challenge. Springer International Publishing, Cham, pp 85–97

Departement Omgeving (2018) Beleidsplan Ruimte Vlanderen. Strategische visie, p 93. https://publicaties.vlaanderen.be/download-file/30641 (last accessed: 1 September 2021)

Desrousseaux M (2017) The French law on biodiversity and the protection of soils. In: Ginzky H, Dooley E, Heuser IL, Kasimbazi E, Markus T, Qin T (eds) International yearbook of soil law and policy 2017. Springer, pp 277–286

Die neue Volkspartei/Die Grünen 2020. Aus Verantwortung für Österreich. Regierungsprogramm 2020–2024. Bundeskanzleramt Österreich, Wien, p 230. https://www.dieneuevolkspartei.at/Download/Regierungsprogramm_2020.pdf (last accessed: 1 September 2021)

Ecoplan, B+S, Hunziker Betatech (2017) Infrastrukturkosten unterschiedlicher Siedlungstypen. Oder: Ist verdichtet und zentral bauen billiger? im Auftrag des Bundesamtes für Raumentwicklung (ARE), Bern, p. 316. https://www.are.admin.ch/dam/are/de/dokumente/infrastruktur/publikationen/infrastrukturkosten-unterschiedlicher-siedlungstypen.pdf.download.pdf/infrastrukturkosten-unterschiedlicher-siedlungstypen.pdf (last accessed: 1 September 2021)

European Commission (2011) Roadmap to a Resource Efficient Europe, Brussels, p. 26. https://eur-lex.europa.eu/legal-content/EN/TXT/PDF/?uri=CELEX:52011DC0571&from=EN (last accessed: 1 September 2021)

European Commission (2016) Next steps for a sustainable European future, Strasbourg, p. 19. https://eur-lex.europa.eu/legal-content/EN/TXT/PDF/?uri=CELEX:52016DC0739&from=EN (last accessed: 1 September 2021)

European Environment Agency (2019a) The European environment - State and outlook 2020. Knowlege for transition to a sustainable Europe, Publications Office of the European Union, Luxembourg, p. 499. https://www.eea.europa.eu/publications/soer-2020/at_download/file (last accessed: 1 September 2021)

European Environment Agency (2019b) Land take in Europe. Indicator Assessment. https://www.eea.europa.eu/data-and-maps/indicators/land-take-3/assessment (last accessed: 1 September 2021)

European Parliament and the Council (2013) Decision No 1386/2013/EU of the European Parliament and of the Council of 20 November 2013 on a General Union Environment Action Programme to 2020 'Living well, within the limits of our planet'. Official Journal of the European Union, 171–200

Ewing R, Hamidi S (2015) Compactness versus sprawl: a review of recent evidence from the United States. J Plan Liter 30:413–432

Federal Statistical Office of Switzerland (2014) Environment. Pocket Statistics 2014, Schweizerische Eidgenossenschaft, Neuchâtel, p. 36. https://www.bfs.admin.ch/bfsstatic/dam/assets/349550/master (last accessed: 1 September 2021)

Flasbarth J (2017) Soils need international governance: a European perspective for the first volume of the international yearbook of soil law and policy. In: Ginzky H, Heuser IL, Qin T, Ruppel OC, Wegerdt P (eds) International yearbook of soil law and policy 2016. Springer, Cham, Switzerland, pp 15–19

Frelih-Larsen A, Bowyer C, Albrecht S, Keenleyside C, Kemper M, Nanni S, Naumann S, Mottershead RD, Landgrebe R, Andersen E, Banfi P, Bell S, Brémere I, Cools J, Herbert S, Iles A, Kampa E, Kettunen M, Lukacova Z, Moreira G, Kiresiewa Z, Rouillard J, Okx J, Pantzar M, Paquel K, Pederson R, Peepson A, Pelsy F, Petrovic D, Psaila E, Šarapatka B, Sobocka J, Stan A-C, Tarpey J, Vidaurre R (2016) Updated inventory and assessment of soil protection policy instruments in EU Member States. Final Report to DG Environment. Ecologic Institute, Berlin, p 462. https://ec.europa.eu/environment/soil/pdf/Soil_inventory_report.pdf (last accessed: 1 September 2021)

Gerber J-D, Hartmann T, Hengstermann A (2018) Instruments of land policy. Dealing with scarcity of land. Routledge, Abingdon, Oxfordshire, UK, p 372

German Federal Government (2002) Perspectives for Germany. Our Strategy for Sustainable Development, p. 345. https://www.bundesregierung.de/Content/DE/_Anlagen/Nachhaltigkeit/perspektives-for-germany-langfassung.pdf?__blob=publicationFile&v=2 (last accessed: 1 September 2021)

German Federal Government (2016) German Sustainable Development Strategy. New Version 2016, p. 252. https://www.bundesregierung.de/Content/DE/_Anlagen/Nachhaltigkeit/2017-06-20-nachhaltigkeit-neuauflage-engl.pdf?__blob=publicationFile&v=2 (last accessed: 1 September 2021)

Grimski D (2019) Tradable land planning certificates to reduce land take: results of a simulation game with communities in Germany. In: Ginzky H, Dooley E, Heuser IL, Kasimbazi E, Markus T, Qin T (eds) International yearbook of soil law and policy 2018. Springer, Cham, Switzerland, pp 131–147

Henger R, Bizer K (2010) Tradable planning permits for land-use control in Germany. Land Use Policy 27:843–852

Henger R, Daniel S, Schier M, Blecken L, Fahrenkrug K, Melzer M, Bizer K, Meub L, Proeger T, Gutsche J-M, Tack A, Ferber U, Schmidt T, Siedentop S, Straub T, Kranz T, Weinhardt C (2019) Modellversuch Flächenzertifikatehandel. Realitätsnahes Planspiel zur Erprobung eines überregionalen Handelssystems mit Flächenausweisungszertifikaten für eine begrenzte Anzahl ausgewählter Kommunen, UBA-Texte 116/2019. Umweltbundesamt, Dessau-Roßlau, p. 191. https://www.umweltbundesamt.de/sites/default/files/medien/1410/publikationen/2019-10-02_texte_116-2019_modellversuch-flaechenzertifikatehandel.pdf (last accessed: 1 September 2021)

Hennig EI, Schwick C, Soukup T, Orlitova E, Kienast F, Jaeger JAG (2015) Multi-scale analysis of urban sprawl in Europe: towards a European de-sprawling strategy. Land Use Policy 49:483–498

Hesse M, Mengs C, Grüttner A (2020) Kosten der Siedlungsstruktur – Herausforderungen für Kommunen. InfrastrukturRecht (IR) – Energie, Verkehr, Abfall. Wasser 17:36–39

Humer A, Sedlitzky R, Brunner D (2019) When does population growth pay off? A case study of suburban land consumption to assess the Lower Austrian infrastructural cost calculator. J Housing Built Environ 34:331–344

IPBES (2018) The IPBES assessment report on land degradation and restoration (laid out version), Montanarella L, Scholes R, Brainich A (eds) Secretariat of the Intergovernmental Science-Policy Platform on Biodiversity and Ecosystem Services, Bonn, Germany. 744 pages. https://doi.org/10.5281/zenodo.3237392 (last accessed: 1 September 2021)

Köck W, Bovet J, Tietz H (2018) Mengensteuerung der baulichen Flächenneuinanspruchnahme - Zur Notwendigkeit eines Flächenzertifikatehandelsgesetzes. Zeitschrift für Umweltrecht 2:67–75

Kowarik I (2011) Novel urban ecosystems, biodiversity, and conservation. Environ Pollut 159: 1974–1983

Le Gouvernement du Grand-Duché de Luxembourg (2019) Luxembourg 2030. 3ème Plan National pour un Développement Durable, Luxembourg, p. 107. https://environnement.public.lu/content/dam/environnement/documents/developpement-durable/PNDD.pdf (last accessed: 1 September 2021)

Marquard E, Bartke S, Font JGI, Humer A, Jonkman A, Jurgenson E, Marot N, Poelmans L, Repe B, Rybski R, Schroter-Schlaack C, Sobocka J, Sorensen MT, Vejchodska E, Yiannakou A, Bovet J (2020) Land consumption and land take: enhancing conceptual clarity for evaluating spatial governance in the EU context. Sustainability 12

Mathey J, Rossler S, Banse J, Lehmann I, Brauer A (2015) Brownfields as an element of green infrastructure for implementing ecosystem services into Urban areas. J Urban Plan Dev 141

Menghini G, Gemperle F, Seidl I, Axhausen KW (2015) Results of an agent-based market simulation for transferable development rights (TDR) in Switzerland. Enviro Plan B-Plan Design 42:157–183

Meub L, Proeger T, Bizer K, Henger R (2016) Experimental evidence on the resilience of a cap & trade system for land consumption in Germany. Land Use Policy 51:95–108

Montanarella L, Pennock DJ, McKenzie N, Badraoui M, Chude V, Baptista I, Mamo T, Yemefack M, Aulakh MS, Yagi K, Hong SY, Vijarnsorn P, Zhang GL, Arrouays D, Black H, Krasilnikov P, Sobocka J, Alegre J, Henriquez CR, Mendonca-Santos MD, Taboada M, Espinosa-Victoria D, AlShankiti A, AlaviPanah SK, Elsheikh EAE, Hempel J, Arbestain MC, Nachtergaele F, Vargas R (2016) World's soils are under threat. Soil 2:79–82

Naumann S, Frelih-Larsen A, Prokop G, Ittner S, Reed M, Mills J, Morari F, Verzandvoort S, Albrecht S, Bjuréus A, Siebielec G, Miturski T (2019) Land take and soil sealing—drivers, trends and policy (legal) instruments: insights from European cities. In: Ginzky H, Dooley E, Heuser IL, Kasimbazi E, Markus T, Qin T (eds) International Yearbook of soil law and policy 2018. Springer, Cham, Switzerland, pp 83–112

Nicolau R, David J, Caetano M, Pereira JMC (2019) Ratio of land consumption rate to population growth rate-analysis of different formulations applied to Mainland Portugal. Isprs Int J Geo-Inf 8:10

Nuissl H, Schroeter-Schlaack C (2009) On the economic approach to the containment of land consumption. Environ Sci Policy 12:270–280

Österreichische Bundesregierung und Landesregierungen (2010) Österreichische Strategie Nachhaltige Entwicklung (ÖSTRAT) – ein Handlungsrahmen für Bund und Länder, p 11. https://www.bmk.gv.at/dam/jcr:f18a3fe4-5704-4732-b7e7-a433b3e08e5a/OeSTRAT_2010.pdf (last accessed: 1 September 2021)

Prokop G, Jobstmann H, Schönbauer A (2011) Overview of best practices for limiting soil sealing or mitigating its effects in EU-27. Technical Report - 2011 - 050. European Commission - DG Environment, Environment Agency Austria, p 231. https://op.europa.eu/en/publication-detail/-/publication/c20f56d4-acf0-4ca8-ae69-715df4745049/language-en (last accessed: 1 September 2021)

Reese M, Bedtke N (2015) Was ist "Nachhaltigkeit" und was ist "nachhaltige Wasserwirtschaft"? Allgemeine Nachhaltigkeitskonzeptionen und Ableitungen für die Wasserver- und Abwasserentsorgung. In: Gawel E (ed) Die Governance der Wasserinfrastruktur. Duncker & Humblot, Berlin, pp 153–195

Ronchi S, Salata S, Arcidiacono A, Piroli E, Montanarella L (2019) Policy instruments for soil protection among the EU member states: a comparative analysis. Land Use Policy 82:763–780

Ruimte Vlaanderen (2016) Adviesnota Verhandelbare Ontwikkelingsrechten p. 44. https://archief-algemeen.omgeving.vlaanderen.be/xmlui/bitstream/handle/acd/229959/Adviesnota_verhandelbare_ontwikkelingsrechten_2016.pdf (last accessed: 1 September 2021)

Schweizerischer Bundesrat (2012) Strategie Nachhaltige Entwicklung 2012-2015, p. 70. https://www.eda.admin.ch/dam/eda/de/documents/aussenpolitik/umwelt-wissenschaft-verkehr/ARE_SNE_d_web.pdf (last accessed: 1 September 2021)

Schwick C, Jaeger J, Hersperger A, Cathomas G, Muggli R (2018) Zersiedelung messen und begrenzen. Massnahmen und Zielvorgaben für die Schweiz, ihre Kantone und Gemeinden. Haupt Verlag, Bern

Siedentop S, Bizer K, Bovet J, Henger R (2013) Handelbare Flächenzertifikate - ein praxistaugliches Instrument? In: Meinel G, Schuhmacher U, Behnisch M (eds) Flächennutzungsmonitoring V: Methodik, Analyseergebnisse. Flächenmanagement Leibniz-Institut für ökologische Raumentwicklung, Dresden, pp 11–20

Swiss Federal Council (2016) Sustainable Development Strategy 2016-2019, p. 76. https://www.are.admin.ch/dam/are/en/dokumente/nachhaltige_entwicklung/publikationen/strategie_nachhaltigeentwicklung2016-2019.pdf (last accessed: 1 September 2021)

TAP Network (2018) SDG Accountability Handbook: A Practical Guide for Civil Society, in: Cardinal, N., Romano, J., Sweeney, E. (Eds.), Transparency, Accountability & Participation (TAP) Network, p. 100. https://sdgaccountability.org/about-the-handbook/ (last accessed: 1 September 2021)

United Nations (2012) The future we want. Resolution adopted by the General Assembly, New York. http://www.un.org/ga/search/view_doc.asp?symbol=A/RES/66/288&referer=http://sustainabledevelopment.un.org/futurewewant.html&Lang=E (last accessed: 1 September 2021)

Wallonian Government (2019) Raumentwicklungsschema. Eine territoriale Strategie für die Wallonie, p. 163. http://lampspw.wallonie.be/dgo4/site_amenagement/amenagement/sdt (last accessed: 1 September 2021)

Regulating Large-Scale Farmland Investments in Low Income Countries ('Land Grabbing'): Appraising Different Modes of Transnational Governance

Till Markus

Abstract This article investigates the governance potential of different policies and laws addressing the issues arising out of large-scale farmland investments in low income countries. To this end, it will interpret their development on the whole as a tentative effort by a set of different actors to forge an effective governance system. It is argued here that the development of these laws and policies has largely been defined by three key features: (1) forming a better understanding of the complex issues arising out of large-scale farmland investments, (2) reframing these issues, the responsibilities of involved actors, and the purpose of regulation, and finally, (3) flexibly shifting modes of governance in view of ineffective national and international laws and institutions. All three strategies have contributed to establishing a set of transnational guiding principles and standards of care that will substantially increase legal accountability and have the potential to effectively change investment practices.

1 Introduction

Over the last decade, large-scale investments in plantation agriculture in low-income countries have raised public, political, and legal concern.[1] It is feared that such investments negatively affect some of the world's least fortunates' rights and livelihoods as well as the environment—particularly fertile soils, forests, and fresh water resources. Critics have termed this kind of investment practice as "land grabbing".[2]

[1] For introductions and overviews see contributions in Montilla Fernández (2017); Kugelman and Levenstein (2013); Demissie (2015); Cotula (2013); see also contributions in Allan et al. (2012).

[2] See, for example, Borras et al. (2017); European Parliament (2014); The Economist (2011); Joan Baxter (2010); Knaup and von Mittelstaedt (2009).

T. Markus (✉)
Helmholtz Centre for Environmental Research - UFZ, Leipzig, Germany
e-mail: till.markus@ufz.de

H. Ginzky et al. (eds.), *International Yearbook of Soil Law and Policy 2020/2021*,
International Yearbook of Soil Law and Policy 2020,
https://doi.org/10.1007/978-3-030-96347-7_5

In response, some target countries have developed an array of national programs, plans, and laws to govern these types of investments in an attempt to mitigate the specific issues in their respective national contexts. In addition, several international non-binding instruments have been developed to guide policy and law makers, investors, and other involved actors towards fairer and more sustainable investments. Already in 2009 the UN Special Rapporteur on the Right to Food proposed "core" or "minimum" principles that aim to safeguard human rights in the context of foreign large-scale investments in land.[3] In 2012 the United Nations Food and Agricultural Organization (FAO) adopted the "Voluntary Guidelines on the Responsible Governance of Tenure of Land, Fisheries, and Forests in the Context of National Food Security".[4] It includes a section specifically addressing farmland investments.[5] In 2014 the United Nation Committee on World Food Security endorsed the more general "Principles for Responsible Investment in Agriculture and Food Systems".[6] At the time of writing the International Institute for Unification of Private Law (UNIDROIT) together with the FAO and the International Fund for Agricultural Development (IFAD) have been in the process of developing another more specific guidance instrument, i.e. the "Legal Guide on Agricultural Land Investment Contracts".[7]

It is argued here that the development of these instruments can be interpreted as a tentative search by actors involved in and affected by farmland investments to better govern the issues arising out of such business practices and to forge an effective governance strategy in response. Three key features have defined this process. First, over the years, scientific and administrative experts have substantially broadened the understanding of the current drivers of large-scale farmland investments and their negative effects in target countries. Second, international organizations involving stakeholders and experts have reframed and reinterpreted the issues arising out of large-scale farmland investments and changed the overall narrative about who bears responsibilities and why investments need to be regulated. Third, in view of frequently ineffective international and national legal instruments and institutions, governance efforts have shifted more towards alternative informal regulatory mechanisms. These more effective mechanisms still fulfill basic normative functions which are usually performed by formal law.

---

[3] De Schutter (2009).

[4] Voluntary Guidelines on the Responsible Governance of Tenure of Land, Fisheries, and Forests in the Context of National Food Security (2012). In addition, the FAO proposed in 2010 the Principles for Responsible Agricultural Investment that Respect Rights, Livelihoods and Resources. These Principles were criticized by various actors and then basically replaced by the two other instruments mentioned here, both developed by the FAO. On the criticism, see De Schutter (2011).

[5] Tenure Guideline, Sec. 12.

[6] The Committee on World Food Security (2014).

[7] The Guide was launched on 24 September 2021 after this article had been submitted. The Guide is available at https://www.unidroit.org/wp-content/uploads/2021/10/ALICGuidehy.pdf. Accessed 2 March 2022.

Reflecting upon these three features will help to better understand the current status as well as the potential of the evolving governance system addressing the problems caused by large-scale farmland investments. In particular it will clarify the broad variety of governance strategies, forms, and mechanisms available and required to solve these issues and call attention to the complex interplay and dependencies between legal and non-legal instruments.

The following article will briefly review available information about the most important characteristics of large-scale farmland investments, including some of the key associated problems (Sect. 2). It will then provide an overview of the different legal regimes that govern them (Sect. 3.1), present a typology of the most important formal governance instruments, i.e. target countries' national laws on foreign farmland investments (Sect. 3.2.1), and provide a review of empirical findings regarding the challenges with their implementation (Sect. 3.2.2). In addition, it will introduce pertinent international guidance instruments developed to mitigate issues and conflicts growing out of large-scale farmland investments (Sect. 3.3). Finally, it will assess national and international instruments and interpret their development mainly from transnational governance perspective in order to better identify those actors and actions that can help to further improve investment practices (Sect. 4). It will then draw some conclusions (Sect. 5).

2 Large-Scale Farmland Investments: Quantity, Quality, Trends, and Issues

Assessing quantities and specific characteristics of large-scale farmland investments in low-income countries has proven challenging. This is owed inter alia to a lack of transparency of investment flows and the lack of a broad number of systematic and representative studies, as well as to conceptual challenges and outdated information.[8] Despite these challenges, there seems to be a general consensus that the number of global large-scale farmland investments surged in 2005, with a slight levelling out around 2012, but then has continued to increase ever since, though at a somewhat slower pace.[9] Several estimates figure that large-scale farmland investments have taken control over at least 45 million hectares of land, whether through the acquisition of concessions or long-term leases.[10] The latest numbers provided by the Land Matrix Project conclude that land controlled through respective contracts summed up to 43.6 million hectares in April 2016. In addition, intended deals

[8] Liao et al. (2016); Oya (2013).

[9] Nolte et al. (2016); Schoneveld (2014a); Cotula et al. (2014); Deininger et al. (2011); De Schutter (2009).

[10] White et al. (2012).

included roughly another 20.2 million hectares.[11] Of all regions on the globe, Sub-Saharan Africa is clearly the main target area, followed by South-East Asia, Latin America, South-East Europe, and Oceania.[12] While in Sub-Saharan Africa, areas controlled through investments make up approximately 9.6% of areas harvested across the subcontinent, in Cambodia, for example, they totalled 36% of the country's agricultural land.[13]

Increasing investments have been driven and made possible by a number of key factors, including, for example, a globally growing demand for agriculture products and energy (which, in turn, have been particularly fuelled by the 2008 global food crisis, BRIC-Countries' economic growth and their increasing demand for food, and the increased use of biofuels as a result of several countries' commitments to promote renewable energies).[14] In addition, trade liberalization, rising prices for land and agricultural commodities, and anticipated growth in demand for food, water, and energy, have all attracted speculative investments.[15] Growing supply risks and volatile food prices on world markets, as well as increasingly stringent quality, safety, and traceability requirements on large food markets such as the European Union (EU) or the U.S. have also motivated the food processing, trading, and retailing industries to engage more directly in farming in low income countries.[16] Some have pointed out that transnational investments in farmland have been—at least in part—motivated by interests in securing access to fresh water.[17] Last but not least, these investments have often been propelled by low income countries' hopes that inviting foreign direct investment in agriculture would help them to promote economic development in rural areas, of "unused" land, to create employment, to modernize their agricultural sectors, to create foreign exchange, and ultimately to alleviate poverty.[18]

In principle, these positive effects and opportunities are possible, provided certain conditions are being met and effective institutional mechanisms are in place.[19] Several issues, however, have repeatedly been observed in the course of

[11] Nolte et al. (2016).

[12] Nolte et al. (2016); Cotula (2013); see also Deininger et al. (2011).

[13] See Schoneveld (2014a); Davis et al. (2015). Such numbers can be somewhat misleading with a view to understanding the overall trend in land investments. For example, land acquisitions in Ethiopia, Ghana, and Tanzania between early 2005 and mid 2012 were estimated to account for 1.1% to 1.9% of land suitable for rain-fed agriculture. See Cotula et al. (2014).

[14] Deininger et al. (2011); Aanseuw et al. (2012); De Schutter (2011); Mackintosh and Burgess (2008).

[15] Kugelman and Levenstein (2013), Chapter 1.

[16] Cotula (2014–2015), pp. 181–182; UNCTAD (2009), pp. 105–106.

[17] Smaller and Mann (2009); De Schutter (2009).

[18] Liao et al. (2016), pp. 131–132; De Schutter (2011).

[19] Deininger et al. (2011).

investments[20] and they have been criticized on various grounds. Though issues and critiques differ regarding specific cases and countries,[21] they may be roughly grouped into three sets.

First, in a broad number of cases, target countries' local communities' land tenures have lost out against those acquired by foreign investors. In many countries these land tenures are based on traditional use rights which are often not sufficiently recognized and protected under formal national law and thus become marginalized in the process of large-scale land acquisitions.[22] Even where national laws recognize and protect customary use rights to some extent, these laws often lack full implementation.[23] Accordingly, in a large number of cases investments have led to evictions of rural populations from the lands traditionally used by them.[24]

Second, it has been observed that excluding small-scale farmers and or other traditional users from their land and replacing them with large-scale farms has in several cases reduced the overall environmental status of the respective areas.[25] In particular, turning smallholder farmland or relatively "unused" land (which in fact is often used for a variety of activities including, for example, gathering wood, foraging for food, hunting, grazing etc.) into industrial-type farmland has led in many cases to accelerated deforestation, reduction of biodiversity, water pollution, and loss of fertile soils.[26]

Third, there have been general political and macro-economic concerns. Large-scale farmland investments will usually introduce financially powerful (often foreign and globally active) investors such as governments (acting directly or through state-run enterprises), agri-businesses, the food processing and retailing industry, and private equity funds into hitherto undeveloped local or national land markets. It is feared that this will be mainly to the detriment of local people and communities who cannot compete in these asymmetrical power relations. For those poor rural populations that are already food insecure this could have extremely harsh consequences. Adopting such an investment strategy to develop farmland would also accelerate the commodification of land and water which neglects their cultural importance and disconnects rural communities and populations from their traditional ways of living and earning their livelihoods.[27] In addition, foreign large-scale farmland investments are inherently designed to systematically draw food resources

---

[20] For a comprehensive general introduction into the challenges of African land policies and the position of large-scale farmland investments, see AU-ECA-AfDB-Consortium (2010), pp. 1–41 (5–12). See also references provided in Cotula (2014–2015), pp. 182–183.

[21] Kazimbazi (2018), pp. 315–330; Chiziane (2018), pp. 331–346; Ubink (2007), pp. 215–248; Neef et al. (2013), pp. 1085–1103.

[22] Cotula (2014–2015), pp. 182–183; German et al. (2013b), pp. 1–18.

[23] German et al. (2013b), pp. 1–18; Schoneveld (2013).

[24] Montemayr (2012), pp. 135–152; Golay and Biglino (2013), pp. 1630–1650.

[25] D'Odorico and Rulli (2012), pp. 324–325.

[26] Davis et al. (2015), pp. 772–775; Messerli et al. (2014); Borras and Franco (2012); German et al. (2013a).

[27] De Schutter (2011), pp. 252, 256–258.

and economic benefits away from target countries, again, particularly to the detriment of their relatively poor rural populations.[28] It has also been argued that these kinds of investments will introduce a type of agri-business that is simply less sustainable and in most cases less effective in reducing poverty than improved small-scale farming.[29]

3 Efforts to Govern Large-Scale Farmland Acquisitions Through Law: Regimes and Instruments

The following subsection provides an overview of the different legal regimes that govern large-scale farmland investments (I). It will then present a typology of target countries' national legal instruments (II.1), and provide an overview of their empirical findings regarding the challenges in implementing these national measures (II.2).

3.1 The Legal Frame

Different types of regimes and laws can apply to large-scale farmland investments: formal or informal (e.g. general or specific binding international or national state law, non-binding international guidance, or private standards), and sectoral or cross-cutting (e.g. investment and trade law or human rights or environmental law). They can also apply at different levels: the international, the national, and local level. To this day, no binding international treaty has been adopted that specifically regulates large-scale farmland investments. Only non-binding guiding instruments have been negotiated under the auspices of different international organizations. Non-specific or general binding international law, however, can become directly applicable: on one hand investment practices may infringe internationally guaranteed human rights of local communities where host countries' governments would not guarantee certain minimum standards in the course of the acquisition process (e.g. right to food, right to self-determination, specific indigenous peoples' rights, or rights of agricultural workers).[30] On the other hand, investments may be protected against national policies or judicial decisions encroaching on rights safeguarded under investment treaties.[31] Both binding regimes, however, do not proactively and

[28] De Schutter (2011), pp. 249–279; German et al. (2013a).

[29] Large-scale plantations "have a higher productivity per active laborer, although their productivity per hectare is lower" compared to small-scale farms. De Schutter (2011), p. 260.

[30] Golay and Biglino (2013), pp. 1630–1650; Borras Jr. et al. (2017).

[31] See on the issues associated with investor protection, for example, Slater (2015); Douglas (2013), pp. 415–444; Tienhaara (2009).

systematically, let alone comprehensively regulate large-scale farmland investments. They would be applicable only as a reflex and on a cases-by-case basis to specific issues arising throughout the investment processes. At present, specific rules and standards governing farmland investments can primarily be found at the national level, i.e. in target countries' national laws. They are thus the most important source of formal law governing farmland investments. Private regulatory efforts and standard setting for certification schemes or the financing sector have, to this day, only played a marginal role.[32] It will become clearer throughout the following passages that this might change in the near future.

3.2 *National Law Instruments in Low Income Countries*

The following sub-sections 1 and 2 aim to illustrate the status of the most important source of formal laws governing farmland investments, i.e. the different legal instruments existing in target countries' national laws. To this end it will largely draw on the analytical and empirical findings of the seminal works by Laura German and George Schoneveld, who—sometimes alone, sometimes together and sometimes in collaboration with other researchers—systematically investigated the design, content, and implementation practice of several national policies and laws governing land acquisition in different African countries (e.g. Ethiopia, Ghana, Mozambique, Nigeria, Tanzania, Zambia).[33]

3.2.1 Typology

To better understand the legal underpinnings of large-scale land acquisitions, German and Schoneveld (together with Mwangi) have identified parameters that allow the exploration of how the law and other institutional mechanisms support different dimensions of customary land rights during the process of land acquisition.[34] Parameters are structured sequentially, following the key stages in the land acquisition process. Notably, these parameters focus primarily on the protection of customary land rights, addressing thus the most contentious aspect of land investments, i.e. the laws regulate access to and use of land. While considering the effects of introducing an industrial type of agri-business in little developed agricultural economies, these parameters indirectly also cover—at least to some extent—ecological aspects (Table 1).

---

[32]See Schoneveld (2014a).

[33]German et al. (2013b), pp. 1–18; Schoneveld et al. (2011), p. 10; German and Schoneveld (2012), pp. 467–491; Schoneveld and Schete (2014); Schoneveld (2014b), pp. 147–162. See also Schoneveld (2013).

[34]For a slightly modified set of parameters see also Schoneveld (2014a).

Table 1 Laws in host countries supporting customary land rights in the process of land acquisition (German et al. 2013b[a])

Parameter	Description
Provisions to protect customary rights	Legal provisions to protect customary rights—whether through formal titling or recognition of existing systems of land occupation and tenure; mechanisms to ensure local rights to land and other natural resources are safeguarded during the negotiation process; (...)
Types and durations of land rights afforded to investors	Nature of land rights that may be acquired by investors (e.g. usufruct, leasehold, or freehold) and conditions (e.g. duration and renewability)
Government programs and actions for promoting and/or guiding land allocation	Government initiatives for identifying suitable and/or available land for particular types of uses; mechanisms for identifying land available for large-scale investments within customary areas; and sector-specific initiatives for promoting land-based investment
– The role of intermediaries – Mechanisms for local representation – Compensation mechanisms	– Legislated role of government agencies or other actors in regulating, mediating, or facilitating the negotiation process – Legislation that specifies mechanisms for representation of "local communities" or customary rights holders in the negotiation process – Legal provisions that specify the level, type, and distribution of compensation to be paid in cases of customary land alienation
Envisioned process for consulting customary land users about investment and land alienation	Legislated steps and processes though which customary rights holders are informed, consulted, or given decision authority over land alienation and its terms, including three related sub-parameters:
Impact mitigation requirements	Legislation requiring that negative socio-economic impacts of large-scale investments be mitigated by project proponents
Monitoring	Legislation requiring the monitoring of social impacts and, where stipulated, the specification of social dimensions or indicators to be monitored
Dispute resolution	Legally recognized mechanisms recourse for aggrieved parties

[a]Parameter explanations are slightly shortened

3.2.2 Empirical Findings: Design and Implementation Issues[35]

Most of the countries investigated by German and Schoneveld were found to have some formal laws in force that recognize and protect customary land rights. Mechanisms and procedures intended to offer recognition and protection vary. Regarding foreign investments, most countries generally forbid the sale of farmland. They do, however, allow investors to acquire leasehold titles for varying periods of time, ranging from, for example, 25 years for biofuels projects in Tanzania to 99 years in Ghana and Zambia, as well as Tanzania for non-biofuel projects. Some countries make such long-term leaseholds conditional upon a positive evaluation of prior performance of investments and investor compliance (e.g. Zambia). All studied countries foresee the development of plans or programs for identifying land suitable for investment. The processes for drawing up these plans and programs often include some form of zoning practice for the purpose of separating investor farmland from other areas. These processes often involve—to varying degrees—consultations with local communities (either with community representatives or the communities themselves). Several countries have some legislation in force that aims to ensure compensation (e.g. for the loss of farmland, settlements, or common pool resources), requires some sort of impact assessment and mitigation (socio-economic or environmental impact assessment), demands implementation-monitoring, and provides for dispute settlement or grievance mechanisms (including informal mechanisms as well as tribunals specialising in land conflicts (Zambia) or disputes regarding the environmental impact assessments, or the misconduct of chiefs (Ghana)).

In most cases studied, however, these different sets of rules have faced an array of implementation challenges and have often not allowed farmland investments to generate the potential benefits without inducing at least some of those negative effects for local communities mentioned above. Mechanisms and procedures under national law which are intended to recognize and protect traditional land rights, de facto, grant very diverse and often low levels of protection. Even though most countries generally forbid the sale of farmland to foreign investors, long-term leaseholds reaching up to 99 years have similar effects, at least from the perspective of traditional land users: these long-term contracts effectively disconnect local people from their land as well as from their traditional ways of living and earning their livelihoods. In addition, none of the procedures or programs in place for identifying land suitable for investment systematically applies any social or ecological criteria; rather, these programs and procedures focus mainly on farming-efficiency (e.g. proximity of the respective land to water, markets, and infrastructures etc.). While countries like Ghana, Mozambique, Tanzania, Ethiopia, Nigeria, and Zambia all have laws requiring some form of involvement of communities, the quality and depth of such consultations has often been rather limited in practice. In most cases there was a lack of information about the positive and negative effects of the projects as well as a substantial power-asymmetry between negotiating parties

[35] Also based on studies by German, Schoneveld, and others. See Fn. 33.

(e.g. a low legal literacy and bargaining power at the local level). In many cases, government agencies and community representatives struggle with using the powers vested in them to balance out this asymmetry and to protect local communities' customary land rights or interests. There is even evidence that some have used their position to promote investment contracts, often for their own personal gain. Compensation measures also often suffer from bad implementation. As a result, respective practices vary widely, are often contentious, and are mostly granted to the government (including local councils) instead of local communities. In particular, laws concerning environmental or socio-economic impact assessments, monitoring requirements, and dispute resolution have been weakly implemented. Despite some good practices, in most countries, investors have not been asked to carry out impact assessments, and government agencies and administrations often do not monitor project implementation. This is often due either to political concerns that a strict implementation would potentially obstruct development, or due to weak technical, administrative, or financial capacities. Similarly, though formal conflict mechanisms exist, in many cases they are difficult to access for local people due to a lack of funding of these institutions and an overall poor understanding of rights and judicial means by local people.

3.3 Governing Large-Scale Farmland Investments Through International Guidance?

Chiefly due to the absence of binding international treaties systematically addressing issues arising out of large-scale farmland investments as well as host countries' severe implementation challenges regarding their national laws targeting these issues, a set of quite distinct non-binding international instruments has been developed. They pursue different objectives and represent a different kind of governance strategy. The following four subsections will briefly outline and explain these four instruments.

3.3.1 Minimum Principles to Address the Human Rights Challenge (2009)

In 2009 the UN Special Rapporteur on the Right to Food, Olivier De Schutter, issued a report containing a set of core principles and measures to address human rights challenges posed by large-scale land acquisitions (Minimum Principles).[36] The report aimed to assess how investments could impact human rights (i.e. right to

[36]The Rapporteur insisted that these principles are "minimum principles". According to him, this means that "large-scale investment in land will not necessarily be justified even though it may comply with the various principles listed." A/HRC/13/33/add.2 (2009), para. 9. Notably, the

food, right to self-determination, specific indigenous peoples' rights, or rights of agricultural workers) and to ensure that states contribute to their realization. The Rapporteur explicitly argued that, in his opinion, "(l)arge-scale farmland investments can work for the benefit of all parties concerned. But that presupposes that an appropriate institutional framework is in place (...)."[37] In addition, according to him, potential benefits of investments can be achieved through business models that do not include any change to the rights over the land.[38] Based on this, he then goes on to recommend and explain the following 11 principles (which are often, again, further explained or complemented with more specific suggestions) (Table 2).

3.3.2 FAO's Guidelines on the Responsible Governance of Tenure of Land, Fisheries, and Forests

The full title is: Voluntary Guidelines on the Responsible Governance of Tenure of Land, Fisheries, and Forests in the Context of National Food Security (Tenure Guide). They were adopted in 2012 by the FAO's Committee on World Food Security (CFS) and seek to improve the governance of tenure of land, fisheries and forests to achieve food security for all.[39] The Guidelines are based on two premises: First, the eradication of hunger and poverty, as well as the sustainable use of the environment depend "in large measure on how people, communities and others gain access to land, fisheries and forests."[40] Access, in turn, is largely defined through systems of tenure. Second, these systems of tenure are increasingly under pressure due to the combination of a growing world population demanding resources on one hand and reduced availability of land, fisheries, and forests on the other.[41] Strengthening responsible governance of tenure can thus contribute to reducing this pressure and achieving food security.

The Guidelines are voluntary in nature and they address both governments and private entities.[42] They lay down some general principles, such as that states should "respect legitimate tenure right holders and their rights", protect them "against arbitrary loss of their tenure rights", and "provide access to justice to deal with infringements."[43] Where states recognise or allocate tenure rights, they should also establish safeguards to avoid infringing or extinguishing tenure rights of others,

Rapporteur argues that in cases where the investor is a private entity, it is also the responsibility of its home state to ensure that these obligations are met, A/HRC/13/33/add.2 (2009), para. 33.

[37] A/HRC/13/33/add.2 (2009), para. 33.

[38] A/HRC/13/33/add.2 (2009), para. 9.

[39] Tenure Guide Sec. 1.1.

[40] Tenure Guide, preface, p. iv.

[41] Tenure Guide, preface, p. v.

[42] Including "all persons concerned to assess tenure governance and identify improvements and apply them." Tenure Guide, Secs. 2.1. and 2.3.

[43] Tenure Guide, Sec. 3.1 Nr. 1–4, also Secs. 4.5, 4.9.

Table 2 Core Principles and Measures to Address the Human Rights Challenge[a]

Principle 1	The negotiations leading to investment agreements should be conducted in a fully transparent manner, and with the participation of the local communities whose access to land and other productive resources may be affected. (. . .) In considering whether or not to conclude an agreement with an investor, the host government should always balance the advantages of entering into such an agreement against the opportunity costs involved, in particular when other uses could be made of the land available, which could be more conducive to the long-term needs of the local populations concerned (. . .).
Principle 2	In general, any shifts in land use can only take place with the free, prior and informed consent of local communities concerned. (. . .)
Principle 3	In order to ensure that the rights of local communities will be safeguarded at all times, states should adopt legislation protecting these and specifying in detail the conditions according to which shifts in land use, or evictions, may take place, as well as the procedures to be followed. Moreover, states should assist local communities in obtaining collective registration of the land they use, in order to ensure that their rights will enjoy full judicial protection. (. . .)
Principle 4	The local population should benefit from the revenues generated by the investment agreement. (. . .) depending on the circumstances, arrangements under which the foreign investor provides access to credit and improved technologies for contract farming (. . .) may be preferable to long-term leases of land or land purchase, (. . .).
Principle 5	Host States and investors should establish and promote farming systems that are sufficiently labour intensive to contribute to employment creation. (. . .) investment agreements should contribute to the fullest extent possible to reinforcing local livelihoods options and in particular provide access to a living wage for the local population involved (. . .).
Principle 6	Host States and investors should cooperate in identifying ways to ensure that the modes of agricultural production shall respect the environment, and shall not accelerate climate change, soil depletion, and the exhaustion of freshwater reserves. (. . .)
Principle 7	Whichever the content of the arrangement, it is essential that the obligations of the investor be defined in clear terms, and that the obligations are enforceable, for instance by inclusion of pre-defined sanctions in cases of non-compliance. For this mechanism to be effective, independent and participatory ex post impact assessments should be made at pre-defined intervals. (. . .)
Principle 8	In order to ensure that they will not increase food insecurity for the local population, (. . .) investment agreements should include a clause providing that a certain minimum percentage of crops produced shall be sold on local markets (. . .)
Principle 9	(. . .) impact assessments should be conducted prior to the completion of the negotiations, in order to highlight the consequences of the investment on the enjoyment of the right to food though (a) local employment and incomes (. . .), (b) access to productive resources of the local communities (. . .), (c) the arrival of new technologies and investments in infrastructure; (d) the environment, including soil depletion, the use of water resources and genetic erosion, (e) access, availability and adequacy of food. (. . .)
Principle 10	(. . .) States shall consult and cooperate in good faith with the indigenous peoples concerned in order to obtain their free and informed consent prior to the approval of any project affecting their lands or territories and other resources (. . .)
Principle 11	Agricultural waged workers should be provided with adequate protection and their fundamental human and labour rights should be stipulated in legislation and enforced in practice, consistent with the applicable ILO instruments. (. . .)

[a]A/HRC/13/33/add.2 (2009), Annex

including those currently not protected by formal law.[44] They should also promote policies and laws to provide recognition to such informal tenure.[45] Notably, the Guidelines also highlight that "non-state actors including business enterprises have a responsibility to respect human rights and legitimate tenure rights."[46]

Section 12 of the Guidelines specifically addresses investments in tenure rights, including foreign large-scale investments. This set of Guidelines begins with generally highlighting the importance of "responsible public and private investments" in land, fisheries, and forest resources with a view to improving food security and argue that "responsible governance of tenure" encourages right holders to make such investments.[47] The section then emphasizes the importance of smallholder producers in this regard as well as their significant contribution to food security, nutrition, poverty eradication and environmental resilience.[48] In addition, it demands that transactions in tenure rights should be transparent, that responsible investments safeguard against dispossession of legitimate tenure right holders, that they not create environmental damages, and that they respect human rights.[49]

Particularly with a view to large-scale investments,[50] states should provide "transparent rules on the scale, scope and nature of allowable transactions" and "should define what constitutes large-scale transactions (...) in their national context."[51] In addition, states are required to "provide safeguards to protect legitimate tenure rights, human rights, livelihoods, food security and the environment from risks that could arise from large-scale transactions in tenure rights."[52] It then suggests that such safeguards "could include introducing ceilings on permissible land transactions and regulating how transfers exceeding a certain scale should be approved, such as by parliamentary approval."[53] Most notably, it then suggests that

> States should consider promoting a range of production and investment models that do not result in the large-scale transfer of tenure rights to investors, and should encourage partnerships with local tenure right holders.[54]

Furthermore, investments should be made upon effective prior consultation with those whose tenure rights could be affected, including indigenous people in particular.[55] "States should also strive to make provisions for different parties to conduct

[44] Tenure Guide Sec. 7.1.

[45] Tenure Guide Sec. 10.1.

[46] Tenure Guide, Sec. 3.2.

[47] Tenure Guide, Sec. 12.1.

[48] Tenure Guide, Sec. 12.2.

[49] Tenure Guide, Secs. 12.3, 12.4.

[50] On large-scale investments in tenure rights, see particularly Tenure Guide, Secs. 12.5, 12.16, 12.10, 12.14 and 12.15.

[51] Tenure Guide, Sec. 12.5.

[52] Tenure Guide, Sec. 12.6.

[53] Tenure Guide, Sec. 12.6.

[54] Tenure Guide, Sec. 12.6.

[55] Tenure Guide, Secs. 12.7–12.9.

prior independent assessment on the potential positive and negative impacts that those investments could have on tenure rights, food security and the progressive realization of the right to adequate food, livelihoods and the environment."[56] They should also "ensure that existing legitimate tenure rights and claims, including those of customary and informal tenure, are systematically and impartially identified, as well as the rights and livelihoods of other people also affected by the investment, such as small-scale producers."[57] The Guidelines also remind contracting parties to inform all relevant persons,[58] effectively monitor implementation and impacts of agreements,[59] take corrective action where necessary, and provide grievance mechanisms.[60] Finally, they also call upon states who invest or promote investments abroad to ensure that their conduct is consistent with the protection of legitimate tenure rights, the promotion of food security, and their existing obligations under international law (including human rights), and with due regard to voluntary commitments (. . .).[61]

3.3.3 FAO Principles for Responsible Investment in Agriculture and Food Systems (2014)

The FAO issued in 2014 a guidance document addressing the topic of "responsible investments in agricultural and food systems" (Farmland Investment Guide). Its scope is thus a lot broader than that of the Tenure Guide which it basically integrates through cross-referencing in its Sect. 5. The principles are based on several assumptions. First, food security requires and will continue to require an increase in responsible investments in agriculture and food systems.[62] Second, smallholders, including family farmers, play a vital role in investment in agriculture and food systems; responsible investments thus include "priority investments in, by, and with smallholders (. . .)".[63] Third, investments in agriculture and food systems can produce multiplier effects including a contribution to overall economic development. To achieve this, additional investments in public goods are required and investments' "viability also depends on well-functioning ecosystems and sustainable use of natural resources."[64]

Declaredly, the Farmland Investment Guide aims to "promote responsible investment in agriculture and food systems that contribute to food security and nutrition,

[56]Tenure Guide, Sec. 12.10.

[57]Tenure Guide, Sec. 12.10.

[58]Tenure Guide, Sec. 12.11.

[59]Tenure Guide, Sec. 12.14.

[60]Tenure Guide, Sec. 12.14.

[61]Tenure Guide, Sec. 12.15.

[62]Farmland Investment Guide, paras. 1–3.

[63]Farmland Investment Guide, para 4.

[64]Farmland Investment Guide, para. 6.

thus supporting the progressive realization of the right to adequate food in the context of national security."[65] It seeks to "address the core elements of what makes investment in agriculture and food systems responsible", identify "key stakeholders (...) engaged in agriculture and food systems", and "guide" their actions.[66] The principles are voluntary in nature and their scope is global, i.e. they "have been developed to be universally applicable."[67] Notably, the Farmland Investment Guide addresses the following stakeholders:[68]

- States;
- Inter-governmental and Regional Organizations;
- Financing Institutions, Donors, Foundations and Funds;
- Research Organizations, Universities, and Extension Organizations;
- Smallholders and their Organizations;
- Business Enterprises, including Farmers;
- Civil Society Organizations;
- Workers and their Organizations;
- Communities;
- Consumer Organizations

Principles included in the Farmland Investment Guide are meant to be complementary and not all principles are relevant for every investment.[69] Responsible investments in agriculture and food systems should adhere to the following principles (Table 3).

The Farmland Investment Guide explains each of these general principles in greater depth.

With a view to large-scale farmland investments the Farmland Investment Guide generally highlights states'responsibility to "ensure, to the extent possible, that actions related to responsible investment in agriculture and food systems both at home and abroad, are consistent with their existing obligations under national and international law, and international agreements related to trade and investment, with due regard to voluntary commitments (...)."[70] Accordingly, the Guide highlights the extraterritorial responsibility of investors' home states, making sure to include their committments under international human rights treaties. They also mention states' responsibilities to support smallholders, including those that are family farmers.[71] Particularly where states themselves or through controlled enterprises engage in

[65] Farmland Investment Guide, Sec. 10.

[66] Farmland Investment Guide, Sec. 11.

[67] Farmland Investment Guide, Secs. 12, 15.

[68] Farmland Investment Guide, Sec. 16.

[69] Farmland Investment Guide, Sec. 17.

[70] Farmland Investment Guide, Sec. 33.

[71] Farmland Investment Guide, Secs. 36, 39.

Table 3 Principles for Responsible Investment in Agriculture and Food Systems

Principle 1	Contribute to food security and nutrition
Principle 2	Contribute to sustainable and inclusive economic development and the eradication of poverty
Principle 3	Foster gender equality and women's empowerments
Principle 4	Engage and empower youth
Principle 5	Respect tenure of land, fisheries, and forests, and access to water (referencing to the FAO's Voluntary Guidelines on the Responsible Governance of Tenure)
Principle 6	Conserve and sustainably manage natural resources, increase resilience, and reduce disaster risks
Principle 7	Respect cultural heritage and traditional knowledge, and support diversity and innovation
Principle 8	Promote safe and healthy agriculture food systems
Principle 9	Incorporate inclusive and transparent governance structures, processes and grievance mechanisms
Principle 10	Assess and address impacts and promote accountability

investments, they are called upon to act in accordance with the principles.[72] Finally, states are responsible to establish monitoring, assessment, and reporting systems.[73]

Notably, the Farmland Investment Guide then outlines more specifically the roles and resposibilities of all different stakeholders involved in investment practice (see above). Financing institutions and other funding entities are explicitly identified as powerful actors. They are encouraged to apply the principles "when formulating their policies for loans and grants (. . .)."[74] They should also "take appropriate measures so that their support to investors does not lead to violations of human and legitimate tenure rights (. . .)."[75] They are also encouraged to develop "innovative financial mechanisms and insurance tools in support of investment in agriculture, especially appropriate solutions for smallholders, including those that are family farmers, that consider a long-term development perspectve."[76] With regards to business enterprises, the guidelines highlight not only their responsibility to abide by any applicable national and internal law but also to "act with due dilligence to avoid infringing on human rights."[77] They are also encouraged to "conduct equitable and transparent transactions (. . .)", and to "use a range of inclusive business models." In addition, "processors, retailers, distributors, input suppliers, and marketers are encouraged to inform and educate consumers about the sustainability of products and services (. . .)."[78]

[72]Farmland Investment Guide, Sec. 42.
[73]Farmland Investment Guide, Sec. 43.
[74]Farmland Investment Guide, Sec. 45.
[75]Farmland Investment Guide, Sec. 45.
[76]Farmland Investment Guide, Sec. 45.
[77]Farmland Investment Guide, Sec. 50.
[78]Farmland Investment Guide, Sec. 51.

3.3.4 UNIDROIT Legal Guide on Agricultural Land Investment Contracts

At the time of writing a rather different type of guide had been developed by the International Institute for the Unification of Private Law (UNIDROIT). UNIDROIT together with the FAO and IFAD submitted in June 2019 a zero-draft of a Legal Guide on Agricultural Land Investment Contracts (UNIDROIT Guide). It aims to provide "detailed guidance to improve such contracts so that they promote respect for tenure rights, equitable access to land and responsible agricultural investment (...)."[79] It is also explicitly based on the assumptions that there is growing need for responsible investment in farmland and claims to be consistent with the "international consensus reflected in the principles and standards for land tenure, agricultural investment and related areas (...)."[80] In accordance with these principles, the UNIDROIT Guide declares that it does "not promote the large-scale transfer of tenure rights and excludes sales of agricultural land from its scope" and points out their potential negative social, environmental and economic effects.[81] It highlights both formal laws as well as other principles, standards and practices with which parties should comply in order to support responsible and sustainable investment.[82] Notably, the UNIDROIT Guide identifies as its target audience a broad set of actors, including "legal professionals involved in the preparation, negotiation, implementation and review of agricultural land investment contracts, including for investors, governement, legitimate tenure right holder and local communities."[83] It also mentions that it might be useful for "other actors (...), including legislators, policymakers, judges, arbitrators, notaries, public-interest legal service organisations, community organisations, law societies and international development organisations."[84] The Guide suggests that investment contracts can be tools to help realise the benefits of investments and minimize negative impacts.[85] Notably, it argues that "where gaps exist (in the laws), the agricultural land investments contract may be a means to fill in some or all of those gaps while the State works towards filling them."[86] Contracts could also provide for effective grievance and dispute resolution mechanisms, especially in cases where the relevant judicial systems do not provide for them.[87]

---

[79] UNIDROIT Guide (zero draft), preface 1. The Guide was finally launched on 24 September 2021 after the paper had been submitted, see above Note 7.

[80] UNIDROIT Guide (zero draft), preface 1, intro 1.

[81] UNIDROIT Guide (zero draft), preface 5.

[82] UNIDROIT Guide (zero draft), preface 6.

[83] UNIDROIT Guide (zero draft), preface 9.

[84] UNIDROIT Guide (zero draft), preface 9.

[85] UNIDROIT Guide (zero draft), intro, para. 5.

[86] UNIDROIT Guide (zero draft), intro, para. 5, see also Sec. 1.2.

[87] UNIDROIT Guide (zero draft), intro, para. 5.

The Guide is structured in six chapters:

- Chapter 1: The Legal Framework
- Chapter 2: Parties, Contractual Arrangement, Due Dilligence and Formation
- Chapter 3: Rights and Obligations of the Parties
- Chapter 4: Contractual Non-Performance and Remedies
- Chapter 5: Transfer and Return
- Chapter 6: Grievance Mechanisms and Dispute Resolution

Currently at 144 pages, the zero-draft's different sections systematically provide criteria for developing, negotiating, or assessing investment contracts. Due to its length, it cannot be adequately represented here. Just to illustrate the Guide's governance approach and "regulatory style", its recommendations about the contracting parties' rights and obligations regarding the environment should be explained in greather depth. After presenting environmental obligations in international and national law and highlighting, again, the individual contracts' gap filling function,[88] the Guide goes on to systematically address, non-exhaustively, several environmental issues that it recommends parties to address in their contracts:[89]

- assessing environmental impact,
- preventing pollution,
- protecting water,
- preventing soil degradation,
- conserving biodiversity and safeguarding ecosystem services,
- mitigating and adapting to climate change,
- adopting sound management of waste,
- monitoring and reporting environmental protection,
- achieving project closure and restoration of the environment,
- addressing failure to comply with environmental obligations,

For illustrative purposes, the Guide's proviso on "soil degradation" is presented here.

> Preventing soil degradation: The control of soil quality should be addressed by parties to ensure responsible agricultural land investment. The need to prevent soil erosion and pollution can be achieved by adopting crop rotation, intercropping practices, (...) and agro-ecological farming options that can improve soil biodiversity. Desertification issues should also be considered. The adoption of contractual provisions that require the restoration of drylands and degraded lands, especially when agricultural land investments occur in Africa, constitutes best practice encouraged by the UN Convention to Combat Desertification CD).[90]

The Guide's following chapters spell out information and recommendations on other aspects that parties need to address in their contracts in order to make investments

[88]UNIDROIT Guide (zero draft), Secs. 3.109–3.113.

[89]UNIDROIT Guide (zero draft), Secs. 3.114–3.124.

[90]UNIDROIT Guide (zero draft), Sec. 3.118.

responsible, i.e. on potential cases of contractual non-performance, on the quality and the procedures of of how to transfer and return rights in land, as well as on grievance mechanisms and dispute resolution.

4 Assessing State and Potentials of the Existing Governance System

It is argued here that the existing governance mechanisms addressing farmland investments can best be understood by highlighting three central features that have defined the process leading up to its current design. The system as it stands is mainly the result of a continuous effort to better understand the complex issues arising out of large-scale farmland investments, reframing and reinterpreting these issues, the responsibilities of actors involved, and the purpose of regulation, and finally, flexibly shifting modes of governance in face of absent effective state made laws and institutions.

4.1 Increasing Knowledge

The surge in large-scale farmland investments in low income countries around 2005 has sparked critical debates about their potential negative effects. At that time, there were only a few empirical or theoretical assessments that had systematically investigated the specific characteristics of this "new wave" of investments.[91] Accordingly, most assessments or comments drew their conclusions either from older studies on agricultural investments or from more general positive or critical theories about the effects of trade and investment liberalisation. Over the last 10 to 15 years, however, knowledge about these new types of investments has substantially expanded. Empirical studies in particular provided a much broader understanding of the size, quality, and the effects of these investments. Three specific key characteristics characterizing the new investments were identified: First, unlike prior investments, the "new wave" of acquisitions was not mainly following a Ricardian logic regarding a comparative advantage through trade in crops; they were instead largely driven by governments' or commercial operators' (geo)strategical interest in securing access to food, energy, and water supplies.[92] Investors did not mainly intend to link local production to global markets, in which they had lost their trust to reliably

[91] For an historical perspective on farmland investments see De Schutter (2011), pp. 250–255; Byerlee (2013); Cotula (2013).

[92] See, for example, Sebastian and Warner (2014).

provide for food and water resources, but rather to circumvent them by tightening control over supply chains.[93] In addition, the surge in land-deals was at least partly fuelled by speculative investments.[94] Second, while there were some projects that had generated benefits for investors and host countries, empirical studies clearly indicated that most large-scale projects created strikingly similar negative effects, particularly in cases where land rights were being transferred for longer periods of time. Third, governmental or quasi-governmental institutions in most target countries had all too often failed to implement national laws that were supposed to prevent these negative effects. Being the only source of formal law that directly and systematically addressed these issues, this situation left large-scale farmland investments *de facto* largely unregulated.

4.2 Reframing the Issue and Developing New Narratives

In response to this growing awareness of the recent surge of investments, scientific and legal experts have engaged incrementally in reinterpreting the problem (in judicial terms, reframing "the perception of what is the case").[95] In the beginning, as De Schutter has convincingly explained, the issues created by large-scale farmland investments were perceived rather narrowly or one-dimensionally. Generally speaking, actors and observers assumed, at least to some extent, that large-scale farmland investments could eventually generate economic gains for investors, target states, and possibly local communities alike, if only target countries (and maybe investor countries) would effectively intervene to prevent investments' negative side effects.[96] This influential interpretation was deeply rooted in (a) classical or neo-classical economic thinking about the positive effects of liberalisation of trade and investment, as well as based on (b) an overly optimistic assumption about the effectiveness of governmental institutions in target states and the regulatory ambition of investor home countries. As negative effects of large-scale farmland investments and target countries' implementation deficits became increasingly obvious, this narrow perspective was substantially broadened and the theoretical underpinnings regarding investments' economic value, the possibilities for governing them, as well as their overall legitimate purpose were newly interpreted. The result is that the following three narratives were added to the powerful trade-liberal economic perspective on farmland investments.

[93] Smaller and Mann (2009), pp. 1–2. See also De Schutter (2009), para. 2.

[94] Kugelman and Levenstein (2013), Chapter 1.

[95] Regarding the cognitive dimension of framing legal issues in international negotiations and relations, see relations, see Markus et al. (2018), pp. 502–509. https://doi.org/10.1093/icesjms/fsx201; Markus (2013), pp. 145–165. See also Wendt (1999).

[96] De Schutter (2011), pp. 255–263.

First, throughout the process of developing international guidance a narrative about universal values was added to the overall picture, i.e. regarding investments' possible effects on human rights. The first important step in this regard was the publication of the UN Special Rapporteur's Minimum Principles. These principles, in effect, subjected large-scale farmland investments to a categorially different valuation system. From now on investments had to be not only economically viable but also in compliance with human rights. Notably, both the Tenure Guide and the Responsible Investment Guide reinforced this approach by explicitly pointing out that non-state actors also have a responsibility to respect human rights. The process of complementing economic rationality by normative premises went on as guidance instruments evolved. All guidance instruments following the Special Rapporteur's Minimum Principles highlighted the importance of pursuing other political and legal objectives such as global food security, well-functioning ecosystems, and sustainability of resources.[97] Particularly the UNIDROIT's Guide reinforced this broadened normative perspective on investments by explaining possible effects on human rights and calling on contracting parties to respect human rights and environmental laws as well as to consider other non-binding international guidance in their contractual agreements in order to carry out responsible investments.

A second aspect about large-scale farmland investments has also been added to the classical investment paradigm through the development and adoption of international guidance documents, i.e. the overall purpose of farmland investments. As a fundamental premise all guidance documents accepted that global food security requires an increase in investments in agriculture and food systems. To contribute to food security, however, these investments have to be "responsible." Whether investments are responsible or not depends on their contribution to promoting food security and avoiding social conflicts and other negative side effects. Against the background of empirical findings on their effects, a distinction was drawn between smallholder and large-scale investments, particularly those involving the transfer of land rights. This distinction was first prominently introduced into the discourse by the UN Rapporteur's Minimum Principles. The UN Rapporteur took this fundamental step when he began to explicitly prioritize some investment practices over others, e.g. when he suggested "alternative business models" to "long term leases of land or land purchase."[98] This distinction was explicitly picked up in the Tenure Guide, which called upon states to consider promoting a range of production and investment models that do not result in the large-scale transfer of tenure rights to investors, and instead encourage partnerships with local tenure right holders. Most notably, the UNIDROIT Guide explicitly declares that it does "not promote the large-scale transfer of tenure rights and excludes sales of agricultural land from its scope" and points out their negative social, environmental and economic effects.[99] According to the international guidance documents, investments pursue a legitimate purpose when

[97] Guidelines for Responsible Investment, para. 6; see also Tenure Guidelines, Sec. 12.6.

[98] Principle 4, A/HRC/13/33/add.2 (2009), Annex.

[99] UNIDROIT Guide (zero draft), preface 5.

they contribute to food security and social and cultural cohesion, and promote stewardship for land instead of (geo)strategically "grabbing resources" or simply providing economic returns for a few investors.

The third aspect added to the overall interpretation of the complex phenomenon of large-scale farmland investments by the international guidance instruments was that not only target or investor states but all stakeholders involved in this commercial practice share responsibilities regarding their effects. While the Special Rapporteur's Minimum Principles were still mainly directed at target or investor states, the Tenure Guidelines already addressed both states as well as private entities.[100] The Guide for Responsible Investment took it one step further and systematically identified "key stakeholders" engaged in "agriculture and food systems" (states, inter-governmental organisations, financing institutions, donors, foundations, funds, research organizations, business enterprises, consumer organizations, civil society organizations, etc.), and established some minimum standards to "guide" their actions.[101] Finally, specifically addressing potential contracting parties of investment treaties (but all other stakeholders as well), the UNIDROIT Guidelines elaborate extensively and in much greater depth their responsibilities.

4.3 *Flexibly Shifting Modes of Governance*

This paper has made clear that issues arising out of large-scale farmland investments in low income countries are neither comprehensively nor effectively governed by formal state-based laws. International law does not systematically and proactively address investments to avoid or mitigate their negative effects; target countries' regulations all too often lack effective implementation, and investor countries have—to this day—widely refrained from taking legislative action to direct investments carried out by themselves or by private entities acting on their behalf or under their jurisdiction. Accordingly, there appears to be a "vacuum" or a "gap" of formal state-based law which is capable of pacifying the highly contentious issues evolving around large-scale transnational investments in land. Such a "vacuum" or "gap" does not mean, however, that large-scale farmland investments have remained entirely unordered or ungoverned.[102] It is argued here that the different guidance instruments taken together may well be understood and described as an emerging order of indirect or informal regulation. Such regulations may not yet have binding force in a positivistic or formal legal sense but they may still have a large potential to change

[100] Tenure Guidelines, Sec. 2.1., Sec. 2.3.

[101] Guide for Responsible Investment, Sec. 11.

[102] Whether there is a gap within the law or whether an issue remains ungoverned largely depends on the perspective adopted. A narrow positivistic-legalistic approach would focus observations on the formal state-made law. A more sociological perspective would look for other mechanisms or institutions that generate or promote order. It depends on whether the main source for social order is the state or society.

the behaviour of actors involved, both from a sociological and a legal perspective. For various reasons that will be outlined here, the existing guidance can be expected to effectively direct investment practices and also to be recognized and authorized at some point by legislators, administrations, and courts when developing or applying formal laws on large-scale farmland investments.[103] This section will highlight the social functions of the different guidance documents as well as their relationship with formal sources of law. It will interpret the development of international guidance instruments as a process in which traditional modes of state-based governance have been tried, but in view of their ineffectiveness they have been complemented by alternative, more informal modes of governance. This approach may be seen as a second best choice or a preliminary or proto-type regulatory step. Such an understanding would, however, underestimate the independent momentum of these transnational governance mechanisms emerging outside, but not independently from, the formal system of state-based laws and institutions.[104]

To restate, the development of standards was initiated by international organizations, i.e. state-based institutions. The first guidance document by the UN Special Rapporteur mainly addressed states, reminding them of their human rights obligations under international law regarding farmland investments. In response to a growing awareness of largely dysfunctional national and international laws, another complementary governance strategy was adopted. In addition to the Minimum Principles, two more sets of guidelines were developed, and these created a general non-binding normative frame that provided orientation about what qualifies investment practices as responsible or irresponsible. Together these non-binding guidance instruments define a rough consensus regarding minimum standards, the legitimate purpose of investment practices, the value of large- and small-scale investments, and certain responsibilities of specific actors (state and non-state actors). In a final step this consensus is operationalised in a guidance document providing standards for designing the content of investment contracts. These standards explain how a responsible investment should be carried out and what contracting parties have to consider when they develop their contracts accordingly. It must be stressed here that investment contracts are not international treaties but bilateral agreements between a competent governmental authority in a target country and an investor that binds only these two parties.

By designing a standard for these kind of contracts, based on the normative frame provided in the international guidance documents, the FAO and IFAD together with UNIDROIT aim to use investment contracts as some sort of transmission belt for

[103] These types of rules and processes are currently addressed by the discourse on "transnational law" or in the case of environmental law "transnational environmental law" (or the "transnationalization of environmental law"). For an overview of the history of terms, concepts, and current interpretations see Dilling and Markus (2018), pp. 179–206. See also Hayvaerts (2018).

[104] See, for example, the social scientific research into the transnational aspect of economic governance. Eberlein et al. (2013); National, inter- and supranational law can form various relationships by initiating, recognizing, integrating, limiting or modifying transnational norms. See, for example, Teubner and Fischer-Lescano (2006).

transferring international non-binding standards into the formal legal order that governs investment contracts. Of course, the investment contracts are not general binding law, but they do regulate the investment practice in a binding fashion, i.e. through agreement between contracting investors and the target countries. Where these agreements integrate international standards (including international law) in order to be "responsible", they can bridge the regulatory gap existing in the formal law.[105]

Though non-binding, the entire set of internationally concerted guidance on farmland investments will most likely have strong effect on the investment practice. From a sociological perspective, these guidance instruments will direct stakeholders' actions on a cognitive level. They will force investment activities through defining what is a responsible and irresponsible practice. So doing, they substantially reduce complexity by providing pathways to potential parties that indicate a pre-defined corridor of negotiable options for business, particularly by exluding those options which are most likely to produce severe conflicts. Commercial investors may not be driven by an ethical motvation to conduct responsible investments, but they still will acknowledge that the guidance on responsible investments aims at preventing potentially costly conflicts in target countries.[106] Reducing the potential for conflicts will benefit investors by reducing reputational, security and financial risks. In economic terms, using the UNCITRAL standard as a reference point for negotiating investment agreements will allow them to substantially lower transaction costs, not only during the negotiation phase, but perhaps even more by avoiding conflicts and dispute settlement later. Moreover, contracting partners in target countries are better informed about the multiple aspects they need to consider if they want to act in line with international law, non-binding standards, and expert knowledge, in addition to generating benefits and avoiding conflicts. Target states' authorities are thus empowered in their capacity as contracting parties. The UNIDROIT Guide is thus informative and normative at the same time. The same applies, however, to all other actors involved in farmland investments, including all commercial and non-commercial actors (e.g. financial institutions, civil society organizations etc.).

Apart from the normative force resulting mainly from a self-interest in reaping the benefits of reduced complexity, lower transactions costs, and a decreased potential for conflicts, investments still take place in the "shadow of law" or "the shadow of the state".[107] By identifying, defining, and assigning reponsibilities of or to different stakeholders, the guidance documents substantially increase legal accountability. Where actors involved in large-scale farmland investments would extensively fall behind the internationally agreed standards and continue to provoke conflicts on the

[105] The Guide notes this on several occasions. See, for example, UNIDROIT Guide (zero draft), intro para. 5, para. 1.18., para. 1.2., para. 1.6., para. 3.113., para. 3.81., para. 5.5. etc.

[106] On the rationality of economic actors in a functionally differentiated society, see particularly Luhmann (1989).

[107] Instructive Abbott and Snidal (2009). The term "in the shadow of the law" goes back to Mnookin and Kornhauser (1979).

ground, they would risk that governments will eventually adopt and enforce more stringent and effective types of formal regulations, nationally or internationally. In addition, international human rights or national courts may be willing at some point to respond to particlarly severe violations of these standards by declaring certain investment practices as violating human rights or certain damages suitable for compensation.[108] Finally, the standards developed by UNIDROIT seem to provide target countries with enough room to develop their national laws without running the risk of being sued for damages before investment arbitrations courts. Different courts may thus use international guidance, particularly the UNIDROIT Guide, for orientation in such cases. Where governments, their administrations, or courts use these standards as reference points for law-making or decision-making, they would connect and integrate these non-binding standards with or into their formal legal orders.

5 Conclusions

The specific issues arising out of large-scale farmland investments in low income countries over the last 15 years remind us that neither food nor land are regular commodities.[109] The food crisis in the years 2007–2008 has made very clear that many people in the world live under food insecure conditions and that several countries depend on food imports and are thus vulnerable to volatile world market prices for food products. Additionally, the increasing commodification of farmland in low income countries has revealed how vulnerable smallholder farmers and other small-scale users of rural farmland are to increasing competition over farmland. Developing farmland in low income countries can be one step toward improving food security as well as living conditions in these countries, particularly in the face of growing populations and increasing pressures on soils, water, and biodiversity. To avoid negative effects and to develop farmland sustainably, however, investments must be carried out (very) responsibly, i.e. they must respect human rights and existing systems of tenure, prioritize smallholders, consider ecologic requirements, etc. Given their overall negative effects on people and land, (transnational) large-scale farmland investments are often problematic with a view to promoting sustainability, particularly where they involve transfers of land rights.

Governing these types of investments, however, has proven difficult. Binding international law is still non-existent and national laws in target states often suffer from weak implementation. In the absence of effective formal binding law, an alternative informal, non-binding order has emerged that has the potential to

[108] The UNIDROIT Guide explicitly mentions "legislators, policymakers, judges, arbitrators" as possible reviewers of investment contracts whose work can be informed by the international guidance. See UNIDROIT Guide, preface 9.

[109] Already in 1944 Karl Polyani has pointed out the grave consequences that might arise if the value of natural resources is solely defined by the market mechanics, see Polyani (1994), pp. 73 et seqq.; see also De Schutter (2011), pp. 252–253, 274.

substantially improve investment practices. Its normative force results from its provision of information about what makes an investment practice responsible as well as economically and socially beneficial for parties involved. It especially empowers target countries by allowing them to better develop their farmland in accordance with their own priorities and autonomous interests. For all actors involved in investments it reduces the overall complexity of farmland development by providing guidance as to which investments are compatible with international and national binding and non-binding standards as well as expert advice. Guidance instruments thus lower their transactions costs, particularly by reducing the risks for conflicts. As such the approach is not only persuasive in an ethical sense but all parties will be able to benefit in the long-term if they comply with these standards. In addition, by identifying responsibilities, the international guidance instruments increase legal accountability. Grossly neglecting guidance standards and creating conflicts in target countries may lead governments to adopt and even implement more stringent measures at the national or international level or courts to qualify certain practices as violations of human rights or to be eligible for compensation.

Pursuing such an alternative governance approach requires further effort. Guidance instruments must be advertised and explained to actors involved in farmland investments, particularly those negotiating, drafting, implementing, or reviewing investment contracts, including "investors, governments, legitimate tenure right holder and local communities"[110] or "legislators, policymakers, judges, arbitrators, notaries, public-interest legal service organisations, community organisations, law societies and international development organisations."[111] The UNCITRAL Guide seems to be a priceless tool in this regard. At best it will develop over the years to come and integrate future experiences, best practices, and judicial or legislative decisions. Now that the UNIDROIT Guide is finalized, all actors involved in transnational farmland investments can and should revisit their investment contracts as well as their policies and laws in order to assess, revise, and improve them. Failing these standards will become increasingly difficult to justify.

References

Aanseuw W, Boche M, Breu T, Giger M, Lay J, Messerli P, Nolte K (2012) Transnational land deals for agriculture in the global south: analytical report based on the land matrix database. CDE/CIRAD/GIGA, Bern/Montpellier/Hamburg

Abbott K, Snidal D (2009) The governance triangle: regulatory standards institutions and the shadow of the state. In: Mattli W, Woods N (eds) The politics of global regulation. Princeton University Press, Princeton, pp 44–88

Allan T, Keulertz M, Sojamo S, Warner JF (eds) (2012) Handbook of land and water grabs in Africa; Foreign direct investment and food and water security. Routledge, London/New York

[110] UNIDROIT Guide (zero draft), preface 9.

[111] UNIDROIT Guide (zero draft), preface 9.

AU-ECA-AfDB-Consortium (2010) Supporting land policy development and implementation in Africa. Addis Ababa
Baxter J (2010) The Great African Land Grab. Le Monde Diplomatique
Borras SM Jr, Seufert P, Backes S, Fyfe D, Herre R, Michéle L, Mills E (2017) Land grabbing and human rights – the role of EU actors abroad. FIAN, Heidelberg
Borras SM, Franco JC (2012) Global land grabbing and trajectories of agrarian change: a preliminary analysis. J Agrar Change 12:34–59
Byerlee D (2013) Are we learning from history? In: Kugelman M, Levenstein SL (eds) The global farms race – land grabs, agricultural investment, and the scramble for food security. Island Press, Washington D.C., pp 21–44
Chiziane EA (2018) The design of the political and legal framework on soil conservation in Mozambique: deeply Unfisinished. In: Ginzky H, Dooley E, Heuser IL, Kasimbazi E, Markus T, Qin T (eds) International yearbook of soil law and policy. Springer, pp 331–344
Cotula L (2013) The great African land grab? Agricultural investments and the global food system. ZED Books, New York
Cotula L (2014–2015) 'Land Grabbing' and international investment law: towards a global reconfiguration of property? In: Bjorklung AK (ed) Yearbook on international investment law and policy. Springer, pp 177–214
Cotula L, Oya C, Codjoe EA, Eid A, Kakraba-Ampeh M, Keeley J, Kidewa AL, Makwarimba M, Seide WM, Nasha WO, Asare RO, Rizzo M (2014) Testing claims about large land deals in Africa: findings from a multi-country study. J Dev Stud 50:903–925
D'Odorico P, Rulli MC (2012) The land and its people. Nat Geosci 7:324–325
Davis KF, Yu K, Rulli MC, Pichdara L, D'Odorico P (2015) Accelerated deforestation driven by large-scale land acquisitions in Cambodia. Nat Geosci 8:772–775
De Schutter O (2009) Large-scale land acquisitions and leases: a set of minimum principles and measures to address the human rights challenge. Report of the Special Rapporteur on the right to food, United Nations General Assembly A/HRC/13/33/add.2
De Schutter O (2011) How not to think of land grabbing: three critiques of large-scale investments in farmland. J Peasant Stud 38:249–279
Deininger K, Byerlee D, Lindsay J, Norton A, Selod H, Stickler M (2011) Rising global interest in farmland: can it yield sustainable and equitable benefits? Agriculture and rural development. World Bank, Washington, DC. Available via http://documents.worldbank.org/curated/en/998581468184149953/Rising-global-interest-in-farmland-can-it-yield-sustainable-and-equitable-benefits. Accessed 27 Sept 2021
Demissie F (2015) The new scramble over Africa's farmland: an introduction. In: Demissie F (ed) Land grabbing in Africa – the race for Africa's farmland. Routledge, London
Dilling O, Markus T (2018) The transnationalization of environmental law. Oxford J Environ Law 30:179–206
Douglas Z (2013) The enforcement of environmental norms in investment treaty arbitration. In: Dupuy PM, Viñuales JE (eds) Harnessing foreign investment to promote environmental protections – incentives and safeguards. CUP, Cambridge, pp 415–444
Eberlein B, Abbott KW, Black J, Meidinger E, Wood S (2013) Transnational business governance interactions: conceptualization and framework for analysis. Regul Gov 8:1–21
European Parliament (2014) Addressing the Human Rights Impacts of "Land Grabbing". Brussels
Food and Agriculture Organization of the United Nations (2012) The Voluntary Guidelines on the Responsible Governance of Tenure of Land, Fisheries and Forests in the Context of National Food Security, Rome
German L, Schoneveld GC (2012) Biofuels in sub-Saharan Africa: review of the early legal and institutional framework for biofuel investments in Zambia. Rev Policy Res 29:467–491
German L, Schoneveld G, Mwangi E (2013b) Contemporary processes of large-scale land acquisitions in sub-Saharan Africa: legal deficiency or elite capture of the rule of law? World Dev 48: 1–18

German LA, Achten WMJ, Guariguata MR (2013a) Environmental impacts. In: Kugelman M, Levenstein SL (ed) The global farms race – land grabs, agricultural investment, and the scramble for food security. Island Press, Washington, Chapter 5
Golay C, Biglino I (2013) Human rights responses to land grabbing: a right to food perspective. Third World Q 34:1630–1650
Hayvaert V (2018) Transnational environmental regulation and governance: purpose, strategies and principles. CUP, Cambridge
Kazimbazi E (2018) Soil protection law in Uganda. In: Ginzky H, Dooley E, Heuser IL, Kasimbazi E, Markus T, Qin T (eds) International yearbook of soil law and policy. Springer, pp 315–330
Knaup H, von Mittselstaedt J (2009) Foreign investors snap up African farmland. Spiegel International
Kugelman M, Levenstein SL (eds) (2013) The global farms race – land grabs, agricultural investment, and the scramble for food security. Island Press, Washington
Liao C, Jung S, Brown DG, Agrawal A (2016) Insufficient research on land grabbing. Science 353: 131
Luhmann N (1989) Ecological Communication. University of Chicago Press, Chicago
Mackintosh J, Burgess K (2008) Hedge funds muck in down on the farm. Financial Times, London
Markus T (2013) Changing the base: legal implications of scientific criteria and methodological standards on what constitutes good marine environmental status. Transnatl Environ Law 1:145–165
Markus T, Hillebrand H, Hornidge K, Krause G, Schlüter A (2018) Disciplinary diversity in marine sciences: the urgent case for an integration of research. ICES J Mar Sci 75:502–509. https://doi.org/10.1093/icesjms/fsx201
Messerli P, Geiger M, Dwyer MB, Breu T, Eckert S (2014) The geography of large-scale land acquisitions: analysing socio-ecological patterns of target contexts in the global south. Appl Geogr 53:449–459
Mnookin RH, Kornhauser L (1979) Bargaining in the shadow of the law: the case of divorce. Yale Law J 88:950–997
Montemayr RQ (2012) Regional perspectives: Asia. In: Kugelman M, Levenstein SL (eds) The global farms race – land grabs, agricultural investment, and the scramble for food security. Island Press, Washington, pp 135–152
Montilla Fernández LT (2017) Large-scale land Investments in least developed countries. Springer, Heidelberg
Neef A, Touch S, Chiengthon J (2013) The politics and ethics of land concessions in rural Cambodia. J Agric Environ Ethics 26:1085–1103
Nolte K, Chamberlain W, Giger M (2016) International land deals for agriculture. Fresh Insights from the Land Matrix: Analytical Report II. CDE/CIRAD/GIGA/University of Pretoria, Bern
Oya C (2013) Methodological reflections on land "grab" databases and the land "grab" literature "rush". J Peasant Stud 40:503–520
Polyani K (1994) The great transformation. The political and economic origins of our time. Beacon Press, Boston
Schoneveld G (2013) The governance of large-scale farmland Investments in sub-Saharan Africa – a comparative analysis of the challenges for sustainability. Eburon Academic Publishers, Delft
Schoneveld GC (2014a) The geographic and sectoral patterns of large-scale farmland investments in Sub-Sahara Africa. Food Policy 48:34–50
Schoneveld GC (2014b) Politics of the forest frontier: negotiating between conservation, development, and indigenous rights in Nigeria. Land Use Policy 38:147–162
Schoneveld GC, Schete M (2014) Modernizing the periphery: citizenship and Ethiopia's new agricultural investment policies. In: Kaag M, Zoomers RB (eds) Land grabbing beyond the hype. Zed Books, London, pp 18–35
Schoneveld GC, German L, Nukator E (2011) Land-based investment for rural development? A grounded analysis of the local impacts of biofuel feedstock plantations in Ghana. Ecol Soc 16:10

Sebastian AG, Warner JF (2014) Geopolitical drivers of foreign investment in African land and water resources. In: Demissie F (2015) Land grabbing in Africa, The race for Africa's Rich Farmland. Routledge, London, Chapter 2
Slater TL (2015) Investor-state arbitration and domestic environmental protection. Wash Univ Global Stud Law Rev 14:131–154
Smaller C, Mann H (2009) A thirst for distant lands: foreign investment in agricultural land and water. IISD, Winnipeg
Teubner G, Fischer-Lescano A (2006) Regime-collisions: the vain search for legal Unity in the fragmentation of global law. Mich J Intl Law 25:999–1046
The Committee on World Food Security (2014) Principles for Responsible Investment in Agriculture and Food Systems. Rome
The Economist (2011) The surge in land deals – When others are grabbing their land. https://www.economist.com/international/2011/05/05/when-others-are-grabbing-their-land. Accessed 27 Sept 2021
Tienhaara K (2009) The expropriation of environmental governance: protecting foreign Investors at the Expense of public policy. CUP, Cambridge
Ubink M (2007) Tenure security: wishful policy thinking or reality? A case from Peri-urban Ghana. J Afr Law 51:215–248
UNCTAD (2009) World Investment Report 2009. New York/Geneva
UNIDROIT/FAO/IFAD (2021) Legal Guide on Agricultural Land Investment Contracts (ALIC), https://www.unidroit.org/work-in-progress/agricultural-land-investment. Accessed 27 Sept 2021
Wendt A (1999) Social theory of international politics. CUP, Cambridge
White B, Borras SM, Hall R, Scoones I, Wolford W (2012) The new enclosures: critical perspectives on corporate land deals. J Peasant Stud 39:619–647

Land Use Policies as Drivers of Land Cover Change in Cameroon

Tassah Ivo Tawe and Ndifon Franshua Becha

Abstract The inextricable bond between land resources and human existence presents evidence of severe threat that seems to undermine the very existence of the human race due to policy failures in land governance. This chapter presents some land governance and policy failures that usher a leeway for land cover changes and environmental degradation in Cameroon. The chapter notes that land resources remain a key asset for the development of the country and that the changes in the land cover stem from unregulated land use practices, which is an expression of policy failures in land governance. Therefore, the future development of the country will be further exposed to threats of vulnerability should policy failures continue to persist with respect to land use and land cover governance. It is recommended that for a developing economy like that of Cameroon to thrive, there is a need to fortify sustainable land management through regulatory policies as a development prerequisite and strategy to minimize expenditure in tackling environmental and social-economic problems.

1 Introduction

Land use policies are key determinants to the social and economic wellbeing of our communities. These policies and decisions play a vital role in fostering sustainable land stewardship.[1] According to Batcha,[2] when the user of a land parcel decides to employ his or her resources toward different uses, land use changes occur, producing

[1] Briassoullis (2003).
[2] Batcha (2015).

T. I. Tawe (✉)
Department of Economics and Environmental Science, National Centre for Education, Ministry of Scientific Research and Innovation, Yaoundé, Cameroon

N. F. Becha
Department of Environmental Science, University of Buea-Cameroon, Buea, Cameroon

H. Ginzky et al. (eds.), *International Yearbook of Soil Law and Policy 2020/2021*, International Yearbook of Soil Law and Policy 2020,
https://doi.org/10.1007/978-3-030-96347-7_6

both desirable and undesirable consequences. In this case, policies and decision-making correlate with variations in land uses. Therefore, appropriate policies and strong institutions significantly determine the nature and dimensions of land uses, which have a bearing on the social and economic welfare of communities.

Most African countries south of the Sahara lack credible regulatory and institutional frameworks that are charged with exercising firm control over land uses. In some cases, if such regulatory organs and institutions exist, they are often flawed in implementation. Most often, prompt feedbacks from farmers usually proceed directly to policy negotiations and implementation, which in turn affect or influence the action carried out by the user of the land without prior consideration and understanding of the root causes of the problem.[3] This is sometimes because of the dimension of the policies in place, which in some cases assume that human and natural systems can be treated independently. This situation raises some compelling questions regarding the credibility of policy makers in African countries south of the Sahara. As the population of Sub-Saharan African countries is projected to reach 1.4 billion persons by 2030,[4] the region's land as a natural capital is expected to experience increasing pressure to meet the needs of its population. With this growing concern, it is therefore important to examine some of the deficiencies that exist in the regulatory instruments with the view of establishing an enabling environment for policies that are more effective both at regional and sub-regional level.

Identifying the root causes of changes in land use are primordial and requires an understanding of how various factors interact in specific contexts to influence decision making on land uses. Growing variations in land uses (which affect global climatic cycles and human welfare) make it necessary to focus attention in understanding the land use decision making actors as their decisions and policies often shape and direct the user of land resources. For decision making and policy implementation on land uses to be effective and sustainable, these decisions should take place exclusively at the local level (though at varying scales), and most importantly, from the individual's landuse preferences so as to have an understanding of the complex ways in which people make decisions.[5]

This chapter is aimed at examining some of the land use regulatory instruments in Cameroon and in which way land use changes shape and orientate policy makers' decisions. In order to address this issue, the chapter attempts to determine if land use policies and decisions are a cause to the present land cover changes and environmental constraints in Cameroon. It also seeks to examine how the role of institutions, State agencies and structures determine the nature of land cover change and the vulnerability of the land user in the face of change. To better appreciate the changes in land uses, it becomes imperative to examine a collection of some domestic land use policies and decisions as. Understanding the complexities of and how these

---

[3]Tassah and Atemkeng (2019), p. 182.

[4]Policy outcome document, Workshop "Implementing Land Degradation Neutrality in Africa (LDN): Means, Legal Instruments and Institutional Challenges? "14 to 15 August 2018. P.3.

[5]Ibid 3.

decisions at different levels and in which way are made and how they influence land use decisions and the sustainability and adaptive capacity of the people.

2 The Human and Physical Environment

Cameroon is a country between West and East Africa and stretches from the Gulf of Guinea to the Lake Chad Basin.[6] It has a total land surface of 472,710 km^2 and a population of 26,545,863 million people.[7] With a growth rate of 2.59%, the population is expected to increase to 50,000,000 people by 2050. This means that almost 600,000 people are added to the population per year.[8]

The physical environment of the country is varied with forests, mountains, water bodies, savannah and deserts.[9] The forest covers about 20 million hectares, making it the second largest in Africa with identified potential of 300 marketable timber species, of which 60 are currently exploited. The area for commercial logging covers a little more than 7 million hectares while community forests represent more than 2 million hectares and the council forests cover 1.8 million hectares.[10] According to the African Development Bank Group[11] as cited by Ruppel C. and Katharina R-S,[12] 10% of the Congo Basin forest is found in Cameroon. covering 41.3% of the national territory.

The economy of the country rests very much on the extractive industry or agriculture constituting the major sector for employment and a source of livelihood for more than 70% of the population.[13] The observed trend in the population growth rate of the country exerts growing pressure on the natural capital, and therefore, requires effective regulations and implementation policies to guide exploitation measures and to enhance sound environmental quality, which is one of the goals of the Sustainable Development Goals (SDGs), namely SDG15.3, which seeks to enhance land degradation neutrality world by 2030.

Cameroon's land use policies are determined by the legislative organ and enshrined in the constitution. To better appreciate the land use policies and decisions and how they shape and direct land cover changes and their impacts on the environment, this study assumes historical "roots" of these policies as they have a close bearing to the present day land use policies in Cameroon and the enhanced challenges faced by land users.

---

[6] Ruppel and Katharina (2018), p. 6.

[7] https://worldpopulationreview.com/countries/cameroon-population/, access on the 12/07/2020.

[8] https://worldpopulationreview.com/countries/cameroon-population/, access on the 12/07/2020.

[9] https://worldpopulationreview.com/countries/cameroon-population/, access on the 12/07/2020.

[10] Cf MINFOF & WRI (2017) cited by Ruppel and Katharina (2018), p. 7.

[11] African Development Bank Group (2015).

[12] Ruppel and Katharina (2018), p. 6.

[13] Lotsmart and Fonjong (2004), pp. 13–29.

3 Historical Overview of Some Land Use Policies/Decisions in Cameroon

Prior to independence, the rush for territories by colonial powers in the African continent for the exploitation of raw materials and other natural resources for their industries in Europe marked the beginning of land use changes in Africa and Cameroon in particular. The major competing colonial powers in Cameroon during this period were the Germans, British and the French. The Germans effectively took control over the territory on the 12th of July 1884 through the signing of the Germano-Douala Treaty between the native traditional rulers and the German traders.[14] Environmental protection during this colonial era did not play a major role, but interestingly, a German imperial decree of 4th April 1900 existed in which the German Governor in Cameroon was authorized "... for the purpose of protecting the forest ... that illegal logging which was in violation of the existing regulations could be ordered to reforest the deforested areas."[15] The colonial occupants had the power to use and dispose of all that constituted the country's wealth, inclusive of its natural resources. Batcha[16] gives a general pattern of the use of land and investments by the Germans, which was mainly for the exploitation of raw materials for their home based industries in Europe. The exploited raw materials include rich natural resources such as mineral, fisheries, and wildlife amongst other resources. This exploitation processes also lead to control over agricultural and industrial activities.

German occupation in Cameroon came to an end after the First World War, which then resulted in the scramble for Cameroon by the British and the French. The scramble for Cameroon by these two colonial powers finally resulted in the separation of the territory by 1919, with the East given to the French while Southern Cameroons was given to the British. The Milner-Simon agreement finalized the partition of Cameroon by putting in place the boundary for the Anglo-French Partition.[17]

The British land use policies in Southern Cameroons centered on crop production—mainly tropical cash crops through the plantation system as well as livestock production in open rangelands. The livestock policy in Southern Cameroons was based on a strategy to ensure the actual practice of livestock grazing. This policy was noted to favour livestock activities as compared to the cultivation of food crops for home consumption. This policy was designed such that livestock could benefit from funds from the "Native Treasury".[18] According to Boutrais,[19] this policy instituted

[14]Ruppel and Katharina (2018), p. 6.

[15]Reichsanz. Nr. 108 vom 5.Mai 1900, Kol. B1. S. 365. Ruppel and Katharina (2018), p. 7.

[16]Batcha (2015).

[17]Batcha (2015).

[18]Batcha (2015).

[19]Boutrais (1995).

Table 1 The "JANGALI" tax in British Southern Cameroons 1949–1952

Year	1948	1949	1950	1951	1952
Tax due (pounds)	18.700	19.600	20.600	22.600	28.900

Source: Boutrais, Hautesterres d'elevage au Cameroun, Vol. 1 Paris, Ostrom, 1995. P.119

the payment of cattle tax or the "jangali" by the local people. Statistics indicate that this tax witnessed an increase over the years during the colonial period (Table 1).

The livestock policy, according to Batcha,[20] had adverse effects on the livestock sector as it limited and controlled the number of cattle owned by a native who depended on income gained from cattle trade.

In 1943, the British also instituted the "Grazing rules" in an effort to control cattle movement and the use of pasture lands. This was mainly due to the fact that cattle movements up and down the slopes were causing problems to the farms located in the valleys. The grazing rules helped to halt some of the problems of crop destruction and conflicts among the natives. The economic crisis that rocked the world in 1929 affected the agro-pastoral sector in terms of crops and cattle prices as some archival sources show that the cattle tax of 1931 was considerably low as compared to the 1930 period, which was due to invasion of locust that destroyed the rich savannah vegetation.[21] The encouragement of agriculture saw the creation of coffee cooperatives and palm oil mills in some parts of the country. It should be noted that the policies that were instituted in British Cameroons were similar to those instituted in French Cameroon, though with slight variations in conceptualization and implementation.

4 Some Legislative and Regulatory Policies on Land Use After Independence in 1960

The legislative and regulatory framework on land use was instituted to guide the system of land tenure and State land in Cameroon. *The Ordinance N°. 74-1 of 1974* gives the State the ownership of land within the territory of Cameroon. However, the State guarantees all natural persons and corporate bodies in possession of landed properties the right to freely enjoy and dispose of such lands. The interest in the formulation of this policy was based on access, ownership, withdrawal, exclusion and alienation of specific lands. The policy gives the State, as the guardian of all lands, the right to intervene and ensure the rational use of land when need arises. The framework guiding the administration of land in Cameroon distinguishes three types of landed properties, which are:

[20] Batcha (2015).

[21] Batcha (2015).

- The national land property
- The public land property
- The private land property

According to Article 14 of the Ordinance of 1974?,[22] national land constitutes land which when entering into force is not classed into the public or private property of the State and what does comprise of land as defined by *Art.2.* of Ordinance *N°. 74-1* of *1974*. These lands, under *Art. 16(1)* shall be administered by the State through consultative bodies presided over by the administrative authorities and comprising representatives of the traditional authorities. National land comprises of two categories *(Art. 15 (1) and (2):*

- Land occupied with houses, farms and plantations and grazing lands manifesting human presence and development.
- Land free of effective occupation

Pursuant to this ordinance therefore, in particular concerning the problem areas in agro-pastoral activities and national lands pursuant *Art. 17(2),* it is seen that customary communities members and any person of Cameroon nationality, peacefully occupying or using land in category *(1) Art. 15* shall continue to occupy or use the said land. In this case, they may apply for a land certificate in accordance with the terms of the decree provided for in *Art. 7*. This shows clearly why there are conflicts over land use and ownership amongst the agro-pastoralist in Cameroon. As the cattle graziers (Mbororo) have claim over the occupation and use of land this has posed conflicts with the native indigenes especially in the savannah zones in Cameroon.

The public property land is guided by *Ordinance N°. 74-2 of 1974* and *77-2 of 1977*. The public land in Cameroon *Art.2. (1) of 1974* shall comprise all personal and real property, which by nature or intended purpose should be available either for the direct use of the public or for public services. Public property is divided into natural and artificial properties, wherein natural public properties consist of coastland waters, sub-soil and airspaces. The artificial public properties comprise all motor ways, highways, tracks and commercial seaports, monuments and communication lines. Finally, public properties shall not be liable to private appropriation.

4.1 Private Land Property

According to *Art. 2 of Ordinance N°. 74-1*, these categories of land shall be subject to the right of private property.

- Registered land
- Free hold lands
- Lands acquired under the transcription system

[22] Article 14 of the ordinance of 1974.

- Land covered by a final concession
- Land entered in the "Grundbuch"

Private lands are administered by the state provincial or divisional lands service, which ensures the legal ownership of certificate of occupancy by private land owners. It is further noted in *Art. 77-(1)* of *Ordinance N°. 77-1 of 10 January 1977* that the terms and conditions of the issue of land certificates and the rules relating to the cancellation of the same shall be fixed by decree. In view of this, issues arising from land ownership and occupancy in the *Ordinances of 1974 and 1977* indicate that the State did notanticipate the problems of "common land uses" otherwise known as Common Property Resource (CPR), wherein the land is communally owned such as grazing field occupancy, which now breed a significant number of conflicts. The policies equally failed to make provision for women and minority groups such as the cattle breeders to access and own land freely.

4.2 Institutional Framework Charged with Formulating and Implementing Land Use Policies

These are government ministerial departments and non-governmental organizations (NGOs) that are involved in formulating and administering land use/cover policies and decisions.. It is important to examine these institutions, especially in their outreach activities and strategies, which affect land use/cover change.

In Cameroon, there are about seven state institutions directly involved in land use decisions, while other state ministries are also involved indirectly to the changes occurring on land. These ministries include, the Ministry of Land Tenure and State Property (MINDAF), the Ministry of Agriculture and Rural Development (MINADER), the Ministry of Forestry and Wildlife (MINFOF), the Ministry of the Environment and Nature Protection (MINEP), Ministry of Livestock, Fisheries and Animal Industries (MINEPIA) amongst a host of others. It is important to note that the various outreach activities of these state institutions determine, to a large extent, the forms of land use/cover changes in all parts of the national territory.

4.2.1 MINADER

The Ministry of Agriculture and Rural Development is charged with the responsibilities of promoting agricultural development and farmer wellbeing especially in the rural areas of Cameroon. However, state policies instituted by MINADER in the domain of agriculture and rural development have far reaching implications on land use/cover changes in the country. According to Fonjong, agricultural policies instituted by the State immediately after independence in the 1960s had, as its primary focus, the aim to revitalize the country's economy in the aspect of cash crop production. The main driver of this policy was the government's vision of generating income through cash crop production. In line with this policy, the government embarked on a series of agricultural reforms. As noted by Article 14 of the Ordinance of 1974,[23] the first five years from (1961–1966) was captioned the "farmers' year" and the second plan (1966–1971) was declared as "the farmers' plan". In this respect, agriculture benefited from institutional developments with the creation of research institutions, agricultural schools and financial institutions in almost all parts of the country to address the specific needs of farmers. All these strategies were administered by the agricultural department at the provincial, district and village levels. This covered the rural environment, the farmers' and grazing land (communal land) and stakeholders' land. Likewise, the grazing and farming laws of 1962 prescribed the acquisition of farming, grazing and movements permits. This was to check on environmental deterioration through the various forms of activities carried out. Due to this agricultural policies, rapid transformations took place resulting in the land cover changes, as most of the forests and grazing lands were transformed into cultural/artificial land covers such as small holders palms, coffee, and cocoa plantations.

It should be noted that these beneficial moves carried out in the agricultural sector did not last, as the shocks of the economic crisis that rocked the world did not spare Cameroon. In order to contain the hardship brought in by the economic crisis of the 1980s, the government adopted new liberalization policies in the agricultural sector. Thus, farmers were left on their own and all subsidies provided to the farmers were withdrawn by the state. This policy resulted in counter reactions, such as the abandonment of most farms by farmers since they could not buy the required crops and equipment. In line with this policy of liberalization, many coffee farms were abandoned while some were burnt down. The frustration faced by farmers in the cultivation of tree/cash crops such as coffee forced a majority of them to resort to the cultivation of food crops such as maize, yams, cocoyam, banana, plantains, and potatoes amidst a host of other food crops. However, the practices adopted in the cultivation of these food crops by the local farmers did not account for the preservation of forests and water resources. Most ecosystems and landscapes were destabilized with many assuming fragile status—some beyond restoration due to the adoption of very unsustainable practice such as the felling down of trees, bush burning and tilling. This resulted in rapid land use and land cover changes.

From the 1990s, the government through the MINADER again resorted to the revitalization of the agricultural sector, which saw the distribution of improved seedlings, hybrid species of animals, the provision of agricultural loans and the creation of roads from farm to market. With the assistance of some of these institutions, the agricultural sector started gaining steam and more pressure was mounted on the land cover.

In 2006, the agricultural sector experienced some new policies that resulted in enormous impacts on land use/cover. The 2006 agricultural policy had as objective to:

[23] Article 14 of the ordinance of 1974.

- Ensure the security and the basic needs of the nation
- Contribute to economic development by enriching knowledge on exportation and employment
- Increase revenue generation from agricultural products
- Ameliorate the living conditions of the rural dwellers and
- Ensure the utilization and sustainable management of capital with an emphasis on agricultural production.

The driving force behind this ambitious policy was to ameliorate the living conditions of the rural population through the development of agriculture. In terms of implementing these programmes, a top down approach is the main strategy used as information dissemination in natural resource management is carried out mostly by government agents who are often somewhat regarded as strangers in most rural communities of the country. In this case therefore, the implementation of these land use policies have been met with stiff resistance from the population who prefer their traditional methods of land exploitation (which are absolutely destructive to the environment).

4.2.2 Ministry of Livestock, Fisheries and Animal Industries (MINEPIA)

The Ministry of Livestock, Fisheries and Animal Industry is charged with the responsibilities of formulating policies that concern livestock management, fisheries and general pastoral land. This ministry works closely with other State ministries (such as MINADER) when taking decisions regarding the management of sectorial policies/decisions regarding, in particular the changes experienced in the livestock sector. This dates back to the colonial days where livestock grazing was principally free grazing due to the abundance of grazing land. Cattle were traditionally managed with very little or no medical care and with no strict regulatory framework in place. Cattle grazing during this period was, therefore, more of nomadic grazing with the practice of transhumance.

From the 1960s to 1990, the livestock land use decisions were guided by a series of laws.[24] Law *N°. 11 of 1962 of the United Republic of Cameroon* provided guidelines for grazing, farming and movement permits. These permits were issued by competent authorities in areas where pasture land was found. The laws allowed farmers to occupy parts of pasture land whenever they deemed it necessary for cultivation. The deficiencies in this law have resulted in most of the conflicts observed between the graziers and the farmers especially in the western highlands of Cameroon.

Ordinance N°. 74-1 of 6th July 1974 defines the ownership of all land and equally defines the use and function of pasture land. In this respect, the law clearly

[24] Batcha (2015).

determines the control rights over pasture land. According to *law No. 76/420 of 14th September 1976 and decree No. 78/263 of 3rd September 1973* defined the modalities on how state land is accessed and used. However, this was geared toward zoning land use activities. It equally defines the nature of livestock system. During this period, free grazing was the most common form of grazing because vast expanse of pasture land could not be allocated as private property for individual use.

From 1990, MINEPIA and its livestock sector was dealing with actions geared toward reinforcing professional animal production institutions. It equally addresses the producers by stimulation respective organizations and improving the extension systems and services. Grazing activities were still predominantly free grazing. The MINEPIA cabinet management strategy 2000 lays emphasis on the need to address the problems arising from the illegal status of communal grazing land and their improper use. The 2000 strategy was to look into the major challenges of the livestock sector. This strategy was focused towards ensuring food security, a sustainable management of natural resources and the competitiveness of the livestock products and poverty alleviation.

In ensuring food security and poverty alleviation, the ministry had a strategy to aid small producers in the face of intensification of production, such that they could have a place among the big firms. This led to the formation of producer's organizations in the rural areas and their activities were being followed up by the MINEPIA extension workers. Equally, another aspect of the MINEPIA cabinet strategy 2000 is on the management of natural resources. In this aspect, the strategy was focused towards the maximum use of production potentials in a sustainable way. Some of the strategies included the following:

- Fight against track degradation and the invasion of rangelands by non-adapted plants (such as the bracken fern)
- Organizing concerted management of grazing land.
- Promotion of fodder plants cultivation (like Bracharia and ensilage).

From this strategy, it shows that laws protecting grazing lands clearly defined the limits of grazing. However, these limits are not respected by the local farmers in most cases due to lapses in implementation and the absence of strong and credible institutions. However, MINEPIA in its outreach activities, especially in the western highlands of Cameroon, has done a lot in the organization of concerted management of grazing land. Grazing land permits are being issued today to graziers upon request. This permit gives access for graziers to occupy a certain portion of pasture land for a given period of time. At the same time, due to the hilly topographic landscape in which some of the grazing is done, some farmers have been forced to occupy hilly and fragile ecological zones, gallery and montane forest areas. This situation poses a problem as they sometimes invade the cattle grazing zones. This analysis shows that MINADER and MINEPIA do not work jointly in this area. In this case therefore, the execution of land use/cover policies and decisions do not reflect the original aims of the policies and decisions made. Likewise, the poor execution of policies and decisions have sometimes resulted in conflicts over land use. Finally, the cabinet strategy 2000 has been revised and efforts are being made to improve on the activities on the ground and also on the regulatory framework guiding the formulation and execution of the said decisions.

5 Forestry and Environmental Policies (MINEP and MINFOF)

The forestry and environmental policies and decisions which are either directly or indirectly decreed are promulgated premised on the necessity to protect the environment. These policies provide a guide to the user by ensuring and providing sanctions wherever necessary. However, the role of international environmental policies aimed at ensuring environmental protection is necessary, such that one can better appreciate the way they aimed at ensuring environmental protection in the country. As aforementioned, the government of Cameroon immediately after independence adopted the five-year development plan. This plan had objectives that were to be realized within a certain timeframe. However, before and during the period between 1961 and 1991, environmental policies and forestry regulations existed such as:

- Decision ***N°. 1/10/1961*** rules on hygiene and sanitation in the entire national territory of Cameroon, Decree ***No. 76/157 of 1976*** regulating the management of the private property of the State, and Decree ***N°. 77/139 of May 12, 1977*** creating the national standing committee on Man and the biosphere. This committee was placed under the Ministry of Economic Affairs and Planning with the aim of educating the entire Cameroonian public on environmental issues and concerns.

However, over the period dating back from the 1960 to 1980, these environmental issues were not clearly spelt out. Most of the environmental concerns with regard to land use/cover issues were managed by the then Federal Ministry of Agriculture. Also, during the period between 1981 and 1991, the agricultural development plan in Cameroon were ambiguous and the issue of environmental protection was still found wanting. During this period, forestry policies and laws such as ***Law No. 84/797 of July 17, 1984*** created the sub-department of human settlement and the environment, which was charged with the responsibilities of the national environmental policy and reporting of up to date information on the entire national territory on environmental issues. By 1990, a lot more had been done to improve on the environmental concern of Cameroon as the ongoing fight against environmental degradation was a major priority. With regard to this, a legal framework was established to check on environmental issues. The Ministry of Environment and Forest Protection (MINEF) was created by ***decree No. 92/069 of April 9, 1992***. According to this law, Section 5 (3) prescribes that MINEF has the responsibility to control the exploitation of forest resources. Also, ***Law No. 96/12 of August 5, 1996*** provides a regulatory framework for the management of the environment. ***Sections 2(1)*** of this law states that “the environment constitutes a national common heritage”. Further, ***section 2(2)*** emphasizes the place of this ministry in environmental protection. The 1996 law of forest and the environment as laying emphasis on sustainable development and further noted that the forest and environmental policies in Cameroon face many challenges such as rapid transformations and modifications with a high decreasing trend.[25] Forest exploitation for timber and the consequent effects of degradation have been continuing despite the forest regulatory policies that have been put in place.

In the year 2000, the Cameroon government saw the need to protect the rapid degradation of the environment. As a consequence, a regulatory framework for environmental management necessitated the creation of the Ministry of the Environment and Nature Protection (MINEP). This ministry is concerned with exercising full control over environmental issues in Cameroon and equally works in synergy with the other ministries. This ministry is taking enormous efforts to put in place an effective regulatory instrument focused on protecting the environment. This ministry adopted the 1973 accord of September 12, 1973 by which an inter-state committee was created which developed a framework for the fight against desertification. However, it is observed that some deficiencies exist in the execution of environmental policies and decisions as there were many actors and ministerial departments involved in land management and environmental issues. Thus, there are many problems which arise from the level of taking decisions and actions with respect to realities on the ground.

6 Analysis on the Local Regulatory Framework in Land Use Policies

The local regulatory framework in Cameroon geared toward land management is guided by ***decree N°. 76-166 of 27th April 1976,*** which provides conditions for the management of all national lands and lays emphasis on the role of traditional authorities over the control of land in Cameroon. In this respect, land use polices are managed on two fronts: the local administrative level of land tenure and use, and the individual or household decision making. From field investigations, land use is guided by a plethora of traditional laws which are varied based on importance and interpretations to the local community.

To trace the evolution of these traditional laws, it is important to look at the historical evolution which is rooted in the colonial legacy of the country. According to Fonjong et al.,[26] individual land rights rarely existed during the pre-colonial days as land was collectively owned by the community or family but allows for individual use according to need. Succession over such land followed the family lineage and when a user of the land died, the male heirs of the family assumed natural control

[25]Batcha (2015).

[26]Fonjong et al. (2013).

over the land. In cases where there was no successor, the land simply reverted to the common pool for redistribution. However, the land was collectively owned by the natives who had acquired it either through first settlement or conquest. The various chiefdoms were the managers and custodians of all land. During this period, however, written land titles were not part of the process of ownership given that the customary jurisprudence knew no writing.[27]

This form of land ownership (with its strengths and weaknesses) were only distorted following the signing of the Germano-Douala Treaty in 1884, which formed the legal basis for German colonial rule in Cameroon. The German administration of 1884 claimed all land that were not effectively occupied, and therefore, forced the natives to small reserves, and introduced the system of land registration, which granted the security of land titles. With the departure of the Germans in the First World War, the British became the new colonial powers in Southern Cameroons through the separation of the territory by the two colonial powers that took control over the entire national territory. The British control over land in Southern Cameroons gave the natives only right of occupancy on native land, which was not equal to full ownership. However, after independence, post-colonial Cameroon introduced the 1974 Land Ordinances to regulate land ownership and registration, which were largely rooted from the ideologies of the former colonial masters. The 1974 Land Ordinances attempted to nationalize all land without land certificates whether or not such lands were effectively occupied. ***Ordinance N°. 74-1 of 6th July 1974*** established the roles that governed national lands. Those with land certificates[28] were, by this new law, given 10 years (if in urban areas) and 15 years (if in rural areas) to convert such lands into proper land certificates, or else their lands will be taken as national land. This is in accordance to Sections ***14(3) of Ordinance N°. 74-1 of 6th July 1974***. This policy therefore put the natives in complete disarray. Since the 1974 Ordinances did not completely capture the interest of the natives who owns the land and what land ownership is, and therefore, put the natives at a juncture of confusion.

Field investigations in present day Cameroon indicates that the majority of the local people still believe that traditional authorities are the primary owners of the land, and therefore, pay more allegiance to traditional methods of land ownership than to statutory provisions. This situation arises because ***Section 1(1) of Decree N°. 76/165 of 27th April 1976*** is based on the fact that the land certificate issued by the State is the only official document recognizing land ownership.[29] It is further noted that this decree and other statutes failed to explain how land is acquired in the first place before seeking registration. As a result, land ownership in every day transaction is not based on land ownership as defined by the law but more oriented by customs that discriminate against women in the process of land ownership. From field investigations and interviews in some of the rural areas in Cameroon, a majority

[27] Fonjong et al. (2013).

[28] Fonjong et al. (2013).

[29] Fonjong et al. (2013).

of the local populations expressed complete ignorance about the existence of land certificate.

As observed in the field, customary laws discriminate against women in the process of land ownership and gives preference to the male. Women constitute the main actors in food crop production and agents of environmental transformations and land cover changes. Field investigations indicate that women have just access to land and not ownership rights as defined by the customary provisions of the land.[30] As we are dealing with a patrilineal society, women's right to land is, therefore, based on their relations either as wives, daughters, sisters or in-laws.

Based on the above observations on land ownership and equally on the traditional regulatory framework, environmental concerns and land cover changes are on the rise since women, who constitute the main actors of food crop production, do not have the right to own land. In this respect, these women have been using the most unsustainable methods in land exploitation and simultaneously occupying marginal lands in the process of food crop production, which have a minimal resilience threshold to stressors. This has resulted in rapid changes in land use/cover changes as they fear that they can be evicted from such lands at any time by the real owners.

7 Conclusion

Humanity for the rest of our existence will have an inextricable interaction with land resources for obvious but diverse reasons. However, the humanity-land interaction in the world and Cameroon in particular presents evidence of indelible footprints of land resources degradation as a direct consequence of policy failure in land governance. The social-environmental cost due to land cover changes is already too heavy for individuals, communities, regions and the State of Cameroon to bear. Land cover change is largely blamed for soil impoverishment, decline in ecosystem services, climate change and variability and declining water resources. Perhaps, the scale of these social-environmental problems are too difficult to be reversed as they include the inconsistencies in land related regulations, gross ineffectiveness in the implementation of regulations by concerned institutions, the growing dominance of customary tendencies in gaining access to and ownership to land, increasing population pressure with the absolute hope by many to make a living from exploiting land persist. Land resource remains a key asset for the development of Cameroon. Unregulated land use practices, and thus land cover changes, in the country have been observed as an expression of policy failures. Therefore, it is evident that the future development of Cameroon will be exposed to further threats of vulnerability should policy failures persist in land use and land cover governance. In other words, should current trends in land cover changes continue without being regulated by policies, most of the nation's landscapes will be degraded in the near future.

[30] Fonjong et al. (2013).

It is incumbent for developing economies to fortify sustainable land management as a development prerequisite and a strategy to minimize expenditures in combatting human-induced environmental and social-economic problems. The government of Cameroon must demonstrate a strong political will to carry out fundamental adjustments in the existing land policy to reflect the realities of the society and the future development orientation of the country. Main possibilities should include introducing measures to reduce pressure on the land resource, enforce gender equity with regards to access and ownership to land, devise a harmonizes mechanism to regulate customary land tenancy options, support initiatives towards land classification with prescribed uses, promote initiatives for the creation of individual and community forest at local levels, as well as enforce restrictions in areas designated as State lands.

References

African Development Bank Group (2015) Cameroon, joint 2015-2020 country strategy paper and country portfolio performance review (CPPR) report, Abidjan, African Development Bank, http://www.afdb.org/fileadmin/uploads/afdb/Document/Project-and-operations/Cameroon__06_2015pdf. Accessed 12 July 2020

Batcha RNE (2015) Land use land cover change and food security in the Santa- Babadjou region: Western Highlands, Cameroon. Unpublished Ph.D Thesis, University of Yaoundé 1, Cameroon

Boutrais (1995) Hautesterres d'elevage au Cameroun, Vol. 1. Paris, Ostrom, p 119

Briassoulis H (2003) Analysis of land-use change: theoretical and modeling approaches. The web book of regional science. West Virginia University, Regional Research Institute, Morgantown, http://www.rri.wvu.edu/WebBook/Briassoulis/contents.htm (January 2005)

Fonjong L et al (2013) Land tenure practices and women rights to land in Anglophone Cameroon: implications for sustainable development. (IDRC)-2013

Fonjong LN (2004) Changing fortunes of government policies and its implications on the application of agricultural innovations in Cameroon. Nordic J Afr Stud 13(1):13–29

http://worldpopulationreview.com/countries/cameroon-population/, Accessed 12 July 2020

MINFOF/Ministry of Forestry and Wildlife and World Resources Institute (2017) Forestry domain of Cameroon, Yaoundé

Policy outcome document, Workshop "Implementing Land Degradation Neutrality in Africa (LDN): Means, Legal Instruments and Institutional Challenges? " 14 to 15 August 2018. P.3

Ruppel C, Katharina R-S (2018) Cameroon in a nutshell-human and natural environment, historical overview and legal setup. Environmental law and Policy in Cameroon, p 6

Tassah IT, Atemkeng AN (2019) An assessment of participatory management of forest resources by communities neighbouring the Mount Cameroon Forest Reserve. Afr J Soc Responsib Sustain Manage 1(1):2019

The Protection of Soil Under Cameroonian Law: The Place of Investors

Christopher Funwie Tamasang

Abstract Increased pressure on soil and threats to its health which are driven by the main land-based investors including; the growing awareness of its multipurpose functions in addressing global environmental and socio-economic challenges have triggered more research in Cameroon to enhance the protection of this key non-renewable resource. This chapter sets out to investigate whether there is an enabling soil legal and institutional environment for the control of land-based investors with respect to the impacts of their activities on soil health. The chapter argues that the regulatory environment for the protection of soil in Cameroon is not comprehensive enough for the control of land-based investors regarding the huge impacts their activities have on soil health. The chapter also establishes that there are numerous, but fragmented, selective and isolated legal attempts to prevent soil degradation and protect for the same across sectors in Cameroon without expressly defining the contours of environmental responsibility and sanctioning them. The chapter concludes by recommending a comprehensive legal framework that specifically articulates the responsibility of 'land-based investors' vis-à-vis soil protection and equally reinforces the administrative control and sanction of investors who violate regulations that are in force.

1 Context and Rationale for Soil Protection

As a key component of the Earth, soil provides crucial services for agricultural production, animal habitation, plant growth, biodiversity, filters and cleans tens of thousands of cubic kilometres of water each year and is; fundamental for regulating the climate as a major reservoir for carbon that helps regulate emissions of greenhouse gases (GHGs), all of which are essential for achieving the United Nations'

C. F. Tamasang (✉)
University of Yaounde II, Soa, Cameroon

H. Ginzky et al. (eds.), *International Yearbook of Soil Law and Policy 2020/2021*, International Yearbook of Soil Law and Policy 2020,
https://doi.org/10.1007/978-3-030-96347-7_7

Sustainable Development Goals (SDGs).[1] Helming and others,[2] believe that soil constitutes a nexus of multiple SDGs, citing another group of researchers who identified direct or indirect contributions of soils to be as many as 13 of the 17 SDGs and that a fundamental role of soils exists for at least four of the 13 SDGs.[3]

As a key non-renewable natural resource, the critical importance of soils has been echoed at the UN level. For instance, during the 2015 World Soil Day (WSD), the UN Secretary-General Ban Ki-moon while calling on all to promote sustainable soil management through proper soil governance and sound investments, referred to soil as: 'a truly solid ground for life' in line with that year's WSD theme entitled: "Soils a solid ground for life".[4] Roosevelt F.D. is known to have asserted that "the nation that destroys its soil destroys itself."[5] Soil has also been considered as 'the living skin of Earth'[6] and a 'foundation for all terrestrial life',[7] or a "melting pot for all life on earth" or better still a "melting pot for all natural resources and human activities".[8] Tita[9] noted that "human life depends on how seriously the soil is protected since it is the backbone of agriculture". These assertions all seek to demonstrate the importance of soil for humankind and the environment, but disappointingly, the protection of soil has long been given little attention. The world's soils are being poisoned, stripped, and degraded more than ever before due to increased demand for food and natural resources as a result of demographic and economic pressures. These factors are driving massive soil transformation and degradation worldwide. In the meantime, the 2030 Agenda instituting the SDGs adopted by the United Nations General Assembly (UNGA) in September 2015 has placed a mammoth responsibility on soil to provide food, water, energy, protect biodiversity, fight climate change, etc. These ever growing demands reflect the centrality of soils in global development and environmental politics. Target 15.3 of the 2030 Agenda on Land Degradation Neutrality mirrors the increasing awareness that land especially soil as a natural resource, constitutes a natural capital that underpins wider sustainable development. This crucial utility of soil requires everyone to protect and promote the multi-functionality of soil by ensuring that all land

---

[1] Hou et al. (2020, p. 729); FAO.

[2] Helming et al. (2018, p. 1).

[3] The four SDGs include: arable soils account for the largest part of global food provision (SDG 2); soils are the basis for bio-based renewable energy production to ensure energy security (SDG 7); the storage capacity of soils for organic carbon is paramount for climate change mitigation (SDG13); and the capacity for water purification and retention, nutrient and matter cycling, and the habitat function of soils are essential for maintaining the terrestrial environment and biodiversity (SDG 15).

[4] FAO.

[5] Notably Stroud (2019).

[6] Wyatt (2008, p. 165).

[7] Wyatt (2008, p. 165).

[8] Tamasang et al. (2019, p. vi).

[9] Tita (2006, p. 24).

users, land-based investors[10] in particular, employ sustainable methods that can enable soils to deliver their multiple goods and services.

The above notwithstanding, soil degradation continues to occur in many places worldwide, including Cameroon. The majority of soils in Cameroon are in only fair or poor condition and in most cases are getting worse than are improving. The degradation of soil in Cameroon is due to natural factors such as soil erosion, salinization, compaction, acidification, but more importantly is due to chemical pollution orchestrated mostly by human activities especially via investors. In Cameroon healthy soils are degrading due to unsustainable land investments such as intense chemical use in agricultural processes, mining, industrial development, etc. It has been rightly asserted[11] that healthy soil especially the topsoil, is home to billions of thriving micro-organisms which allow crops to be optimally produced, but as the topsoil is physically eroded or degraded, its biological matter correspondingly declines, and therefore its capacity to produce optimal yields is compromised. The time is therefore ripe for land-based investors to create positive impacts by improving on strategies to protect, restore and cultivate healthy soil. In order to achieve the SDGs by 2030, the African Vision 2063 and the 2035 Emergence Agenda for Cameroon, the use of soils must seek to ensure healthy food, people, nature and climate.

1.1 Soil and Land Distinguished

Land and soil are usually considered synonymous. Of course, *prima facie*, they may appear closely linked to each other and probably inseparable but on a deeper look, they are different. The distinguishing elements are hard to pinpoint but do exist. Land matters deal with tenure and property rights, but soil is spoken about more in terms of possession, ecology and responsibility. Again, the mention of land tenure raises a lot of concern among people but this is not usually the case with soil which is critical for investment and human survival. This reaction may be explained by the fact that land has, from time immemorial, been a highly valued and tradable commodity unlike soil.[12] Furthermore, as water is indispensable for human survival, so too is the soil which succinctly put is "a melting pot of all natural and human activities". This description is perhaps justified by the Biblical provision that "from dust and unto dust we shall return".[13] Soil from this perspective can therefore be

[10]The expression land-based investors is used in this chapter to refer to investors who use land to carry out their projects which damage the soil in numerous ways.

[11]Stroud (2019).

[12]Fonjonga et al. (2010, p. 156, 157); Belaunde et al. (2010, p. 18).

[13]See the Book of Genesis (chapter 3: verse 19); Belaunde et al. (2010, p. 18); Fonjonga et al. (2010, p. 156).

considered as being the point of entry as well as the point of outlet of human and natural phenomena.

1.2 Soil Protection and Investment Nexus

Soil protection is increasingly commanding unanimity for reasons of its importance as highlighted above. On the other hand, there is also unanimity in recent times on encouraging investment in any country across the world including Cameroon. But who is an investor? An "investor" as defined by the Investment Charter of Cameroon[14] means "any individual or a corporate body of Cameroon or foreign nationals, resident or non-resident, whose activity involves acquiring assets with a view to earning interest". The 2017 law[15] to amend and supplement the 2013 law on private investment incentives also defines an "investor" as a resident or non-resident Cameroonian or foreign natural or legal person that acquires assets in the conduct of business for profit. These definitions *prima facie* appear to give the impression that investors are only interested in profit making. Even if investors impact negatively on the environment and soil in particular, they do bring to their host countries some positive impacts which cannot be undermined. Investment creates employment opportunities; contributes to revenue generation through payment of taxes, royalties, etc.; contributes to Gross National Product; and to socio-economic development in respect of their corporate social responsibility and local content obligations. Unfortunately, the impact of investment projects on soil is usually enormous as investors are more often concerned with generating profits at the detriment of any other societal achievements including soil sustainability. This is more disturbing for this is short-sighted on the part of the investors especially where the investments are of long term character. In this case, the sustainability component of the project which is largely dependent on the sustainability of soil is relegated to the footnote. As a matter of fact, this then becomes problematic.

1.3 So What Is the Real Problem?

The discourse on soil protection often sounds utopic as it is not usually perceivable what really constitutes soil. Soil has many meanings but in its traditional meaning,

[14]See Section 3 of Law No. 2002/004 of 19 April 2002 Instituting the Investment Charter of the Republic of Cameroon and amended by Law No. 2004/020 of 22 July 2004 to Amend certain Provisions of Law No. 2002/4 of 19 April 2002 to Institute the Investment Charter of the Republic of Cameroon.

[15]Section 3 new (8) of Law No. 2017/015 of 12 July 2017 to Amend and Supplement Some Provisions of Law No. 2013/0 of April 2013 to Law Down Private Investment Incentives in the Republic of Cameroon.

soil is the natural medium for the growth of plants, habitat for animals and a base for human activities. Soil and its sustainable management is a cross-cutting issue and has linkages or interfaces with general environmental protection, land use and ownership, agriculture, grazing, the extractive industry, industrial activities, industrial waste disposal, climate change, etc. Sadly, there is no comprehensive legislation on soil protection in Cameroon. In the absence of such a comprehensive soil legislation[16] and because of its cross-cutting nature, sectoral policy and legal instruments such as the Framework Law on Environmental Management[17] and its enabling instrument[18] dealing directly with soil and subsoil, environmental and social impacts assessment (ESIA) legislation, agriculture, mining, petroleum, water and other sectoral legislation lend support to soil protection in Cameroon. The relevant prescriptions of these sectoral policy and legal instruments therefore act as instruments of control for land-based investors whose activities may constitute threats to soil health. The problems that must therefore be addressed and answered in this chapter is whether the existing legal prescriptions purporting to lend support to soil protection clearly establish the responsibility of land-based investors to protect the soil. Additionally, do such investors comply with the relevant soil protection provisions? Is there an effective implementation and enforcement of the relevant soil protection provisions vis-à-vis land-based investors?

The Cameroonian land tenure legislation which is a corollary of soil protection law guarantee ownership which gives landowners of the three land categories the right to exclusive possession and use of their land, but remains silent on the question of their responsibility to sustainably manage the land/soil they own. Investors are known for the negative impacts of their activities on soil quality and health and failure by the land tenure laws to take cognizance of soil concerns which constitute an inherent flaw of the Cameroonian land tenure laws. This legal gap may give investors the liberty to unsustainably use land without due regard to the imperative of the protection the soil. Given that land and soil are inextricably linked and in the absence of a comprehensive soil law, one should expect to see comprehensive and unambiguous provisions apportioning responsibility to owners and users of the three categories of land in Cameroon to ensure the protection of soil. The problem is equally serious given that the Cameroonian land law which also deals with the use of soil is void of any provision on the environmental responsibility of land owners and/or users such as investors. It is true that in this regard, reliance can be had in the

[16]Comprehensive soil legislation here refers to one that clearly envisages all potential threats to soil health and clearly establishes and apportions visible responsibility for soil protection for all and, in our case, for land-based investors.

[17]Law No. 96/12 of 5 August 1996 relating to environmental management.

[18]The legislative drafting process in Cameroon is one that is significantly inspired by the Continental legal system. Such a legal tradition practices a law-making process where Parliament or any other Organ of the State entrusted with powers to make laws, crafts a piece of legislation by setting out the main issues of the law and allow subsidiary or delegated legislators to lay down details to complete such a legislation in terms of a better understanding and implementation of the law in question.

Constitutional provision in the Preamble requiring that: "Every person shall have a right to a healthy environment. The protection of the environment shall be the duty of every citizen. The State shall ensure the protection and improvement of the environment".[19] Since soil is an integral part of the environment, this environmental protection provision may be interpreted to impose such responsibility on everyone, including investors to protect soil. Although these Constitutional pronouncements are relevant, they are vaguely worded and will equally require a specific legal text on soils to enhance a clearer understanding, interpretation and application that assigns visible responsibilities for soil protection for all and, in our case, for investors. The problem is equally serious because the *locus standi* to sue for the enforcement of the Constitutional provisions is restricted to certain categories of individuals,[20] who may not be environmental protection advocates or may simply not give priority for the need to protect the environment especially soil.

A *prima facie* reading of an overarching legislation regarding the responsibility to protect the environment especially soil under Cameroonian law notably, the Environmental and Social Impact Assessment (ESIA) Law,[21] gives the impression that all promoters of projects (usually investors) are targeted by the law. While this appears to be the case from the substantive provisions of law provided, the problem as far as protecting the soil is concerned is towards the effective compliance with the prescriptions on this legislation by investors. The problem is accentuated by the fact that the administration in charge of the environment and other sectoral administrations do not conduct regular controls on project sites in order to identify and sanction investors violating environmental and soil laws.

1.4 Some Key Theoretical Underpinnings on Soil and Soil Degradation

The discussion on soil protection law in Cameroon with regard to activities of investors that cause soil degradation cannot be complete without invoking some key theoretical underpinnings on soil[22] and soil degradation.[23] The first is the theory

---

[19]Paragraph 26 of the Preamble of Law No. 96/06 of 18 January 1996 to amend the Constitution of 2 June 1972, amended and supplemented by Law No. 2008/001 of 14 April 2008.

[20]They include the President of the Republic, the President of the National Assembly, the President of the Senate, one-third of the members of the National Assembly or one-third of the Senators, Presidents of regional executives. For details, see Article 47(2) of the Cameroonian Constitution.

[21]Decree No. 2013/0171/PM of 14 February 2013 fixing modalities for carrying out Environmental and Social Impact Assessment.

[22]In Cameroon, Soil is characterised by topography, amount of rainfall and natural occurrences.

[23]The main threats to soil degradation in Cameroon include mining and industrial operations; unsustainable agricultural practices; demographic growth rate and urbanization; weather and climatic factors; wildfire for farming, hunting, and cattle raring; topographic constraints; and land-grabbing.

of collective goods which holds that a pure collective good has three properties: non-excludability (nobody can be excluded from the consumption benefit), non-rivalry in consumption (one person's consumption does not impair that of another) and externalities (negative externalities which are related to the degradation of the good and positive externalities which are related for instance, to protection of the collective good).[24] Soil, as an example of an environmental good, has the properties of a collective good which means that in as much as everyone, including an investor, is entitled to use it, such users have the responsibility not to damage but to protect it. According to this theory, environmental problems such as soil degradation emerge when users exploit environmental goods without contributing to their maintenance or conservation.[25] Investors as economic agents have caused and continue to cause soil degradation through their investments. This requires the legal control of investors in order to ensure the protection of soil in Cameroon for present and future generations.

Another theory underpinning soil and soil degradation is the theory of social cost which holds that economic agents must bear the full costs of their actions.[26] In the context of this discussion, the theory of social cost explains soil degradation as the result of investors' use of unsustainable production methods and consequently, it is their responsibility to make good on the negative impacts of their activities on soil. Thus, there is the social cost to prevent soil degradation, to pay reducing degradation of the soil, for repairing or compensating for soil degradation, to pay for restoration and rehabilitation of degraded soil. It is the considered opinion of this author that the respect of the principles and ideas elucidated in the above theories in the context of soil degradation could enhance the legal protection of soil against the activities of investors in Cameroon. Thus, the respect of the principles embedded in the theories of collective goods and social cost in implementing the legal prescriptions on the control of investors is critical for soil protection in Cameroon.

The important questions this chapter attempts to answer are: how do investors contribute to soil degradation in Cameroon? Is there a comprehensive soil protection legislation including the investors' role that protects against soil degradation? Do investors comply with the relevant soil protection law in Cameroon? Is there an enabling institutional environment for the effective implementation and enforcement of the policy and legal prescriptions for soil protection? And lastly, what are some of the existing opportunities for, and challenges faced in the effective implementation of soil protection law and policy from an investor perspective?

[24] Maiangwa et al. (2007, p. 790).

[25] Maiangwa et al. (2007, p. 790).

[26] Maiangwa et al. (2007, p. 790).

2 How Are Investors a Source of Soil Degradation in Cameroon?

Apart from natural causes, soil degradation in Cameroon is facilitated by several actors with varying degrees of intensity, in different spheres of influence. Investors stand out as the most prominent actors responsible for soil degradation in Cameroon through their activities which are mostly detrimental to soil health. Their role as actors of soil degradation in Cameroon is heavily felt in the extractive industry (more specifically in the mining sector) and also in the agricultural sectors, more likely within the large scale agro-industrial establishments as opposed to the industrial operations.

The mining sector in Cameroon for example, has witnessed the presence of many investors especially multinational corporations who unfortunately, are not usually preoccupied by the concerns of soil protection as they are usually in no hurry to obtain environmental authorisation before launching their activities. Moreover, the investors care little about rehabilitating or reinstating the exploited sites which become environmentally degraded including; the degradation of soil. More critically is the fact that such investors make illegal use of chemical products, such as mercury, which leads to soil toxicity. The chemicals destroy the attendant biodiversity of the sites including soil biodiversity as well as polluting the water used by the surrounding populations. The mining of mineral deposits in Cameroon has led to the excavation of large quantities of earth materials from the mining sites, exposing the soils and rendering the landscapes desolate especially when the exploited mining sites are not restored, reinstated or rehabilitated.

Dallou and others[27] for instance, conducted an assessment of the level of heavy metals pollution in soil samples collected around gold mines located in Betaré Oya and Batouri ((Kambélé)) in Eastern Cameroon and analysis results showed that soils examined in the gold mining areas are polluted by heavy metals such as iron, manganese, arsenic, zirconium and lead. The METALICOM Company carries out gold mining in the areas. A wide range of large-scale mining concessions have been initiated in the last few years by Chinese companies[28] and the rapid development of this phenomenon constitute to a danger to soil health and its attendant biodiversity.

[27] Dallou et al. (2018, p. 212 and 216).

[28] Examples of such mining concessions include the East Batouri mining search permit (PCMRC S. A.), Bandongwe mining search permit (PCMRC S.A.), Garoua Boulai South mining search permit (PCMRC S.A.), Mvangan mining search permit (PCMRC S.A.), Sangmelima South mining search permit (PCMRC S.A.), Bangbel West mining search permit (CMRC S.A.), Batouri West mining search permit (Longsheng Cameroon S.A.) Garoua Boulai mining search permit (Longsheng Cameroon S.A.), Lobéke East mining search permit (Longsheng Cameroon S.A.), Sangmelima mining search permit (Longsheng Cameroon S.A.), Beck mining search permit (Société Coast Investments Int. (SCII), Bougma mining search permit (SCII), Messeng mining search permit (SCII), Ngoundi mining search permit (SCII), Akonolinga North mining search permit (FMRC), Boulou mining search permit (FMRC), Mpmpwe mining search permit (FMRC). Mayers et al. (2019, p. 16).

Small-scale Chinese miners are also reported[29] to be highly involved in small-scale and artisanal mining of gold in the East Region which apart from operating illegally,[30] often do not rehabilitate the mining sites, resulting in landslides and the death of local miners.

Investors in large-scale agriculture also constitute principal actors of soil degradation in Cameroon. The large scale agro-industrial establishments in the coastal lowlands of the southwest, littoral and south regions of Cameroon make use of chemical inputs such as fertilizers and pesticides, but also tillage and deforestation with acute implications on biodiversity loss and the quality of soils in the regions where they are applied. Furthermore, their activities on soil degradation are huge because they acquire large expanses of land e.g., Cameroon Development Corporation (CDC), PAMOL and Tole Tea Estate in the Southwest region of Cameroon, SOCAPALM and HEVECAM[31] in the South Region of the country. All banana plantation investors for instance, make use of Chlorpyrifos—an insecticide that is put in polyethylene bags that are used to protect the bananas. These bags, if not recycled, may cause soil hazards when dumped in the environment as the chemicals in the bags can be washed into soils and rivers thereby killing living organisms that have the effects of destroying the soil fertility and its attendance biodiversity.[32] Most plantation investors also use herbicides that kill herbs thereby leaving the soil bare and leading to exposed erosion. When rain falls, the topsoil and its fertility is washed away by soil erosion. Soil degradation is also associated with the clearing and destruction of forests and other forms of vegetation, exposing the topsoil to frequent rain storms. This always results to the leaching of the soil and runoff causes soil erosion.[33] Plantations also involve the clearing of large areas of lands, mostly forests that include the loss of vegetation cover leading to erosion and consequently, soil degradation. It is reported[34] that the rubber industry mostly controlled by Chinese investors[35] is an intensive user of phosphate, nitrogen and potassium fertiliser—with the associated dangers of chemical pollution and toxicity of water and soil. While there is some evidence of planning by the companies' social and environmental

[29] Mayers et al. (2019, pp. 47–51).

[30] In March 2014, the Divisional Mining Delegate for Lom and Djerem in the East Region went on a tour of mining sites under her jurisdiction and ame across five Chinese mining companies—Gipo Mining, Ging Ma SA, d'Or de lion, Danshang Ming and Fuhaï Ming operating illegally. Mayers et al. (2019, p. 51).

[31] SOCAPALM and HEVECAM are the French acronym For Cameroon Palm Oil Company German Rubber Company respectively.

[32] Tita (2006, p. 26).

[33] Tita (2006, p. 26).

[34] Mayers et al. (2019, p. 45).

[35] Examples of Chinese companies (investors) involved in the rubber sector are Hevecam – Halcyon Agri Co. Ltd. 52,607 ha and Sudcam – Halcyon Agri Co. Ltd. 58,931 ha. Rubber production is the objective of recent Chinese company acquisitions of about 112,000 ha of land, at two sites in the departments of Ocean and Dja Lobo by the above mentioned companies. See Mayers et al. (2019, p. 42, 43 and 45).

responsibility plans to mitigate these effects, there is little evidence of effective implementation.[36] Given that agriculture is the mainstay of Cameroon's economy,[37] land acquisitions for agro-industry by agro investors is increasing in number and scale with serious implication for soil health. With the entry of new investors, especially oil palm plantation companies in Cameroon and the interest shown by existing companies to expand their operations and increase production, there will be significant land-use changes with serious negative impacts on soils in the near future. Attention must therefore be given, in granting large areas of land, to the avoidance of any additional soil degradation.

The activities of most industrial operators in Cameroon do not "promise good" for the health of the soil. Industrial operations are associated with the discharge of effluents and other wastes into the environment without considering the health of soil. The increasing contamination of soil in Cameroon is partly attributed to industrialisation processes[38] which in most cases have preceded ESIA requirements. Industrial operators therefore remain some of the main actors of soil degradation in Cameroon. The impacts of industrial operations on soils are particularly felt around the Douala industrial zone.

The increased role of investors across many economic activities in Cameroon therefore leads to a spiral in the effects of soil degradation. As the demand for land from the mining and agro-industrial sectors increases and the government pursues its policy of encouraging foreign investment in these sectors in the country, so there is an acute risk of soil degradation.

Other actors of soil degradation in Cameroon are small-scale farmers who use unsustainable farming methods such as the slash-and-burn method also called swidden,[39] bush burning and *"the ankara system"*[40] especially in the Western Highlands of Cameroon; livestock farmers who burn most of the grazing lands in order to allow regeneration for their livestock, with direct implications for soil

[36]Mayers et al. (2019, p. 45).

[37]Agriculture contributes 43% of GDP, and providing employment to 70% of the working population in Cameroon. See Mayers et al. (2019, p. 42).

[38]Fogwe et al. (2001, p. 8).

[39]Slash-and-burn farming includes several phases: (i) clearing of a portion of forest; (ii) burning of the plant debris; (iii) cultivation of the land for a generally brief period; (iv) leaving the land to fallow for a generally long period of time. When returns from land diminish, they shift cultivation to another plot and allow the previous parcel to lie fallow for some years to regain fertility. About 80% of rural population in Cameroon practices it. Slash and burn agricultural practice is detrimental to soil. This type of agriculture affects vegetation and soil carbon stocks.

[40]The *"ankara system"* is a traditional farming method practiced by farmers in the Western Highlands of Cameroon with the aim to improve soil fertility and increasing crop production. Here, dry plant waste is put in the middle of beds, then partly covered with soil and then burnt. Then, seeds are planted on the portions. During burning, too much smoke is being emitted, which indicate loss of carbon to the atmosphere with all what this entails for climate change and also reducing soil fertility in the long run.

degradation;[41] hunters who use fire for hunting; urban sprawls as a result of rapid increase in the population with corresponding urbanization in the rural areas with significant effects on vegetation cover; topographic factors; and climatic factors although not actors, are some of the factors responsible for soil degradation in Cameroon.

3 Current Soil Regulatory Mechanisms in Cameroon: Focus on Investors

In this chapter, the regulatory mechanisms for soil protection as regards investors in Cameroon are considered under two main heads: the enabling legal environment and enabling institutional environment.

3.1 *Enabling Legal Environment*

The enabling legal environment for the control of investors regarding soil degradation in Cameroon can be conveniently catalogued in principles and concepts that draw from international and domestic regulatory frameworks.

3.1.1 International Legal Environment: The Principles and Concepts on Soil and Investments

Domestic legal and policy instruments are usually enacted with respect to international commitments subscribed to by Cameroon. It seems there is no specific international soil instrument at present despite a proposal for a 'Convention on sustainable use of soil' made early in Tutzing, Germany in 1998.[42] Notwithstanding this huge legal gap at the international level with respect to soil protection, international environmental legal and policy instruments to which Cameroon subscribes, have generated numerous principles and concepts that lend support to soil protection which investors are expected to comply with. With the above in mind, an enabling international legal environment for soil protection and the control of investors is afforded by principles and concepts emanating from international environmental legal and policy instruments. Cameroon subscribes to some international principles

[41] Extensive grazing in the Western Highlands is also a major contributor to soil degradation. Livestock grazing in the region is done using the free-range technique over communally owned lands. This technique of grazing alongside the non-respect of the carrying capacity has given rise to over grazing, soil compaction and soil degradation.

[42] Boer et al. (2016, p. 56).

and concepts which offer support to soil protection. The relevant ones examined below:

3.1.1.1 Land Degradation Neutrality Concept and Its Implication for Investors Control for Soil Protection

The 'Land Degradation Neutrality'[43] (LDN) Concept: The objective of LDN is to maintain and enhance the natural land and its associated ecosystem services. The United Nations Convention to Combat Desertification (UNCCD) provides for the international framework of the development and implementation of the LDN initiative which largely seeks to protect soil as the principal component of land.[44] In this regard, the Science-Policy Interface of the UNCCD led to the development of the "Scientific Conceptual Framework for Land Degradation Neutrality" which provides a conceptualised scientific paradigm for planning, implementing and monitoring the LDN Concept at national or regional levels.[45] Actions to achieve the LDN include sustainable land management practices that prevent or mitigate degradation, and all efforts to reverse degradation through rehabilitation or restoration of degraded lands. The implementation of the LDN Concept requires efforts to avoid further net loss of the land-based natural capital[46] such as soil.

Crucially important is this LDN concept that it has also been prescribed in Goal 15 and its Target 15.3[47] of the 17 Sustainable Development Goals (SDGs).[48] Other Goals relevant for soil protection include Goal 12 which sets outs to ensure sustainable consumption and production patterns especially as such production relates to land; Goal 13 which encourages everyone to take urgent action to combat climate change and its impacts. In line with Goal 13, soil is one of the carbon sinks and reservoirs whose protection and conservation is critical for climate change mitigation. Investors are particularly responsible for soil-based carbon emissions and are therefore specifically targeted by this goal to protect soil. Goal 13 therefore provides a global policy framework for soil protection in the context of fighting climate change. Some of the SDGs therefore constitute a panacea for fighting soil

[43] 'Land Degradation Neutrality' was defined at 12th Conference of the Parties (COP) of the United Nations Convention to Combat Desertification (UNCCD) as: "a state whereby the amount and quality of land resources necessary to support ecosystem functions and services and enhance food security remain stable or increase within specified temporal and spatial scales and ecosystems. Decision 3/COP.12, UNCCD, 2015.

[44] Cowie et al. (2017, p. 25).

[45] Cowie et al. (2017, p. 25).

[46] Orr et al. (2017, p. 3).

[47] Goal 15 sets out to protect, restore and promote sustainable use of terrestrial ecosystems, sustainably manage forests, combat desertification, and halt and reverse land degradation and halt biodiversity loss. Target 15.3 urges States to strive to achieve a land degradation neutral world by 2030 constitutes the most critical target for soil protection.

[48] The 17 SDGs are catalogued in the Global 2030 Agenda for Sustainable Development.

degradation as they provide a policy framework for investors to comply with. Furthermore, COMIFAC's[49] Sub-regional Action Plan for Combating Land Degradation and Desertification (SAP-CLDD), known by its French acronym as PASR-LCD lends support to the achievement of the LDN and consequently to soil sustainability. In this light, investors are most targeted as they are the greatest users of land and their activities which will definitely frustrate the achievement of the LDN objective if this concept is not properly integrated in the domestic regulatory framework in Cameroon.

3.1.1.2 The Principle of Inclusive Growth and Sustainable Development and Its Implication for Investors Control for Soil Protection

The principle of inclusive growth and sustainable development is one that helps people and, in the context of soil protection, investors to contribute to and benefit from economic growth and development with minimal impact on the environment, especially with regards to soil. The principle finds policy backing under the African Union (AU) Vision 2063 commonly known as Agenda 2063: The 'Africa We Want'.[50] Although none of the 7 Aspirations for the 'Africa We Want' specifically mentions soil. Aspiration 1 which sets out to achieve "A prosperous Africa based on inclusive growth and sustainable development" implies soil protection which is the major natural capital for investment in Africa. In fact, soil constitutes the principal means and resources to drive development as iterated under Paragraph 10 of Aspiration 1. More critical is the fact that investors are the principal actors that will drive the much expected inclusive growth and sustainable development through their land-based investments such as agriculture, mining and industrial operations that can produce negative externalities by way of soil degradation. Some of the objectives[51] of the Green Wall for the Sahara Initiative (GWSI),[52] notably are: to slow the advance of the Sahara Desert, enhance environmental sustainability, control land degradation, promote integrated natural resources management, and conserve biological diversity also lend support to soil protection. The Principle of inclusive

[49] The Central African Forests Commission.

[50] Adopted in September 2015 as Africa's master plan for transforming Africa into a global powerhouse of the future. See African Union Website, available at https://au.int/en/agenda2063/overview, accessed on 05 June 2019.

[51] The Green Wall for the Sahara Initiative, a programme initiated by AU and developed by the AU Commission in collaboration with the Food and Agricultural Organization, United Nations Environment Programme (UNEP), UNCCD, etc., was launched in December 2006. African Heads of State and Government in their Summit in January 2007 adopted the Decision on the implementation of the initiative. The programme covers a wide group of countries, including Cameroon.

[52] See Commission of the African Union, The Green Wall for the Sahara Initiative, available at http://iriportal.ldeo.columbia.edu/portal/server.pt/gateway/PTARGS_0_2_1208_0_0_18/AU%20-%20The%20Green%20Wall%20for%20Sahara%20Initiative%20%2D%2D%20In%20brief.pdf, accessed 23 May, 2019.

growth and sustainable development is also reflected in all of the 17 SDGs, which, like Vision 2063 and GWSI, provides a policy framework for sustainable soil management and use. Investors are expected to consider them when making investments plans and decisions.

3.1.1.3 The Concept of Improved and Sustainable Soil Management and Implication for Investors' Responsibility

The Concept of Improved and Sustainable Soil Management is simply a clarion call on every one and more importantly on all investors whose investments have negative soil impacts to ensure that such investments do not cause harm to soil, but improve soil quality. The Regional Implementation Plan for the African Soil Partnership (AfSP) adopted in 2016 recognises the importance of soil as the main resource base for many people in Sub-Saharan Africa (SSA). AfSP calls for the strong support of national governments as well as national and regional entities involved in natural resource management to contribute to achieving the common goal of improved and sustainable soil management. The implementation of the AfSP in SSA is structured on five main pillars[53] which are all relevant to sustainable soil management. The implementation plan of AfSP also recognises six main areas[54] of action as priority in SSA which are also relevant for sustainable soil management. National governments as well as national and regional entities are some of the land-based investors. It is therefore expected of them to carry out their investment with the goal of achieving improved and sustainable soil management.

[53] Pillar 1 promotes sustainable management of soil resources for soil protection, conservation and sustainable production; Pillar 2 encourages investment, technical cooperation, policy, education, awareness and extension in soil; Pillar 3 promotes targeted soil research and development focusing on identified gaps, priorities, and synergies with related productive, environmental, and social development actions; Pillar 4 enhances the quantity and quality of soil data and information: data collection (generation), analysis, validation, reporting, monitoring and integration with other disciplines; Pillar 5 anchors on the harmonisation of methods, measurements and indicators for the sustainable management and protection of soil resources.

[54] Addressing soil degradation as a major factor in food insecurity and making country and region-wide efforts to rehabilitate degraded lands and change them to productive assets through community participation (Relates to Pillar 1); guiding and enabling the implementation of sound and sustainable soil management practices and the restoration of soil health at all levels towards food security (Relates to Pillar 1); developing, updating and disseminating updated and harmonised national and regional soil resource information addressing all user needs including soil fertility information and making the best use of science available to increase soil productivity (Relates to Pillars 4 and 5); addressing climate change and developing resilience towards climate change adaptation (Relates to Pillars 1 and 3); developing and implementing training and capacity building programmes in all soil applications for existing and new generations of experts in soil science and land management, taking into account the gender balance (Relates to Pillar 2); establishing linkages and networks with other national, regional and global initiatives that impact soil health (Relates to Pillars 1 and 2).

3.1.2 Domestic Legal Environment

In the absence of a comprehensive soil law in Cameroon, relevant provisions on soil protection are embedded within cross-cutting instruments for general environmental protection such as ESIA law, the Cameroonian Constitution and the 1996 Framework Law on Environmental Management[55] and every other relevant enabling instrument[56] on soil protection and in relevant sectoral policy and legal instruments such as agriculture, land tenure,[57] forestry,[58] water,[59] mining,[60] industrial

[55]Law No. 96/12 of 5 August 1996 relating to the framework law on environmental management.

[56]Decree No. 2011/2584/PM of 23 August 2011 Establishing the Modalities for the Protection of the Soil and Subsoil.

[57]Ordinance No 74/2 of 6 July 1974 to establish the rules governing state lands and Ordinance No 74/1 of 6 July 1974 to establish the rules governing land tenure as complemented by several other enabling instruments such as: Decree No. 76/166 of 27April 1976 to establish the terms and conditions for the management of national lands; Circular No. 001/CAB/PM of 1 April 2014 relating to the provisions applicable to investors on access to land in Cameroon; Decree No. 2003/418/PM of 25 February 2003 to establish the rates for indemnities to land owners whose lands have farm lands and planted trees have been destroyed on grounds of public utility; Decree No. 87/1872 of 16 December 1987 to enable the application of law No. 85/009 of July 4, 1985 on Expropriation on grounds of Public Utility; Decree No. 214/32111 of 29 September 2014 to establish the minimum prices applicable to transactions on land that falls under the private property of the State; Decree No. 2014/3210/PM of 29 September 2014 to establish the conditions for the acquisition of BAUX and the modalities for payment of land taxes in economic areas; Decree No. 2014/3209/PM of 29 September 2014 to establish the minimum amounts for annual dues for the occupation of attachments to state lands; Decree No. 2015/3580/PM of 11 August 2015 to establish the modalities of registration and the regime of insurance and guarantees applicable to concensus and baux domaniaux; Decree No. 2016/3058/PM of 28 July 2016 to establish the rules for the use of land and construction; Arrête No. 00832/Y.15.1/MINUH/D000 of 20 November 1987 to establish the bases for calculating value of buildings that are expropriated on grounds of public utility; Circular No. 001 of 22 March 1994 to establish the minimum amounts for the sale of national lands; Decree No. 95/678/PM of December 18, 1995 to institute an indicative land use framework; and Law No 85/09 of 4 July 1985 on Expropriation for public utility purposes and the compensation regime.

[58]Law No. 94/001 of 20 January 1994 to lay down forestry, wildlife and fisheries regulations; Decision No. 135/D/MINEF/CAB of 26 November 1999 establishing the procedures for the classification of the forests in the permanent forest areas of the Republic of Cameroon.

[59]Law No. 98/005 of April 14, 1998 on the Water Code; Order No. 2001/165/PM of 08 May 2001 specifying the modalities of protection of surface water and underground water against pollution.

[60]Law No. 2016/017 of 14 December 2016 repealing Law No. 001 of 16 April 2001 establishing the Mining Code and its modification by Law No. 2010/011 of July 29 2010.

operations,[61] petroleum[62] and gas,[63] investments,[64] and other sectoral legislation. The relevant prescriptions for the control of investors that cause soil degradation are fragmented and dispersed across these policy and legal instruments. The relevant provisions prescribe, the mode of access to soil, the manner of dealing with and the use of soil by all land users including investors, demonstrating the strengths and weaknesses of such policy and legal prescriptions. The principal ones are examined in the sub sections that follow.

3.1.2.1 Domestic Policy Prescriptions for the Control of Investors for the Purpose of Soil Protection

At the national level, there is ample evidence of the incorporation into national policy of international principles and concepts on soil protection in most cases with emphasis on investors. In the first place, the strength of the 1996 National Environmental Management Plan (NEMP)[65] is that its five priority areas[66] have important implications for sustainable soil management and protection especially as it provides for the restoration of degraded land and improvement of soil fertility. Sadly, one of its weaknesses is that it does not contain any reference as to the control of investors whose activities impact seriously on soil through the use of chemicals that have the effect of causing soil degradation. Moreover, the NEMP is a policy instrument with only directives that do not impose binding obligations on land users such as investors. It is therefore important to urgently update the NEMP and thereafter

[61] Order No. 002/MINEPDED of October 15, 2012 fixing the specific conditions for the management of industrial waste (toxic and / or hazardous); Decree No. 2012/2809/PM of 26 September 2012 fixing the conditions for the sorting, collection, transport, recovery, recycling, treatment and final disposal of waste.

[62] Law No. 2019/008 of 25 April 2019 establishing the Petroleum Code; Law No. 99/013 of 22 December 1999 to institute the Petroleum Code.

[63] Law No. 2012/006 of 19 April 2012 to institute the Gas Code.

[64] Law No. 2002/004 of 19 April 2002 Instituting the Investment Charter of the Republic of Cameroon; Law No. 2004/020 of 22 July 2004 to Amend certain Provisions of Law No. 2002/4 of 19 April 2002 to Institute the Investment Charter of the Republic of Cameroon; Law No. 2013/0 of April 2013 to Law Down Private Investment Incentives in the Republic of Cameroon; and Law No. 2017/015 of 12 July 2017 to Amend and Supplement Some Provisions of Law No. 2013/0 of April 2013 to Law Down Private Investment Incentives in the Republic of Cameroon.

[65] The NEMP established in March 1996 forms part of the policy framework for environmental protection and its overall objective is to guide the development of strategies and actions for environmental protection and the rational management of resources to contribute to sustainable development. It is a general framework of reference for the various sectoral environmental management actions in Cameroon. See the REDD Desk website available at http://theredddesk.org/countries/plans/national-plan-environmental-management-cameroon, accessed 25 August 2020.

[66] Its five priority areas include participatory land use management, sustainable management of natural resource, restoration of degraded land and improvement of soil fertility, capacity building and concerted management of shared resources at the sub-regional level.

craft an instrument that establishes the obligations of all carrying out activities especially investors likely to degrade the soil to take steps to avoid soil degradation and restore degraded soil.

The Nationally Determined Contributions (NDCs) of Cameroon initially established as the Intended Nationally Determined Contributions (INDCs)[67] in the lead up to the 2015 Paris Climate Change Agreement[68] identifies soil as a critical component in achieving Cameroon's carbon emission reduction pledge. As an important carbon sink which is a critical natural solution in mitigating climate change, the strength of the NDC is that it provides for the promotion and adoption of better soil management practices in order to enhance soil productivity through restoration for instance.[69] The weakness of the NDC regarding soil protection however lies in the fact that it fails to sound a warning to investors regarding the impact of their activities on soil health.

For the time being, the National Action Plan for the Fight Against Desertification (NAP-FAD)[70] is considered a government policy document within the framework of the fight against desertification in general and the protection of soil in particular. The overall goal of the NAP-FAD is to reverse the desertification/degradation trends in order to fight against poverty and encourage sustainable development.[71] One of the major axis of the action plan is based on the restoration of degraded lands and the improvement of soil fertility in order to ensure that the fertility of marginal soil is improved; degraded soils are restored; and techniques of soil restoration mastered and used by the population. Although this is progressive in terms of soil protection, the weakness of this action pan is its failure to expressly establish the land and soil degradation responsibility of investors as major actors of soil degradation.

The National Biodiversity Strategy and Action Plan II (NBSAP II) provides an enhanced vision and a new orientation for interventions to reverse the trend of biodiversity loss in Cameroon.[72] The NBSAP II lends support to soil protection in the sense that soil is an inherent part of biodiversity as well as the major part of its

---

[67] INDCs is an expression used under the UNFCCC for reductions in greenhouse gas emissions that all countries that signed the UNFCCC were asked to publish in the lead up to the 2015 UN Climate Change Conference held in Paris, France in December 2015. On 28 September 2015, Cameroon submitted its INDCs to the Secretariat of the UNFCCC, in which it pledged a 32% reduction in carbon emissions by 2035 compared to business-as-usual levels, taking 2010 as the reference year and conditional upon international support in the form of financing, capacity building and transfer of technology.

[68] Adopted on 15 December 2015, signed in New York on 22 April 2016, entered into force on 04 November 2016.

[69] See the INDC of Cameroon, pp. 4 and 5.

[70] Upon ratifying the UNCCD in 1994, the Government committed itself to working out a NAP-FAD—a coherent framework for the control of desertification. See GEF/ANCR et al. (2007, p. 34).

[71] Republic of Cameroon (2012, p. 43).

[72] Republic of Cameroon (2012, p. 43).

foundation.[73] Like the NAP-FAD, the NBSAP II disappointingly fails to establish the responsibility of investors regarding the degradation of biodiversity especially soil degradation.

The Readiness Plan Idea Note (R-PIN) of 2008 and the REDD+ Readiness Preparation Proposal (R-PP) of 2013, both prepared in view of the national implementation of the REDD+[74] initiative in Cameroon indirectly offer their support for soil protection by identifying some of the common causes of deforestation, forest degradation and soil degradation. A major weakness of these documents is that they fail, like others, to impose responsibility on investors who through mining and large-scale agricultural projects cause deforestation, forest degradation and soil degradation. Cameroon's National REDD+ Strategy of 2018 provides for the implementation of a program on landscape restoration and management for climate change resilience in the Northern Regions through the restoration of degraded soil by setting up private plantations for the production of wood energy; regaining soil fertility and water catchment protection through agroforestry techniques.[75] One of the recommendations of the Strategy is regaining soil fertility which is critical for soil protection. The Strategy is however silent regarding the role of investors in the restoration of degraded soil and regaining their fertility.

The 1995 Indicative Land Use Framework or Zoning Plan[76] is relevant to soil protection as it makes provisions for the permanent forest estates (PFEs)[77] which are meant to remain forested such as protected areas which are relevant for the protection of the soil covered by such areas.[78] The Indicative Land Use Framework is believed to be inaccurate because of outdated information and poor consultation with the local

[73] Hannam and Boer (2004, p. 1).

[74] REDD+ stands for reducing emissions from deforestation and forest degradation and the role of conservation, sustainable management of forests and enhancement of forest carbon stocks in developing countries. The initiative has been conceptualized and operationalized as a PES scheme to reduce the conversion of forests to alternative land uses within the framework of the implementation of the UNFCCC with the objective of mitigating climate change through reducing net emissions of GHGs through enhanced forest management in developing countries. See the REDD Desk website at https://theredddesk.org/what-redd.

[75] Cameroon's National Strategy for reducing emissions from deforestation and forest degradation, sustainable management of forests, conservation of forest and enhancement of carbon stocks, 2018.

[76] The indicative land use framework or zoning plan developed and adopted in 1995, aimed at the distribution of land use. As land or soil constitutes a multiple-use resource requiring the harmonisation or coordination of users' interests, which are sometimes environmentally unsustainable and detrimental to soil quality.

[77] See Articles 1(2) and 6(2) of Decree No. 95/678/PM of December 18, 1995 to institute an indicative land use framework. In addition to Decree No 95/678, there are additional texts addressing zoning and land tenure, including: Decision No. 135/D/MINEF/CAB of 26 November 1999 establishing the procedures for the classification of the forests in the permanent forest areas of the Republic of Cameroon; Ordinances No. 74/2 and No. 74/1.

[78] See the REDD desk Website, available at http://theredddesk.org/countries/policies/indicative-land-use-framework-cameroon, accessed on 25 August 2020.

communities.[79] This is probably why some investors have been engaged in conflicts with local communities for taking over and degrading their lands and their attendant biodiversity, including soil. There is therefore the urgent need to put in place a robust land use plan that clearly apportions and indicates the different land uses in the country and the corresponding responsibility of all land users especially investors whose activities are prone to degrading the soil.

Other important policy documents which are not only relevant for soil protection in Cameroon, but promote substantial investments on land likely to cause soil degradation, are the 2035 Vision, the Poverty Reduction Strategy Papers (PRSPs) 2003–2010 and the GESP 2010–2020. The weakness of these policy documents regarding soil protection is that they equally fail to establish the responsibility of investors to protect soil notwithstanding the enormous pressure that Cameroon's vision of becoming an emerging country will put on soil. In fact, all the above policy documents, which have a strong focus on ensuring the modernisation of the agricultural, extractive and industrial sectors place a huge burden on soil without establishing the corresponding responsibility of investors of these sectors to protect soil and maintain its health which is the foundation of sustainable development. The domestic policy frameworks have their strengths, but equally their weaknesses in terms of soil protection in Cameroon.

3.1.2.2 Domestic Legal Prescriptions for the Control of Investors for the Purpose of Soil Protection

Relevant national legal provisions for soil protection and the control of investors can be found in various legal sources, including, the investment laws, the Constitution, the Framework Law on Environmental Management, ESIA laws, land tenure law, natural resources laws, water management law, investments laws, etc. In conformity with the Investment Charter[80] of the Central African Economic and Monetary Community (CEMAC), Cameroon like other Member States is required to enhance its legal environment to attract investors and investments. The fundamental investment laws are also oriented towards the promotion of the overall sustainable development strategy aimed at increased and sustainable economic growth, job creation in all sectors of the economy and social well-being of the people.[81] This in effect creates an enabling legal environment for investors especially foreign investors who are desirous of investing in Cameroon or who are already operating in the country. In this regard, the bulk of the Cameroon investment law establishes certain rights, liberties, obligations and powers for investors. With respect to the investor's obligation to protect the soil, the Investment Charter provides for "the

[79] See Forest Legality Initiative Website, available at http://www.forestlegality.org/risk-tool/country/cameroon-0, accessed on 25 August 2020.

[80] Regulation No. 17/99/CEMAC – 020-CM-03 Relating to the CEMAC investment Charter.

[81] Section 6 of the Cameroonian Investment Charter.

safeguard of the ecological environment and the rational exploitation of natural resources of the soil and sub-soil, for sound and sustainable development".[82] A similar obligation is also echoed in the law on private investment incentives,[83] though in an ambiguous manner. Another provision[84] of the law also incentivizes investors by way of tax credit if they invest in combating pollution which is critical for soil protection. This later provision may however produce little or no positive impact in terms of protecting the soil since it does not specifically encourage investors to invest in protecting soil.

As the *Grundnorm*, the 1996 Cameroonian Constitution addresses important national issues in a very general manner leaving the details to be addressed by specific legislation. Surprisingly, the drafters[85] of the 1996 Constitution failed to specifically envisage and address the concern of soil protection which is the cornerstone that supports all forms of life on Earth. However, in the absence of a specific Constitutional provision dealing with soil protection in Cameroon, reference to sustainable management of natural resources,[86] to the right to a healthy environment and to the protection of the environment[87] may be interpreted to relate equally to soil. Such an analogy is considered reasonable as soil constitutes the springboard of all natural resources and environmental protection. This analogy notwithstanding, it may be apt to argue that soil being a crucial multi-purpose resource, requiring to be expressly articulated in the Constitution. Moreover, the Constitution addresses the issues of environmental protection and sustainable management of natural resources in a rather sketchy manner and in a less conventional way, that is, in the Preamble. One expects that such crucial issues which have implications for soil protection ought to be given special attention and therefore be addressed in specific articles, as is the case in the Constitution of other countries[88] and not merely be mentioned as declarative statements in the Preamble. That notwithstanding, Article 65 of the Constitution unambiguously makes the Preamble part and parcel of the Constitution thereby giving it the force of law and making its provisions implementable. The violation of such prescriptions is tantamount to a violation of the Constitution for which an action can lie. One therefore expects that legislation inspired by the Constitution will deal with the question of soil in more details, including in particular the responsibility of investors.

---

[82] Section 2 of the Cameroonian Investment Charter.

[83] See Section 1(2).

[84] See Section 8 (new) (1) of Law No. 2017/015 of 12 July 2017 to Amend and Supplement Some Provisions of Law No. 2013/0 of April 2013 to Law Down Private Investment Incentives in the Republic of Cameroon.

[85] Legislator.

[86] Paragraph 3 of the Preamble and Article 26(2)(d)(5) of Law No 96/06 of January 20 1996 as amended.

[87] Paragraph 26 of the Preamble of the Constitution.

[88] See the Constitution of Kenya, Articles 69, 10, 60, 22.

The protection of soil and subsoil is envisaged first and foremost in the 1996 Framework Law on Environmental Management. The relevant soil protection provision[89] which investors are expected to comply with provides that *(1) the soil and subsoil as well as the limited renewable or non-renewable contained therein, shall be protected against any form of degradation and jointly managed rationally by the competent administrations; and in (2) an enabling decree of this law, prepared in collaboration with the administrative units concerned, shall lay down: the specific conditions for the protection and fight against desertification, erosion, loss of arable land and pollution of the soil and its resources by chemicals, pesticides and fertilizers; the list of fertilizers, pesticides and other chemical substances whose use shall be authorised or encouraged in agriculture; the authorised quantities and the terms and conditions for their use so that the substances do not endanger soil quality or other receptor environments.* We hail the fact that the enabling instrument of this specific provision of law is in force. Unfortunately, the effective implementation of the Law and the Decree like many other instruments is still problematic as some unauthorised chemicals[90] in the form of pesticides, herbicides and fertilizers are still used for agriculture and mercury in mining, thereby degrading the soil and subsoil by way of pollution. There is therefore the need to enhance the implementation of the law against investors and other users of unauthorised chemicals and pesticides.

Another relevant soil protection provision[91] of the Framework Law which investors are expected to comply with requires holders of a mining or quarrying permit to rehabilitate exploited mining sites. There is good reason to commend the intentions of the legislator regarding this provision as soil is usually left exposed after mining activities have taken place and therefore rehabilitation measures need to be applied. Regrettably, delegated legislators have delayed the legislative drafting process as the enabling instrument for the application of this provision is yet to be promulgated in spite of the fact that mining activities continue to take place, exposing the soils to pollution and erosion. In the absence of such an enabling instrument, the implementation of the law becomes practically difficult if not impossible. This legal shortcoming could be availed by drafting the enabling instruments at the same time with the main law or drafting a complete law detailed enough for understanding and application.

---

[89] Article 36 (1) and (2).

[90] This is disturbing as the Community Regulation of CEMAC has harmonized the importation of chemicals and fertilizers. Community Regulation on pesticides and fertilizers—the Common Provisions on the Homologation of Pesticides in the CEMAC Zone sets out to bring together experiences and expertise of the Member States for the evaluation and homologation of pesticides in order to ensure their rational and judicious use, as well as the protection of human health and the environment; and the experimenting, authorisation, importation, exportation, transportation, transit, stocking, marketing, use, control and elimination of active materials and formulated products of pesticides within the states of the CEMAC zone.

[91] Article 37.

Other relevant aspects of the Framework Law that offer support to soil protection which investors are expected to comply with include the provisions dealing with: the allotment and management of land for agricultural, industrial, urban or other uses, as well as prospecting, research or exploitation of subsoil resources likely to endanger the environment;[92] the protection of nature, the preservation of animals and plant species and their habitats, the maintenance of biological balances and ecosystems and the conservation of biodiversity and genetic diversity against all causes of degradation;[93] the rational management of natural resources;[94] the sustainable use of biodiversity;[95] the exploitation of biological and genetic resources;[96] the recognition and protection of historic, archaeological and scientific sites as well as sites with special panoramic beauty;[97] the rational and ecological exploitation of mining resources and quarries in line with environmental considerations;[98] and the protection of land against erosion and prevention and fighting against desertification.[99] Soil forms the bed rock of nature, natural resources, and biodiversity. A reading of the above provisions relating to these aspects suggests that sufficient measures are in place to protect soil. The above provisions may lead one to safely conclude that the 1996 Framework Law adequately provides for soil protection both directly and indirectly albeit with some legal hitches by way of the unavailability of some enabling instruments to ease the implementation of the relevant soil protection provisions as identified above.

The most relevant and specific soil protection piece of legislation in Cameroon is Decree No. 2011/2584/PM of 23 August 2011 on the Modalities of Protection of Soil and Subsoil. It is one of the enabling instruments of the 1996 Framework Law on Environmental Management in matters of soil protection. The Decree stipulates that: "any activity relating to the exploitation of the soil must be done in a manner as to avoid or reduce erosion of soils and desertification."[100] Worthy of note is the fact that this Decree specifically takes into account land that is suitable for agricultural activities. To this effect, the protection of arable land is envisaged in Article 5 which forbids any activity which degrades or modifies the quality and/or the structure of arable lands or contributes to the loss of such lands. The Decree[101] prohibits the exploitation of zones with high risk of erosion. The Decree[102] also prohibits any activity that degrades or modifies the quality and/or structure of arable

[92] Article 38.
[93] Article 62.
[94] Article 63.
[95] Article 64.
[96] Article 65.
[97] Article 66.
[98] Article 67.
[99] Article 68.
[100] Article 3.
[101] Article 4(1).
[102] Article 4(2).

lands or which contributes to the loss of these lands. The Administration in charge of Environment is tasked with overseeing the implementation of this legal text by ensuring its respect as need be, and impose administrative sanctions in cases of violation. On the issue of environmental impact assessment (EIA) which conditions the realisation of all development projects on the conduct of EIA as envisaged in the 1996 Framework Law on Environmental Management, the Decree has retaken the provisions in order to make this prescription emphatic for a better protection regime for soil and subsoil. This is what seems to be the purport of Article 9 which is to the effect that "all physical or moral persons, private or public who undertake agricultural exploitation and practice intensive use of fertilizers and/or pesticides or apparatuses that condition the soil, are obliged to carry out in a regular manner an evaluation of their impacts on the environment in accordance with regulations in force" and Article 10 which provides that "all physical or moral persons desirous to manufacture fertilizers and/or pesticides on the national territory are obliged to carry out environmental and social impact assessment in conformity with regulations in force".

From the foregoing provisions, one can safely assert that the Decree is comprehensive enough with regards to the protection of soil and subsoil. This is even more plausible given the fact that the Decree makes ESIA mandatory for any project likely to cause harm to soil and the use of the word physical or moral person points to the activities of investors. The challenge remains that of compliance with the Decree by investors and effective implementation and enforcement by the relevant administrations. Other soil protection provisions requiring compliance by investors are scattered across sectoral legislations mostly dealing with environmental protection.

Regarding investors in the mining sector, the point has been highlighted above, namely that, although the mining legislation in Cameroon addresses soil degradation and provides for soil protection, it also contains some gaps that encourage soil degradation. The relevant provisions of mining law imposing responsibility on investors who in this context are the multinational investors in the majority to protect the soil seem to be the following: Land owners are entitled to compensation stemming from, among others things, damage caused to the natural surface of the land.[103] The mining law also subjects any mining and quarry operation to comply with the laws and regulations in force relating to sustainable environmental protection and management; and subjects the granting of all mining and quarry permits and titles apart from the non-industrial mining license, to the prior conduct of an ESIA, a hazard and risk assessment and provision of an environmental management plan as provided for by the laws and regulations in force in matters relating to the protection and sustainable management of the environment.[104] A further provision[105] of the mining law provides for the restoration, rehabilitation and closure of mining and quarry sites. The mining law also requires holders of mining and quarry titles—

[103] Article 123.

[104] Article 135(1) and (2).

[105] Article 136.

multinational investors in the majority—to prevent hazards and disasters, to prevent or minimise the discharge of waste in the open, to protect fauna and flora, to promote or maintain the general health of the population, to reduce waste, to dispose non-recycled waste in such manner as to ensure safety of the environment, after informing and receiving the approval of the authorities in charge of mining and the environment, and to manage waste in accordance with the laws and regulations in force.[106] The mining law further requires mining investors to ensure the rehabilitation and closure of mining and quarry sites and proper management of waste to ensure safety of the environment; reduce waste; prevent geo-hazards and geo-disasters and also requires holders of mining or quarry permits to dismantle, in accordance with standard rules, any mining plant liable to payment of duties and taxes.[107] Another important provision[108] of the mining law requires buildings, outbuildings, wells, galleries and all structures in general built and used for the mining or quarry activities to be secured, in accordance with conditions set out in the Environmental and Social Management Plan (ESMP) and the mining sites rehabilitation programs at the end of mining or quarry activities. The mining law also makes provisions for the monitoring or supervision and control of mining activities.[109]

By imposing obligations on mining investors, the above provisions of the mining law demonstrate the good intentions of the legislator to maintain the health of the soil either directly or through the protection of man and the environment. Notwithstanding the strength of the Mining Code regarding soil protection, it also has a weakness that acts as a potential driver of soil degradation. The weakness lies in the fact that the Code is not being applied due to the absence of an enabling instrument which is aimed at enhancing the effective implementation of this law. For example, Article 199 provides that the conditions for supervising and controlling mining, quarry, and spring water, mineral and thermo-mineral water tapping activities, as well as the mining of geothermal deposits, and appointing inspectors and assistant inspectors of the ministry in charge of mines shall be laid down by regulation. Such regulatory instruments are yet to be promulgated in Cameroon. This is one of the challenges in the legislative drafting process in Cameroon, especially regarding natural resources and environmental management in general and soil protection in particular. A possible explanation for this pitfall is that the Executive Power of the State which is in charge of preparing and issuing subsidiary legislation (regulatory instruments) sometimes deliberately delays the crafting and issuance of regulatory Acts in order to frustrate the implementation of laws passed by Parliament some of which provisions may work against the Government agenda or it may be the usual Government inertia. Again, the making of the law could have been a condition from international monetary institutions like the World Bank or the International Monetary Fund for a negotiation of much needed loan for development projects in this area. The

[106] Article 137.

[107] Article 138.

[108] Article 140.

[109] Article 196.

non-application of the Mining Code for want of an enabling instrument could be remedied as suggested by Tamasang and others,[110] by either enacting a complete and detailed law that does not require an enabling instrument or the law and its enabling instrument could be drafted at the same time so as to facilitate the understanding and the immediate implementation of the law.

Concerning the regulation of investors in industrial development regarding the impacts of such development on soil, it is good to know that Cameroon at present is characterised by a low level of industrialisation. However, with the Government's ambitious goal of becoming an emerging economy by 2035, industrial development has a major role to play in achieving this goal as outlined in the Vision 2035 and the GESP that serve as the country's compass for emergence by 2035. The government's ambition may mean increased pressure on soils. Cameroon has elaborated and validated its National Industrialization Plan aiming at translating into practical terms Cameroon's ambition of becoming an emerging economy by 2035. The Plan focuses on three key areas referred to as "industrial sanctuaries" consisting of Agro-industry; Energy and Digital and five industrial pillars consisting of Forest and Timber; Textile/Confection/Leather; Mining/Metallurgy/Iron and steel industry; Hydrocarbons/Petrochemicals/Refining; and Chemical/Pharmacy. These targets all have implications for soil degradation although to varying degrees of intensity. For instance, the agro-industry makes use of chemical inputs such as fertilizers, herbicides and pesticides, but also tillage and deforestation with serious implications on the quality of soil and its attendance biodiversity in the regions where they are applied. Also, mining involves the use of chemical products, such as mercury, destroys the attendant biodiversity of the sites including soil biodiversity. Again, Hydrocarbons, Petrochemicals, and Refining processes, when recklessly managed, can cause severe damage to soil in the regions where such processes take place. This therefore, requires strong regulation with a particular objective to protect the soil. At the moment, there is no specific legislation governing the industrial sector in Cameroon that protects the soil. The sector for the time being is regulated through a plethora of laws and regulatory instruments governing other sectors. The relevant soil protection provisions of sectoral legislations governing industrial developments and wastes include:

The 1996 Framework Law on Environmental Management regulates the treatment of special industrial waste which, by reason of its properties, is dangerous and may not be deposited in storage facilities receiving other categories of waste.[111] The same law provides for the regulation of classified establishments, factories, workshops, depots, building sites and, in a general way, the industrial, craft or commercial installations that may be dangerous to health, safety, public health, agriculture, nature and the environment in general, or disadvantages for the convenience of the

[110] See Tamasang et al. (2019, p. 80).

[111] Article 47(3).

neighbourhood.[112] In the same spirit, the 1994 Forestry Law[113] forbids the dumping of any toxic product or industrial waste in national forests as well as in public waterways, in lakes and in the sea without obtaining the prior permit of the government, and their proper treatment. An enabling instrument[114] to the 1996 law calls on every generator and/or operator of the field of industrial waste (toxic and/or dangerous) to provide at the end of the semester to the administrations in charge of the environment, a declaration containing a summary of the information of different manifestos. The same instrument[115] requires any operator of an installation that annually generates more than two tons of such waste to provide the administration responsible for the environment with a management plan for such waste. Every carrier of such waste is obliged to use a manifest of traceability of the waste according to the form in force.[116] To reinforce and ease the practical implementation of the foregoing provisions, sworn agents of the administration in charge of the environment are urged to prescribe samples and analyses to check the conformity of the loading to the manifest.[117] The requirements to be met by the generator or shipper of such waste before and during transportation and the conditions under which such waste is transported are also spelled out in the enabling instrument.[118] The enabling instrument is quite firm when it provides that any breach of one or more of the obligations, incurs the responsibility of the operator involved.[119] These are all relevant to soil protection from the perspective of investors' role in the sense that strict observance of the provisions will prevent the discharge of toxic and/or dangerous waste into the environment, which lends support to soil protection.

The Prime Ministerial Order[120] fixing the conditions for the sorting, collection, transport, recovery, recycling, treatment and final disposal of waste also applies to industrial waste. The Order subjects the collection, transportion and storage of industrial waste (toxic and/or dangerous) to obtaining an environmental permit issued by the administration in charge of the environment.[121] The Order also requires the transport of such waste to be accompanied by a manifest of traceability of the waste issued by the administration in charge of the environment.[122] The Order further requires every generator, collector, transporter or destroyer of industrial

[112] Article 54.

[113] Article 18.

[114] Order No. 002/MINEPDED of October 15, 2012 fixing the specific conditions for the management of industrial waste (toxic and/or hazardous). Article 2(1).

[115] Article 3.

[116] Article 4.

[117] Article 5.

[118] Articles 6 to 9.

[119] Article 10.

[120] Decree No. 2012/2809/PM of 26 September 2012 fixing the conditions for the sorting, collection, transport, recovery, recycling, treatment and final disposal of waste.

[121] Article 9.

[122] Article 10.

waste (toxic and/or dangerous) to keep a register in which it records the type, nature, quantity, characteristics of danger and origin of the hazardous waste which it has produced, collected, stored, transported, recovered or disposed of.[123] This register is subject to the control of the administration in charge of the environment.

Despite the existence of these regulatory instruments which indirectly govern the industrial sector, one still finds effluent or liquid and gaseous discharge of industrial origin, treated or untreated discharged directly or indirectly into the environment in some major cities such as Douala, which may be harmful to soil health. For instance, when toxic and/or dangerous waste is discharged into the environment, it may cause soil hazards as the waste can be washed into soils and rivers thereby killing living organisms that have the effects of destroying the soil fertility and its attendance biodiversity. In the absence of a comprehensive policy and regulation that balance industrial developments with the exigencies to protect soils; such developments and spillages may have negative impacts on the environment causing soil degradation. There is therefore the need to develop an appropriate legal instrument that will govern industrial development in Cameroon. Another way of availing this legal gap is for the government to mainstream soil concerns in a comprehensive manner in such sectoral legislations.

In the domain of investment in the water sector, the Water Code and its enabling instrument have envisaged a number of investment related activities that could impact negatively on soil health by destroying the soil fertility and its attendance biodiversity. For instance, the Code[124] prohibits the spills, discharges, jets, infiltrations, landfills, spreading, direct or indirect deposits in the waters of any solid, liquid or gaseous material and, in particular, industrial, agricultural and atomic wastes that are liable to: alter the quality of surface or groundwater or seawater within territorial boundaries; harm public health and aquatic and sub-marine fauna and flora; and jeopardise the economic and tourist development of the regions. The operationalization and effective implementation of the Water Code in relation to soil protection from an investor's perspective are enabled by a Prime Ministerial Order.[125] Consequently, the Water Code addresses soil related concerns probably not very directly but by implication as soil is not expressly mentioned. The danger of such implied prescriptions is that their application may be completely ignored or, if considered, may fall to be interpreted depending on the whims and caprices of those who are called upon to apply the law. And of course knowledge and interest on the subject matter (soil protection) are important considerations for an interpretation in favour of soil protection, especially from an investor's perspective. However, as observed by Tamasang and others,[126] what matters here is that the water legislator has demonstrated the importance of soil in the water management regime given that

[123] Article 11.

[124] Article 4(1).

[125] Order No. 2001/165/PM of 08 May 2001 specifying the modalities of protection of surface water and underground water against pollution.

[126] See Tamasang et al. (2019, p. 53).

water has its source from the soil. The Water Code and its enabling instruments therefore seem to have captured all the implications that the exploitation of water may have on soil. Thus, since water has its source from the soils, sustainable exploitation of water by implication means the protection of soil.

The current Petroleum Code also takes on board concerns relating to soil protection. For example, it[127] requires holders of petroleum permit to carry out petroleum operations in such a manner as to ensure, under all circumstances, the conservation of natural resources, in particular hydrocarbon deposits, and due protection of essential features of the environment, including soil. In this respect, investors with petroleum permits are required to take all the necessary measures to preserve the safety and protection of the environment and ecosystems including soil. The Code[128] also requires holders of petroleum contract which are of course, investors, to carry out ESIA with respect to the laws and regulations applicable in matters of environmental protection in order to evaluate the direct or indirect impacts of petroleum operations on the ecological balance of the surrounding areas, the people's living environment and quality of life and the environment in general. Compared to the repealed 1999 Petroleum Code,[129] the present code does not contain innovations in terms of a soil protection paradigm as its provisions have no direct reference to soil apart from general environmental protection provisions. One expects that the new code to contain clear provisions imposing responsibility on investors in the sector to protect soil. The importance of soil is therefore not well captured by the new Petroleum Code. Another main shortcoming is, as others, the lack of enabling instruments which would probably have articulated in greater details which could touch on the protection of soil.

Furthermore, within the framework of environmental protection by investors, the Gas Code[130] indirectly protects the soil from all related gas activities by obliging the operator to comply with environmental protection and safety regulations in force, as well as with internationally accepted environmental protection and safety standards. Another gas law[131] equally provides an indirect protection of soils within the context of environmental and ecosystem protection.

Finally, perusing through the fragmented and selective pieces of Cameroonian land tenure legislation, one can hardly find any express anchor point for soil protection and more specifically a provision that imputes responsibility on investors to protect soil. While there is ample evidence of national policy and legal prescriptions about soil protection, the 1996 Framework Law on Environmental Management has gone a step further to put an assent on the application of international Conventions duly ratified by Cameroon in keeping with Article 45 of the

[127] Article 91.

[128] Article 92(1).

[129] Law No. 99/013 of 22 December 1999 to institute the petroleum code. See Articles 82 and 83 directly addressing soil protection.

[130] Article 30.

[131] Article 11 of Law No. 2011/25 of 14 December 2011 on the valorisation of associated gases.

Constitution. Cameroon has actually ratified all, if not most, of the relevant international conventions on soil and soil related issues. The country is in the implementation phase of national legislation and international conventions, but with obvious challenges which could be overcome.

3.2 Enabling Institutional Environment

Institutions constitute part of the soil protection regulatory mechanism in Cameroon as they are responsible for the effective implementation and enforcement of the law. The relevant institutions are considered in the paragraphs that follow. Institutions are tasked with overseeing the implementation and enforcement of law by ensuring its respect, and impose administrative and judicial sanctions in cases of violation.

The first is the Ministry of State Property, Surveys and Land Tenure (MINDCAF). Since the drivers of soil degradation such as mining, agricultural projects and industrial operations begin with the acquisition of land, the role of MINDCAF can be crucial for soil protection for the simple reason that land is the primary factor of investment. According to Article 1 of Decree No. 2012/390 of 18 September 2012 organising MINDCAF, it is responsible for elaborating and implementation of government policy related to land tenure, land registration and land survey. Such policy includes the elaboration of legislative and regulatory texts in its different areas of competence, management of the government's private and public property, make propositions for the attribution of State lands, elaboration and conservation of cadastral plans, etc. In this regard, MINDCAF can make the acquisition of private property of the state or private lands by investors subject to undertakings to the implementation of soil protection measures.

The second is the Ministry of Agriculture and Rural Development (MINADER). Given that poor agricultural practices can result in soil degradation such excessive use of chemical inputs, the role of this Ministry is critical for the control of investors for the purpose of soil protection. In this regard, MINADER is responsible for elaborating, implementing and evaluating government policies in agriculture and rural development.[132] MINADER also has the mandate of developing standards and rules applicable to the agricultural sector and identifying and preparing projects and programs for investments in the sector. This mandate therefore confers on MINADER the power to control the activities of investors in large-scale agricultural projects with positive implications for soil protection.

The third is the Ministry of Environment, Protection of Nature & Sustainable Development (MINEPDED). By its organigramme of 2012,[133] MINEPDED has as

[132] Decree No. 2005/118 of 15 April 2005 organizing the Ministry of Agriculture and Rural Development.

[133] See Decree No. 2012/431 of 1 October 2012 creating the Ministry of Environment, Protection of Nature & Sustainable Development.

principal missions *inter alia*: the definition of modalities and principles for the sustainable management of natural resources; monitoring of environmental compliance in the implementation of large projects; information and sensitisation of the public to incite their participation in the management, protection and restoration of the environment and nature; negotiation and implementation of international conventions and treaties related to the protection of environment and nature in liaison with the Ministry of External Relations; and others.[134] Since soil degradation is a sub set of environmental degradation, MINEPDED ensures that ESIA, Strategic Environmental Evaluation and others are properly conducted. MINEPDED is the co-ordinator Ministerial department for monitoring the respect of environmental commitments and meting out administrative sanctions to project implementers violating environmental standards. The role of MINEPDED can therefore not be overstated in matters of soil protection and engaging the responsibility of investors to avoid, reduce and reverse soil degradation.

The Ministry of Economy, Planning and Regional Development (MINEPAT) is in charge of developing, monitoring and coordination of the government's economic policy; of the elaboration of norms and rules of regional planning and control of their application.[135] By dint of these roles, MINEPAT is another institution that can implement regulations to prevent investors such as industrial operators from carrying out activities that degrade soil by establishing and implementing industrial and economic activities in a manner that includes soil conservation practices.

The mission of the National Consultative Commission for the Environment and Sustainable Development created by the Framework Law on Environmental Management is also critical for soil protection. The Commission has the mission of ensuring a wide participation of all stakeholders in the development and implementation of environmental management policies and strategies. Such participation can create an avenue for concerned actors or advocates of soil protection to bring pressure to bear on investors to comply with soil protection measures, but also on implementation and enforcement authorities to be more proactive in their functions.

Although an Inter-ministerial Committee in charge of Environmental Protection exists in Cameroon, its organisation and functioning regarding soil protection is weak. The competencies are shared by several ministerial departments that face a substantial under-staffing which hinders the implementation and enforcement of measures to protect the soil across sectors. There is therefore, the need to further develop its implementation and enforcement mechanisms in relation to soil protection.

---

[134] Other missions include: the definition of environmental management measures in collaboration with other ministries and concerned specialised institutions; the elaboration of sectoral master plans for the protection of the environment in liaison with other concerned ministries; coordination and monitoring of all interventions by regional and international organizations in the areas of environmental and natural resources in liaison with the Ministry of External Relations and concerned administrative bodies.

[135] Article 1(2) of Decree No 2008/220 of 4 July 2008 relating to the organization and functioning of MINEPAT.

The role of the courts and other administrative review procedures for lodging complaints related to soil degradation is crucial for soil protection. Adequate access to justice is a critical component for effective or sustainable environmental management and consequently, soil protection. Poor judicial mechanism can undermine soil protection efforts. There is an absence of an environmental law court in Cameroon or at least a chamber in each court level that deals with environmental matters and consequently with soil issues. Generally, environmental justice is not well institutionalised in Cameroon. Judges and other court officials are not well informed about the importance of soil protection. Any proper environmental justice system must be widely known, accessible, and cheaper and should be able to be able to provide timely results.[136] Avenues such as judicial forums, administrative procedures, traditional authorities and arbitration guaranteed under relevant environmental, natural resources and related laws,[137] are all available for settling disputes, but are ill-adapted for enforcing soil protection provisions. These avenues together with their personnel need to be strengthened for proper enforcement of soil protection provisions. Under international law,[138] governments have a number of duties and obligations to ensure adequate access to the essential components of access to justice, namely: access to effective judicial proceedings, or the absence of economic obstacles preventing citizens from accessing the courts.[139] Access to an appropriate environmental justice mechanism increases the accountability of investors or project implementers and public authorities as it provides an opportunity to challenge illegal acts of public authorities and investors that violate soil protection law. Such a mechanism needs to be transparent, independent and accountable and allow for concerned or soil protection advocates to institute actions against investors for violating soil protection law. Measures that can be taken to enable access to justice regarding soil protection include *inter alia*:

- legal standing or clear rules that enable soil protection advocates to initiate litigations to enforce soil protection related legal provisions and to enshrine such important requirement in the Constitution;
- developing programs to raise awareness among the general public of extended *locus standi* right to initiate actions to enforce the violation of soil protection provisions; and

[136] See Costenbader (2009, p. 112); Chapman and Wilder (2014, p. 28).

[137] The 1996 Framework Law on Environmental Management, 1994 Forestry Law, 1998 Water Law, 2003 Biosafety Law and the 1974 Land Tenure laws.

[138] Reference to the right of access to justice under international law includes: Principle 10 of the Rio Declaration on Environment and Development; Articles 2, 9, 14, 26 and 50 of the International Convention on Civil and Political Rights and its First Protocol; Articles 8, 11, 13, 20, 28, 32 and 40 of the UN Declaration on the Rights of Indigenous Peoples; Article 15 of the Convention for the Elimination of All Forms of Discrimination Against Women; Articles 9, 12 and 14 of the ILO Convention No. 169; Articles 7, 8 and 10 of the Universal Declaration of Human Rights; Article XVI of the 2003 African Convention on the Conservation of Nature and Natural Resources.

[139] Denier et al. (2014, p. 129).

- providing legal aid and services to advocates of soil protection which could include exemption for costs, holding judicial proceedings in local languages or engaging the services of interpreters, etc.

Procedural rights such as full and effective public participation, access to information and access to justice are keys to ensuring and enhancing sustainable soil management and protection.[140]

From the above, it is advocated that the institutional environment for soil protection in Cameroon has its strength as well as its weaknesses. Thus, as laws have their strength and gaps or challenges regarding the control of investors for the purpose of soil protection so too are the institutions.

4 Opportunities for, and Challenges to the Implementation of Soil Protection Regulations: Investors' Perspectives

Cameroon has in place an arsenal of policy and legal instruments constituting of an enabling legal environment that lends support to the protection of soil, including an enabling institutional structure for implementation and enforcement, albeit with some challenges with respect to investors that must be addressed for soil to be effectively protected in the country. The legal instruments also offer opportunities for investors that range from tax concessions through compromise arrangements, compensation awards to arbitration. These opportunities and challenges are considered in the sub sections that follow.

4.1 Opportunities

4.1.1 Tax Concessions: Exoneration from Custom Duties

An opportunity for investors to protect soils is afforded by the Investment Charter, by the application, using tax credits, of a consistent mechanism tailored to foster for instance, the protection of the natural environment especially the soil.[141] Such an opportunity is also afforded by the 2013 law to lay down Private Investment Incentives in the Republic of Cameroon[142] which offers to extend the benefit of some tax and customs exonerations to investors whose projects have been duly assessed to be of some importance to sustain development. Investors who invest in projects for the restoration of degraded land for instance, can benefit from such tax and custom exonerations as such projects are critical for sustainable soil

[140] Tamasang et al. (2019, p. 64).

[141] Section 34.

[142] Section 11.

management. The 2017 amendment[143] to the 2013 law also incentivizes investors by way of tax credit if they invest in combating pollution which is critical for soil protection.

4.1.2 Compensation Award (Annual Prizes) or Tax Rebate for Green Investment

Annual compensation awards or prizes for green investment or tax rebate constitutes yet another opportunity available for investors to carry out sustainable soil management practices. For this to happen, the investor must produce a report of the green investment that have been carried out in the company and such report must be verified by the Environmental Ministry through investigation and control conducted by the Environmental Inspectors who are sworn officials and conferred the status of Judicial Police Officers with special competence.[144] The investigation and control referred above to could include eliciting information from communities peripheral to the company's activities to affirm or disaffirm the reporting of its activities. The media is also be a useful source for the Environmental Inspectors to lead them to the conclusion that the company has been reporting.

4.1.3 Compromise (Transaction) and Arbitration

Compromise is another opportunity available to investors that may offer help to soil protection. Under the relevant legal provision,[145] compromise applies at the initiative of the investor. It is an out-of-court-arrangement offering an opportunity to a defaulter (investor) who has been caught in violation of the law, to pay a minimum amount in sanction for an environmental harm (such as soil degradation) caused by the said defaulter or his employee within the scope of vicarious liability. The law however emphasizes that the amount payable for such a compromise should not be less than the minimum amount of fine prescribed by law. Unfortunately, in practice, this is usually not respected by the administration in charge of the environment.

---

[143] See Section 8 (new) (1) of Law No. 2017/015 of 12 July 2017 to Amend and Supplement Some Provisions of Law No. 2013/0 of April 2013 to Law Down Private Investment Incentives in the Republic of Cameroon.

[144] Judicial Police Officers with special competence are entitled to handle issues specifically related to their respective areas of competence. Pursuant to Article 80 of the CPC, they are civil servants with the prerogatives of JPOs recognised by special texts. They exercise these prerogatives under the conditions and restrictions defined by these texts. They are staff of certain departments (e.g. the Ministry of Environment and Nature Protection, the Ministry of Forestry and Wild Life, the Custom department, etc.

[145] See Article 91 of the 1996 Framework Law on Environmental Management.

The law[146] also allows parties to an environmental dispute the choice to settle such a dispute by a joint agreement reached through arbitration. The advantage of both processes is that they save the investor of the cumbersome, expensive and lengthy court procedures.

4.2 *Challenges*

4.2.1 The Challenge of Compliance, Implementation and Enforcement of Soil Protection Related Provisions

Within the framework of soil protection, some of the challenges regarding investors include; inadequate compliance with the law on soil protection by investors and ineffective implementation and enforcement of soil protection provisions. Investors in the agricultural, mining and industrial sectors do not seem to comply with the relevant soil protection provisions identified in the above discussion especially those requiring the conduct of ESIA. Again, the administration in charge of the environment control project implementation in order to ensure compliance by investors or project promoters with the law in force (commonly referred to in Cameroon as *cahier de charge*). The first challenge in this direction is that the controls are not regular. The second challenge is that such controls are usually tainted with corrupt practices as the promoter may offer the payment of bribe to the environmental controllers/inspectors thereby making them to be biased in their reports. This, as a matter of course, does not augur well for soil protection. Furthermore, where a report identifies violators, the administrative or judicial sanctions meted out are usually not sufficient enough to have a deterrent effect on violators. Most of the cases that go to court usually take a lot of time and most often, the sanctions are not in favour of the administration for at least two reasons. The first one is that judges are not sufficiently schooled on this relatively grey area of the law and secondly, even where they are, some will yield to corrupt practices from all powerful investors whether multinational or national corporations.

Again, legal provisions requiring users or investors practicing intensive use of fertilizers and/or pesticides or apparatuses that degrade the soil seem not very much practical because if read as it stands, all user of pesticides would have to conduct an ESIA. This seems not to be clear. However, the Decree on ESIA specifically mentions "intensive use" which may be interpreted to mean that users who do not practice intensive use are not required to carry out ESIA. Nevertheless, their activities may still lead to soil degradation if the total summation of their small-scale projects is considered. To avoid this scenario, the law would have imposed strategic environmental assessment or environmental notice and these could at least serve dissuasive purposes.

[146] Article 92 of the 1996 Framework Law on Environmental Management.

4.2.2 Absence of Enabling Instruments for Effective Implementation of Some Relevant Soil Protection Legislation

Another shortcoming or challenge regarding the control of investors is the fact that the implementation of the relevant soil protection provisions like most pieces of legislations in Cameroon are ineffective without the support of enabling instruments. In other words, most of the enabling instruments need to complete and ease the implementation of some relevant soil protection provisions, notably the relevant provisions of the Framework Law on Environmental Management, the Petroleum Code, the Mining Code, etc. lack these supporting instruments which makes their implementation difficult if not impossible. These give room for powerful investors to twist the law and force implementation in their favour, thereby hampering soil health.

4.2.3 Lame Collaboration and Coordination among Centralised and Decentralised Territorial Collectivities

MINEPDED which is the administration that spearheads the application of environmental law encounters many challenges in its mission of ensuring soil protection. In effect, in the organisation chart of many other public administrations, one finds some internal services of such administrations that are specifically in charge of environmental concerns that carry out their missions without necessarily concerting or collaborating with MINEPDED. This paradigm of functioning generally leads to soil degradation in Cameroon. For example, in matters of mineral exploitation or any other industrial activity that affects the soil, MINEPDED requires that all such sites should be put in their initial state or be restored. This requirement is seldomly respected by industrial operators and MINEPDED is handicapped to take any effective measures against the defaulting operators because the industrial activities carried out on these sites has been authorised by another Ministerial Department. This requires the need for frank collaboration and through coordination among centralised and decentralised territorial collectivises.

5 Conclusion and Recommendations

5.1 Conclusion

The following conclusions are drawn from the above discussion: First and foremost, investment is undeniably the driving force of any economic development and consequently, the economic growth of any country including Cameroon. Soil degradation in Cameroon is caused, apart from natural factors, by a range of actors, amongst which investors (multinational or national) occupy the frontline. The role of investors as actors of soil degradation in Cameroon is much felt in the extractive industry (more specifically the mining sector) and the agricultural sectors than in industrial operations. Secondly, sectoral activities such as agriculture, mining, industrial development, etc. are the main drivers of soil degradation in Cameroon. In the absence of a comprehensive soil protection law in Cameroon, soil protection relevant provisions are fragmented and scattered across sectoral legislations which also promote and encourage activities that drive soil degradation. Thirdly, the soil protection provisions constitute benchmarks for investors in the various sectors to observe and ensure that their activities and operations do not cause harm to the soil. Unfortunately, for the most part, such provisions are usually not complied with by investors as they are not usually preoccupied with the concerns of soil protection. Fourthly, non-compliance with the relevant soil protection provisions is also weakly implemented and enforced in Cameroon. Fifthly, the institutional environment for soil protection exists, albeit with operational challenges such as government inertia, corruption and bribery, and an administrative bottle neck that do not work favourably for the implementation and enforcement of soil protection legal provisions. In spite of these challenges, there are also opportunities for investment offered by the law.

5.2 *What Is the Way Forward?*

There is a strong need to address soil degradation and protection concerns in private investment and land tenure regulations and policies by inserting provisions that either oblige investors to care for soil or provisions that incentivise investors to carry out eco-friendly or green investments that protect the soil. In fact, there is the need to introduce a specific and comprehensive soil protection legislation for Cameroon. Such a comprehensive soil protection legislation must include appropriate regulations that provide for soil degradation taxes that discourage harmful practices by investors and that obliges investors whose activities affect soils or are likely to have negative impacts on soil quality, to invest in research development and extension, to develop and test sustainable soil management technologies and practices.

Furthermore, a comprehensive soil protection legislation could grant appropriate incentives/subsidies through financial instruments which support investor to implement sustainable soil management practices. Subsidies/incentives for instance, can enhance sustainable agricultural practices by encouraging investors to replenish soil quality through the implementation of sustainable or organic agricultural practices. It should also envisage a comprehensive education and awareness programme on the specific types of soil conservation measures that can be applied, and the circumstances of their use, for the different types of land use activities in order to achieve the first two points. Such a legislation should lay emphasis on the fact that soil is a common good and that there is a social cost for investing in the soil.

Additionally, a comprehensive soil protection must set standards for the degree of soil contamination at various levels; establish the responsibility to protect the soil for land owners and land users, especially investors; decentralized soil survey, inventory, sustainable use, restoration, sustainable management and monitoring responsibilities; provide for coercive administrative measures and other sanctions for soil investors.

A specific soil protection legislation must establish an institutional framework with very clear and precise missions including coordination of soil protection activities, and must contain an exhaustive list of prohibited activities that degrade the soil and its health by way of an annex.

References

Belaunde S, Cortes M, Hogstad J, Ku E, Nascimento K, Trzcinski L (2010) Land, legitimacy and governance in Cameroon. Institute for Research and Debate on Governance and Columbia University School of International and Public Affairs

Boer BW, Ginzky H, Heuser IL (2016) International soil protection law: history, concepts and latest developments. In: Ginzky H, Heuser IL, Qin T, Ruppel OC, Wegerdt P (eds) International yearbook of soil law and policy. Springer, pp 49–72

Chapman S, Wilder M (2014) Defining the legal elements of benefit-sharing in the context of REDD+, REDD+ law project working paper available at http://www.4cmr.group.cam.ac.uk, Accessed 9 Sept 2015

Commission of the African Union, The Green Wall for the Sahara Initiative, available at http://iriportal.ldeo.columbia.edu/portal/server.pt/gateway/PTARGS_0_2_1208_0_0_18/AU%20-%20The%20Green%20Wall%20for%20Sahara%20Initiative%20%2D%2D%20In%20brief.pdf, Accessed 23 May 2019

Costenbader J (ed) (2009) Legal frameworks for REDD: design and implementation at the national level. IUCN, Gland

Cowie LA, Orr BJ, Castillo Sanchez VM, Chasek P, Crossman ND, Erlewein A, Louwagie G, Maron M, Metternicht GI, Minelli S, Tengberg AE, Walter S, Welton S (2017) Land in balance: the scientific conceptual framework for land degradation neutrality. Environ Sci Pol 79:25–35

Dallou GB, Ngoa EL, Abdourahimi, Bongue D, Saïdou, Ndjana NIIJE, Kankeu B, NMG K (2018) Environmental pollution by heavy metals in the gold mining region of East Cameroon. Am J Environ Sci 14(5):212–225

Denier L, Korwin S, Leggett M, MacFarquhar C (2014) The little book of legal frameworks for REDD+. Global Canopy Programme, Oxford

FAO, Soils are endangered, but the degradation can be rolled back, available at http://www.fao.org/news/story/en/item/357059/icode/, accessed on 25 August 2020

Fogwe ZN, Ndifor F, Lambi CM, Etame RME (2001) Industrial water pollution: the case of the Ndogbong Industrial District, Douala (Cameroon). In: Lambi CM (ed) Environmental issues: problems and prospects. Bamenda, Unique Printers, pp 7–22

Fonjonga L, Sama-Lang I, Fon FL (2010) An assessment of the evolution of land tenure system in Cameroon and its effects on Women's land rights and food security, PGDT 9. Koninklijke Brill NV, Leiden, pp 154–169

GEF/ANCR, UNEP & Cameroon's MINEP (2007) Building Cameroon's Capacity to ensure Synergy between Environmental Conventions, Capacity building Action Plan of Cameroon in the implementation of international environmental Conventions. The National Coordination of the ANCR-NCSA Process, Yaounde

Hannam I, Boer B (2004) Drafting legislation for sustainable soils: a guide. IUCN, Switzerland, p 100

Helming K, Daedlow K, Hansjürgens B, Koellner T (2018) Assessment and governance of sustainable soil management. Sustainability 10:4432. https://doi.org/10.3390/su10124432
Hou D, Nanthi S, Bolan DCW, Tsang MBK, O'Connora D (2020) Sustainable soil use and management: an interdisciplinary and systematic approach. Sci Total Environ 729:138961, Available at https://doi.org/10.1016/j.scitotenv.2020.138961, Accessed 25 August 2020
Maiangwa MG, Ogungbile AO, Olukosi JO, Atala TK (2007) Land degradation: theory and evidence from the North-West Zone of Nigeria. J Appl Sci 7:785–795
Mayers J, Nguiffo S, Assembe-Mvondo S (2019) China in Cameroon's forests: a review of issues and progress for livelihoods and sustainability, Research Report. IIED, London
Orr BJ, Cowie LA, Castillo Sanchez VM, Chasek P, Crossman ND, Erlewein A, Louwagie G, Maron M, Metternicht GI, Minelli S, Tengberg AE, Walter S, Welton S (2017) Scientific conceptual framework for land degradation neutrality, A Report of the Science-Policy Interface. United Nations Convention to Combat Desertification (UNCCD), Bonn
Republic of Cameroon (2012) National Biodiversity Strategy and Action Plan – Version II – MINEPDED
Stroud J (2019) Soil Erosion and Degradation: Opportunity Amidst the Loss. https://bonnefield.com/2019/04/23/soil-erosion-and-degradation-opportunity-amidst-the-loss/, Accessed 25 August 2020
Tamasang CF, Effala C, Tassi I (2019) Mapping out Options for a Model Legislation for Sustainable Soil Management in Africa, GIZ, UBA and DROP PROJECT, Country Report for Cameroon
Tita JA (2006) Social and Environmental issues in Cameroonian Banana Production, Master of Science Dissertation. Chalmers University of Technology, Göteborg, Sweden
Wyatt AM (2008) The dirt on international environmental law regarding soils: is the existing regime adequate? Duke Environ Law Policy Forum 19:165–207

Sustainable Land Management Through Social Innovation in Land Tenure

Larissa Stiem-Bhatia, Serah Kiragu-Wissler, and Anna Kramer

Abstract Landscape restoration plays a key role in mitigating the impacts of climate change, providing food and energy for growing populations, and protecting biodiversity. When rolling out large-scale land restoration programmes it is crucial—both from economic and social justice standpoints—to secure and protect legitimate claims to land. Yet, the implementation and enforcement of land policy is weak in many countries.

This chapter presents two cases, from Kenya and Burkina Faso, in which women's access to land was secured through alternative, locally driven approaches. TMG Research and partners piloted social innovations that give farmers secure land use rights through formal leases and intra-household tenure arrangements. The success of these innovations depended on strong community participation, facilitation by trusted and neutral civil society organisations, and the alignment of customary and statutory land tenure systems.

A variety of approaches is needed to address the nexus of land tenure security and soil restoration. The inclusion of innovative, community-driven approaches within land and soil policy frameworks and development programmes is particularly effective where the implementation of broader land reforms is weak.

1 Introduction

In recent years, the urgency of preserving soil and land resources has increasingly been reflected in the global political agenda. *Land Degradation Neutrality* is enshrined in the 2030 Agenda for Sustainable Development as Sustainable Devel-

L. Stiem-Bhatia (✉) · S. Kiragu-Wissler · A. Kramer
TMG Research gGmbH, Berlin, Germany
e-mail: Larissa.stiem-bhatia@tmg-thinktank.com; serah.kiragu-wissler@tmg-thinktank.com

H. Ginzky et al. (eds.), *International Yearbook of Soil Law and Policy 2020/2021*,
International Yearbook of Soil Law and Policy 2020,
https://doi.org/10.1007/978-3-030-96347-7_8

opment Goal (SDG) 15.3,[1] and this decade has been declared as the "UN Decade of Ecosystem Restoration". Concomitantly, the Intergovernmental Panel on Climate Change (IPCC) Special Report on Climate Change and Land emphasises the pivotal role of soils in climate change mitigation.[2] It is estimated that restoring 350 million hectares of land will sequester 13 to 25 gigatons of greenhouse gases.[3]

Restoration of large areas of land is indispensable to the fight against climate change, desertification, food insecurity, and other challenges of the twenty-first century. At the same time, high population growth rates, expanding cities and the promotion of large-scale agricultural investments to meet growing food demands put additional pressure on scarce land resources. Secure land rights and tenure become increasingly important, both to safeguard legitimate rights and to provide incentives for sustainably managing land resources. The positive relationship between land tenure security and investments in sustainable land management has been confirmed by a large body of literature.[4] In many non-industrialised countries and emerging economies, smallholder farmers do not have secure land rights. Causes of land tenure insecurity include lack of land ownership documents, short-term or oral land use arrangements, lack of transferability, and the risk of expropriation. Vulnerable and disadvantaged groups—especially women—are particularly affected by insecure access to land. Farmers who fear losing their land have little incentive to invest in soil protection in the long run. This is notably the case for sustainable land management (SLM) practices, such as agroforestry and terracing, that only provide returns on investment after several years.

Nowadays, it is widely acknowledged that land tenure security is a prerequisite for sustainable land management. According to the IPPC Special Report on Climate Change and Land, "insecure land tenure affects the ability of people, communities and organisations to make changes to land that can advance adaptation and mitigation".[5] The 14th Conference of the Parties to the United Nations Convention to Combat Desertification (UNCCD), held in New Delhi in 2019, adopted a decision on land tenure which recognises that "responsible land governance is a fundamental component of sustainable land management and is important to addressing desertification/land degradation and drought".[6]

Despite the recognition of tenure security as a premise for landscape restoration, many projects and programmes fail to embrace land tenure issues. An analysis of SLM-related programmes in southwestern Burkina Faso revealed that only three out

[1] SDG 15.3: "By 2030, combat desertification, restore degraded land and soil, including land affected by desertification, drought and floods, and strive to achieve a land degradation-neutral world".

[2] IPCC (2019).

[3] See: https://www.unenvironment.org/news-and-stories/press-release/new-un-decade-ecosystem-restoration-offers-unparalleled-opportunity. [accessed on 25 August, 2021].

[4] cf. Goldstein and Udry (2008); Higgins et al. (2018).

[5] IPCC (2019), p. 31.

[6] See: https://www.unccd.int/sites/default/files/sessions/documents/2019-11/26-cop14.pdf. [accessed on 25 August, 2021].

of ten programmes had integrated land tenure into their design and implementation, while insecure access to land, at the same time, was a major impediment to post-project sustainability.[7]

Many countries have progressive land policies which recognise gender issues and legitimate claims to land. However, they are often weakly enforced. According to the World Bank, three quarters of the world's population, most of whom are already disadvantaged, lack access to formal systems to register and safeguard their land rights.[8] Implementation and enforcement of land policies is impeded by a lack of resources (financial and human/technical expertise), low awareness of land policies, and corruption, among other factors.[9]

The international land community acknowledges that statutory land regulation is not the only way to secure land rights. The *continuum of land rights*,[10] which includes customary and legitimate[11] claims to land, as well as secondary rights such as access, use and cultivation rights often held by women, need to be considered in any effort to secure land tenure.

Large land reform efforts are not always required in order to strengthen the tenure security of legitimate rights holders. There are also many effective small-scale options for addressing land tenure in soil restoration programmes. This chapter presents case study research from Burkina Faso and Kenya, where locally embedded social innovations to secure land use rights have been developed. These innovations are based on community-driven approaches, and specifically address women and land-constrained farmers. The research was conducted in the context of a global transdisciplinary research project in five countries, implemented by the Berlin-based Think Tank TMG Research gGmbH and partners. It served to accompany a global programme entitled "Soil Protection and Rehabilitation for Food Security" (2015–2022) under the Initiative "One World, No Hunger" by the German Federal Ministry for Economic Cooperation and Development (BMZ).[12] The project aimed to address structural hindrances smallholder farmers face in sustainable land management; land governance was a key research topic.[13]

---

[7] Koudougou and Stiem (2017).

[8] Enemark et al. (2015).

[9] Paradza (2011); Zevenbergen et al. (2013).

[10] For the concept of "continuum of land rights", see UN Habitat (2008).

[11] The *Voluntary Guidelines on the Responsible Governance of Tenure of Land, Fisheries and Forests in the Context of National Food Security* (VGGT) emphasise the need to recognise and protect legitimate land tenure rights. There is no formal definition of legitimate land rights, as the notion of legitimacy is highly contextual. However, the VGGT refer to legitimate land rights in the context of informal tenure rights that are regarded as socially legitimate at the local level.

[12] More information about this research project can be found here: https://soilmates.org/. More information about the GIZ-led global programme can be found here: https://www.giz.de/en/worldwide/32181.html.

[13] The idea of this accompanying research project was born at the Global Soil Week, a platform bringing together a diverse range of actors to initiate and strengthen policies and actions on sustainable soil management and responsible land governance. The results of the accompanying

2 The Role of Social Innovation in Securing Legitimate Access to Land

Case study research in West and East Africa confirms the need for flexibility in order to account for the fluid nature of tenure relations in land governance. A review of case studies on land governance in Niger, Burkina Faso, and Mali finds that process-oriented, participatory approaches focusing on local dialogue and alignment with customary tenure systems are key in moving towards responsible, gender equitable land governance.[14] Gaynor Paradza, who analysed local land governance initiatives in Rwanda, Uganda, and Kenya, concludes that top-down provision of legal land frameworks cannot alone ensure women's equal access to land.[15]

Social innovations in land tenure, based on community-driven approaches, can provide alternative and effective options for securing access to land.[16] The concept of social innovation emerged in the 1990s in the field of business and management studies. In the early 2000s, the study of social innovation became increasingly popular in the social sciences, and particularly in the area of rural development. Shifting attention towards social innovations was, inter alia, a response to classical innovation theory, which focuses mainly on technological and material aspects. The social dimension of innovations is based on the idea that the diffusion and adoption of innovations depend on social contexts and require a change in behaviours and attitudes.[17] According to Stefan Neumeier from Germany's Federal Research Institute for Rural Areas, Forestry and Fisheries, "if the new form of collaborative action becomes accepted and is implemented by the majority of participating actors and beyond, as it proves to be superior to 'traditional' forms then one can speak of a social innovation.".[18] In the field of rural development, social innovation is often discussed in the context of austerity politics that affected public spending and provision of public services. James Phills et al. define social innovation as "*the process of inventing, securing support for, and implementing novel solutions to social needs and problems*".[19] The authors further claim that social innovation becomes specifically relevant where markets have failed, or where policies are inefficiently implemented.

Social innovation is not only concerned with *what* is to be achieved (outcome), but specifically with *how* the solutions are developed (process). The process dimension (e.g., mobilisation and collaboration of actors, mutual learning, communication

research project were regularly discussed at the Global Soil Week and provided lessons learnt for a broad range of stakeholders from the partner countries and beyond. More information can be found here: https://globalsoilweek.org/.

[14] Hughes (2014).

[15] Paradza (2011).

[16] Paradza (2011); Stiem-Bhatia and Koudougou (2018).

[17] Bock (2012).

[18] Neumeier (2017), p. 35.

[19] Phills et al. (2008), p. 36.

processes) and the outcome dimension (e.g., change in attitudes and behaviours, new practices) are interlinked. Both of these dimensions are critical to the success of social innovation.[20] For example, while one intended outcome of an innovation process may be the acquisition of new skills, this outcome may also result from the nature of the process itself and the opportunity it presents actors to develop solutions jointly with others, changing their attitudes and perceptions. A solution developed through a locally embedded process is likely to benefit from greater social acceptability, and therefore a higher chance of eventual success.

In the context of land governance, social innovations, as collaboratively developed solutions, can offer a pragmatic way to secure land rights for marginalised groups based on social legitimacy where the appropriate institutions are lacking or have insufficient resources. Social legitimacy, in this context, means that the concerned stakeholders have a sense of ownership of the solutions due to their involvement in developing them. Inspired by the concept of social innovation, the hypothesis of the aforementioned research project was that effective collaboration in research design and implementation between multiple stakeholders will generate innovations that are better tailored to the context and socially legitimate, and therefore more likely to be adopted by the intended users.

3 Case Study: Intra-Household Land Use Agreements in Burkina Faso

Burkina Faso, a landlocked country in the Sahel region, is severely affected by land degradation and desertification. Deforestation, inappropriate farming practices, demographic pressures and recurrent droughts have taken a toll on the country's natural resources. About 34% of its arable land area is considered degraded. Sustainable management of land resources as well as land tenure security are among the six priorities of the country's National Programme of the Rural Sector (2016–2020). However, despite the government's efforts to protect natural resources, land degradation increased by an average of 469,000 ha per year between 2002 and 2013.[21] Weak institutional capacities, incoherence of sectorial policies, insufficient monitoring, and weak law enforcement hinder the effective protection of natural resources.[22]

About 86% of the Burkinabé population depend directly on agriculture, yet smallholder farmers often lack the enabling environment to sustainably manage their land. Rural services, such as access to agricultural advisory services, micro-finance schemes and support in securing and formalising land rights, are often unavailable.

[20] Neumeier (2017).

[21] Government of Burkina Faso (2018a).

[22] Government of Burkina Faso (2018b).

In relation to land governance, women are in a much weaker position than men. They make up half of the agricultural labour force, but typically face discrimination in land allocation, control and ownership.[23] Under customary land governance regimes, women have secondary rights to land and access land through male family members. Traditional authorities, so-called *chefs de terre*, oversee land allocation at the village level.[24] Women's access to land is commonly renegotiated after each agricultural season, and household heads can revoke women's land use rights at any time.[25] Villagers refer to this phenomenon as *forced rotation*. Consequently, many women refrain from making long-term investments in land productivity, such as soil fertility measures, which in turn limits their potential to enhance agricultural production, income and financial autonomy.

Over the past two decades, many land policies in Burkina Faso have been reformed with the aim of achieving tenure security through formalisation. The most recent land policy reform was Decree 034-2009 on Rural Land Tenure, which was passed in 2009.[26] This policy recognises and strengthens customary land rights as well as men and women's equal right to land. However, implementation and enforcement advance slowly.[27] According to census data from 2010 and 2012, only 4% of land is held under formal rights.[28]

In an effort to provide incentives for smallholder farmers, specifically women, to invest in soil restoration, the Burkinabé expert network Groupe de Recherche et d'Action sur le Foncier (GRAF)[29] and TMG Research developed a social innovation to secure land use rights for women.[30] This innovation was developed in the village of Tiarako, Satiri municipality, southwest Burkina Faso, and involved a range of stakeholders.

3.1 Description of the Study Site

The study site, Tiarako village, is situated around 45 km northeast of Bobo-Dioulasso and has approximately 2000 inhabitants. The village population is very homogenous with around 90% originating from the area and belonging to the *Bobo* ethnic group. Other ethnic groups living in Tiarako include the *Mossé*, *Bwaba* and *Fulani*. Subsistence farming and livestock rearing are the main economic activities. The most commonly planted crops are cotton and maize (mostly grown by men) and

[23] Bambio and Bouayad Agha (2018); Koudougou et al. (2017).

[24] Linkow (2016).

[25] Kevane and Gray (1999).

[26] Government of Burkina Faso (2009).

[27] Hughes (2014).

[28] Bambio and Bouayad Agha (2018).

[29] See: www.graf-bf.org.

[30] cf. Stiem-Bhatia and Koudougou (2018).

sesame, beans, and groundnuts (mostly grown by women). Declining soil fertility is a serious issue threating agricultural productivity in Tiarako. Causes of declining soil fertility include intensive farming, overuse of chemical fertilisers and pesticides, loss of vegetative cover, and unfavourable climatic conditions. Traditional soil restoration practices include contour ploughing, construction of earthen bunds, vegetative contours, intercropping and crop rotation, mulching, composting, and the use of animal dung as fertiliser. The practice of leaving land fallow has been largely abandoned due to demographic pressure. Insecure land tenure is a major impediment to investing in soil restoration, especially for women. Other obstacles include poor access to agricultural advisory services, equipment, inputs, credit, and markets.

3.2 *The Research Process*

The research process focused on the development of a methodology to secure land use rights for women. It was conducted between February 2017 and June 2019 and built on preceding studies led by TMG and GRAF which analysed sustainable land management projects[31] and a gender-focused study of the nexus between SLM and land tenure.[32] The hypothesis underlying this research project was that legitimate land use agreements, based on consensus-building within the community and families, can be an interim and preparatory step towards increased equality in land tenure security for women.

The research process was conducted in multiple phases (see Fig. 1). In the conceptual phase, various people and organisations with a stake in land governance and agriculture were brought together. These stakeholders included representatives of the village community, the municipal administration, agricultural extension officers, project implementers, and other resource people. During the joint design of the research process, activities and objectives were defined and roles and responsibilities clarified.

The research process in the village began with awareness raising about the land policy framework in general and on men and women's equal rights to land in particular, as well as outlining the social and economic benefits of women's secure access to land.

In the next phase, the current tenure system was analysed jointly with the local community to better understand customary tenurial relationships and practices. Further, the facilitating civil society organisation, GRAF, conducted a census of all farming households interested in participating in intra-household land rights transfers; these recorded the primary land rights holder (who retained the right to

[31] cf. Koudougou and Stiem (2017).

[32] cf. Koudougou et al. (2017).

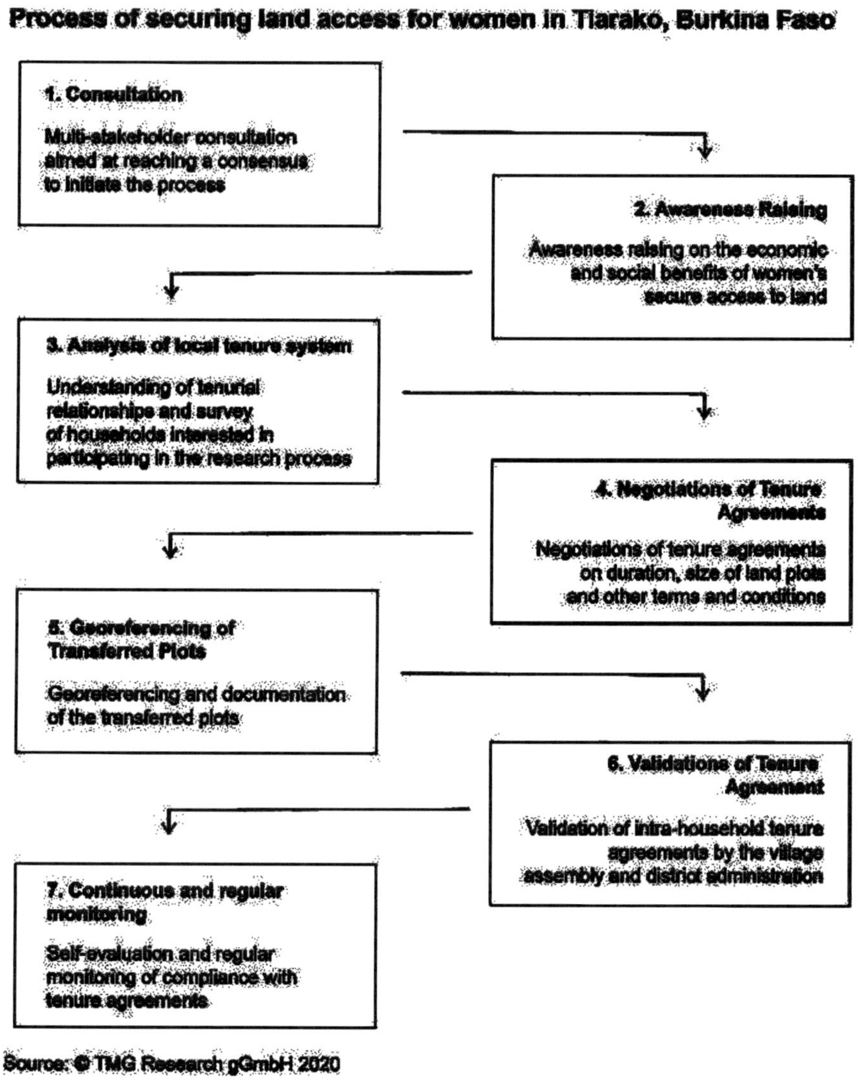

Fig. 1 Illustration of research process in Tiarako

transfer land),[33] the head of the family farm/household (who retained the right to control the land),[34] the beneficiary women, and the size of the agricultural holding (s).

This phase was followed by facilitated negotiations about the land to be transferred to women. Negotiations and dialogues took place at different levels; for example, between heads of clans and heads of households, as well as between heads of households and women beneficiaries. The size of the land holdings, duration of land use, and conditions of tenure agreements formed the main topics of negotiation. GRAF facilitated these negotiations where necessary. The intra-household tenure agreements were later validated in a village assembly, in which representatives of the municipal administration, public extension services and GRAF participated.

Once all stakeholders consented to the tenure arrangements, the plots to be transferred were georeferenced through participatory GPS recording. Public agricultural extension agents provided technical support.

Finally, the plots were officially transferred in a village assembly on eyewitness account by the whole village community and other participating stakeholders, such as the mayor. During this assembly, 145 land rights holders, mostly men, had transferred long-term land use rights, mostly of unlimited duration, to 228 women, individually and collectively, to 189 parcels or a total of 412 ha.

The negotiated tenure arrangements were recorded in a document presented and collectively signed in a Village Assembly. This record of tenure arrangements is not a legal document but rather an accord validated by the village assembly and recognised by the municipal administration. In the event that an accord is contested, the village conciliation committee will mediate the dispute. The issuing of formal land titles to women was not possible within this pilot process. Local institutions responsible for issuing land titles did not exist in the region.

After the land rights transactions were completed, GRAF and TMG organised local multi-stakeholder dialogues to reflect on the process and mobilise the participating actors to support women in farming the newly received land. Support for women in making the best use of the fields to which they were granted secure access and long-term use rights was important, as women farmers not only face the obstacle of land tenure insecurity, but also frequently lack access to equipment, credit, training, and other inputs.

Continuous monitoring after conclusion of the research process served to assess the effects of the intra-household land rights transfers to women (*outcome dimension* of the social innovation) and to reflect on the research process itself in an *ex post*

[33]The primary land rights holder here was the head of clan or *chef de terre*, who has the right to allocate plots of the clan's communal land to families. He has the authority to approve land rights transactions within and across families.

[34]The head of the family farm (mostly equates with the head of the household) has the right to control the land (i.e., decide on the use of land), but must seek consent from the head of clan or *chef de terre* if the family decides to transfer land rights. Therefore, the head of the family farm/household is referred to as the secondary land rights holder.

manner (*process dimension* of the social innovation). Tenure relations, perception of tenure security, agricultural investment behaviours, and social dynamics within the community and families were analysed based on focus group discussions and individual interviews with farmers and other key informants.[35]

3.3 Effects and Outcomes of the Land Rights Transfer Process

3.3.1 Changed Attitudes to and Perceptions of Gender Roles

One key observation made during the discussions with farmers, both men and women, was the shift in attitudes about women's right to land and their role in farming. Most of the men who were interviewed acknowledged women's legal right to secure access to land. Previously, the men were reportedly reluctant to cede control over land to women, as they feared that women could become increasingly independent, thereby undermining patriarchal authority. The shift in mindset from preserving men's control over land towards granting stronger land use rights to women was associated with a positive recognition of women's role in the family farm business. The fact that women may be better able to invest in farming, enhance productivity and therefore increase the family's overall income was a convincing argument for men to recognise not only the *de jure* but also the *de facto* value of women's secure and equal right to land.

3.3.2 Increased Tenure Security

Eighteen months after the land rights transfers were concluded, land tenure security is perceived to have increased compared to the baseline situation. During follow-up discussions, both men and women farmers referred to the *forced rotation* phenomenon as the old method, which has been abandoned.

> Now the women have their own fields and no longer do they have to ask for plots of land, left to right, season after season. (focus group discussion with female farmers, Tiarako.)

[35]The monitoring of the research process took place between March 2018 and February 2020. Qualitative data provided an in-depth understanding of the process and outcomes of this research project. Data were collected through multi-stakeholder workshops (2), focus groups discussions (26), individual farmer (16) and expert interviews (6), and informal communication with farmers, village leaders, the mayor and other resource people. Local perceptions of farmers, extension workers and municipal administrators helped us to understand how the research process resulted in the development of socially legitimate tenure agreements between women and men. Observation as *a participant* in and *evaluator* of the research process complemented the data (*ex-post* self-evaluation). Document analysis and literature reviews completed the data triangulation.

None of the respondents, except for one household,[36] reported reversal of land rights transactions. Thanks to the clarification of land rights within families and larger clans, land conflicts have reportedly declined at the community level.

3.3.3 Incentives for Adoption of Soil Protection Measures

The outcome monitoring further showed that increased tenure security provided incentives for female farmers to invest in farming and SLM in some cases: Some female respondents stated during focus group discussions that they were better organised in female farmer groups than before. In a few cases, the men helped the women cultivate their fields, as per their commitments during the multi-stakeholder dialogues.[37] Several men, however, refrained from supporting women in farming their fields, as they were waiting for external support. Many of the women were similarly awaiting further project support, such as the provision of free fertilisers and seeds, and therefore did not invest more than previously in their farmland. A rapid assessment of harvests in February 2020 showed that those women who organised themselves to invest more in soil fertility, for example, by participating in compost training courses and joint management of field preparation, could achieve considerably higher yields compared to women who were not supported by their peers or family or waited for external support.

3.3.4 Integration of Social Innovation Into Statutory Land Governance and GIZ Soil Projects

Both the municipal administration and GIZ have shown strong interest in scaling up the social innovation to secure women's land use rights. At present, the innovation, with financial support from the GIZ global soil programme, is being tested in villages in a similar socio-cultural context to the pilot site, as well as in a different context (i.e., with a higher proportion of migrants). At policy level, discussions are being held with the Department for Rural Land Tenure under the Ministry of Agriculture to explore how the social innovation can contribute to more effective application of the land law 034-2009.

[36] During the focus group discussions, one household was found to have had ulterior motives for participating in the research process. The woman who was supposed to benefit from land allocation never even saw the land she was allegedly given.

[37] During the final multi-stakeholder workshop (*atelier de bilan*) to conclude the process in Tiarako, men who had transferred land use rights to women publicly committed to support the women in making best use of the fields they received in the context of the land rights transfer process.

3.4 *Reflection on the Process*

Our research experience shows that facilitated negotiations between men and women, and consensus-building at village level, led to increased land tenure security and enhanced incentives for SLM, to the benefit of women. Several process characteristics were key in achieving these outcomes.

The process was organised to maximise communities' participation in transparent and honest discussions which secured their trust. In a variety of settings, including multi-stakeholder workshops, village assemblies, focus group discussions, and informal meetings, participants discussed details such as the type of tenure arrangements (permanent or temporary) and the geographical locations of transferred fields. Involved from the beginning of the initiative, able to determine its design and implementation, and encouraged and enabled to assume key roles, participants developed a sense of ownership and trust over the process. Trust in the process and its outcomes was further increased by endorsement of the land rights transfers by traditional village leaders, village committees and the district administration, as well as close support from GRAF experts who speak the local language, and other local partners.

The men in the village played a crucial role too. Participation reached a tipping point after customary leaders were convinced of the value of the process. The time invested by the GRAF experts in discussing with the villagers the important contribution women can make to the family's income and well-being was important in convincing the men. Once they were persuaded of the prospective economic benefits, the men were more willing to break away from patriarchal practices, though they tended to assert that as heads of the farmthey should suggest the terms and conditions of the land tenure agreements. Most insisted that the provision of secure land use rights to women should be contingent on their continued membership of the family. In return the maintenance of norms related to the traditional bond of marriage, many men accepted women's demands for a larger area of allocated land, which would allow them to produce for the market as well as for subsistence. Crucially, men's openness to new ideas of managing and controlling land within the family was also ensured by the support and endorsement of the municipal administration and customary authorities.

> The project has been explained, but there have always been doubts. Seeing the authorities on the ground gave legitimacy to the project. (Interview with male farmers, Tiarako)

Another important condition for creating legitimate land tenure agreements that enhance gender equality was the blending of customary practice and gender-progressive ideas. Due to the locally-led nature of the research process, the method developed in Tiarako is responsive to the local context and the fluid nature of tenure relations. While respecting the land law, the method acknowledges women's differentiated tenure situations. Terms and conditions of tenure agreements were negotiated between customary land rights holders (e.g., heads of clans), heads of households, and female family members in an iterative manner. The analysis of the prevailing local tenure system ensured that existing tenurial relationships and practices were considered.

Embedding the social innovation in socio-cultural realities, such as the need for endorsement of traditional leaders, was important in ensuring the social legitimacy of the tenure arrangements. Villagers' decision-making power over the process and the consensus reached between them further reinforced the legitimacy of these arrangements.

4 Case Study: Community-Led Land Lease Guidelines in Western Kenya

Kenya is a country of high agricultural potential, but soil fertility is declining. An estimated 40% of its total land area is degraded.[38] A rapidly expanding population has led to sub-division of land into small and uneconomic holdings, many of less than a hectare in size, while continuous cultivation to meet increasing food demands has exhausted soil fertility. Soil erosion due to heavy rainfall and overexploitation along with increasing soil acidity have reduced agricultural productivity. Promoting sustainable land management to restore soil fertility is a priority for the government. The Kenya Strategic Investment Framework for Sustainable Land Management acknowledges that "land degradation remains a major threat to the provision of environmental services and the ability of smallholder farmers to meet the growing demand for food and income".[39]

Insecure access to land has been identified as a major obstacle to sustainable investment in soil protection. This affects particularly tenant farmers with short-term and oral leases, many of whom are women and young farmers.[40] A senior advisor for GIZ's soil programme notes in this regard:

> Some of the project participants only get land for a season because they lease for one season or two. They will not take up the activity of planting trees. It's labour intensive, you hire people to do it for you, so if you are not guaranteed [a return on the investment], or it's not your piece of land, then you would not fully implement that part of the project. You find that those beneficiaries would just do the bare minimum, not think in the long term. (Flora Ajwera, GIZ Kenya, personal communication)

Land leasing is best understood in the context of Kenya's land policy framework and land tenure system—the National Land Policy 2009. This sets out the terms and conditions under which rights to land and land-based resources are acquired, retained, used, disposed of, or transmitted. Individuals or corporations acquire land either through purchase, inheritance, gifting or leasing. The purchase option is

[38] UNDP (2019).

[39] Ministry of Environment and Natural Resources (2016).

[40] Kiragu and Flohr (2016).

limiting for many resource-poor people, particularly women.[41] Inheritance is often shaped by customary rules which favour men while women are left to access land through marriage.[42] In Kenya, only around 1% of title owners are women, and 5% of registered land titles are held by women and men jointly.[43] As the population has increased, leasing has become a viable option. Nyangena notes the growing trend whereby land is accessed through formal and informal rental markets.[44] The legal formal process of transferring land user rights requires that parties (lessor and lessee) enter into an agreement in the presence of a witness, preferably an advocate of the High Court of Kenya. Formal leasing is common between large landlords and tenants who have a strong financial base. Poorer land-constrained households tend to opt for informal lease arrangements due to the high cost associated with entering a formal lease. These informal leases are often short-term, and offer little motivation to invest in soil and land conservation.

4.1 Description of the Study Site

The study site, comprised of Isukha Central and Isukha West wards, is located in Shinyalu sub-county of Kakamega County. The two wards have a combined population of about 60,000 inhabitants (or 14,000 households). Kakamega is the fifth most densely populated rural county in Kenya, with 618 persons per square kilometre.[45] The population of the study site is ethnically homogenous, comprising the Isukha community, a sub-community of the larger Luhya community. Mixed farming, comprising crop cultivation and livestock rearing, is the main economic activity. Other economic activities include small-scale mining and brick-making. For household use, the community mainly grows maize, beans, sweet potatoes, bananas and vegetables. Sugarcane and tea are the most common cash crops.

In Kakamega County, a growing population has put great pressure on the land. The average land holding per family is just 0.6 ha.[46] This land is typically farmed continuously with little replenishment of soil nutrients, resulting in sub-optimal agricultural productivity. Even where land is available, access is unequally distributed between family members due to customary inheritance laws that favour male household heads.

Consultations with stakeholders in Kakamega County, facilitated by TMG Research, revealed that land leasing is a common way resource-poor households

[41] Women, while not the only social grouping disadvantaged in terms of land tenure, were the focus of the social innovations analysed in the context of this chapter.

[42] Kameri-Mbote (2002).

[43] Musangi (2017).

[44] Nyangena (2010).

[45] Government of Kenya (2019).

[46] County Government Kakamega (2018).

access land.[47] Since most lease arrangements are informal, there are no statistical records showing the extent of land leasing. The communities also noted that many lease arrangements were not secure for tenants. The unregulated nature of the land lease market poses numerous problems for tenant farmers. Landowners may breach informal lease agreements made without witnesses, or damage crops without compensation, and arbitrarily change boundaries. In these conditions it is difficult for farmers to invest in improving soils. Consequently, both agricultural production and sustainable rural development suffer. Lease tenure was thus considered by stakeholders as an important element of land tenure rights, especially for women and youth.

4.2 The Research Process

The research project in Western Kenya focused on developing a methodology to secure land access for land-constrained farmers through written lease agreements. It was part of the research for the GIZ Global Soil Programme and succeeded previous studies on smallholder farmers' challenges in adopting soil protection practices.[48] TMG Research partnered with a grassroots women's organisation—Shibuye Community Health Workers (CHW)—to lead this research process in collaboration with governmental representatives, community leaders and project implementers, such as GIZ. The process was conducted between April 2017 and April 2018 and can be divided into seven main steps (see Fig. 2).

The process of developing land lease guidelines (LLGs) began with kick-off workshops. These workshops, which incorporated local institutional stakeholders including the Land Control Board, the county ministries of agriculture and lands, and the area administration, ensured that the process received buy-in from local stakeholders and did not contravene any governmental regulations on land.

The next engagement targeted community members and sought not only validation of the LLG idea, but also to clarify roles and responsibilities of community members in the process of developing the lease guidelines. Diverse representation—comprising men, women, youth and single-headed households—was key in ensuring broad community ownership of the process and outcome.

In the next step, community-level committees were formed to draft the LLGs. The content of the LLGs included the terms and conditions of the same with the aim of serving both the landowner and the tenant. The terms and conditions were based on the experiences and challenges the communities previously faced in land leasing. In total, about 380 persons directly participated in developing the community LLGs.

The consolidated draft of the LLGs was then shared with the institutional stakeholders for their approval. The aim of doing so was to ensure that the proposed

[47] Lusweti (2017).

[48] For more information, see Kiragu and Flohr (2016); Mburu and Kiragu-Wissler (2018).

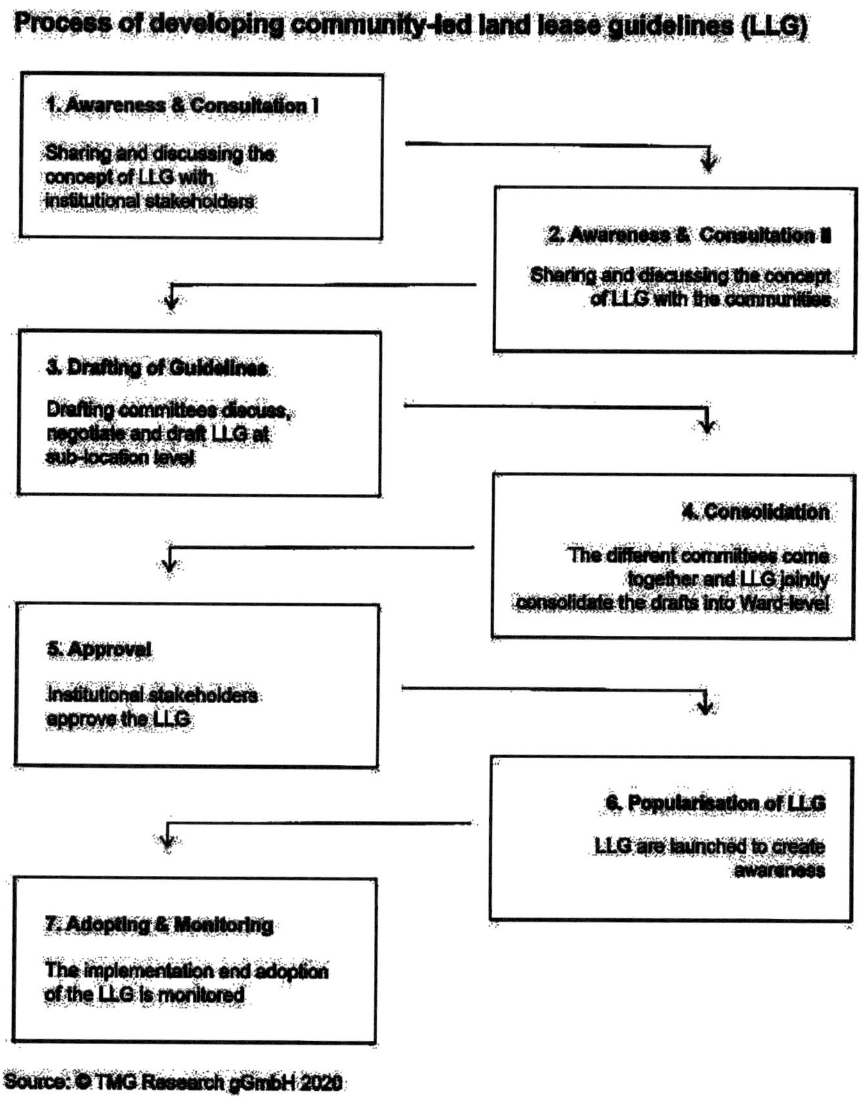

Fig. 2 Illustration of the research process in western Kenya

LLGs received the recognition and support of the formal institutions as some of them would be involved in their implementation. After being validated by the institutional stakeholders, the LLGs were launched in large community meetings, followed by awareness campaigns. Once the community had approved the LLGs, community committees were formed to monitor the adoption of land lease forms and to mediate conflicts.

The forms for leasing land, based on the LLGs, are available at the local assistant chief's office. Since the launch of the LLGs in 2017, over 200 (recorded) households have used them, establishing a new precedent in a context in which informal agreements were previously the norm.

Following the launch of the LLGs, TMG and Shibuye CHW assessed the results of their implementation. This was done to reflect on how the development process had influenced the adoption of the LLGs, as well as to assess effects of the adoption. The assessment was conducted between March 2018 and February 2020, when changes in behaviour, land conflicts and local governance structures, as well as the adoption of land lease forms, were analysed.[49]

4.3 *Effects and Outcomes of the Process to Develop Land Lease Guidelines*

4.3.1 Enhanced Land Tenure Security and Reduced Land Disputes

Broad and direct engagement of the local stakeholders contributed to increased awareness as to how to securely lease land and deal with any conflicts. Focus group discussions with land owners and tenants revealed that the presence of several witnesses when signing the lease forms, including local government administrators, and the detailed terms and conditions of the leasing arrangement give the leasing parties a stronger sense of security and legality. Landowners perceive that their interests are protected, i.e., that their land will be used as per the agreement. Land lessees perceive assurance that they will reap the benefits from farming the leased farm without problems like those they experienced before.

The assessment of the adoption of the lease guidelines further revealed a reduction in the number of land disputes. Six local administrators reported that they had not received any complaints over leased land since the introduction of the LLGs. Likewise, members of the land lease committees observed a reduction in conflicts relating to land leases. The lease agreements were found to enhance transparency by formalising the lease arrangement and stipulating detailed terms and conditions. The presence of witnesses when the agreements were signed also boosted confidence in the lease arrangement and enhanced transparency.

[49]The monitoring of the research process combined qualitative and quantitative research methods. Data were collected through records of land lease agreements, multi-stakeholder workshops (2), focus groups discussions (26), individual farmer (16) and expert interviews (6), and informal communication with farmers, village leaders, the mayor and other resource people. Local perceptions helped us to understand how the research process resulted in the development of socially legitimate tenure agreements between women and men. Observation as *a participant* in and *evaluator* of the research process complemented the data (*ex-post* self-evaluation). Document analysis and literature reviews completed the data triangulation.

4.3.2 Enhanced Land Access for Women

The lease agreements enjoyed high resonance with female farmers. A notable 40% of leases signed have women as the principal tenant, of which 45% are widows. In a society where widows' access to land is limited by a cultural patriarchy, the lease guidelines provide a new opportunity for such women to securely access land.

> Women were not allowed or never had the courage to go and lease land because it was believed that women didn't have a say in issues concerning land. With the sensitization meetings and encouragement to men that women also have a bigger role in taking care of the families, at least men have allowed women to participate freely in farming. Women are now free to lease land. (Key informant interview with Shibuye CHW volunteer)

4.3.3 Increased Adoption of Soil Protection Measures

The LLGs include agreements on SLM measures, allowing tenant farmers to practice SLM while assuring the landowner that assets such as trees will not be destroyed. The value of the LLGs has been well recognised by the GIZ Global Soil Programme. In the study site in Kakamega County, farmers accessing land through the land lease agreements now benefit from training on conservation agriculture provided by the GIZ project. Based on their training and experience practising conservation agriculture on the leased plots, farmers have taken up composting, making liquid organic manure, push and pull technology, as well as planting of nappier and vetiver grass and leguminous cover crops, such as *mucuna*, for two to four seasons.

4.3.4 Policy Integration of Land Lease Guidelines

The LLG development process and outcomes received positive recognition by Kakamega County policy makers. The Minister for Agriculture and the County Governor have acknowledged the LLGs as an effective strategy to ensure that more people gain access to land.

To scale up the lease guidelines, Shibuye CHW is collaborating with Kakamega County Government for broad public review of the LLGs by communities in other regions of the county. This process aims to come up with harmonised, county-wide, and county government-backed lease guidelines.

4.4 Reflection on the Process

The experience in Kakamega County showed that joint/inclusive and community-driven research, facilitated by a community-based organisation, led to innovative approaches to land leasing. Due to a range of factors, the land lease guidelines responded to local needs and were accepted by the local community. These success factors are analysed below.

The assessment study showed that community members were willing to adopt the LLGs thanks to their strong participation in developing them. In doing so, community members drew on their experiences of the issues faced when leasing land. Consideration of these very context-specific challenges ensured that the LLGs responded to actual local needs.

> There could be a problem, but since the LLG were made locally, we have accepted and owned [them] fully. I have faith in them because we participated in drafting them. (Focus group discussion, land tenant, Virhembe sub-location)

The research process of developing the LLGs contributed to enhancing existing local land tenure systems while empowering the community to actively engage in land governance processes. Viable, sustainable implementation of the LLGs depended on this engagement.

Communities are also better able to deal with land disputes thanks to their involvement in the process of developing the LLGs. The LLGs have been instrumental in curbing fraud due to their detailed character and the presence of witnesses. During the process of developing the LLGs, participants complained about how land leases could be fraudulently turned into sale agreements. The communities noted that they lacked clarity on leasing arrangements. Leasing parties negotiated the leasing deal the way they knew best—orally or on a loose sheet of paper, and rarely with witnesses. The LLGs have filled this gap.

> Before the booklet [land lease guidelines] was developed and adopted, there was a lot of confusion in that the parties involved in leasing land had to go to the village elder who didn't know how to write, read and even didn't know how to frame the agreement. The agreement from the village elder wasn't authentic because anyone would manipulate and change the agreement (Key informant interview with Assistant Chief, Mukomari sub-location).

Critical to the success of the process was the presence of the grassroots women's organisation Shibuye CHW as its facilitator. Shibuye CHW, whose members belong to the community, had extensive knowledge of the local context, and is respected locally. These characteristics gave them legitimacy as a convener and facilitator of dialogue between formal institutions and the community.

5 Discussion

Many countries have progressive legal frameworks when it comes to land policies. Often, these explicitly specify and guarantee women's rights to land. However, the enforcement of such legislation and the implementation of related projects is frequently lacking. Since enforcement advances slowly in many countries while pressure on land resources increases, alternative approaches are urgently needed to secure and safeguard land tenure, specifically the legitimate rights of women and other vulnerable groups to contribute to the creation of enabling conditions for smallholder investment in soil protection.[50]

The case studies from Kenya and Burkina Faso show how innovative approaches can be jointly developed with local communities to address the problem of weak enforcement of land policy frameworks. Although the monitoring assessment provides only a snapshot of a continuously evolving process, the findings show that local multi-stakeholder dialogues and collaboration, facilitated by trusted civil society organisations, can strengthen legitimate land use rights for vulnerable groups. The two case studies provide insights into the factors that enabled social innovation in land tenure to emerge, as well as their outcomes, which are discussed below.

5.1 Civil Society Organisations as Bridging Agents

Although civil society has gained a foothold in multi-stakeholder governance, civil society and community-based organisations (CBOs)[51] are often overlooked in innovation processes. Yet, their role can be particularly pivotal in the context of grassroots innovations. Due to their independent, non-governmental status, civil society organisations can bridge various types of knowledge, experiences, and interests between governmental actors, village communities, project implementers, and researchers.

The civil society organisations (CSOs) GRAF and Shibuye CHW played a key role in navigating the research process in complex socio-cultural contexts. Thanks to their existing ties with the local communities, and their ability to understand local customs and governance regimes, they enjoyed a high degree of trust in the communities. The CSOs were able to mobilise both governmental actors and the local community and win their support for the innovation process. Muok and Kingiri found that community participation is key to the success and sustainability of grassroots innovations.[52] Previous research by GRAF and TMG shows that weak involvement of local institutions constitutes a major obstacle to the development of sustainable rural development projects in western Burkina Faso.[53]

In dealing with politically sensitive topics, in some instances, CSOs have an advantage over governmental actors. Even though CSOs do not have a mandate to enforce land policies, in both the Kenyan and Burkina Faso cases they were perceived as neutral and unbiased, supporting the government in the process of

---

[50]TMG Research (2019).

[51]Civil society organisations and community-based organisations are used interchangeably in this context because both are non-governmental organisations representing civil society. Community-based organisations here are understood as civil society organisations at community level.

[52]Muok and Kingiri (2015).

[53]Koudougou and Stiem (2017).

securing land access. This was important in the facilitation of land governance processes, which are typically characterised by complex power dynamics, vested interests, and disputes among different interest groups.[54] Perceived as a neutral process facilitator, the CSOs were in an ideal position to achieve consent among different stakeholders. In an analysis of CSO-facilitated local land governance initiatives in Rwanda, Uganda, and Kenya, Paradza found that CSOs, despite lacking statutory authority, were able to provide a platform for "unbiased dispute resolution" in land conflicts, particularly for women.[55]

While the state retains a preeminent role in implementing and enforcing land policies, trusted and neutral civil society organisations provide a crucial platform for social innovation in land tenure.

5.2 *Towards Gender Equality Through Social Innovation*

Customary land tenure systems provide one explanation for gender imbalances in access to land. In many cultures, customary institutions are patriarchal and grant women only secondary land rights through male family members. However, statutory law may also privilege male-dominated land governance institutions which discriminate against women in land registration and administration.[56] Ignorance of gender-progressive land policies further exacerbates patriarchal law enforcement. Where land ownership is formalised and secured, this is often in the name of the male family head, neglecting secondary land use rights of women family members.[57] The disadvantaged tenure situation of most women, and patriarchal control of land, is likely to persist even with the implementation of gender-progressive land policies.

Empowerment of marginalised groups and enhanced equality are desired goals in developing social innovations.[58] Moulaert et al. note that "[social innovation] refers to those changes in agendas, agency and institutions that lead to a better inclusion of excluded groups and individuals in various spheres of society at various spatial scales".[59]

The research experience in Burkina Faso and Kenya demonstrates how social innovation can improve the tenure situation of vulnerable groups. Facilitated negotiations, and consensus-building at household and community levels, led to legitimate land rights transfers to the benefit of women and land-constrained farmers. Due to the locally-led nature of the research process, the social innovations were responsive to the local context and fluid nature of tenure relations. For example, they

[54] Paradza (2011).
[55] Paradza (2011), p. 11.
[56] Deininger et al. (2017); Namubiru-Mwaura (2014).
[57] Zevenbergen et al. (2013).
[58] Bock (2012); Moulaert et al. (2005).
[59] Moulaert et al. (2005), p. 1978.

acknowledged women's differentiated tenure situations, such as secondary land rights held by married women, or widows' disadvantaged position in accessing land in Kenya.

Securing land access for women is not only crucial in terms of social justice but also for the protection of natural resources. Ali et al. found that the uptake of soil protection measures tripled among female-headed households after their access to land had been secured through formalisation of land rights.[60]

5.3 Assuring Legitimacy of Land Tenure Agreements

Under customary law, the legitimacy of land tenure agreements is based on mutual trust and respect between those who govern and those who are governed by the tenure system. Social cohesion within the community, shared values, and ancestral linkages further underpin the legitimacy of customary tenure systems.

In both cases studied, respect for traditional practices of land allocation, and knowledge of the local tenure system, were key to achieving legitimacy. Traditional authorities, such as the *chefs de terre* in Burkina Faso, who have a mandate to allocate and transfer communal land played a central role in the research process. In Kenya, the local institutions mandated to implement and oversee the land lease guidelines were involved in drafting them. Convincing the local leaders of the innovation process was the first step before the research process was rolled out in the community. Local actors' endorsement of and support for the process was crucial to the success of the social innovations. In the context of pro-poor land administration, Zevenbergen et al. regard co-management of land tenure by community leaders as a risk-avoidance strategy.[61]

Another important factor in seeking legitimacy is the inclusion of witnesses in the land transfer process. In Burkina Faso, land transactions were carried out in a village assembly with the participation of local leaders. In Kenya, witnesses oversaw the signing of land lease agreements, which reportedly prevented fraudulent practices in land leasing.

5.4 Alignment of Customary and Statutory Land Tenure Systems

Progressive land policies usually recognise customary land systems. However, due to their weak enforcement, legal protection is often not assured. The social innovation processes in Burkina Faso and Kenya were developed to increase tenure security

[60] Ali et al. (2014).

[61] Zevenbergen et al. (2013).

with the aim of providing incentives for sustainable land management, enhanced agricultural production, and household food security. In the land governance community, it is acknowledged that statutory law alone cannot provide land tenure security. Reflecting on the social innovation process in Burkina Faso, GRAF founding member S. Sanou remarked that land titles do not automatically ensure social acceptance within a community:

> Someone can have a land title but not be able to exploit the field simply because, on the village level, the people do not agree with him or her managing this field. (S. Sanou, land policy expert, pers. comm., Bobo-Dioulasso, Burkina Faso)

Many scholars agree that statutory and customary land law need to be aligned. Their alignment is important to acknowledge local customary leadership and conflict resolution systems, which are key elements in the legitimacy of tenure agreements, as discussed above.[62] This is important to avoid top-down approaches that do not respect existing tenure systems, as well as to ensure that existing local tenure systems are protected by law. Additionally, a tenure system that builds on local governance models, as opposed to centralised, top-down land administration processes, is more likely to be transparent.

Safeguarding people's customary and legitimate rights to land is also crucial to avoid rights violations. Experiences from previous large-scale restoration projects, such as REDD+, provide critical lessons. Restoration projects are sometimes based on the false perception of vast tracts of land lying "idle". Land may be home to many people, but this fact might not be recognised in statutory land use plans if those people's rights are not formally recognised. Ambiguous land tenure status and a lack of community participation can prevent REDD+ initiatives from generating positive outcomes and improving livelihoods.[63]

Considering the above, the hypothesis underlying this research project was that legitimate land use rights agreements, based on consensus-building within the community and families, can support the implementation of existing land policies. The *continuum of rights*, specifically secondary land use rights, and securing access to land for people without land ownership, were core considerations of the research projects in Kenya and Burkina Faso. Given weak institutional capacities for statutory law enforcement, the social innovations sought to translate legal provisions into practice. For example, the Rural Land Law in Burkina Faso makes specific provision for access to land for women and other vulnerable groups, such as nomadic herders and youth, in local land charters. However, communities are largely unaware of these provisions. In Kenya, the community-led land lease guidelines, piloted by Shibuye CHW and TMG Research, followed juridical procedure by considering the role of witnesses in setting up lease agreements. In both case studies, statutory land governance institutions were involved in the development of the social innovations around land tenure, thereby ensuring that the process complies with the statutory

---

[62] Zevenbergen (2013).

[63] Awono et al. (2014); Stiem and Krause (2016).

legal framework. The fact that, in both Kenya and Burkina Faso, governmental actors are seeking to institutionalise and scale up the social innovations shows the potential to align alternative approaches to securing land, based on customary practices and innovative ideas, with existing legal frameworks.

5.5 *Tackling Land Governance for Soil Restoration*

The social innovations in Kenya and Burkina Faso were developed to provide incentives for sustainable land management. Often, soil restoration programmes shy away from tackling land rights issues. An evaluation conducted by the World Bank-financed watershed management projects, which were implemented between 1990 and 2004 across different regions and prioritised SLM, contained very few references to legal land governance frameworks or to land tenure in general.[64] Similarly, a review of experiences of the TerrAfrica Strategic Investment Programme (SIP) on SLM in Sub-Saharan Africa found that the majority of SIP projects disregarded land tenure.[65] The complexity of responsible land governance as well as the short-term nature of the SIP projects were the reasons given for this neglect.[66] The relatively high transaction costs associated with helping decentralised governmental agencies regulate and administer land tenure at local level was another obstacle.[67]

The monitoring of the outcomes of the social innovations in Kenya and Burkina Faso revealed that secure access to land, especially for women and other vulnerable groups, provided investment incentives for SLM. In Burkina Faso, secure land use rights, combined with composting training, resulted in higher agricultural production. In Kenya, farmers were able to enhance their farm size through lease agreements. This enabled them to benefit from SLM training as model farmers. Further, the lease agreements provided a basis for the adoption of SLM measures without leading to conflicts between landowners and tenants. Previously, the implementation of certain SLM measures, such as tree planting, was constrained when only oral lease arrangements existed.

[64] World Bank (2006).

[65] FAO and TerraAfrica (2016).

[66] Ibid.

[67] Paradza (2011).

6 Concluding Remarks

Tenure security is not a silver bullet for widespread adoption of soil protection measures by smallholder farmers. Many obstacles to their adoption exist. These include lack of or limited access to locally responsive agricultural extension services, dependency on unsustainable agricultural inputs, unaffordable credit schemes, or limited access to markets. However, neglecting individuals' or communities' legitimate claims to land when rolling out large-scale restoration programmes is not an option—neither from an economic nor a social justice standpoint.

Progressive land policies have been adopted on paper in many countries. However, the enforcement and implementation of such policies remain weak. Legal reforms often do not translate into more secure land tenure, especially for women and other vulnerable groups. They continue to be in a disadvantaged situation with regard to land tenure and have therefore little incentive to invest in sustainable land management practices that only pay off after several years.

Addressing these insufficiencies has been at the core of TMG's research projects in Burkina Faso and Kenya. Facilitated by civil society organisations and involving strong partnerships with local actors, social innovations have strengthened legitimate claims to land and enhanced the tenure situation for women and other land-constrained farmers. These innovations provide evidence for the potential of local and effective approaches to tackle land tenure issues, such as insecure access to land for vulnerable groups, within soil restoration projects. The fact that these have been integrated into a global soil programme testifies to this potential.

Ever-increasing pressures on land resources due to growing demand for food, energy, housing, as well as restoration for biodiversity preservation and climate change mitigation, require sustainable development pathways that do not compete with one another. Safeguarding peoples' legitimate right to land is a core element of such pathways. Innovative, community-driven approaches with the aim of aligning customary and statutory land tenure systems need to be further explored where land reform policies have not delivered the intended results. It is necessary to explore other options to address the nexus between tenure security and soil restoration. The inclusion of innovative, community-driven approaches in land and soil policy frameworks and development programmes provides a viable and effective alternative in situations where the implementation of broader land reforms is lacking.

References

Ali DA, Deininger K, Goldstein M (2014) Environmental and gender impacts of land tenure regularization in Africa: Pilot evidence from Rwanda. J. Dev. Econ. 110:262–275. https://doi.org/10.1016/j.jdeveco.2013.12.009

Awono A, Somorin OA, Eba'a Atyi R, Levang P (2014) Tenure and participation in local REDD+ projects: Insights from southern Cameroon. Environ. Sci. Policy 35:76–86. https://doi.org/10.1016/j.envsci.2013.01.017

Bambio Y, Bouayad Agha S (2018) Land tenure security and investment: does strength of land right really matter in rural Burkina Faso? World Dev. 111:130–147. https://doi.org/10.1016/j.worlddev.2018.06.026

Bock BB (2012) Social innovation and sustainability; how to disentangle the buzzword and its application in the fi eld of agriculture and rural development. Stud. Agric. Econ. 114:57–63. https://doi.org/10.7896/j.1209

County Government of Kakamega (2018) Kakagema County Integrated Development Plan 2018–2022. https://kakamega.go.ke/download/county-integrated-development-plan-2018-2022/ [last accessed 28 August, 2021]

Deininger K, Savastano S, Xia F (2017) Smallholders' land access in Sub-Saharan Africa: a new landscape? Food Policy 67:78–92. https://doi.org/10.1016/j.foodpol.2016.09.012

Enemark S, Bel KC, Lemmen C, McLaren R (2015) Fit-for-purpose land administration, FIG/World Bank. https://doi.org/10.1080/00050326.1978.10441468

FAO, TerraAfrica (2016) Informing Future Interventions for Scaling-up Sustainable Land Management

Goldstein M, Udry C (2008) The profits of power: land rights and agricultural investment in Ghana. J. Polit. Econ. 116:981–1022

Government of Burkina Faso (2009) LOI N° 034-2009/AN PORTANT REGIME FONCIER RURAL

Government of Burkina Faso (2018a) PROGRAMME DE DEFINITIONS DES CIBLES DE LA NEUTRALITE EN MATEIRE DE DEGRADATION DES TERRES (PDC/NDT)

Government of Burkina Faso (2018b) Deuxième Programme National du Secteur Rural (PNSR II)

Government of Kenya (2019) Kenya Population and Housing Census Volume I: Population by County and Sub-County. https://www.knbs.or.ke/?wpdmpro=2019-kenya-population-and-housing-census-volume-i-population-by-county-and-sub-county [last accessed 28 August, 2021]

Higgins D, Balint T, Liversage H, Winters P (2018) Investigating the impacts of increased rural land tenure security: a systematic review of the evidence. J. Rural Stud. 61:34–62. https://doi.org/10.1016/j.jrurstud.2018.05.001

Hughes O (2014) Literature review of land tenure in Niger, Burkina Faso, and Mali. Context and Opportunities, Baltimore, USA

IPCC (2019) Climate Change and Land: an IPCC special report on climate change, desertification, land degradation, sustainable land management, food security, and greenhouse gas fluxes in terrestrial ecosystems, IPCC Intergovernmental Panel on Climate Change. https://doi.org/10.4337/9781784710644

Kameri-Mbote DP (2002) Gender Dimensions of Law, Colonialism and Inheritance in East Africa

Kevane M, Gray LC (1999) A woman's field is made at night: gendered land rights and norms in Burkina Faso. Fem. Econ. 5:1–26. https://doi.org/10.1080/135457099337789

Kiragu S, Flohr A (2016) Sustainable Land Management in Western Kenya – Lessons Learnt and Future Directions. Potsdam, Germany. https://doi.org/10.2312/iass.2016.012

Koudougou S, Stiem L (ed., 2017) La Gestion Durable des Terres au Burkina Faso: une analyse d'expériences de projets dans le Houet, le Tuy et le Ioba. Potsdam, Germany. https://doi.org/10.2312/iass.2017.006

Koudougou S, Stiem-Bhatia L, Bary H, Tall F (2017) Genre, foncier et gestion durable des terres au Burkina Faso. Potsdam, Germany. https://doi.org/10.2312/iass.2017.024

Linkow B (2016) Causes and consequences of perceived land tenure insecurity: survey evidence from Burkina Faso. Land Econ. 92:308–327. https://doi.org/10.3368/le.92.2.308

Lusweti I (2017) Land leasing rapid appraisal in Isukha West ward, Shinyalu sub-county, Kakamega County, Kenya. Internal TMG Report. Unpublished

Mburu J, Kiragu-Wissler S (2018) Sustainable Land Management in Western Kenya: An Analysis of Project-Basd Interventions. Berlin, Germany

Ministry of Environment and Natural Resources (2016) Kenya Strategic Investment Framework for Sustainable Land Management 2017–2027. http://www.environment.go.ke/wp-content/

uploads/2018/08/KSIF-Kenya-Strategic-Investment-Framework-on-SLM-2017-2027.pdf [last accessed 28 August, 2021]
Moulaert F, Martinelli F, Swyngedouw E, González S (2005) Towards alternative model(s) of local innovation. Urban Stud. 42:1969–1990. https://doi.org/10.1080/00420980500279893
Muok BO, Kingiri A (2015) The role of civil society organizations in low-carbon innovation in Kenya. Innov. Dev. 5:207–223. https://doi.org/10.1080/2157930X.2015.1064558
Musangi P (2017) Women, land and property rights in Kenya, Paper prepared for presentation at the "2017 World Bank Conference on Land and Poverty."
Namubiru-Mwaura E (2014) Land Tenure and Gender: Approaches and challenges for strengthening rural women's land rights (No. 7 92761). Women's Voice and Agency Research Series
Neumeier S (2017) Social innovation in rural development: identifying the key factors of success. Geogr. J. 183:34–46. https://doi.org/10.1111/geoj.12180
Nyangena W (2010) Determinants of household participation in land rental markets in Rural Kenya. University of Nairobi, School of Economics, Nairobi, Kenya
Paradza G (2011) Innovations for securing women's access to land in Eastern Africa. The International Land Coalition
Phills JA Jr, Deiglmeier K, Miller DT (2008) Rediscovering social innovation. Stanford Soc Innov Rev
Stiem L, Krause T (2016) Exploring the impact of social norms and perceptions on women's participation in customary forest and land governance in the Democratic Republic of Congo—implications for REDD+. Int. For. Rev. 18:110–122. https://doi.org/10.1505/146554816818206113
Stiem-Bhatia L, Koudougou S (2018) Innovations for women's land access. Rural 21 52
TMG Research (2019) Global Soil Week 2019: Creating an Enabling Environment for Sustainable and Climate-Resilient Agriculture in Africa. Berlin, Germany. https://globalsoilweek.org/wp-content/uploads/2019/09/gsw_report_edit_v3-1.pdf [last accessed 28 August, 2021]
UNDP (2019) Kenya [WWW Document]. Hum Dev Reports. URL http://hdr.undp.org/en/countries/profiles/KEN [last accessed 28 August, 2021]
World Bank (2006) Sustainable Land Management: Challenges, Opportunities, and Trade-offs. The International Bank for Reconstruction and Development/ The World Bank, Washington, DC
Zevenbergen J, Augustinus C, Antonio D, Bennett R (2013) Pro-poor land administration: principles for recording the land rights of the underrepresented. Land Use Policy 31:595–604. https://doi.org/10.1016/j.landusepol.2012.09.005

Part II
Recent International Developments on Soil Governance

UNCCD COP 14: Mirroring Soil and Land's Growing Relevance at the Interface of Climate and Biodiversity; Discussion Focus on Drought and Land Tenure

Judith Rosendahl, Alexander Erlewein, and Antje Hecheltjen

Abstract This chapter presents the discussions and outcomes of the UNCCD's 14th Conference of the Parties (COP), held in New Delhi, India in September 2019. It highlights the most relevant decisions and discusses their significance in light of the Convention's development. The record number of participants mirrored the growing relevance of the Convention. Especially the decisions on the most debated agenda items, drought and land tenure, have given new impetus to UNCCD's work. Overall, the entanglement of soil and land with climate and biodiversity and thus the synergies between the three Rio Conventions were emphasised. Land Degradaton Neutrality (LDN) remains key to the Convention, but the LDN target setting process did not gain further momentum. The discrepancy between the UNCCD's formally restricted mandate for drylands and its de facto removal through LDN as a global goal remains the key source of tension and impediment for the Convention. In the eve of the UN Decade for Ecosystem Restoration, the question of how to mobilise financing for LDN is coming increasingly to the fore.

1 Introduction

1.1 Overview of Characteristics of UNCCD

The United Nations Convention to Combat Desertification (UNCCD), together with the United Nations Framework Convention on Climate Change (UNFCCC) and the Convention on Biological Diversity (CBD), forms the group of three so-called Rio Conventions resulting from the United Nations Conference on Environment and

---

J. Rosendahl (✉) · A. Erlewein · A. Hecheltjen
Deutsche Gesellschaft für Internationale Zusammenarbeit (GIZ) GmbH, Department Sector and Global Programmes, Bonn, Germany
e-mail: judith.rosendahl@giz.de; alexander.erlewein@giz.de; antje.hecheltjen@giz.de

H. Ginzky et al. (eds.), *International Yearbook of Soil Law and Policy 2020/2021*, International Yearbook of Soil Law and Policy 2020,
https://doi.org/10.1007/978-3-030-96347-7_9

Development (UNCED) held in Rio de Janeiro in 1992.[1] The UNCCD was adopted in 1994, entered into force in 1996 and has been ratified by 197 countries.[2] The Convention was established mainly in response to pressure from African states and aims primarily to combat land degradation in drylands (= desertification) and to combat the effects of drought. Accordingly, it was long considered an "Africa Convention" or "Desertification Convention" and remained politically significantly less relevant than the UNFCCC and CBD. Yet, UNCCD is not only the sole legally binding international agreement in the soil/land sector, but also explicitly links environment and development.

Meanwhile, it is recognized that, while drylands are most severely affected, soil and land degradation is a widespread problem in all parts of the world and that soil and land play a fundamental role in climate protection and in the preservation of biodiversity, ecosystem services and human well-being.[3] This is reflected in the increased attention paid to soils and land in recent years by development and environmental policy, culminating in the inclusion of soil and land degradation in the Agenda 2030 for sustainable development. Target 15.3 of the Sustainable Development Goals (SDGs) calls for land degradation neutrality (LDN) for every country by 2030, meaning that zero net land degradation should be achieved by applying a response hierarchy of avoiding, reducing and reversing land degradation.[4] The UNCCD is the custodian agency of SDG 15.3, thus officially entrusted with measuring and reporting on progress towards this target and has strongly focused the convention process on it in recent years. However, with LDN as a global goal and center of UNCCD's strategy 2018-2030, the UNCCD addresses land degradation as a global problem and thus transcends its mandate which is still legally restricted to drylands, generating an area of tension and constant source of dissent among the Parties to the Convention.[5]

1.2 *Overview of the 14th Conference of the Parties*

The UNCCD biennially holds Conferences of the Parties (COPs) as their supreme decision-making body, during which also the two subsidiary bodies for reviewing

---

[1] Full name of UNCCD: United Nations Convention to Combat Desertification in those Countries Experiencing Serious Drought and/or Desertification, particularly in Africa.

[2] All 193 UN Member States, the Cook Islands, Niue, the State of Palestine and the European Union. United Nations Treaty Collection: https://treaties.un.org/Pages/ViewDetails.aspx?src=TREATY&mtdsg_no=XXVII-10&chapter=27&clang=_en#1 (last accessed on 9 September 2021).

[3] United Nations General Assembly (2012), para 205.

[4] "By 2030, combat desertification, restore degraded land and soil, including land affected by desertification, drought and floods, and strive to achieve a land degradation-neutral world", United Nations General Assembly (2015), p. 24.

[5] Boer et al. (2017), p. 61 f.

implementation (CRIC)[6] and for scientific advice (CST)[7] meet in parallel. After COP 13 was held in Ordos, China, in September 2017, India took over the COP presidency from China and hosted the 14th COP in New Delhi from 2 to 13 September 2019. As presidency, India displayed its commitment to environmental policy and announced, among other things, that it would increase its land rehabilitation targets to 26 million hectares. Ibrahim Thiaw, the new Executive Secretary of the Convention since 2019, presented himself and his priorities for the Convention. He emphasised the need to focus more strongly on the implementation of the convention and to link it even better with the other two Rio Conventions, UNFCCC and CBD. The approximately 6000 participants made this COP the largest in the history of the UNCCD. As the decision-making body of the Convention, this Conference of the Parties concluded with 36 decisions that will shape the UNCCD process.

Following this overview of UNCCD and the 14th Conference of the Parties, this chapter presents and discusses the most important agenda items. These are Land Degradation Neutrality (LDN), drought, and land tenure. For each topic, we (i) provide background information and (ii) summarize the discussions and negotiations, before (iii) analyzing the decisions. The chapter concludes by synthesizing the findings and providing an assessment and outlook. Further COP 14 decisions that go beyond the scope of this article deal with, among others, the issues of sand and dust storms, displacement and migration, gender equality, the Convention's financial performance and budget, it's communication plan, and the involvement of civil society organizations.[8]

[6]Committee for the Review of the Implementation.

[7]Committee for Science and Technology.

[8]The increased global relevance of the issue of Displacement and migration is mirrored in the Convention. COP 13 (decision 28) firstly takes up the positive role that measures taken to implement the Convention can play to address DLDD as one of the drivers that causes migration. COP 14 took this up, stressing the multiple ecological, social and economic benefits of measures to combat DLDD. The "3S initiative" (sustainability, stability, security) is one concrete initiative in the frame of which some African Parties dispose of external financing to implement measures countering DLDD for migration mitigation. Decision 22/COP.14 invites Parties to various activities at the national level, among other fostering renewable energy, promoting the development of small- and medium-sized enterprises in rural areas, encouraging responsible and sustainable investment by the private sector in land restoration, and supporting the 3S initiative. The Secretariat is requested to support Parties with the above-mentioned activities and to support relevant regional and international cooperation, as well as to strengthen cross-sectoral cooperation with relevant agencies and stakeholders.

2 Thematic Foci: Land Degradation Neutrality, Drought and Land Tenure

2.1 *Land Degradation Neutrality (LDN)*

2.1.1 Background

The UNCCD has long suffered from the absence of a unifying and quantifiable target. Neither there was an agreement on how to measure desertification consistently, nor a framework to set quantitative targets for reducing the degradation of land and soils. This finally changed in 2015 when COP 12 took up the vision of a *land degradation neutral world* as stipulated in Sustainable Development Goal 15.3. Two years later COP 13 completed this re-orientation of the Convention and made LDN the centerpiece of the UNCCD Strategic Framework for 2018–2030. LDN is understood as no net loss of land resources.[9] In other words, a balance between new degradation and restoration of past degradation that can be achieved through avoiding, reducing and reversing land degradation. In order to make progress measurable, the Scientific Conceptual Framework for Land Degradation Neutrality sets out three indicators: land cover (land cover change), land productivity (net primary productivity) and carbon stocks (soil organic carbon).[10] Based on this concept, the UNCCD started an LDN Target Setting Programme that aims to bring down the global vision of a *land degradation neutral world* to more tangible, albeit voluntary, national targets.

2.1.2 Discussions and Negotiations

At COP 14 Parties reaffirmed the orientation of the Convention towards LDN, particularly praising the success of the LDN Target Setting Programme. As of 2020 123 states have so far committed themselves to voluntary national targets for achieving LDN and 84 states have in fact formulated these targets, even though in different ways ranging from rather general objectives to concrete targets that are specific in terms of land use class and quantification. Interestingly, many countries went far beyond drylands when setting LDN targets and identifying priority areas. This shows once more that land and soil degradation is a major challenge for rural development in all climatic zones and not limited to arid, semi-arid and sub-humid areas on which the UNCCD focuses traditionally. Overall, the number of participating countries has clearly exceeded initial expectations. With a growing number of

---

[9] LDN has been defined by the Parties to the Convention as "A state whereby the amount and quality of land resources, necessary to support ecosystem functions and services and enhance food security, remains stable or increases within specified temporal and spatial scales and ecosystems" [Decision 3/COP.12, in UNCCD (2015)].

[10] Orr et al. (2017), p. 92 ff.

countries that have set LDN targets, discussions at COP 14 underlined that the real challenge lies in their implementation. Several countries urged that it is at this stage where bold targets will have to be translated into policies and projects that actually improve the way land is managed on the ground. A number of positive country examples were presented in the COP sessions and side events, demonstrating that in some countries LDN created significant momentum to address desertification and soil degradation. Key success factors emerging from these discussions included integration of LDN targets in existing policies for sustainable land management and restoration initiatives as well as intersectoral cooperation between environmental and agricultural agencies. At the same time, it became also clear that LDN target setting in several countries did not go far beyond the office of the respective national UNCCD focal point. In terms of financial support, it was recognized that a number of multi- and bilateral agencies have joined the LDN journey and scaled up investments in sustainable land management and landscape restoration. However, numerous Parties called for stronger support to implement LDN targets, making it clear that LDN does not fully live up to expectations in terms of mobilizing funding for sustainable land management and rehabilitation.

2.1.3 Decision and Follow-Up

The decision taken mandates the Secretariat to continue to support the Parties in the implementation of LDN, including the identification of project and financing opportunities, as well as the provision of concepts and guidelines. Parties renewed UNCCD's mandate to use LDN indicators for reporting and reaffirmed the commitment for harmonization with Agenda 2030. Furthermore, the decision "invites Parties who have not yet done so, to formulate voluntary land degradation neutrality targets [...] and invites Parties that have committed to voluntary land degradation neutrality targets to implement measures to accelerate their achievement, as appropriate".[11] With regard to implementation the decision highlights the role of integrated land use planning and the creation of an "enabling environment for the achievement of land degradation neutrality, including the responsible governance of land and tenure security, the engagement of stakeholders and the improved access of smallholders to advisory and financial services".

[11] Decision 3/COP.14, in UNCCD (2019b).

2.2 Drought

2.2.1 Background

Mitigating the effects of drought is enshrined as an objective in the text of the Convention.[12] For a long time, however, the topic was not the focus of the Convention. This changed with the adoption of the Strategic Framework 2018–2030 at COP 12, when drought mitigation and management became one of the five strategic objectives. The expected impact is to reduce ecosystems' vulnerability through sustainable land and water management and to increase communities' resilience to drought.[13]

The nexus of land and drought, climate change, natural resource-based conflict, environmentally induced migration, and security is well recognized.[14] Africa is particularly affected. Therefore, African country Parties have long been fighting for a higher priority of drought under the Convention, culminating in their call for a binding drought protocol in the Windhoek Declaration for Enhancing Resilience to Drought in Africa in 2016.[15] A drought protocol would be binding under international law and would be one of several possible legal instruments. Other forms, both binding and non-binding under international law, include annexes and treaty amendments, principles, declarations, decisions and standards.[16]

2.2.2 Discussions and Negotiations

The proposal of a drought protocol was once again highly debated in negotiations at COP 14. African Parties considered a protocol being the best way to mobilize resources for drought mitigation. Traditional donor countries continued to oppose the proposal. Common European position is that the added value of a protocol is unclear and would involve many years of negotiations. Several delegates pointed out that energy and resources would be better invested in addressing today's urgent needs rather than in lengthy protocol negotiation processes. They preferred to make use of and strengthening existing mechanisms such as early warning systems and examine how these can be used more efficiently. As a compromise, they proposed an informal intersessional process to elaborate the modalities of assistance to Parties affected by drought.[17]

---

[12] UNGA (1994).

[13] Annex to Decision 7/COP.13, in UNCCD (2017a).

[14] Mapedza et al. (2019), Reichhuber et al. (2019).

[15] African Member States and Parties to the United Nations Convention to Combat Desertification (UNCCD), Ministers, Heads of Delegation and Experts, attending the High Level Meeting of the first African Drought Conference (ADC) (2016).

[16] Decision 16/COP.14, in UNCCD (2019b).

[17] IISD Reporting Services (2019).

The negotiation process on drought was also characterized by the concern to avoid duplication of efforts and to maximize synergies through cooperation with relevant institutions. Different intergovernmental institutions deal with drought from different perspectives, for example: the World Meteorological Organization (WMO) from a meteorological perspective, the United Nations Office for Disaster Risk Reduction (UNDRR) from a disaster risk management perspective, the Food and Agriculture Organization (FAO) from an agricultural, social and food security, and social protection perspective, the United Nations Framework Convention on Climate Change (UNFCCC) from a climate change adaptation perspective, the World Health Organization (WHO) from a health perspective or the Global Water Partnership (GWP) from a water perspective. UNCCD's perspective is that of land-based interventions for drought management and mitigation, stressing the point that sustainable land management is an essential building block of drought mitigation.[18]

2.2.3 Decisions and Follow Up

Finally, Parties agreed on a workable compromise between a binding protocol and an informal process, namely the establishment of an intergovernmental working group on effective policy and implementation measures for addressing drought under the UNCCD (IWG). The 30 government representatives and experts, who form the IWG, are to draw up proposals for policy, advocacy and implementation measures to address drought under the Convention more effectively.[19] They will present their final report by the next COP, which will serve as a basis for upcoming COP decisions on drought.

In addition to establishing the IWG, the Parties agreed on further actionable measures to mitigate the effects of drought. They mandated the Secretariat, the Global Mechanism and UNCCD bodies to build on the results of UNCCD's drought initiative in cooperation with relevant stakeholders and institutions.[20] This includes (a) upgrading the drought toolbox to boost the resilience of people and ecosystems to drought, (b) designing and testing innovative, gender-responsive and transformative drought mitigation approaches, and (c) continuing with the development and implementation of national drought preparedness plans.[21] As of June 2020, more than 70 countries are in the process of drafting national-level plans of action, which indicates that drought preparedness is perceived as a pressing issue at national level.[22]

---

[18]Decision 17/COP.14 in UNCCD (2019b) and Reichhuber et al. (2019).

[19]Decision 23/COP.14 in UNCCD (2019b).

[20]The drought initiative was implemented for the biennium 2018–2019 based on decision 29/COP.13.

[21]Ibd.

[22]15 exemplary national drought preparedness plans are available online at https://knowledge.unccd.int/drought-toolbox/page/drought-planning (last accessed on 9 September 2021).

This also explains the strong commitment of the Parties to drought monitoring and reporting. In the course of the negotiation process, the text of the draft decision became more binding: while the draft negotiation text stated rather generally that "reporting is required", the final COP decision reads that affected country parties are "requested" to report on at least one of the three drought indicators. The agreed indicators are with increasing complexity: "trends in the share of the drought-affected country in the total land area" (indicator of level 1), "trends in the share of the drought-affected population in the total population" (indicator of level 2), and "trends in the degree of vulnerability to drought" (indicator of level 3). As with LDN monitoring, the Secretariat is requested to (a) provide affected country Parties with national estimates from global datasets as default data for validation (for level 1 and level 2 indicators) and (b) prepare methodological good practice guidance and provide capacity-building and technical assistance to affected country Parties (including on level 3 indicator).[23]

2.3 *Land Tenure*

2.3.1 Background

This COP for the first time dealt with land tenure as a separate issue and agenda item.

In recent years, the Convention has more and more considered and acknowledged that implementing measures to combat desertification, land degradation and drought (DLDD) is not only or primarily a technical but in essence also a governance issue.

COP 13 decisions firstly register mentions of land governance, namely the scientific framework for LDN that recommends the adoption of the Voluntary Guidelines on the Responsible Governance of Tenure of Land, Fisheries and Forests in the Context of National Food Security (VGGT), the recommendation from civil society organizations to include land rights in LDN measures and the SPI work programme for the biennium 2018/2019.[24]

The work by the SPI started in 2016 with the development of their draft work programme for the biennium 2018–2019 by inviting Parties and other UNCCD stakeholder to submit potential topics, by reviewing inputs and identifying priorities based on criteria.[25] As a result, the work programme contained in a COP 13 decision

---

[23]Decision 11/COP.14, in UNCCD (2019b).

[24]UNCCD (2019a), p. 3; UNCCD (2017b); The VGGT are a non-binding document under international law, which was developed in intergovernmental negotiations involving civil society, the private sector and academia and adopted by the FAO Committee on World Food Security in 2012. The VGGT collect, combine and develop existing governance standards for land management and use agreed language of other international negotiations. Since then, work is underway to implement them and they have become the central reference document in this area and have imprinted UNCCD's discourse; Decision 15/COP.13, in UNCCD (2017a, c).

[25]Ibid.

states as one of two objectives to provide refined guidance for LDN implementation, with one deliverable being a report providing "science-based evidence of how land-based interventions can improve the well-being and livelihoods of people while simultaneously enhancing environmental security".[26] The report, along with a policy paper entitled "Shaping an Enabling Environment for Land Degradation Neutrality" was published in September 2019 in the eve of COP 14.[27] It established "enabling environment" as a concept, defined as "a conducive institutional, policy, regulatory and financial setting for progress to be made towards LDN" and identifies four dimensions: science-policy interface; financial; institutional; and policy and regulatory.[28] Subsumed under "institutional", land governance is presented as a critical cross-cutting element that covers land-use planning, land-based interventions and land tenure. Land tenure security is described as fundamental for the attainment of LDN, "...because a) it enables land managers to be effective agents towards sustainability; b) lack of tenure security can cause degrading land management practices; c) many land-based interventions (...) cannot work under insecure land tenure conditions".[29]

2.3.2 Discussions and Negotiations

Initiated by Germany, the European Union and its Member States developed a joint position towards land tenure in the context of UNCCD, contained in a Non-Paper that outlines several recommendations for Parties' commitments as possible elements for a UNCCD draft decision. This EU Non-Paper strongly builds on and makes reference to the VGGT. Following an EU proposal, land tenure was included in the official agenda of COP 14 as (the only) "new and emerging issue". Parties for the first time were requested to take a formal decision on the relevance and recommended modes and principles of land tenure in the context of fighting DLDD and achieving LDN. The UNCCD's "Note by the Secretariat" containing background information and the draft decision text on land tenure made a similarly strong case as the SPI and the EU Non-Paper.[30] It gives an overview of causal linkages between tenure security and land degradation, outlines positive impacts of tenure security, examines the integration of the VGGT into LDN and addresses the question whether SDG indicators on land tenure can usefully be integrated in UNCCD reporting, before presenting the draft decision text.[31]

[26] Ibid.

[27] Verburg et al. (2019).

[28] UNCCD/Science-Policy Interface (2019), p. 1.

[29] Ibd. p. 4.

[30] Decision 20/COP.14, in UNCCD (2019b).

[31] 1.4.2 "Proportion of total adult population with secure tenure rights to land, with legally recognized documentation and who perceive their rights to land as secure, by sex and by type of tenure", 5.a.1 "Women's ownership of agricultural land" and 5.a.2 Proportion of countries where the legal framework (including customary law) guarantees women's equal rights to land ownership and/or control.

It was against this backdrop that statements and negotiations started at COP 14. Statements made by Parties in the plenary session ahead of the negotiations underlined the sensitivity of the issue. Several Parties expressed concerns about the draft decision text, while others were supportive. Critical Parties were concerned about their national sovereignty in this area of negotiation, stressed the different country-specific contexts, or requested a voluntary nature of any wording of the decision. Some were also concerned about a possible expansion of the scope of the convention, that the issue would draw attention away from core issues of the convention or add additional reporting obligations. Among the supportive Parties welcoming the integration of VGGT into LDN, some acknowledged land tenure as one barrier to achieving LDN while others pointed out a need for technical assistance and funding.

In the course of the COP it transpired that, together with drought, land tenure was one of the most controversial issues. Contentious points implied whether certain aspects of the VGGT should be singled out or not, the principle of free, prior and informed consent, and the linguistic graduation of what Parties should—voluntarily—do.

2.3.3 Decision and Follow Up

The eventually adopted decision *notes* the importance of land tenure and of the VGGT for the implementation of measures against DLDD, *encourages* Parties to follow the VGGT, taking into account the principles of implementation, and highlights several specific aspects such as *encouraging* Parties to recognize legitimate tenure rights, including customary rights or *encouraging* Parties to provide effective, timely, and affordable access to justice and transparent dispute resolution mechanisms. Compared with the draft, it contains stronger wording at diverse points in the text. Immediate actions points implied in the decision are (i) the exploration of options for the integration of existing SDG indicators relevant to land governance in the UNCCD reporting process, (ii) the exploration of options on how to promote awareness-raising on responsible land governance for combating DLDD and (iii) the development of a Technical Guide on how to integrate the VGGT into the implementation of the Convention and LDN. For the elaboration of the Technical Guide, UNCCD and FAO will hold an online consultation in 2020 whose findings will again feed into the ongoing debate at CRIC 19 and COP 15.

3 Conclusions

The increased attention paid to soils and land in recent years by development and environmental policy is reflected in the growing relevance of UNCCD, culminating in the inclusion of soil and land degradation in the Agenda 2030 for sustainable development. Since 2015 the Convention has focused its goals and processes on

SDG 15.3 on Land Degradation Neutrality, thus aiming to gain global relevance beyond its original legal mandate for drylands. This ambivalence between the Convention's mandate for drylands and this concurrent de facto removal of this limitation through its custodianship for LDN as a global goal remains a continued source of tension and the main impediment for a higher relevance and impact of the Convention. However, the outcomes of COP 14 are overall predominantly assessed positively. Especially the decisions on drought and land tenure have given new and substantive impetus to UNCCD's work and will shape the Convention's work in the years to come.

With regard to Land Degradation Neutrality (LDN), the Parties reaffirmed the orientation of the Convention towards this goal. In particular, the success of the LDN Target Setting Programme was praised and negotiations focused primarily on the implementation of these targets. The decision mandates the Secretariat to continue to support the Parties in the implementation of LDN. Parties renewed its mandate to use LDN indicators for reporting and reaffirmed its commitment to harmonization with Agenda 2030, while encouraging Parties to integrate LDN into national land use planning and to exploit synergies with the implementation of the other Rio Conventions and Agenda 2030. This decision creates a basis for the continuation of the LDN agenda but falls short of developing it further. Though the updating of LDN targets was proposed, substantial discussions on processes such as stock taking or ambition raising did not take place. Overall, the momentum around LDN that characterized previous COPs was not maintained at the same degree, also because of new priorities of the secretariat's new leadership. Thus, the initial expectation of some Parties and observers that LDN targets may develop a similar dynamic as the target setting processes of the other Rio Conventions did not materialize.[32]

After the Conventions' strong focus on LDN in the past years, debates on how to deal with the issue of drought have been reinforced since COP 13. The Parties decided to continue the initiative to develop national drought plans, as well as to develop corresponding guidelines and cooperation with other organisations. In the negotiations, after African states initially once again called for a drought protocol that would be binding under international law, Parties agreed on the establishment of an intergovernmental working group (IWG) on drought with the mandate to develop proposals for further work on the topic by the next COP. Moreover, Parties agreed to establish a drought monitoring framework that will be effective for the upcoming reporting cycle. Even if a drought protocol could not prevail, first steps towards formalization have been taken with the establishment of the IWG and the new reporting obligation on drought. After having played a minor role in the Convention for years, drought is now becoming an integral part of the Convention process. With the development of national drought preparedness plans in now 70 countries, the process has also developed a stronger dynamic at the national level.

The new agenda item of land tenure centers on the integration of human rights principles in the implementation of LDN. After the LDN scientific framework makes

[32] Bodle et al. (2020).

reference to the VGGT and the SPI's work on an "enabling environment" for LDN reached the conclusion that land tenure security is fundamental for the attainment of LDN, Parties struggled to formulate the relevance of the VGGT for the implementation and corresponding action points. The decision that encourages Parties to follow the VGGT is considered by many experts in UNCCD, intergovernmental and civil society organizations not only as a main outcome and achievement of COP 14, but as a landmark decision that has put the issue firmly on the Convention's agenda. The International Institute for Sustainable Development (IISD) lists it as one positive development hit of global environmental governance in 2019 altogether and UNCCD, after recalibrating its work programme in light of the COVID-19 pandemic, confirmed that land governance will remain a top priority.[33,34]

In the frame of the consultation for the elaboration of an FAO Technical Guide and beyond, UNCCD stakeholders will continue to debate the relevance of including such human rights principles into LDN implementation. While the theoretical relevance has been documented sufficiently by SPI and others, more evidence on the practical relevance and factual situations on the ground will be needed to further the debate and make it meaningful for the implementation of the Convention and ultimately for affected people.

A further recurrent discussion item that was mainly addressed in some of the 144 side events and that points to the bigger picture of activities to counter DLDD is the question of how to promote private sector investment in sustainable land management and land restoration. With the Land Degradation Neutrality Fund, UNCCD has set up a blended finance instrument to leverage private sector investments in SLM and land restoration which had been endowed with 100 Mio USD until mid-2019 and started a first restoration project, after debates on its environmental and social safeguards.[35] Yet, the need to finance sustainable land management beyond this fund in order to achieve the announced political goals is enormous. Estimated annual funding needs for conservation and landscape restoration amount to USD 300-400 billion. Within this massive financing gap, 80 percent of the available funds currently stem from public sources.[36] The UN Decade for Ecosystem Restoration 2021–2030 provides an international framework and further impetus for restoration activities.

[33] Allan et al. (2020).

[34] UNCCD (2020), p. 4.

[35] UNCCD: https://www.unccd.int/actions/impact-investment-fund-land-degradation-neutrality (last accessed on 9 September 2021).

[36] Gichuki et al. (2019), p. 22.

References

African Member States and Parties to the United Nations Convention to Combat Desertification (UNCCD), Ministers, Heads of Delegation and Experts, attending the High Level Meeting of the first African Drought Conference (ADC) (2016) The Windhoek Declaration for Enhancing Resilience to Drought in Africa. Adopted on the 19th August 2016 in Windhoek, Namibia

Allan J, Antonich B, Bansard J, Bhandary RR, Chasek P, Jones N, Leone F, Jungcurt S, Paul D, Tsioumani A, Tsioumani E (2020) The State of Global Environmental Governance. Winnipeg, Canada

Bodle R et al (2020) Improving international soil governance - Analysis and recommendations. Umweltbundesamt Texte 75/2020

Boer BW, Ginzky H, Heuser IL (2017) International soil protection law: history, concepts and latest developments. In: Ginzky H, Heuser I, Qin T, Ruppel O, Wegerdt P (eds) International yearbook of soil law and policy 2016. Springer, Cham

Gichuki L, Brouwer R, Davies J, Vidal A, Kuzee M, Magero C, Walter S, Lara P, Oragbade C, Gilbey B (2019) Reviving land and restoring landscapes: policy convergence between forest landscape restoration and land degradation neutrality. IUCN, Gland, Switzerland. viii + 34pp

IISD Reporting Services (2019) Summary of the Fourteenth Session of the Conference of the Parties to the UN Convention to Combat Desertification: 2-13 September 2019. Earth Negotiations Bulletin - A Reporting Service for Environment and Development Negotiations. Vol. 4 No 290. 16 September 2019

Mapedza E, Tsegai D, Brüntrup M, McLeman R (eds) (2019) Drought challenges: policy options for developing countries. Elsevier

Orr BJ, Cowie AL, Castillo Sanchez VM, Chasek P, Crossman ND, Erlewein A, Louwagie G, Maron M, Metternicht GI, Minelli S, Tengberg AE, Walter S, Welton S (2017) Scientific conceptual framework for land degradation neutrality. A report of the science-policy interface. UNCCD, Bonn, Germany

Reichhuber A, Gerber N, Mirzabaev A, Svoboda M, López Santos A, Graw V, Stefanski R, Davies J, Vuković A, Fernández García MA, Fiati C, Jia X (2019) The Land-Drought Nexus: enhancing the role of land-based interventions in drought mitigation and risk management. A report of the science-policy interface. United Nations Convention to Combat Desertification (UNCCD), Bonn, Germany

UNCCD (2015) ICCD/COP(12)/20/Add.1 Integration of the Sustainable Development Goals and target into the implementation of the United Nations Convention to Combat Desertification and the Intergovernmental Working Group report on land degradation neutrality

UNCCD (2017a) ICCD/COP(13)/21/Add.1. Report of the Conference of the Parties on its thirteenth session, held in Ordos, China, from 6 to 16 September 2017. Part two: Action taken by the Conference of the Parties at its thirteenth session. Addendum

UNCCD (2017b) ICCD/COP(13)/CST/2. The scientific conceptual framework for land degradation neutrality. Synthesis report by the Science-Policy Interface. 28 June 2017

UNCCD (2017c) ICCD/COP(13)/CST/7. Work programme for the Science-Policy Interface (2018–2019), 23 June 2017. Note by the secretariat

UNCCD (2019a) ICCD/COP (14)/20. New and emerging issues: land tenure. Note by the Secretariat. 14 August 2019

UNCCD (2019b) ICCD/COP(14)/23/Add.1. Report of the Conference of the Parties on its fourteenth session, held in New Delhi, India, from 2 to 13 September 2019. Part two: Action taken by the Conference of the Parties at its fourteenth session. Addendum. 23 October 2019

UNCCD (2020) UNCCD and the COVID-19 crisis: land-based solutions for healthy people and a sustainable planet

UNCCD/Science-Policy Interface (2019) Shaping an Enabling Environment for Land Degradation Neutrality. Science-Policy Brief No: 5. September 2019. United Nations (UNCCD), Bonn, Germany

UNCCD: An Impact Investment Fund for Land Degradation Neutrality. https://www.unccd.int/actions/impact-investment-fund-land-degradation-neutrality. Accessed 30 June 2020

UNCCD: Drought Planning: https://knowledge.unccd.int/drought-toolbox/page/drought-planning Accessed 30 June 2020

United Nations General Assembly (1994) A/AC.241/27. Elaboration of an international Convention to Combat Desertification in countries experiencing serious drought and/or desertification, particularly in Africa. Final text of the Convention. Note by the Secretariat. Intergovernmental Negotiation Committee for the elaboration of an international Convention to Combat Desertification in countries experiencing serious drought and/or desertification, particularly in Africa. Final text of the Convention. 12 September 1994

United Nations General Assembly (2012) The future we want. Outcome document of the Rio+20 conference. UN General Assembly Resolution A/RES/66/288, 11 September 2012

United Nations General Assembly (2015) Transforming our world: the 2030 Agenda for Sustainable Development. UN General Assembly Resolution A/RES/70/1, 25 September 2015

United Nations Treaty Collection: United Nations Convention to Combat Desertification in those Countries Experiencing Serious Drought and/or Desertification, Particularly in Africa https://treaties.un.org/Pages/ViewDetails.aspx?src=TREATY&mtdsg_no=XXVII-10&chapter=27&clang=_en#1 Accessed 30 June 2020

Verburg PH, Metternicht G, Allen C, Debonne N, Akhtar-Schuster M, Inácio da Cunha M, Karim Z, Pilon A, Raja O, Sánchez Santivañez M, Şenyaz A (2019) Creating an enabling environment for land degradation neutrality and its potential contribution to enhancing well-being, Livelihoods and the Environment. A Report of the Science-Policy Interface. United Nations Convention to Combat Desertification (UNCCD), Bonn, Germany

Assessment of the African Union, FAO, and UNCCD Roles in Enhancing Soil Governance in Africa Through the Lens of Agriculture Policy Actions

Robert Kibugi

Abstract The governance of soil is important to the African continent, and this is evident when seen through the lens of agriculture. Food production is core to the continents livelihood, and the African Union's (AU) Comprehensive African Agriculture Development Programme (CAADP), implemented nationally is evidence of this. The continent has two treaties with provisions for land and soil conservation, but both have inherent structural weaknesses. African countries continue to implement CAADP, and some have adopted National Agriculture Investment Plans (NAIPs) to drive agricultural transformation. Although this tools focus on enhancing productivity, they also provide a critical opportunity for the AU to cross-fertilize knowledge and lessons from Food and Agriculture Organization of the United Nations (FAO); and the United Nations Convention to Combat Desertification (UNCCD). FAO has developed and is implementing the concept of sustainable soil management (SSM) through which the health and fertility of the soil is of primary importance. This echoes with land degradation neutrality (LDN), which is a key plank of the UNCCD's current strategic plan. The AU through its CAADP programme and, particularly an enhanced Maputo Convention which governance land and soil health including during agriculture, has the requisite legal and policy mechanisms to integrate SSM and LDN into agriculture policy, through sustainable land management strategies. This chapter is an exposition of the poor health, and weak soil governance situation on the African continent and proposes how AU approaches can be enhanced through infusion of knowledge from FAO and UNCCD methods and processes.

R. Kibugi (✉)
Faculty of Law, University of Nairobi, Nairobi, Kenya
e-mail: rmkibugi@uonbi.ac.ke

H. Ginzky et al. (eds.), *International Yearbook of Soil Law and Policy 2020/2021*,
International Yearbook of Soil Law and Policy 2020,
https://doi.org/10.1007/978-3-030-96347-7_10

1 Introduction

This chapter is an assessment of the roles of the African Union (AU), Food and Agriculture Organization of the United Nations (FAO); and the United Nations Convention to Combat Desertification (UNCCD), in the enhancement of sustainable soil governance in Africa. The review is in context of prioritized agricultural actions, especially for smallholders, which are adversely affected by land and soil degradation. The chapter argues that FAO and the UNCCD continue to provide knowledge and methodologies to implement sustainable soil management, and land degradation neutrality (LDN) respectively. In this context, the role of the AU, as the principal treaty body established by African nations, and through which treaties (1968 Algiers, 2003 Maputo) impacting soil governance have been adopted, is reviewed. The AU has also adopted the 2003 Maputo and the 2014 Malabo declarations focused on enhancements in agricultural productivity and food security. Similarly, since 2003, country-level implementation of the Comprehensive African Agriculture Development Programme (CAADP) has been underway, with a current focus on development and implementation of National Agriculture Investment Programmes (NAIP), some receiving technical support from FAO.

The chapter is organized into six sections. Section 2 is an appraisal of the soil management situation in Africa, followed in Sect. 3 by a review of the soil governance mechanisms in context of the AU, including review of treatment by agricultural priorities. Section 4 is an assessment of how the soil management and governance challenges can be buttressed by methodologies and actions driven by the FAO and the UNCCD. Section 5 is the conclusion.

2 Appraising the Soil Management Situation in Africa

In this section, the chapter reviews literature on the status of soil health and fertility in Africa. This review provides an indication on the legal and policy interventions that the African Union, working with FAO and the UNCCD can develop for implementation. Africa contains a great soil diversity with over 60% representing arid, hot or immature soil assemblages.[1] The continent has recorded a long and steady decline in soil health which has negatively affected food productivity in Africa.[2] According to the Soil Atlas for Africa, the soils across the continent face a number of threats which include: loss of soil organic matter; loss of soil carbon; degradation including expansion of decertification in arid and semi-arid areas; climate change; soil sealing; erosion; decline in biodiversity; contamination; compaction; and salinization.[3] When these threats materialize, the result is poor soil

[1] Soil Atlas of Africa (2013), p. 50.

[2] Mokgolodi et al. (2011), pp. 123–131.

[3] Soil Atlas of Africa (2013), p. 149.

health which in turn impacts agriculture, food security, and some risks such as soil sealing cause flooding. Depletion of soil fertility, along with the concomitant problems of weeds, pests, and diseases, is a major biophysical cause of low per capita food production in Africa.[4] Over decades, small-scale farmers have removed large quantities of nutrients from their soils without using sufficient quantities of manure or fertilizer to replenish the soil.[5] A 2016 FAO study on agriculture in East Africa reported that over time, agricultural production will diminish soil carbon stocks unless land is fallowed or soil fertility is restored, attributing the current loss of these carbon stocks to land-use intensification and the conversion of new land for food and fibre production.[6] Reporting on Kenya, the study noted that intensive land uses were also expanding into areas where soil organic carbon stocks are less resilient, or soil conditions are marginal for agriculture such as semi-arid savannahs and grasslands and tropical rainforests which are being converted to arable land at an increasing rate.[7] An earlier, 2015 study, also on Kenya reported that when measured in the surface soils, within a depth of up to 15 cm, the highest soil organic carbon stocks and total nitrogen stocks were in natural forests respectively at 71.6 and 7.1 Megagrams per hectare (Mg Ha).[8] In turn, the lowest soil organic carbon stocks and total nitrogen were in natural forests converted for cultivation at 35.4 and 3.5 Mg Ha respectively.[9] It's important to emphasize that the protection of soil organic carbon is important to prevent degradation and for soil health because it comprises roughly 58% of the organic matter and is most concentrated in the top 20 centimetres (cm).[10] Soil degradation processes, such as in intensified cropping systems, affect soil organic carbon stocks and is one of the most dramatic examples of ecosystem deterioration limiting agricultural productivity and ecosystem health.[11]

Cameroon also experiences soil fertility and soil quality management, which has become a major agricultural production problem.[12] A 2020 study focused in the Tombel area which experiences high agriculture pressure due to increasing population growth and the settlement of large-scale farming companies. It found that 21.5% of the soils had the poorest fertility, and they included umbric Andosols (fragic) (ANuf) and umbric pisoplinthic Plinthosols (haplic dystric) (PLuphd).[13] This class of soils with poor fertility require several years of tillage for the restoration of its properties.[14] The study proposed that farmers should practice the fallow system after

[4]Sanchez (2002), p. 1.

[5]Sanchez (2002), p. 1.

[6]FAO (2016), p. 46.

[7]FAO (2016), p. 46.

[8]Were et al. (2015), p. 125.

[9]Were et al. (2015), p. 125.

[10]Olson (2017), p. 65A.

[11]Olson (2017), p. 64A.

[12]Nguemezi et al. (2020), p. 1.

[13]Nguemezi et al. (2020), p. 7.

[14]Nguemezi et al. (2020), p. 8.

cultivation to maintain soil fertility; to alternate the types of cultures; and promulgate new soil nutrient supply mechanisms from inputs (composts, chemical fertilizers, mineral fertilizers).[15] In Zambia's Luangwa basin, in the Zambezi system, pressure to find more fertile soils has pushed farmers away from depleted, exhausted farmland into more sensitive, hilly landscapes where soil and water run-off are contributing to thousands of tons of soil loss annually.[16]

According to Kihara et al., that land degradation characterized by soil erosion, nutrient depletion, loss of biological diversity and decreasing quality and quantity of water is a major problem facing many countries in sub-Saharan Africa (SSA).[17] About 65% of the agricultural land in SSA is degraded due to poor management practices, which induce declines in soil biological, chemical and physical quality; reducing the capacity of the soil to support crop production and provide other ecosystem service.[18] Thus, the annual nutrient depletion in year 2000 was estimated to reach 38 kg ha, leaving soils with serious fertility and other constraints, such that the degraded lands reduce the annual agricultural productivity by nearly 3%, costing the SSA region about US$ 68 billion annually.[19] Cereal yields in this region hover around 1.5 Metric Tonnes per Hectare (MT^{-1}) as compared with 3^{-1} in Latin America and South Asia, 5 MT ha^{-1} in China, and more than 10 MT ha^{-1} in North America, Europe, and Japan.[20]

Steward et al., caution that future investments in soil fertility improvements must support an equitable green revolution that supports resilient and sustainable livelihoods that both provides the nutrition and economic returns to improve human and social well-being.[21] This, they argue, is important to avoid falling into the same trap as the last decade where productivity in the SSA has increased but the per capita well-being (i.e. food and nutrition security and poverty reduction) has not gone up.[22] Meeting the growing demand for food, while also supporting livelihoods of 70% of the households in the region depending directly or indirectly on agriculture, will thus be a big problem if the current trend of degradation continues.[23] According to Vanlauwe et al., low soil fertility is directly linked to low productivity, food insecurity and rural poverty, which, in turn, can cause rural-to-urban migration, and an increased reliance of rural communities on remittances sent from family members living in towns or abroad.[24] They thus stress that there is an urgent need to reverse soil degradation before it becomes unachievable and many African

[15] Nguemezi et al. (2020), p. 7.
[16] Zambia (2015) NBSAP 2015–2025, p. 8.
[17] Kihara et al. (2020), p. 1.
[18] Kihara et al. (2020), p. 1.
[19] Kihara et al. (2020), p. 1.
[20] Steward et al. (2020), p. 633.
[21] Steward et al. (2020), p. 633.
[22] Steward et al. (2020), p. 633.
[23] Kihara et al. (2020), p. 1.
[24] Vanlauwe et al. (2015), p. 1.

smallholder farmers become locked in a poverty trap.[25] Therefore in order to enhance soil governance, the prevalent problem of soil degradation must be resolved, especially through improving soil health on agricultural lands.[26]

The foregoing assessment, although limited in scope, shows that Sub-Saharan Africa especially, faces a significant challenge of low soil fertility and degradation. This in turn impacts agriculture which is a key means of livelihood. Governance interventions must therefore be adequately innovative to address this dual problem of decline in soil health and agricultural productivity. In the next section, the chapter reviews the soil governance arrangements under the African Union. The review is focused on soil governance in context of agriculture as this is instrumental to livelihoods as seen above.

3 Soil Governance Arrangements Under the African Union

The African Union was established in 2000 as the successor to the Organization of African Unity. The objectives of the AU, according to the Constitutive Act, include among others the promotion of sustainable development; promoting and protecting human rights; promoting cooperation at all levels in order to raise the living standards of Africans.[27] The AU, as the continental body ratified by all fifty-five African nations, occupies a unique position that can be utilized to enhance soil governance, promote sustainable soil management and support productive land uses while protecting soil. In terms of law, there is the 1968 African Convention on the Conservation of Nature and Natural Resources signed in Algiers (also known as the Algiers Convention). This was amended in 2003 the Revised African Convention on the Conservation of Nature and Natural Resources (also known as the Maputo Convention). Under the AU, African leaders have adopted a variety of continent-wide programmes that impact land governance, such as the Comprehensive African Agriculture Programme (CAADP). Additionally, they have adopted, for national implementation, the Maputo, Malabo and Abuja declarations which have impacts on soil management. The AU Agenda 2063 is the overall guiding plan for implementation of programmes.

The discussion in this section, against the background in Sect. 2 on the state of soil degradation and the impact on agriculture and livelihoods, focuses on AU soil governance approaches. This examines treaty law and demonstrate shortfalls and opportunities to enhance soil governance. A review of the AU approach to soil governance in context of agriculture is undertaken, tracking implementation of CAADP through the Malabo Declaration and its adoption at national level through Agriculture Investment Plans (NAIPs).

---

[25] Vanlauwe et al. (2015), p. 1.

[26] Kihara et al. (2020), p. 1.

[27] Constitutive Act of the African Union (2000), article 3.

3.1 Soil Governance Under African Treaty Law

The 1968 African Convention on Nature and Natural Resources (Algiers Convention) requires States to take effective measures for conservation and improvement of the soil with a particular focus on combating erosion and misuse of soil.[28] This treaty prioritization of soil has been carried on by the 2003 Revised African Convention on Nature and Natural Resources (Maputo Convention), which requires Parties to take effective measures to prevent land degradation, and develop long-term integrated strategies for the conservation and sustainable management of land resources, including soil, vegetation and related hydrological processes.[29] The 2003 Convention, in particular, requires the State Parties to adopt measures for the conservation and improvement of the soil, to, inter alia, combat its erosion and misuse as well as the deterioration of its physical, chemical and biological or economic properties. Notably, the 1968 African Convention on the Conservation of Nature and Natural Resources (Algiers Convention) was the pioneer treaty with continental application.[30] It was signed by 46 African States and ratified by 32 of them.[31] This, together with its provisions, demonstrates that a consciousness on soil governance was present in 1968, slightly over a decade after Ghana became the first independent nation. In the preamble, the Heads of States and Governments affirmed their full consciousness that soil, water, flora and faunal resources constitute a capital of vital importance to mankind. The fundamental principle of the Algiers Convention binds the State Parties to adopt the measures necessary to ensure the conservation, utilization, and development of soil together with water, flora and faunal resources in accordance with scientific principles and with due regard to the best interests of the people.[32]

Article 4 of the 1968 Convention is dedicated to soil, requiring Parties to take effective measures for its conservation and improvement and in particular to combat soil erosion and misuse. Parties are required to establish land-use plans based on scientific investigations (ecological, pedological, economic, and sociological) and, in particular, classification of land-use capability. Further, when implementing agricultural practices and agrarian reforms, Parties are required to (a) improve soil conservation and introduce improved farming methods, which ensure long-term productivity of the land; and (b) to control erosion caused by various forms of land-use which may lead to loss of vegetation cover. Parties are also required to take into account the productivity of soils through scientifically based conservation and management plans for the protection, utilization and development of flora.[33]

[28] African Convention (1968), article 4.

[29] Revised African Convention (2003a), article 6.

[30] African Convention (1968).

[31] List of Ratifications to the 1968 African Convention. Online: https://au.int/sites/default/files/treaties/7763-sl-AFRICAN-4.pdf (accessed on 27 September 2021).

[32] African Convention (1968), article 2.

[33] African Convention (1968), article 6.

Important to soil governance, States are required to cooperate where this is necessary to implement treaty provisions; or whenever any national measure under the Convention was likely to affect the natural resources of another African State.[34]

Despite robust soil governance provisions, the Algiers Convention failed to integrate implementation mechanisms, such as a Secretariat or a Conference of Parties; or set up compliance and enforcement mechanisms, making its implementation subpar.[35] These challenges led to adoption of the 2003 Revised Convention under the auspices of the African Union. The 2003 African Convention on the Conservation of Nature and Natural Resources (also known as Maputo Convention) is an amendment of the Algiers Convention.[36] It was adopted on 11 July 2003 by Heads of States and Governments in Maputo, Mozambique and entered into force on 23 July 2016 upon receiving the 16th ratification from Burkina Faso. Liberia provided the 17th ratification in 2014 but deposited the instruments in 2017.[37] Taking into account the challenges of land degradation and soil governance on the African Continent, and in order for its provisions to bring about significant change, the Maputo Convention likely requires more than 17 African countries to ratify.[38]

The provisions of the Maputo Convention on soil augment those of the Algiers Convention. Article 6 provides for land and soil, while the 1968 Convention focused only on soil. Parties are required to take effective measures to prevent land degradation and develop long-term integrated strategies for the conservation and sustainable management of land resources, including soil. They are expected to adopt measures for the conservation and improvement of the soil, including among others, to combat soil erosion and misuse. In order to achieve this, when implementing agriculture practices and reforms, Parties are required to (a) improve soil conservation and introduce sustainable farming and forestry practices, which ensure long-term productivity of the land; (b) control erosion caused by land misuse and mismanagement which may lead to long- term loss of surface soils and vegetation cover; and (c) control pollution caused by agricultural activities, including aquaculture and animal husbandry. The Maputo Convention requires Parties to ensure that non-agricultural land uses, including mining do not result in soil erosion, pollution or other forms of land degradation. In areas affected by land degradation, Parties are required to plan and implement mitigation and rehabilitation measures.

The Maputo Convention, in an enhancement of provisions relative to its 1968 predecessor, provides for a Conference of Parties (CoP), as the decision-making body, and should be convened by the Chair of the Commission of the AU not later

[34] African Convention (1968), article 16.

[35] Kibugi (2018), p. 402.

[36] Revised African Convention (2003a), preamble.

[37] List of Ratifications to the Revised African Convention. Online: https://au.int/sites/default/files/treaties/7782-sl-revised_african_convention_on_the_conservation_of_nature_and_natural_resources.pdf (accessed on 27 September 2021).

[38] Kibugi (2018), p. 403.

than one year after the entry into force of the Convention.[39] At the time of writing, the CoP had not been convened, and therefore there has been no designation of an organization to undertake the Secretariat functions as required by the Convention.[40] The Maputo Convention also establishes a financing mechanism linked to the AU, which was absent in the 1968 version. In this case, the Convention in view of the central importance of financing to its objectives, provides that each Party shall make every effort, within national capability, to ensure that adequate financial resources are available for implementation. Finances for the budget to implement Convention programmes should be drawn from parties' contributions, AU contributions, and other sources. Importantly the manner of contribution, including amounts, are to be assessed by the CoP, which is yet to be convened. The Chair of the AU Commission ought to convene the CoP to kickstart implementation of the Convention. As argued earlier, there is urgent need for advocacy so that more of the 55 African nations to ratify the 2003 Convention because with only 17 ratifications, it would mean other nations are still subject the 1968 Algiers Convention, with all its shortcomings including weaker mechanisms for soil governance and management.

3.2 *African Union Policy Approaches to Prioritizing Soil Management Through Agriculture Investments*

While adopting the Maputo Convention in July 2003, African leaders also adopted the Maputo Declaration on Agriculture and Food Security in Africa.[41] This Declaration resolved to revitalize the agriculture sector through interventions targeted at small scale and traditional farmers in rural areas, including removal of constraints to agricultural production such as soil fertility.[42] The Declaration resolved to urgently implement the Comprehensive African Agriculture Programme (CAADP) and commitment to allocate at least ten percent (10%) of national budget for implementation of agriculture policies.[43] CAADP, also adopted in 2003, is Africa's continental initiative to reduce poverty and eliminate hunger by raising economic growth through agriculture-led development.[44] Pillar 1 of the CAADP advocates for the extension of the area under sustainable land management and reliable water control ecosystems.[45] It calls for increased investment in land and water and makes the point that protecting and improving water and the soil makes good business sense.[46]

[39] Revised African Convention (2003b), article 26.

[40] Revised African Convention (2003b), article 27.

[41] African Union (2003).

[42] African Union (2003), p. 1.

[43] African Union (2003), p. 1.

[44] CAADP (2003), p. 5.

[45] CAADP (2003), p. 12.

[46] CAADP (2003), p. 12.

The Malabo Declaration on Accelerated Agricultural Growth and Transformation for Shared Prosperity and Improved Livelihoods, adopted in 2014 by African Heads of States, is a set of agriculture goals to be attained by 2025 in enhancement of CAADP goals, and the Maputo Declaration.[47] It reiterated the Maputo commitments to allocate at least 10% of public expenditure to agriculture.[48] The Declaration made the commitment to sustain annual agricultural GDP growth of at least 6% and to create job opportunities for at least 30% of the youth in agricultural value chains.[49] Yet, throughout its text, there is no reference to soil, or any of the principles or priorities of promoting soil health. The Declaration nevertheless sets out a commitment to end hunger in Africa by 2025 by at least doubling current agricultural productivity. This is to be achieved, among others, through the necessary appropriate policy and institutional conditions and support systems to facilitate sustainable and reliable production and access to quality and affordable inputs (for crops, livestock, fisheries, amongst others) through, among other things, provision of 'smart' protection to smallholder agriculture. Also, the Declaration commits to ensure that 2025, at least 30% of African farm, pastoral, and fisherfolk households are resilient to climate and weather-related risks. The focus on climate resilience does, implicitly, bring out the likelihood for soil protection to be prioritized through the agricultural investments, but it is not explicit. The focus is on enhancement of agricultural production and this, while critical to livelihoods, equally requires healthy and fertile soils.

CAADP and the accompanying declarations, which focus on enhancing agricultural production remain inadequate. Enhancement of the treaty mechanisms is needed, particularly having more States ratify or accede to the Maputo Convention so that strong soil governance obligations can accompany investments in agriculture. Illustratively, with treaty soil governance obligations, implementation of CAADP Pillar 1, which focuses on extension of the area under sustainable land management and reliable water control ecosystems can equally pay attention to soil. Indeed, Pillar 1 urges that improving and protecting the soil, including through adequate husbandry, makes good business sense.[50] This logic is helpful because by 2019, forty-three (43) out of fifty-six (56) African countries had signed up to the CAADP process and 39 of them have begun to implement CAADP-based National Agricultural Investment Plans (NAIPs), with a small random sample of these reviewed in the next section as part of FAO support to African nations.[51]

This section has reviewed the treaty context and AU approach is governance of soil and attitude in context of agriculture. In the next section, the chapter appraises the roles of FAO and UNCCD, two international organizations, and explores

[47] Malabo Declaration (2014).

[48] Malabo Declaration (2014), part II (2)(a).

[49] Malabo Declaration (2014), part IV (4).

[50] CAADP (2003), p. 12.

[51] African Union Development Agency (2019), online: https://www.nepad.org/caadp/interactive-timeline/43-countries-have-joined-caadp (accessed 27 September 2021).

strategic options through which they can enhance effectiveness of AU treaty or policy means to enhance governance of soil in a sustainable manner.

4 Assessing the Role of FAO and UNCCD in Soil Governance and Window of Opportunity to Support Sustainable Actions in Africa

It is clear that the African continent, which relies heavily on agriculture, continues to have a soil health challenge and inadequate governance and policy approaches adopted by the African union. In this section, the chapter examines how the mandate and experience of FAO and the UNCCD provide a window of opportunity for integration of sustainable soil management into African agricultural legal and policy priorities. The section is divided into two parts, with the first reviewing the role of FAO in soil governance and examining application of the sustainable soil management methodology in NAIPs developed by African nations. The second part appraises the utility of the LDN approach soil governance in African. The section concludes by showing the window of opportunity to integrate it with sustainable soil management through the NAIPs and the National Action Programmes required under the UNCCD.

4.1 Assessing Integration of FAO's Sustainable Soil Methodology in CAADP Implementation Through National Agriculture Investment Plans

The FAO, a United Nations Specialized Agency, has 45 member countries in Africa.[52] It has a broad mandate which includes the conservation of natural resources and the adoption of improved methods of agricultural production.[53] FAO is host to the Committee on Food Security (CFS), a UN intergovernmental body that serves as a forum for review and follow up of food security policies. It is the CFS that approves the Global Strategic Framework for Food Security and Nutrition (GSF), a non-legally binding document.[54] The assessment in this part is focused on lessons from the GSF, and how they can impact prominence of soil governance through NAIPs which are now key tools through which African nations are implementing the AU agriculture agenda under CAADP. The GSF recommends that governments

[52]FAO Member Countries in Africa. Online: http://www.fao.org/africa/countries/en/ (accessed on 27 September 2021).

[53]Constitution of the FAO (2017), article 1.

[54]Committee on World Food Security (2014), p. 6.

should prioritize actions, to secure food security and nutrition, by increasing sustainable agricultural production and productivity; enhancing resilience to climate change; and finding sustainable solutions to the increasing competition for natural resources.[55] Further, the GSF has identified several recommendations that countries could prioritize to achieve food security, nutrition and rural development. Two of these are relevant to this chapter.

The first recommendation is for nations to increase smallholder sensitive investment in agriculture.[56] This GSF recommendation is important for review here because the bulk of investment in agriculture is undertaken by farmers and small-scale food producers themselves, their cooperatives and other rural enterprises. Smallholder farmers, including many rural women, play a significant role in the economic survival of their families, rural and national economy.[57]

Smallholder farms in many countries are significantly smaller than 2 ha, such as in Kenya (0.47 ha) and Ethiopia (0.9 ha) but the average small farm sizes hide significant productivity differences across countries.[58] These differences arise due to soil quality, technology and productive assets, such as irrigation.[59] Unlike Asia where smallholder farms are irrigated, African agriculture is rain fed. In Tanzania, where agriculture contributes towards 28% of the GDP and 73% of the population lives in the rural areas, there are about 3.7 million smallholdings (those smaller than the middle- size farm threshold of 2.2 ha), which make up for 80% of total farms.[60] In Kenya, agriculture contributes about 33% of total GDP, and employs more than 40% of the total population and about 70% of the rural population.[61] Kenya estimates that total employment in agriculture is around nine (9) million persons, in terms of those formally employed (~340,000) and those in informal employment (~8.3 million).[62] These figures, although not representative, show that economic reliance on smallholder agriculture is significant. Therefore, law and policy approaches towards smallholder agriculture should pay heed to the challenges, such as those discussed in Sect. 2, afflicting soil health and fertility in Africa. In order to demonstrate why this is important, this discussion analyses CAADP indicators adopted by the AU and the NAIPs developed by African countries for implementation of CAADP. This is a small randomly selected sample for illustration purposes. To implement the Maputo and Malabo Declarations, the AU has approved a CAADP results framework (2015–2025) with several result areas, through of which are important to this chapter:[63]

---

[55] Committee on World Food Security (2014), p. 11.

[56] Committee on World Food Security (2014), pp. 16–17.

[57] United Nations (2018), p. 3.

[58] Rapsomanikis (2015), p. 5.

[59] Rapsomanikis (2015), p. 5.

[60] Rapsomanikis (2015), p. 5.

[61] Kenya NAIP (2019), p. 19.

[62] Kenya NAIP (2019), pp. 25–26.

[63] African Union (2015), p. 7.

(a) Results area 2.1—which aims to increase agriculture production and productivity
(b) Results area 2.4—which aims to increase resilience of livelihoods and improved management of risks in the agriculture sector
(c) Results area 2.5—which aims to improve management of natural resources for sustainable agriculture.

The indicators for 2.1 are focused on value addition per work; per acre of arable land and yields for priority commodities. The indicators for result area 2.4 which is focused on livelihood and resilience are (i) coverage of social assistance, social protection, social insurance and labour programs, and (ii) existence of food reserves, local purchases for relief programmes, early warning systems and food feeding programmes.[64] Notably, the indicator for results area 2.5 is the share of agriculture under sustainable land management practices, making it the only one of the three result areas described here to integrate sustainable agriculture. Collectively, the indicators show that CAADP implementation, by the AU, still falls short of the focus on soil governance required to target interventions and investments addressing structural causes of agricultural related poverty, including low or declining soil fertility, land degradation, soil erosion, or post-harvest losses. Below, a sample of NAIPs from Ghana, Gambia, Malawi and Kenya are reviewed to demonstrate prioritization of smallholder agriculture interventions:

(a) Ghana—the NAIP reports that smallholder farmers have low levels in the use of technology, such as seed/planting materials, and other inputs, leading to comparatively low yields.[65] It has identified various strategic actions to enhance smallholder farming, including the establishment of sufficient number of fertilizers blending plants to meet the demand for blended fertilizers by smallholder farmers based on the value chain needs:[66]
(b) Gambia—the NAIP recognizes that more than 500,000 smallholder farmers participate in agriculture and natural resources production and most of them are among the poorest and also net food purchasers.[67] The NAIP notes that actions are needed to reduce rate of loss and degradation of natural habitats, facilitate adoption of technologies and innovations that enhance productivity particularly of smallholders. The NAIP, for instance, proposes various actions to support smallholder farmers based on rice and vegetable value chains.[68] The interventions include strengthening governance capacity of rice farmer cooperatives and other farmer organizations; provision/rehabilitation of water control infrastructure and facilities for irrigation/drainage and water harvesting/retention.

[64] African Union (2015), p. 7.
[65] Ghana NAIP (2018), p. 32.
[66] Ghana NAIP (2018), p. 44, 47, 48.
[67] Gambia NAIP (2019), p. 9.
[68] Gambia NAIP (2019), p. 43, 46, 47, 49.

(c) Malawi—the NAIP reports that most smallholder farmers cultivate customary land using hand tools and minimal inputs and technologies.[69] It argues that the commercialisation of smallholder farming will be the main source of growth with the highest contribution to poverty reduction.[70] In this respect, Malawi's NAIP with respect to livelihoods takes a different approach to the social support of the indicators for result area 2.4 above, targeting commercially oriented smallholder farmers with activities geared towards women and youth entrepreneurs in agriculture, to diversify production towards high-value crops and livestock breeds in the priority value chains.[71] It also includes ensuring school feeding programmes source local foods, with diversified food packages.[72]
(d) Kenya—the NAIP has, as a priority area, the total value chain output that comes from smallholders. Across the value chains, total production from smallholders is estimated as following: Maize (75%), potato (83%), Beef (80–90%), Bananas (80%, and Sugar (92%), tea (58%), coffee (63%) while the lowest is flowers (3%).[73] The NAIP aims to raise average annual incomes from KES 465/day to 625/day (~US$ 4.65 to 6.25/day) for approximately 3.3 million small-scale Kenyan farmers.[74] It also aims to increase household food resilience by reducing the number of food-insecure Kenyans in the ASAL regions from 2.7 million on average to zero, while reducing the cost of food and improving nutrition.[75]

Each of the NAIPs reviewed above acknowledge that smallholder agriculture is important and that investments for its enhancement are prioritized by African governments. Considering the findings from Sect. 2 analysis that the soil health situation is precarious, its protection should be concomitant. Lessons can be drawn from FAO's Voluntary Guidelines on Sustainable Soil Management (VGSSM). Under this VGSSM approach, the management of soil is deemed to be sustainable if the supporting, provisioning, regulating and cultural services provided by soil are maintained or enhanced without significantly impairing either the soil functions that enable those services, or biodiversity.[76] Sustainable soil management places a particular concern on securing the balance between the supporting and provisioning services for plant production and the regulating services the soil provides for water quality and availability and for atmospheric greenhouse gas composition.[77] Each of the NAIPs reviewed here acknowledges the technical and financial support received from FAO at country level. The discussion in Sects. 2 and 3 shows there is a gap in

[69] Malawi NAIP (2019), p. 6.
[70] Malawi NAIP (2019), p. 17.
[71] Malawi NAIP (2019), p. 29.
[72] Malawi NAIP (2019), p. 37.
[73] Kenya NAIP (2019), p. 20.
[74] Kenya NAIP (2019), p. xi.
[75] Kenya NAIP (2019), p. xi.
[76] FAO (2017), p. 3.
[77] FAO (2017), p. 3.

soil health and governance and for this reason the NAIPs at national level, and the CAADP at continent level should embrace the sustainable soil management approach. The second GSF recommendation reviewed here aligns with this argument, as it calls on nations to increase agricultural productivity and production in a socially, economically and environmentally sustainable manner. It recommends that yield increases and overall productivity gains in food and agricultural production should be undertaken in the context of a more socially, economically and environmentally sustainable agriculture.[78] This is important in order for investments concerning smallholder agriculture to remain viable and deliver positive social economic and environmental outcomes. It is also very aligned to the methodology of sustainable soil management. Notably, the VGSSM proposes several practices that can aid agricultural yield increases in socially, economically and environmentally sustainable manner:[79]

(a) Minimizing soil erosion soil erosion by water and wind as the most significant threat to global soils and the ecosystem services they provide.
(b) Enhancing soil organic matter content which plays a central role in maintaining soil functions and preventing soil degradation.
(c) Fostering soil nutrition balance, based on crop needs, local soil characteristics and conditions, and weather patterns, through nutrient recycling or additions including mineral (chemical) fertilizers, organic fertilizers and other soil amendments including primary sources (e.g. rock phosphate) and secondary sources (e.g. phosphorus from sewage sludge).
(d) Improving soil fertility through soil conservation practices such as the use of crop rotations with legumes, green- and animal manures, and cover crops in combination with reduced- or no-tillage with attention to reduced herbicide use, as well as agroforestry.
(e) Prevent, minimize and mitigate soil salinization and alkalinization
(f) Prevent and minimize soil contamination—which remains the best way to maintain healthy soils and food safety
(g) Reduce or minimize soil acidification, which is primarily associated with removal of base cations and loss of soil buffering capacity or increases in nitrogen and sulfur inputs (e.g., legume pastures fertilizer inputs, atmospheric deposition).
(h) Preserve and enhance soil biodiversity to maintain core soil functions.
(i) Minimize soil sealing and land conversion which cause a largely irreversible loss of some or all soil functions and the ecosystem services they provide.
(j) Prevent and mitigate soil compaction which reduces soil aeration by destroying soil aggregates and collapsing macropore density, and reduces water drainage and infiltration, generating higher runoff.

[78]Committee on World Food Security (2014), p. 23.

[79]FAO (2017), pp. 6–13.

While the list of sustainable soil management actions proposed by the VGSSM above is long, it has been set out to demonstrate that knowledge of appropriate and ecologically compatible actions for protecting and enhancing soil quality exists. FAO is a key driver in generating this knowledge on sustainable soil management. Looking at the NAIPs, developed by respective countries with some level of FAO technical support, both Ghana and Kenya undertake to carry out soil testing.[80] Kenya also plans to integrate mandatory extension services to explain what fertilizer to use based on soil needs (according to national soil testing results), with compulsory lime vouchers for farmers with acidic soils, and mandatory treatment where the risk of aflatoxin is high.[81] Gambia identifies soils and soil fertility as a problem with infertile and erosion prone soils compounded by unaffordability and/or untimely availability of chemical fertilisers and fertiliser use efficiency.[82] The Gambia NAIP identifies key actions to include improving water-use efficiency of crops through soil fertility restoration and on farm water management; promotion of climate change adaption measures such as soil and water conservation techniques and climate smart-agriculture in the farming system; and disseminating the integration of soil fertility management practices and use of the bio-control product, Aflasafe, for suppression of aflatoxin contamination in maize.[83]

In 2018, the AU published a NAIP Toolkit for Domestication of the Malabo Declaration.[84] Among the thematic priority goals, concern for soil health and fertility is evident under climate change smart agriculture which includes an indicator for rate of adoption of soil fertility management practices.[85] Some of the options in the Toolkit include sustainable soil management approaches such as no-till farming, integrated soil fertility management, water harvesting, drip irrigation, sprinkler irrigation, heat tolerance, drought tolerance, nitrogen-use efficiency and crop protection.[86] Despite the wide knowledge of sustainable soil management options there is nonetheless, no direct reference to sustainable soil management, or soil health in the CAADP Results Framework. Without a deliberate shift in AU agriculture and land policy to mainstream sustainable soil management into continentally driven actions, it is likely there will be further degradation, environmental damage and decline in food security. Efforts to implement the CAADP should be coordinated with activating provisions of the Maputo Conventions on land and soil and recruiting more countries (such as those already implementing CAADP and NAIPs) to ratify or accede to the convention. In this way, the knowledge that FAO has developed can be integrated, but with innovations that respond e to African soil protection needs,

[80] Ghana NAIP (2018), p. 66 & Kenya NAIP (2019), p. 49.

[81] Kenya NAIP (2019), p. 41.

[82] Gambia NAIP (2019), p. 35.

[83] Gambia NAIP (2019), p. 45.

[84] African Union (2018).

[85] African Union (2018), p. 12.

[86] African Union (2018), p. 36.

which are twinned closely with agricultural productivity needs for smallholders, who comprise the bulk of producers.

4.2 Assessing the Potential of the UNCCD and Land Degradation Neutrality Agenda in Enhancing Soil Governance in Africa

In this section, the chapter examines how the UNCCD, and the land degradation neutrality (LDN) agenda, could enhance the role of the AU and complement FAO's sustainable soil management on the African continent. LDN is the state that results when the amount and quality of land resources necessary to support ecosystem functions and services and promote food security remain stable or increase within specified temporal and spatial scales and ecosystems.[87] The pursuit of LDN provides a pathway to attain the target of "no net loss" and aims to achieve the world's resource of healthy and productive land by applying a dual-pronged approach of measures to avoid or reduce land degradation in combination with ways to reverse existing degradation to ensure the losses suffered are balanced by gains.[88] The practices highlighted in the preceding section for enhancing sustainable soil management are therefore quite integral to achievement of LDN.

SDG 15.3 calls on States to combat desertification, restore degraded land and soil, and strive to achieve a land degradation neutral world by the year 2030. Indicator 15.3.1 has been adopted for this SDG target. It will measure the spatial extent (hectares or km^2) expressed as the proportion (percentage or %) of land that is degraded over total land area.[89] This indicator uses a binary (either as degraded or not degraded) quantification based on the analysis of available data for three sub-indicators and it is to be validated and reported by national authorities.[90] The three sub-indicators (Trends in Land Cover, Land Productivity and Carbon Stocks) were adopted by the UNCCD's governing body in 2013 as part of its monitoring and evaluation approach. Notably, a consortium of international organizations comprising UNCCD, FAO, UNEP, UN Statistics Division, UNFCCC and the CBD secretariat are the custodians for this indicator.[91]

The indicator is important to sustainable soil management, as seen in the three sub-indicators that examine land cover, productivity and carbon stocks, all which are key to soil ecosystem services. The attainment of LDN is therefore the most desirable outcome for sustainable soil management, and land productivity especially in socio-economic terms is critical to enhancing the value contribution of

[87] UNCCD (2017), p. 10.
[88] UNCCD (2017).
[89] https://landportal.org/book/sdgs/153/indicator-1531 (accessed on 26 November 2020).
[90] https://landportal.org/book/sdgs/153/indicator-1531 (accessed on 26 November 2020).
[91] https://landportal.org/book/sdgs/153/indicator-1531 (accessed on 26 November 2020).

smallholder agriculture in Africa. In this respect, the priorities of the UNCCD 2018–2030 Strategic Framework are important.[92] Strategic objective 1 aims to improve the condition of affected ecosystems, combat desertification/land degradation, promote sustainable land management and contribute to land degradation neutrality. This, if implemented, could result in enhancement of land productivity and related ecosystem services; increased resilience of affected ecosystems while reducing their vulnerability; and setting and adoption of national voluntary LDN targets including adoption of necessary implementation and monitoring measures. Strategic objective 2 aims to improve the living conditions of affected populations, whose impact could include improvement in food security and adequate access to water for people in affected areas. These strategic objectives, reviewed in context of the AU soil and agriculture agenda, or the focus of NAIPs and sustainable soil management, confirm the close nexus between livelihoods and the ecosystem services provided by soil. In this sense, realization of SDG 15.3 has a likely positive impact on elimination of poverty and hunger. In practice, the UNCCD Strategic Framework is to be implemented by Parties through the National Action Plans (NAPs), in accordance with the Convention. The review of NAIPs in the foregoing section has elucidated that urgent law and policy interventions are needed to improve the livelihoods of smallholder farmers; but that this need to be accompanied by protection of soil. In this context, a framework is required to ensure that NAPs for implementing UNCCD obligations are aligned with NAIPs at national level such that the prioritization of enhanced agricultural production and soil health is concomitant. The AU occupies an important legal and policy role, as a continental body, to midwife the mainstreaming of sustainable soil management, agriculture productivity and land degradation neutrality. As argued earlier, enhancing the ratifications to the Maputo Convention, and activation of its CoP, financing and other treaty mechanisms could catalyse this. Using this reasoning for instance, a methodology could be designed between the AU, FAO and UNCCD Secretariat to guide African nations on harmonizing NAIP and NAP actions in a manner that meets CAADP priorities, and UNCCD commitments. Similarly, the AU should prioritize the review of critical guidelines including the Malabo Declaration and the NAIP domestication toolkit in order to ensure that prioritization of the balance between agricultural productivity, livelihood benefits and soil health is coherent.

5 Conclusion

Africa faces a significant challenge of low soil fertility and degradation. This in turn impacts agriculture which is a key means of livelihood. Investments in agriculture should therefore integrate measures to protect soil. This is because soils are an important part of the ecosystem, providing critical regulating, supporting,

[92] UNCCD (2017).

provisioning, and supporting services to the earth, and humanity. The Algiers Convention, and its 2003 Maputo successor contain robust provisions to enhance sustainable soil management, but these have not been utilized. The AU's CAADP is not robust enough on ecosystem protections including soil such that its prioritization of agricultural productivity could fail. A review of NAIPs, developed by different countries, shows that a national level the focus on soil health as a baseline for success in agriculture is recognized, but critical methodologies such as sustainable soil management and LDN are not systematically taken into account. Indeed, while NAIPs are core to the African agriculture agenda, the NAPs developed under the UNCCD to advance the LDN agenda, which impacts farmlands, are just as important. There is need for a systematic methodology between the AU, FAO and the UNCCD to ensure that deleterious actions do not result from parallel actions aiming to enhance agricultural productivity, soil management and LDN respectively, while indeed all are inherently complementary. The AU, as a continental body with treaties that already have provisions to protect land and soil (e.g., Maputo Convention) can midwife development of a framework to ensure there is integration between NAPs and NAIPs through clear guidelines developed in conjunction with FAO and UNCCD. Enhancing the ratifications to the 2003 Maputo Convention, activation of its CoP, financing systems and other treaty mechanisms could catalyse this process. The mainstreaming of climate change adaptation interventions should be mandatory in all these interventions aimed at halting and/or reversing decline in soil health and farm productivity. This will enhance another critical linkage between NAIPs, NAPs and Nationally Determined Contributions (NDCs). The latter are the main legal and policy tool through which nations, including those in the SSA, implement their commitments under the Paris Agreement to reduce Greenhouse Gas (GHG) emissions, and enhance resilience to adverse impacts of climate change.

References

African Convention (1968). Online: https://au.int/sites/default/files/treaties/7782-treaty-0029_-_revised_african_convention_on_the_conservation_of_nature_and_natural_resources_e.pdf

African Union (2003), Maputo Declaration on Agriculture and Food Security in Africa. Assembly/AU/Decl.7 (II). Online: https://www.nepad.org/caadp/publication/au-2003-maputo-declaration-agriculture-and-food-security

African Union (2015) The CAADP Results Framework 2015-2025. African Union Commission and NEPAD Planning and Coordinating Agency

African Union (2018) AU NAIP Toolkit for Malabo Domestication. African Union Commission and NEPAD Planning and Coordinating Agency.

CAADP (2003) Comprehensive Africa Agriculture Development Programme. African Union

Committee on World Food Security (2014), Global Strategic Framework for Food Security and Nutrition (GSF). FAO Rome.

FAO (2016) Mainstreaming ecosystem services and biodiversity into agricultural production and management in East Africa. FAO, Rome

FAO (2017) Voluntary Guidelines for Sustainable Soil Management Food and Agriculture Organization of the United Nations Rome, Italy

Gambia NAIP (2019) National Agricultural Investment Plan. Online: https://www.gafspfund.org/sites/default/files/inline-files/7.%20The%20Gambia_Investment%20Plan%20%282019-2026%29_1.pdf (accessed on 21 September 2021)

Ghana NAIP (2018) National Agricultural Investment Plan. Online: https://mofa.gov.gh/site/publications/policies-plans/316-national-agriculture-investment-plan-ifj (accessed on 21 September 2021)

Kenya NAIP (2019) National Agricultural Investment Plan 2019-2023. Online: https://www.agck.or.ke/Downloads/NATIONAL-AGRICULTURE-INVESTMENT-PLAN-NAIP.pdf (accessed on 21 September 2021)

Kibugi R (2018) Soil health, sustainable land management and land degradation in Africa: legal options on the need for a specific African soil convention or protocol. In: Ginzky H, Dooley E, Heuser I, Kasimbazi E, Markus T, Qin T (eds) International yearbook of soil law and policy 2017. International yearbook of soil law and policy, vol 2017. Springer, Cham. https://doi.org/10.1007/978-3-319-68885-5_21

Kihara J, Bolo P, Kinyua M, Nyawira SS, Sommer R (2020) Soil health and ecosystem services: Lessons from sub-Sahara Africa (SSA) Geoderma 370:114342. https://doi.org/10.1016/j.geoderma.2020.114342

Malabo Declaration (2014). Online: https://www.nepad.org/caadp/publication/malabo-declaration-accelerated-agricultural-growth

Malawi NAIP (2019) National Agricultural Investment Plan. Online: https://www.scotland-malawipartnership.org/assets/resources/National_Agicultural_Investment_Plan_2018_Final_Signed.pdf (accessed on 21 September 2021)

Mokgolodi NC, Setshogo MP, Shi LL, Liu YJ, Ma C (2011) Achieving food and nutritional security through agroforestry: a case of Faidherbia albida in sub-Saharan Africa. Forestry Stud China 13(2):123–131

Nguemezi C et al (2020) Soil quality and soil fertility status in major soil groups at the Tombel area. South-West Cameroon Heliyon 6. https://doi.org/10.1016/j.heliyon.2020.e03432

Olson KR, Al Kaisi M, Lal R, Morton LW (2017) Soil ecosystem services and intensified cropping systems. J Soil Water Conserv 72(3). https://doi.org/10.2489/jswc.72.3.64A

Rapsomanikis G (2015) The economic lives of smallholder farmers: an analysis based on household data from nine countries. FAO, Rome

Revised African Convention (2003a) Online: https://au.int/en/treaties/african-convention-conservation-nature-and-natural-resources-revised-version

Revised African Convention (2003b) African Convention on the Conservation of Nature and Natural Resources, African Union

Sanchez PA (2002) Soil fertility and hunger in Africa. Science 295(5562):2019–2020

Soil Atlas of Africa (2013) In: Jones A, Breuning-Madsen H, Brossard M, Dampha A, Deckers J, Dewitte O, Gallali T, Hallett S, Jones R, Kilasara M, Le Roux P, Micheli E, Montanarella L, Spaargaren O, Thiombiano L, Van Ranst E, Yemefack M, Zougmoré R (eds) European Union, 2013

Steward ZP, Pierzynski GM, Middendorf BJ, Prasad PVV (2020) Approaches to improve soil fertility in sub-Saharan Africa. J Exp Bot 71(2):632–641. https://doi.org/10.1093/jxb/erz446

UNCCD (2017) The future strategic framework of the Convention ICCD/COP (13)/L.18, 14 September 2017

United Nations (2018) Declaration on the Rights of Peasants and Other People Working in Rural Areas. A/HRC/RES/39/12

Vanlauwe B, Six K, Sanginga N, Adesina AA (2015) Soil fertility decline at the base of rural poverty in sub-Saharan Africa. Nature Plants 1:15101

Were LO, Singh BR, Dick ØB (2015) Effects of land cover changes on soil organic and total nitrogen stocks in the eastern Mau forest reserve, Kenya. In: Lal R, Singh BR, Mwaseba DL, Kraybill D, Hansen DO, Eik LO (eds) Sustainable intensification to advance food security and enhance climate resilience in Africa. Springer International Publishing, Cham, pp 113–133

Zambia (2015) National Biodiversity Strategy and Action Plan 2015-2025. Online: Online: https://chm.cbd.int/api/v2013/documents/DE100323-262E-ADBF-5A36-5FBBDABF895F/attachments/NBSAP%20Report%20for%20PRINTING.pdf (accessed on 21 September 2021)

Part III
Regional/National Reports

Options for Tackling the Challenges of Effective Management of Soils in Africa

Emmanuel Kasimbazi and Hadijah Yahyah

Abstract Soils are fundamental to life on earth but human pressures on soil resources are reaching critical limits especially in less developed countries. Soils are the bedrock of food production, habitat for biodiversity, and many ecosystem services. They are essential for the reduction of greenhouse gas emissions and building resilience to climate change that is made possible by maintaining or boosting their carbon content. This makes effective soil management a critical factor in the attainment of the 2030 Agenda for Sustainable Development as well as the Paris Climate Change Agreement particularly in Africa whose majority rural and urban population depend on agriculture for livelihood and income generation. A lot of effort has been made both globally and regionally to ensure healthy soils by way of legal, policy, and institutional frameworks. Various initiatives on the continent have provided a lead in facilitating sustainable management of soil in many countries. However, soil degradation continues unabated in Africa leading to food insecurity, desertification, climate change, famine and conflicts. Factors such as population growth, pollution, poor agriculture practices, under or overuse of inorganic fertilizers, land tenure, and rights-related issues and climate change are identified as being responsible for continued soil degradation. The Status of the World's Soil Resources report identified ten key threats that hamper the achievement of effective soil management. These threats are soil erosion by water and wind, soil organic carbon loss, soil nutrient imbalance, soil salinization, soil contamination, acidification, loss of soil biodiversity, soil sealing, soil compaction, and waterlogging. This chapter reviews the existing challenges hindering effective/sustainable management of soils in Africa; policy and legal, institutional, and the political economy. It further explores potential strategies that could be employed in enhancing soil management in African countries for better soil health critical in attaining sustainable development.

E. Kasimbazi
Makerere University, School of Law, Kampala, Uganda

H. Yahyah (✉)
Kampala International University, School of Law, Kampala, Uganda

H. Ginzky et al. (eds.), *International Yearbook of Soil Law and Policy 2020/2021*, International Yearbook of Soil Law and Policy 2020,
https://doi.org/10.1007/978-3-030-96347-7_11

1 Introduction

1.1 Effective/Sustainable Soil Management

Healthy soil provides food, stores nutrients for plant life, and delivers essential ecosystem services such as water purification and carbon sequestration. Literature suggests that soil may be the greatest reservoir of biodiversity[1] and the most valuable single natural commodity, worth close to US$4 trillion in 2012.[2] Soil is also critical in building resilience to and combating climate change.[3] In contrast, unhealthy soil can have devastating effects on everything from food security to global business and the quality of our environment. The challenge of nutrient-deficient soils is particularly serious for smallholder farmers who form the majority living in Africa, where the land is incredibly diverse, very depleted, and supports some of the highest population in the world. Soil is an essential natural resource for human survival in the form of livelihood, food production, industrialisation, and most ecosystems without which terrestrial life is virtually impossible.

In the African context, healthy soils are a fundamental building block for productive crop and pastoral agriculture. Soils provide important services to ecosystems and rural and urban economies. These include host nutrients for almost all crops, livestock, and forest production; provide carbon storage hence mitigation of greenhouse gas emissions; host a variety of in-soil biodiversity and retain and purify water resources.

Sustainable development in this context has been defined to mean development that meets the needs of the present without compromising the ability of future generations to meet their own needs for land.[4] We can then infer from this definition to derive a definition for effective or sustainable soil management as that which meets the needs of the present without compromising the ability of future generations to meet their own needs from soil. Thus, soil management is sustainable when it does not alter the capacity of the soil to provide for future needs.

The Revised World Soil Charter 2015 defines sustainable soil management to comprise activities that maintain or enhance the supporting, provisioning, regulating, and cultural services provided by soil are maintained or enhanced without significantly impairing either the soil functions that enable those services or biodiversity. Thus, soil degradation can be defined in contrast to this, as the processes by which soils can no longer maintain the provisioning, supporting and regulating ecosystem services required by current and future generations.[5] Unfortunately, this is at the core of the main challenges faced by the planet and particularly in Africa. It is manifested by rampant soil degradation and declining productivity due to the increasing

[1] Nielsen et al. (2015), p. 83.

[2] Amundson et al. (2015), p. 648.

[3] Lal (2004), p. 1624.

[4] Brundtland Report (1987), para 27.

[5] Tully (2015), p. 6524.

population, inappropriate farming practices, and changing land-use despite the international and regional initiatives that have emphasized the importance of well-functioning soils.

Effective soil management can be threatened by numerous practices including over-cultivation, decreased or increased water abstraction, under or over-fertilization, careless use of biocides, failure to maintain soil organic matter levels, and clearing natural vegetation. These may threaten sustainability in several ways through physical and chemical processes including increased soil erosion, salinisation, desertification, or biological processes e.g., by decreasing soil fertility. When soil management is poor, soil sustainability is often threatened by a combination of these factors. The costs and consequences resulting from soil degradation are borne by society as a whole, including the cost implications of restoration or prevention and mitigation such as damage to infrastructures due to sediment runoff among others.

Soil resources in many agricultural and pastoral landscapes of East African countries of Burundi, Kenya, Rwanda, Uganda, and Tanzania are rapidly being depleted by increased erosion, contributing to widespread land degradation. This threatens food security, water security and livelihood security in the region.[6]

By 2050, the global population is expected to increase from 7.3 to 9.7 billion with Africa having the greatest growth with a doubling of population from 1.2 to 2.5 billion, translating into enormous pressure on African soils.[7] Approximately 25% of Africa's productive lands are degraded, driven by desertification and erosion but mainly due to the loss of nutrients and soil organic carbon under continuous cropping.[8] In 2007, these degraded soils were estimated to affect 485 million Africans and cost the continent nearly US$9.3 billion annually.

This chapter aims to examine the challenges inhibiting the effective management of soils in Africa and further explores possible mechanisms that may facilitate and enhance the use and management of soil in a more sustainable manner. Examples of best practices from various African countries and beyond are relied upon for purposes of benchmarking.

2 International Context of Sustainable Soil Management

Soil performs a fundamental role in human well-being as the foundation of food security, sustainable development, and climate change adaptation and mitigation and as such, a good understanding of effective and sustainable management of soils should start from appreciating the value of the soil.

---

[6] Blaikie and Brookfield (2015), p. 23.

[7] UN (2015).

[8] Stewart (2020), p. 663.

Sustainable soil management is a precondition for sustainable development not only in Africa but the whole world as the degradation of soils is a cause for hunger, migration, and even wars in extreme cases. Sustainable management of soils is therefore to be regarded as a common concern of humankind from a scientific point of view. Considering decisions taken by the international community, this is also politically valid and ought to be paid attention to. For instance, statements and decisions of the Rio+20 outcome document or during COP 12 UNCCD, reaffirm that the sustainable management of soils is politically regarded as a common challenge for humankind.

It is estimated that about 23% of the terrestrial surface is already degraded affecting about 1.5 billion people.[9] Accordingly, the global community has committed itself to the 2030 Agenda for sustainable development. SDG 15 (life on land) target 3 obliges states to try to achieve Land Degradation Neutrality (LDN) by 2030. Considering the centrality of soil to land, this target is the most critical in international law for sustainable management of soils. Other SDGs relevant to SSM are SDG 1 (no poverty); SDG 2 (zero hunger); SDG 3 (good health and well-being); SDG 6 (clean water and sanitation); SDG 8 (decent work and economic growth); SDG 13 (climate action); SDG 14 (life below water); and SDG 17 (partnerships for the Goals).

In addition to the SDGs, the United Nations Framework Convention on Climate Change (UNFCCC), the United Nations Convention to Combat Desertification (UNCCD), the Convention on Biological Diversity (CBD) and the Paris Agreement on Climate Change all deal with soil issues. The development and submission of the Nationally Determined Contributions (NDCs) by the State Parties under the Paris Agreement is directly important for effective soil management and the use of NDCs seems a particularly promising mechanism in the promotion of soil-related actions in Africa. This is more so, because, while under the Paris Agreement, NDCs are voluntary for developing countries, the submission of the NDC targets by each country builds an international legal commitment.

The role of the general principles of international environmental law such as the precautionary principle, the polluter pays principle, common concern for mankind, inter- and intra-generational equity, environmental democracy, the principle of common but differentiated responsibilities and respective capabilities among others cannot be overemphasized. Major challenges such as drought, food insecurity, and water scarcity are all hinged on soil health which is crucial for mitigating climate change.

Further, The Basel Convention on the Control of Hazardous Wastes and their Disposal[10] as well as the Bamako Convention[11] banning import into Africa and control of Transboundary movement and management of Hazardous wastes within Africa deal with chemicals interaction with soil biosecurity concerns. This speaks to

[9] Remus (2016), p. 265.

[10] The Basel Convention (1989).

[11] The Bamako Convention (1991).

policies and measures protecting a nation's food supply and protection of natural resources from biological invasion in Africa.

The UNCCD is the international lead organization for implementation of the LDN objective. Projects on the LDN voluntary target-setting have been launched in around 122 countries.[12] The same lead agency has developed the scientific conceptual framework for LDN which is critical in providing a conceptualized model approach on how best to implement the LDN objective at the regional and national levels.

The Revised African Convention on Conservation of Nature and Natural Resources (2017) provides for land and soil conservation under Article VI. Parties are required to take effective measures to prevent land degradation and develop long-term strategies for conservation and sustainable management of land resources including soil. They are required to adopt measures for improvement and conservation of soils to combat erosion and its misuse. The government has helped farmers in Uganda to identify low-cost inputs that can address soil fertility problems.[13]

3 Key Conceptual and Regulatory Aspects of Soil Management in Africa

Despite Africa being the least affected continent about soils and land degradation, natural and anthropogenic pressures on soils are currently continuously increasing due to a range of factors including poverty, over-exploitation, population growth, and climate change.[14] Africa has experienced a widespread decline in soil quality since the early twentieth century that continues today thereby greatly hindering crop and livestock production across the continent. It threatens food security and environmental sustainability as well as economic growth and well-being, particularly for Africa's poorest and most vulnerable people. It is estimated that erosion and other types of degradation of soil quality have caused the abandonment of roughly one-third of the world's arable land.[15] The first *Status of the World's Soil Resources'* report concluded that the majority of the world's soil resources are in only fair, poor, or very poor condition.[16] The International Food Policy Research Institute (IFPRI) supports these findings.[17]

Sustainable soil management is a cross-cutting issue and has interfaces with i.e., environmental conservation in general, climate change, property ownership, agriculture, chlorophyll influence, public health-pests, wastes, and extractives. There are

[12] Gilbey et al. (2019), p. 4.

[13] Pender et al. (2004), p. 3.

[14] Yahyah et al. (2020), p. 1.

[15] UNCCD (2017).

[16] FAO, ITPS (2015), p. xix.

[17] Nkonya et al. (2016), p. 204.

several regulatory approaches indirectly relevant to soil protection in most African countries. However, regulations are fragmented and do not follow a coherent and consistent policy. So far, sustainable soil management is not a legally established objective and there is, therefore, no clear obligation to restore or rehabilitate degraded land.

In some African countries, environmental legal standards and obligations are not directly implemented and enforced on customary land.[18] Traditional authorities are regarded as custodians of natural resources, including land and soils. Examples (evidence) however demonstrate that local communities due to their internal priorities are not necessarily the best protectors of soils.

3.1 Post-independence Effects

Many of the post-independence issues regarding unsustainable land management can be attributed to the erosion of indigenous social structures during the colonial period, combined with the sudden release from the strict colonial rules into the chaotic new nation-states. For instance, in Tanzania the process of losing the 'indigenous' conservation ethic during colonial time, combined with increasing population pressures and a sudden release from colonial forest enforcement led to uncontrolled exploitation and encroachment in the years following independence.[19]

3.2 Relevance of Land Rights/Tenure in Africa to Sustainable Soil Management

Clarity on land tenure and rights is critical and ought to be seen as a prerequisite for sustainable soil management in Africa. This is also important for medium or even long-term investments.[20] The land rights systems are very different in African states. Thus, the one size fits all approach does not exist. In most states, there are usually three land types: state or public land, private land, and customary/tribal land/community land. Land tenure security in the form of western land titling/registration is often low in Africa. In Uganda, less than 20% of the land is formally registered.[21] There is consensus that an effective land titling system is the key for land security in particular as the economic interest in land is constantly increasing, including for foreign investors.

[18] Cotula et al. (2011), pp. 16–17.

[19] Enfors and Gordon (2007), p. 690.

[20] Wily (2011), pp. 4–13.

[21] Musinguzi et al. (2021), p. 2.

Looking at Kenya, the western highlands are among the most densely populated regions of the Sub-Saharan Africa, with about 40–50% of the population living in poverty. The average farms in the area are 0.5–2 hectares in size.[22] Farming households cultivate maize, beans, and other staple food and cash crops, keep chicken and livestock and grow trees on woodlots for timber and fuel. Farms in the area have medium to high agricultural potential, but most suffer from severe soil degradation.

3.3 Traditional and Ecological Approaches

Traditional elements of increasing nutrient and water use efficiency and building long-term fertility through soil organic matter lie at the heart of this agenda as relates to improved soil, water and biodiversity management, and conservation; decreased or reversed trends in deforestation, nutrient depletion, soil erosion, genetic diversity erosion, and water pollution.

3.4 Climate Change, Adaptation, and Mitigation

There is a strong link between poverty and social-cultural collapse on soil degradation and climate change. Soils being one of the greatest carbon sinks, effective management of such land resources is worth all the protection in the era of climate change effects that continue to affect many Africa countries.

4 Challenges of Effective/Sustainable Soil Management

4.1 Legal

Generally, there is no dedicated, comprehensive, and coherent soil policy in place, which could inform the law. Accordingly, no comprehensive soil protection law has been issued and soil relevant provisions are very much fragmented. At the national level, soil use regulation is contained in land use and environmental legislation. For instance, there is no specific constitutional, policy, or legal tool that deals with soil protection in countries like Uganda, Kenya, and many other African countries. It is only embedded within the environmental, agricultural, and water statutes as well as

[22] Jayne and Muyanga (2012), p. 411.

policies. Soil as a natural resource is thus not explicitly and sufficiently integrated into most African legislation.

There are several challenges in existing international, regional, and even national soil regulation which are very fragmented. First, although the UNFCCC, the UNCCD, and the CBD all deal with soil issues, no coherent concept exists. The Protocol for the implementation of the Alpine Convention of 1991 in the area of Soil Protection is the only legally binding instrument in the world specifically for soil. This means that the majority of the instruments are non-binding and so a State Party can abandon negotiations at any time, walk away without having signed a definitive agreement, and refuse to implement the provisions of the non-binding instrument. The CBD contains provisions mostly targeting conservation and sustainable use to prevent degradation, but the focus is on preventing the degradation of biological diversity more broadly than just on land or soil in this case. At the African continental level, there is a brief mention of land and soil which is not detailed to address sustainable soil management issues.

The land ownership concept in most African states poses a significant barrier to sustainable soil management as land ownership is perceived as entitling persons to an absolute use and even misuse right on the respective land. Although in most African countries land belongs to people collectively as a nation, communities, or individuals, the various land laws confer absolute proprietorship with all rights and privileges on the landowner. This has created the perception that the landowner is not obligated to abide by certain legal requirements concerning the use of his or her land. Similarly, gender inequality is a land rights issue that needs to be addressed in preparing an enabling soil protection environment.

Soil protection laws where available, are mainly on pollution, loss of organic matter, and erosion, lack effective implementation or are insufficiently monitored. Similarly, soil related provisions are not effectively implemented and enforced. Lawmakers face a challenge in drafting effective legislation for soil given the volume of soil science knowledge required, and the many actions involving the use of land resources. Most national laws, for instance Uganda, do not directly focus on the protection of soil.[23] Protecting soil is often a side-effect of the protection of natural resources such as water, land, and biodiversity. There are laws providing for Strategic Environmental Assessment (SEA) and Environmental Impact Assessment (EIA) in the assessment of large-scale projects' impact on various soil functions. However, the soil is usually not the core consideration of a SEA or an EIA.

There is limited jurisprudence specific to soil protection in Africa. While most states are cognizant of environmental degradation and have gone ahead to develop laws and policies to this end, there are hardly any judicial decisions explaining the need to sustainably manage soil resources like is the case with land, water and forests.

[23] Kasimbazi (2019), p. 320.

Finally, land or agriculture laws provide subsidies for special agricultural management processes that have direct guiding effects for land use and its consequences on soil protection. However, the subsidies only last as long as there is money to give to the farmers. If the system of subsidies ceases, farmers easily go back to unsustainable soil management actions.

4.2 *Political, Social, Economic and Institutional Challenges*

In most countries, inadequate budgetary allocations and lack of or poorly articulated regulations on the implementation of policies and coordination remains the greatest challenge facing soil management.[24]

Further, most soil management institutions lack adequate capacity, knowledge, and experience to plan and implement Sustainable Soil Management (SSM) and optimally manage, mitigate and monitor the production and degradation status of the soils especially under intensive cultivation. In that respect, the centralised governments in most of African states have not succeeded in developing or implementing adequate land management strategies to safeguard the soil base in comparison with the local 'indigenous' systems.[25] The lack of national or umbrella organizations for sustainable soil management stakeholders to lead the campaign to promote and create awareness of SSM undermines sustainable soil management efforts and further compounds the challenge.

Insufficient political will towards enactment and introduction of laws and rules as well as their implementation by political parties, leaders or executives has also hindered the effective implementation of SSM policies.[26] Competent institutions are in a relatively weak position, in particular concerning implementation, compliance and enforcement.

Public participation and access to environmental information are not sufficiently developed. Besides, soil information/data at the national level is often inadequate, outdated, not in digital format, and not geo-referenced. Even where the information exists, it is not readily accessible and is inadequately disseminated to key stakeholders such as land users (farmers, livestock holders, foresters) and policymakers, and planners.[27]

Land grabbing is another major issue in many African countries. Land grabbing is generally understood as being an unfair land acquisition. Unfair land acquisition could be both legal and illegal. Foreign investments requiring land are commonly

---

[24] Msangi (2007), pp. 487–499.

[25] Ananda and Herath (2003), p. 348.

[26] Maathai (2011), pp. 1–2.

[27] Bationo et al. (2007), p. 18.

regarded as important and helpful if performed fairly. Hence the need for regulations and additional actions to deal with unfair land acquisition.[28]

Since small scale farmers' access to credit is still very limited, short-term priorities often trump long-term investments in things like soil health. Appreciating the role of farmers' time and risk preferences, and accounting for them is important for the design of effective agricultural programs and policies.[29] Given farmers' shorter-term horizons and risk aversion, we may not see widespread adoption of soil management practices that often deliver benefits in the long run. Designing programs that deliver more immediate benefits and subsidizing the initial investments will be crucial for the requisite buy-in.

The competences are often shared by several ministerial entities. However, the communication and cooperation on legislation, implementation, and enforcement between these entities need to be developed further. Usually, these entities face substantial under-staffing which hinders the desired implementation and enforcement outsides.

The linkages between researchers, farmers, and extension services to optimize information exchange are weak and this hinders knowledge sharing. The top-down planning approach to technical assistance has proven ineffective in the past. Sustainable development can be achieved only when the actual beneficiaries of technical assistance feel they are equal partners and that they, rather than the government, own and drive the process. Sustainability is more likely to be achieved if development is driven from the bottom up and addresses farmers' and community needs and constraints. This is the case in Ethiopia where such an approach has contributed to the weak dissemination of proper land management practices and very poor sustainability of conservation measures, ultimately aggravating land degradation in the country.[30]

Africa's rural poor are heavily dependent on natural resources for survival, have limited access to capital, fertilisers, and technology and thus cannot invest in improved land management. Hence, poverty is a major driver of soil degradation.[31] The investment in sustainable soil management and alternative livelihoods is inadequate. When people lack access to alternative sources of livelihood, they are most likely to exert more pressure on the few resources that are available to them. There is a high rate of poverty especially in third world countries and high dependency on the extraction of natural resources for income. The level of the farmers' incomes affects soil management efforts because farmers with low-income levels are likely to be financially constrained to adopt sustainable soil management measures.[32]

The incentives to the private sector for the provision of sustainable soil management services are non-existent or inadequate. Without incentives, both in the long

---

[28] Vandergeten et al. (2016), pp. 1395–1417.

[29] Shiferaw et al. (2009), pp. 601–619.

[30] Gebreselassie et al. (2016), pp. 404–406.

[31] Wynants et al. (2019), p. 1915.

[32] Ntshangase et al. (2018), p. 9.

and short term, farmers are not able to invest the time and money required to develop new techniques and overcome typical adoption barriers, be they technical, cultural or financial.[33]

5 Conclusion and Recommendations

A review of soil health, its critical roles for human survival and sustainable development, vis a vis its management in Africa reveals that the existing legal, policy and institutional mechanisms have not achieved the objective of sustainable management of soil. Countries do not have sufficient provisions to protect and manage soil in a sustainable way at the international, regional, and national levels. All the existing relevant MEAs have provisions that broadly cover various aspects of soil management which can be expanded to cover specific aspects of soil management or specific laws for SSM can be developed under one of the MEAs such as the UNCCD to anchor important environmental, climate, social, and economic aspects of soil protection.

A Protocol would be ideal after the treaty that specifically deals with institutional mechanisms and sets targets for the achievement of zero land degradation which is at the heart of effective soil management. There is therefore, need to recognize the soil essential element of terrestrial ecology and to include the environmental law concepts and elements needed to achieve the sustainable management of soil which calls for more deliberate attention from African policymakers in their national laws, policies, and strategies toward sustainable soil management.

Regarding land ownership challenges, the police power of the state and the public trust doctrine to regulate land use are founded in the constitution, and this could, with modifications in the legislation, be applied directly to govern soil health. For instance, the 2009 National Land Policy of Kenya has clarified that the dual existence of the freehold and the absolute proprietorship is confusing. Land cover policy in Cameroon[34] is capable of setting incentives for sustainable soil management and all it may need is to be supported by economic instruments and tools to raise awareness for the importance of the maintenance of fertile soils.

Access to land and land property rights need to be fully understood for the establishment of an effective regime on sustainable soil management. Overarching principles for land and soil protection legislation such as prevention, precaution, obligation to rehabilitate, and planning approaches need to be worked out and agreed upon. Stronger involvement of women is required in the planning, decision making, and monitoring of land use activities such as agriculture.

Capacity building will be useful at all levels: government level including legislators, policymakers, executive, enforcement, and the judiciary as well as civil

[33]Garrity et al. (2010), pp. 197–214.

[34]Takoutsing (2015), p. 33.

society organizations and land right holders. It would be particularly effective to train the trainers to increase the knowledge base. Judges and other judicial officials require training and tailored workshops or initiatives to help build the requisite capacity because of the wide jurisdiction of the different courts, such as the Environment and Land Court (ELC) of Kenya. Continuous measures on capacity building are required as opposed to one-off exercises. UNEP, UNCCD, IUCN, and academic institutions could be resourceful to this end.

Restoration of degraded soils, converting marginal agricultural areas to rangeland and forest land, and adoption of recommended management practices have a large technical potential to sequester carbon and off-set anthropogenic emissions, improve the environment, and enhance and sustain soil agriculture productivity.

Drafting, implementation, and enforcement of future soil protection legislation require personnel resources. These resources must be in place if such laws are to be successful. The respective governments ought to be proactive in identifying available funding from the various sources such as under the climate change framework through Global Environment Facility, multinational companies, the private sector and state government funding. Also, the Pan African Parliament's Permanent Committee on Rural Economy, Agriculture, Natural Resources and Environment gathered on the last day of the Sitting of the Permanent Committees in Midrand South Africa, to establish the way forward on development of the Soil Management Model Law for the continent which provides a window of opportunity.

There is a need to clarify the requirements of customary land in various communities. Cultural aspects have to be taken into account as the concept of customary land varies amongst the different communities. It is to be considered that according to customs and traditions land could carry more value than the pure substance, inter alia religious or spiritual values. The procedure of land titling should involve local communities as key stakeholders. The procedure should be simple and cost-effective. For instance, the registration process in Uganda is time consuming and demanding while the registration process in Kenya establishes a deadline that seems to contradict the concept of inclusiveness.

There is a need to develop multi-stakeholder engagement by establishing platforms that bring together all levels, all perspectives and stakes, in particular, if run in a participatory and interdisciplinary manner. Ecologists, environmental economists, soil scientists, lawyers, social workers all need to work together to enhance soil management. The division of labour between central and local entities needs to be clarified to avoid duplication of work and the challenge of "no one being in charge" hence dealing with institutional inertia.

Access to information on the need to sustainably use soil resources could go a long way in sensitizing the different stakeholders especially the rural population. Materials to be disseminated should be adequate for the audience to be addressed taking into account such things as language, culture, and context. Information on ecological background and "economics of land degradation" is of crucial importance.

Embracing new technologies by African countries in soil management is critical. In the past, soil mapping efforts have largely been developed and used for academic purposes. For example, regarding the patterns of chemical nutrients across landscapes and the spatial patterns of soil types. These aspects have produced some useful soil maps, unfortunately, these maps are usually left in dusty stores redundant or for the use of scientific "experts." Modern geoinformatics technology, however, has changed our ability to interact with soil maps and make the outputs relevant to general audiences, policymakers, legislators, or farmers. These technologies are uniquely designed to adapt to the diverse needs of users. However, the adaptation of these new technologies to land uses such as agriculture, especially in Africa, has been slow. Evidence of early applications to soil management has been informative, but further investment across all sectors of the agricultural service economy is needed to bring about system transformation.[35]

There is a need for improved soil conservation measures to reverse the influence of anthropogenic activities coupled with population growth that have negatively affected the landscape. However, this trend is assumed to have changed in recent times in Burkina Faso due to the implementation of Soil and Water Conservation Mechanisms (SWCM).[36] Other African countries could borrow some lessons from this model. Appropriate approaches include: use appropriate tillage practices; Applying organic manures and mineral fertilizers provides essential plant nutrients in the soil for better crop growth; mulching and the use of crop residues, and agroforestry which is the planting of trees or shrubs or protecting the naturally sustaining trees and REDD+ are potentially huge opportunities for Africa in effective and sustainable soil management.

It is our recommendation that national soil legislation is developed in countries that do not have and where they exist the legislation should be improved to ensure that soil ecological and cultural functions and land use are covered. The Protocol on the Implementation of the Convention Concerning the Protection of the Alps of 1991 in the Area of Soil Protection could be used as a basis for the improvement of global and national laws for the soil in the absence of other legal precedents. There is a further need for the development of a specific soil Protocol under the African Convention on Nature and Natural Resources (2017).

The intervention clause of the German Nature Conservation Act serves as a model for implementing the Land Degradation Neutrality concept, in African soil regulatory tools. The German Intervention clause requires three steps of decision making: (1) prevent/avoid, (2) physically offset; and (3) assessing and weighing the competing interests, plus mandatory monetary compensation. This one rule captures the LDN concept and also corresponds to the UNCCD's "LDN response hierarchy". As a precondition for effective use of the German Intervention Clause, data is required for the determination of the baseline and the assessment of degradation and restoration processes. Moreover, indicators must not be too technical and restoration/offsetting requirements must not be interpreted as a "license to degrade". The German prototype emphasizes that planning instruments are required for LDN

[35]Berazneva et al. (2019), pp. 1–8.

[36]Nyamekye et al. (2018), p. 3182.

implementation, in particular, to enforce the neutrality requirement and standards and essentials for sustainable soil management must be area-specific taking into account the different settings in African countries.

There is a need for more deliberate regulatory tools and a stronger partnership in Africa to provide an avenue for soils in terms of its multiple roles and to move toward achieving major outcomes in terms of sustainable soil management and food security on the continent.[37]

References

Amundson R, Berhe AA, Hopmans JW (2015) Soil and human security in the 21st century. 348(6235):1261071. https://doi.org/10.1126/science.1261071

Ananda J, Herath G (2003) Soil erosion in developing countries: a socio-economic appraisal. J Environ Manag 68(4):343–353. https://doi.org/10.1016/s0301-4797(03)00082-3

Bationo A, Kihara J, Vanlauwe B et al (2007) Soil organic carbon dynamics, functions and management in West African agro-ecosystems. Agric Syst 94(1):13–25. https://doi.org/10.1016/j.agsy.2005.08.011. ISSN 0308-521X

Berazneva J, Conrad JM, Güereña DT (2019) Agricultural productivity and soil carbon dynamics: a bioeconomic model. Am J Agric Econ 101(4):1021–1046

Blaikie P, Brookfield H (2015) Land degradation and society. Routledge, p 23

Cotula L, Vermeulen S, Mathieu P, Toulmin C (2011) Agricultural investment and international land deals: evidence from a multi-country study in Africa. Food Secur 3(1):99–113

Enfors EI, Gordon LJ (2007) Analysing resilience in dryland agro-ecosystems: a case study of the Makanya catchment in Tanzania over the past 50 years. LLD 18(16). https://doi.org/10.1002/ldr.807

Food and Agriculture Organization (2015) International Theme Park Services (ITPS). Status of the World's Soil Resources: Main Report; FAO, ITPS: Rome, Italy, ISBN 978-92-5-109004-6

Food and Agriculture Organization of the United Nations (2015) Revised World Soil Charter, 2015

Garrity DP, Akinnifesi FK, Ajayi OC et al (2010) Evergreen agriculture: a robust approach to sustainable food security in Africa. Food Sec 2:197–214. https://doi.org/10.1007/s12571-010-0070-7

Gebreselassie S, Kirui OK, Mirzabaev A (2016) Economics of land degradation and improvement in Ethiopia. In: Economics of land degradation and improvement–a global assessment for sustainable development. Springer, Cham, pp 404–406

Gilbey B, Davies J, Metternicht G, Magero C (2019) Taking land degradation neutrality from concept to practice: early reflections on LDN target setting and planning. Environ Sci Policy 100:230–237

Gro Harlem Brundtland (1987) Report of the World Commission on Environment and Development: Our Common Future. United Nations General Assembly document A/42/427

Jayne TS, Muyanga M (2012) Land constraints in Kenya's densely populated rural areas: implications for food policy and institutional reform. Food Secur 4(3):399–421

Kasimbazi E (2019) Soil protection law in Uganda. In: International yearbook of soil law and policy. Springer, Cham, pp 315–330

Lal R (2004) Soil carbon sequestration impacts on global climate change and food security. Science (New York, N.Y.) 304:1623-7.10.1126/science.1097396

[37]Global Soil Partnership Report, Africa (2015).

Maathai W (2011) The challenges of Africa. Sustain Sci 6(1):1–2. https://doi.org/10.1007/s11625-010-0120-2
Msangi JP (2007) Land degradation management in southern Africa. In: Climate and land degradation. Springer, Berlin, Heidelberg, pp 487–499. https://doi.org/10.1007/978-3-540-72438-
Musinguzi M, Enemark S, Mwesigye SP (2021) Fit for purpose land administration: country implementation strategy for addressing Uganda's land tenure security problems. Land 10:629. https://doi.org/10.3390/land10060629
Nielsen UN, Wall DH, Six J (2015) Soil biodiversity and the environment. Annu Rev Environ Resour 40:63–90. https://doi.org/10.1146/annurev-environ-102014-021257
Nkonya E, Mirzabaev A, von Braun J (eds) (2016) Economics of land degradation and improvement – a global assessment for sustainable development. Springer Open, New York
Ntshangase NL, Muroyiwa B, Sibanda M (2018) Farmers' perceptions and factors influencing the adoption of no-till conservation agriculture by small-scale farmers in Zashuke, KwaZulu-Natal Province. MDPI. Sustainability 10:555. https://doi.org/10.3390/su10020555
Nyamekye C, Thiel M, Schönbrodt-Stitt S et al (2018) Soil and water conservation in Burkina Faso, West Africa. Sustainability 10(9):3182. https://doi.org/10.3390/su10093182
Pender J, Nkonya E, Jagger P, Sserunkuuma D, Ssali H (2004) Strategies to increase agricultural productivity and reduce land degradation: evidence from Uganda. Agric Econ 31:3
Remus P (2016) Drylands extent and environmental issues. A global approach. Earth-Sci Rev 161: 259–278
Shiferaw BA, Okello J, Reddy RV (2009) Adoption and adaptation of natural resource management innovations in smallholder agriculture: reflections on key lessons and best practices. Environ Dev Sustain 11:601–619. https://doi.org/10.1007/s10668-007-9132-1
Stewart ZP, Pierzynski GM, Jan Middendorf B, Vara Prasad PV (2020) Approaches to improve soil fertility in sub-Saharan Africa. J Exp Bot 71(2):632–641. https://doi.org/10.1093/jxb/erz446
Takoutsing B, Ayenkulu E, Tchoundjeu Z et al (2015) Land health surveillance for identifying land constraints and targeting agroforestry intervention in smallholder farming systems in Western Cameroon. ICRAF Working Paper No. 191. Yaounde, World Agroforestry Centre. https://doi.org/10.5716/WP14253.PDF
Tully K (2015) The state of soil degradation in sub-Saharan Africa: baselines, trajectories, and solutions. Sustainability 2015(7):6523–6552. https://doi.org/10.3390/su7066523
United Nations Convention to Combat Desertification (UNCCD) (2017) The Global Land Outlook, pp 192–194. Available at https://www.unccd.int/sites/default/files/documents/2017-09/GLO_Full_Report_low_res.pdf
Vandergeten E, Azadi H, Teklemariam D et al (2016) Agricultural outsourcing or land grabbing: a meta-analysis. Landscape Ecol 31(7):1395–1417
Wily LA (2011) The law is to blame: the vulnerable status of common property rights in sub-Saharan Africa. Dev change 42(3):733–757
Wynants M, Kelly C, Roberts N (2019) Drivers of increased soil erosion in East Africa's agro-pastoral systems: changing interactions between the social, economic and natural domains. Reg Environ Change 19(7):1–13. https://doi.org/10.1007/s10113-019-01520-9
Yahyah H, Ginzky H, Kasimbazi E et al (eds) (2020) Legal instruments for sustainable soil management in Africa. International year of soil law and policy regional perspective, 1st edn. Springer Nature, Switzerland AG. ISBN-10: 3030360032

The European Green Deal: Progress for Soil Protection?

Irene L. Heuser and Julie Itey

Abstract The chapter deals with legal aspects of soil protection in the European Green Deal (EGD). It aims at examining the various legal instruments used in the different policy approaches of the EGD contributing to soil protection, based on the objectives and content of this extensive growth strategy. The selected standards for assessment are the identified soil threats in the EU and the need for soil protection, based on the results of scientific studies of the IPCC and IP-BES, among other institutions. Currently, there is no comprehensive EU directive to address soil threats and soil-related issues in a coherent way. The chapter analyses in detail the EU Climate Law, the Biodiversity Strategy, the Zero Pollution Strategy and the Farm to Fork Strategy, as well as the general approach to a greener economy and a more sustainable society. It raises the question of whether the current mosaic of approaches to EU soil protection can be overcome and replaced by an overall harmonised and comprehensive strategy. The chapter further addresses the rationale for a legal framework extracted from the policy approaches already presented in the to-be-revised 2006 Thematic Strategy for Soil Protection. Additional emphasis is placed on Sustainable Development Goal 15 and its soil degradation neutrality target, on food security and the relationship with climate change mitigation. Finally, the authors explain the necessary conditions for the adequate development of soil protection law at the European level.

The views expressed in this article are personal and do not reflect those of the organisation represented.

I. L. Heuser (✉)
IUCN World Commission on Environmental Law, Kleinmachnow, Germany

J. Itey
The Global Soil Partnership, "Land and Water Division" of the Food and Agriculture Organizations, Rome, Italy
e-mail: Julie.Itey@fao.org

H. Ginzky et al. (eds.), *International Yearbook of Soil Law and Policy 2020/2021*, International Yearbook of Soil Law and Policy 2020,
https://doi.org/10.1007/978-3-030-96347-7_12

1 Introduction

In December 2019, the new President of the European Commission, Ms. Ursula von der Leyen, presented the 2019–2024 roadmap for a greener Europe embodied in the European Green Deal (EGD):[1] a set of policy initiatives with the overarching aim of making Europe climate neutral by 2050. These policy initiatives fall in line with global environmental challenges and more specifically with those tackled in the United Nations 2030 Agenda for Sustainable Development (UN 2030 Agenda) and the Paris Agreement to the United Nations Framework Convention on Climate Change (UNFCCC). The EGD's objectives are to be supported by specific individual agri-environment-climate initiatives, mainly being: the greening of the common agricultural policy, the "farm to fork" strategy, the biodiversity strategy, moving towards "zero pollution" for an environment free of toxic substances, the integration of sustainability in all EU policies, the European Climate Pact and the proposal for an 8th Environmental Action Programme. The initial EGD strategy of the European Commission of 11 December 2019[2] was intended as an ambitious growth programme of the newly appointed Commission. However, this strategy now appears in a new light, as the Covid-19 crisis has reshuffled priorities; whether the EGD will now be pushed into the background or redefined in the course of the discussions at EU level remains to be seen. In any case, it could be a compass for Europe's economic recovery and reorientation as well as towards the achievement of the climate, biodiversity and soil protection goals.[3] In this regard, the first communication on the EGD's intention was promising and the Covid-19 crisis should be an opportunity to emphasise its relevance even more, especially for related issues, such as the dependency of human living on natural resources. This aspect in particular was already dealt with by the EGD but without sufficiently mentioning soil and, above all, without improving its legal protection status.

Currently, there is no comprehensive legal instrument addressing soil protection or the steadily increasing soil degradation in the European Union while the EGD has not announced any specific legislative action to be undertaken in the near future so far. Soils play a fundamental role in the functioning of terrestrial ecosystems[4] by providing and regulating key ecosystem services critical for many global challenges linked to food and water security, biodiversity protection, energy sustainability and climate change mitigation. The variety of functions soils perform is therefore of crucial importance for human life, especially as storage and buffer of pollutants and chemical compounds, as a basis for food and fiber production, cultural heritage, source of raw materials etc.. In this perspective, ensuring healthy soils must be a core aspect that new and future environmental and agricultural policies need to integrate at the global, regional and national levels. Therefore, there are still many pending

[1] COM (2019) 640 final of 11.12.2019, p. 1 ff.

[2] COM (2019) 640 final of 11.12.2019, p. 1 ff.

[3] COM (2020) 380 final of 20.05.2020, p. 1.

[4] Hannam and Boer (2002), p. 10.

questions regarding the EGD's capacity to successfully complete this ambitious green transition, if not enclosed in an adequate EU soil protection framework. Each single EGD's strategy targets, more or less directly, some aspects of soil protection promoting, to à certain extent, sustainable soil use or management. If adequately designed and implemented, such measures could unleash potentials for an adequate sustainable green transition involving a better soil protection across Europe.

Therefore, a closer look at the detailed single EGD's legal proposals and their references to soil protection might reveal how serious the Commission really is about change in the long-term strategy, especially with respect to soils. The following chapter[5] will examine which input the EGD can provide with regard to the general concept of soil protection. It will also consider the specific soil threats which have been identified in the first version of the 2006 EU Soil strategy and raise, for instance, the questions whether soil biodiversity loss can be tackled by the EGD biodiversity strategy or whether the EGD "farm to fork strategy" can effectively address soil threats in agricultural lands. The chapter will further dig into the soil legal status and examine whether soil is still regarded as the "poor cousin" within the European environmental policy framework, by dealing with several concerns. Would general or additional soil protection be needed in the light of the current legal instruments-related challenges at EU level? What is the 2030 and 2050 roadmap for soil? What would need to be done in order to protect and sustainably use our soils, especially as part of this green transition? The chapter aims at addressing these questions about the role of soil as an environmental medium, from a legal point of view, taking into account soils ecological condition and their need for protection in the European Union. It ultimately aims at fostering debate toward the adoption of an adequate soil protection framework at the European level.

2 Current Status of Soils in the EU

Soil is one of our most important natural resources that provides us with vital goods and services.[6] It is mainly considered as a non-renewable resource.[7] Indeed, it's formation is an extremely slow process: Building up fully functioning soil layers of 30 cm requires a period of 1000 to 10,000 years.[8] Therefore soil degradation and its consequences are perceived only with considerable temporal delay due to the large resilience of soil systems to changes.

Europe's soils are characterized by great differences in their physical, chemical and biological properties, with over 320 groups of major soil types (10,000 different

[5] It includes the Green Deal proposals up to the deadline of 26.06.2021.

[6] European Commission (2016), p. 1 ff. (p. 13).

[7] COM (2002) 179 final of 16.4.2002, p. 1 ff. (p. 8).

[8] EEA/UNEP (2000), p. 7.

soil types) reported on the European soil map.[9] In this regard, soil data has been collected for many years by the European Soil Bureau and its Network, located at the Joint Research Centre of the European Commission in Ispra, and a new EU Soil Observatory was launched on 4 December 2020, providing the EU Community with the best available scientific knowledge to support policy making.[10] The following analysis uses data extracted from the 2010 and 2020 Environment State and Outlook Reports as well as from the 2006 EU Soil strategy.

In the European Union, soil resources are dramatically degrading, as presented by the 2020 European Environment Agency's State of the Environment Report.[11] Chemical, physical and biological soil degradation[12] is driven or exacerbated by human activity, especially agricultural and forestry practices, including land use changes (e.g. deforestation, drainage of wetlands and conversion of permanent grasslands into arable land), tourism, urban, industrial sprawl and construction activities with soil sealing, as well as pollution from industrial emissions, poor waste management or contaminants present in fertilisers or sewage sludge that are applied to soil.[13] Unsustainable land uses as well as growing demographic pressure, changes in consumption patterns and increasingly extreme weather events have been identified as the main factors for soil degradation in Europe.[14]

More precisely, Europe's soils suffer from certain threats which have already been identified by the European Commission in the 2006 EU soil strategy: soil erosion, soil contamination, floods and landslides, decline in soil organic matter, salinisation, soil compaction, soil sealing, and soil biodiversity loss.[15] In light of the current and projected global situation, this list should also encompass desertification and soil acidification threats while soil organic matter (SOM) decline should be emphasised even more. Indeed, with respect to soils' role in the climate change mitigation and adaptation context,[16] SOM decline is of particular relevance, especially within those regions in which soil organic carbon accumulation is the highest, namely: peat soils[17] in the northern permafrost region, but also black soils and wetlands.[18] Past trends have shown that 45% of the European mineral soils have a very low to low organic carbon content (0–2%),[19] directly or indirectly affected by the different forms of land degradation. Over the last centuries, the pressure for land

---

[9] Heuser and Montanarella (2016), Fig. 1 on the distribution of major soil groups in Europe.

[10] European Commission, Homepage: https://ec.europa.eu/jrc/en/eu-soil-observatory.

[11] EEA (2020) SOER, p. 115.

[12] EEA (2020) SOER, p. 125.

[13] COM (2006) 231 final of 22.09.2006, p. 1 ff. (p. 3). See Homepage EU COM, DG ENV: https://ec.europa.eu/environment/soil/index_en.htm.

[14] European Commission, Homepage: https://ec.europa.eu/jrc/en/eu-soil-observatory.

[15] COM (2006) 231 final of 22.09.2006, p. 1 ff.

[16] European Commission (2016), p. 1 ff. (p. 9 ff.).

[17] See European Commission (2008) CLIMSOIL, p. 69.

[18] European Commission (2016), p. 39 f.

[19] COM (2006) 231 final of 22.09.2006, p. 1 ff; European Commission JRC (2009), SoCo, p. 2.

in Europe has resulted in the reclamation of large soil organic carbon-rich areas, including peatlands, to make them suitable for agriculture or other land uses.[20]

In addition, soil degradation due to soil erosion is still a widespread issue in Europe, leading to an average soil loss rate of 5 to 10 tons per hectare per year.[21] In 2003, more than 105 million hectares were affected by water erosion and 10–42 million hectares by wind erosion,[22] leading, in the most extreme cases, to desertification. The situation has not reversed ever since, especially not for agricultural and natural grassland area. According to Eurostat, these areas represent over 80% of the areas in the EU estimated to be affected by moderate to severe soil erosion.[23] Currently, soil erosion in Europe affects over 12 million hectares of land and leads to €1.25 billion loss in crop productivity a year.[24]

Other forms of land degradation are of concern as well, since Western Europe is a highly urbanized area, resulting in land-use competition, increased soil sealing and soil contamination.[25] Overall, the EU suffers from different levels of land degradation, with 13 EU Member States (MS) having declared themselves as affected Parties under the United Nations Convention to Combat Desertification (UNCCD).[26] There is increasing evidence that land and soil degradation have major economic consequences, whereas the cost of preventing damage is significantly lower.[27]

The EU legal system task is to create the conditions to protect and sustainably use soils. Soil is mainly understood as protecting soils functions. These include serving as a platform for human activities and landscapes, an archive of our cultural heritage and an habitat and gene pool; furthermore soils store, filter, buffer and transform many substances, including water, nutrients and carbon.[28] Soils therefore play a major role in climate change mitigation and adaptation, agricultural production and food security, nature and biodiversity preservation, and are the foundation of our health and our wealth.[29] In order to address soils' multifunctionality, specific strategies need to be developed. In the European Union, the term "soil protection" is used in a wide sense, including protection from all soil threats.[30]

[20] European Commission (2016), p. 39.

[21] Eurostats (2019).

[22] JRC/EEA (2010), p. 15 and p. 17.

[23] Eurostats (2019).

[24] European Commission (2018), Soil erosion costs European farmers €1.25 billion a year.

[25] COM (2006) 231 final of 22.09.2006, p. 1 ff.

[26] Homepage EU COM, DG ENV: https://ec.europa.eu/environment/soil/index_en.htm.

[27] European Environment Agency (2010), The European Environment State and Outlook, p. 3.

[28] Homepage EU COM, DG ENV: https://ec.europa.eu/environment/soil/index_en.htm.

[29] Homepage EU COM, DG ENV: https://ec.europa.eu/environment/soil/index_en.htm; see also USDA (2021).

[30] See soil threats, Chapter 1.1. (p. 3).

3 European Green Deal: Scope and Relevance for the Legal Protection of Soils

3.1 General Scope and Objectives

The European Green Deal is a growth strategy aiming at implementing a set of ambitious measures across sectors in Europe to undergo a sustainable green transition and becoming the first climate-neutral continent by 2050.[31] It is not a single strategy but rather provides an overarching framework for the implementation of many environmental and climate related targets in key areas.[32] The EGD aims to complement the existing policy framework by providing a long-term guidance. The European Commission and the Member States will launch new initiatives and step up the EU's efforts to ensure current legislation and policies, relevant to the Green Deal, are enforced and effectively implemented.[33]

The EGD is targeting eight key aspects across policy sectors:[34]

1. Increasing the EU's climate ambition for 2030 and 2050
2. Supplying clean, affordable, secure energy
3. Mobilising industry for a clean and circular economy
4. Building and renovating in an energy and resource efficient way
5. A zero-pollution ambition for a toxic-free environment
6. Preserving and restoring ecosystems and biodiversity
7. Farm to Fork: a fair, healthy and environmentally friendly food system
8. Accelerating the shift to sustainable and smart mobility

The European Green Deal intends to review each existing law on its climate merits as well as introducing new legislation on the circular economy, building renovation, biodiversity, farming and innovation.

Figure 1 displays the flagship areas and substantive policy pledges that are part of the EGD framework. Several of these strategies are key to achieve a green transition, especially related to soils. Among them are for example: the EU biodiversity strategy for 2030, the 'farm to fork' strategy for sustainable agriculture (and the references to the CAP reform proposal), the zero pollution action plan and the Climate Law. The framework for implementation will be based on three main policy instruments: a regulatory reform to review and revise existing legislative measures in light of the Green Deal's objectives, policy proposals to develop strategy plans and address specific challenges and MS actions to ensure national policies are in line with the objectives and ambitions of the EDG. A Just Transition Mechanism and a Just Transition Fund are to support the costs associated with the implementation of

[31] European Council (2019), p. 1.

[32] Fetting (2020), p. 5.

[33] COM (2019) 640 final of 11.12.2019, p. 1 ff (p. 4).

[34] Munta (2020), pp. 9–10.

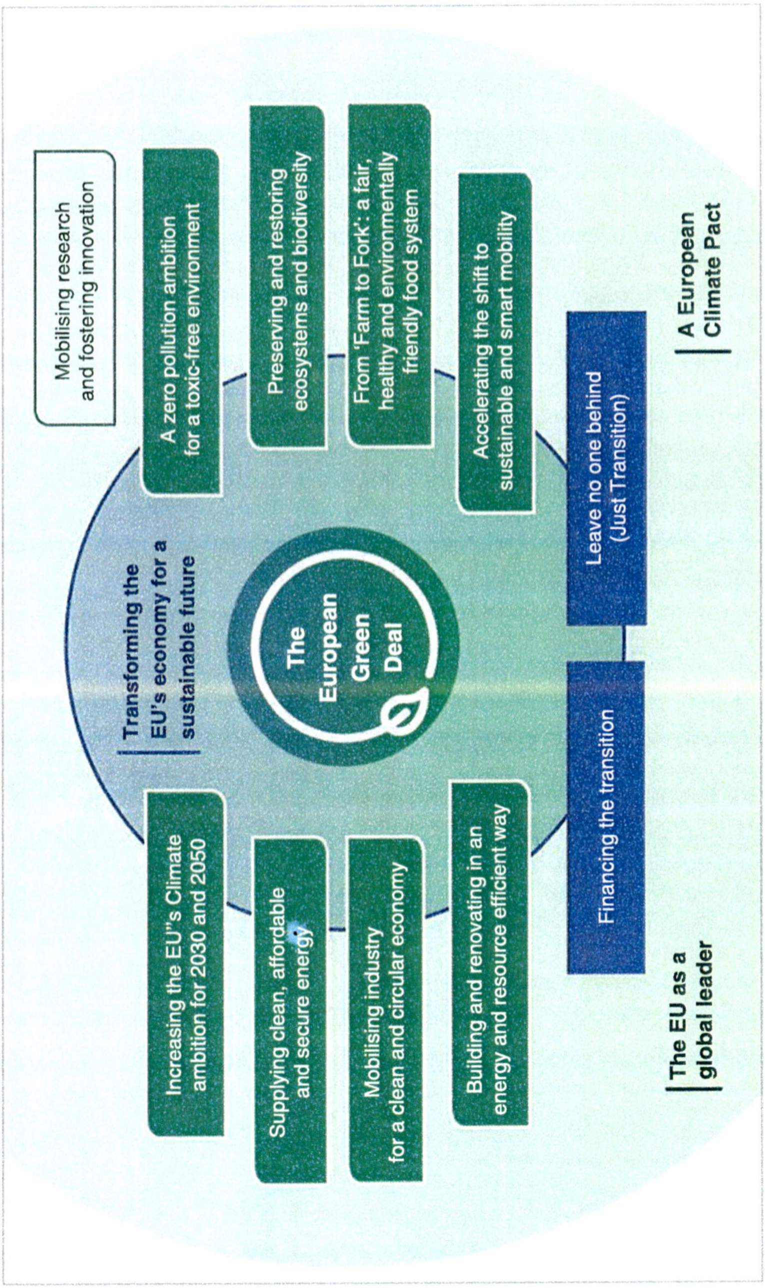

Fig. 1 Elements of the Green Deal (COM(2019) 640 final, p. 3)

these strategies but also aim at providing an incentive for Member States to comply with the 2050 zero net emission target among others.

3.2 *Existence of a Specific EU Soil Protection Law?*

In the European Union, air and water resources have been protected by EU law since many years, and monitoring, assessment and policy frameworks are already in place. In contrast, strategies for soil protection are not adequately developed at European level and there is currently no comprehensive EU legal instrument dealing with soil protection.[35]

Theoretically, chemical, biological and physical soil degradation threats could be tackled by the following objectives, according to Article 191 TFEU (Treaty on the Functioning of the European Union): preserving, protecting and improving the quality of the environment as well as prudent and rational utilization of natural resources. According to Art. 191 para. 2 TFEU (Treaty on the Functioning of the European Union) and the overall objective of the Soil Thematic strategy of 2006, the protection and sustainable use of soils shall be based on the precautionary, preventative and the ‘polluter pays’ principles, to prevent further soil degradation, preserve its functions as well as to restore degraded soil to a level that enables at least its current or intended use, which entails considering the cost of restoration.[36] Any proposed legislation also has to take into account the complexity of environmental situations (for example different soil types) peculiar to the various regions of the EU (Art. 191 para. 2).

The European Commission carried out a consultation for a revised soil thematic strategy till the end of April 2021 with the aim to address soil-and-land-related issues in a comprehensive way to help meeting international obligations such as achieving Sustainable Development Goal (SDG) Target 15.3. on land degradation neutrality and other related SDGs. In reaction to the Commission’s roadmap with the title “New Soil strategy – healthy soil for a healthy life” many participants voted that a soil-specific instrument, such as a soil legislative framework or a European soil (framework) directive[37] or at least specific binding targets as part of the thematic strategy for soil protection, was needed.[38] Input from large and powerful industrial enterprises and other opponents of a soil directive in the consultation process stressed once again—like in the disputes about a Soil Framework Directive—the need for local solutions and reasserted that soils lack cross-border effects (which is

[35] EEA (2020), SOER, p. 131.

[36] European Commission (2006), p. 8.

[37] European Commission (2021), Commission consults on new EU Soil Strategy para. 1–7; Institute for European Environmental Policy (2021), p. 1; Deutscher Naturschutzring (2020), p. 1.

[38] Bundesverband Boden (2020), p. 1.

relevant in the context of the subsidiarity principle).[39] In the same consultation the European Commission further stressed the need for a new soil strategy and soils are essential for achieving the objectives of the EGD, such as climate neutrality, biodiversity restoration, zero pollution, healthy and sustainable food systems and a resilient environment.[40] These goals cannot be achieved without strengthening the legal situation of soil, appreciating their role and the scientific context. Ultimately, a mindset shift, away from the view of soils as productive commodities only, is needed.

In any case, the EGD intends to improve the alarming situation for European soils. Whether the Green Deal will meet the demands placed on policymakers and contribute to a solution that is not merely tokenistic will be seen in the overall design and specific sub-strategies of the EGD and their consideration of soil protection needs.

4 Environmental Provisions of the European Green Deal

The European Green Deal mainly focuses on environmental and climate change aspects; the European Commission intends to transform its economy and society to put them on a more sustainable path and to decouple economic growth from resource use.[41] It aims to protect, conserve and enhance the EU's natural capital,[42] which should also include soils.

4.1 Zero Pollution Action Plan for Soil: Tackling Soil Pollution

In order to create a toxic-free environment, the EU needs to agree on action to prevent, clean and remedy environmental pollution. Soils play an important role in this process, as their storage and buffering capacity allows pollutants to be accumulated to a high degree. Therefore, the entry of pollutants into soils and their ecological consequences can only be perceived with considerable delay. This is a major contrast to the effects of air and water pollution which usually become noticeable after a relatively short period of time and affect human beings more directly.

---

[39] Federation of German Industries (2020), p. 1.

[40] European Commission, Directorate-General for Environment, Commission consults on new EU Soil strategy (2021), para. 2.

[41] COM (2019) 640 final of 11.12.2019, p. 1 ff. (p. 2).

[42] COM (2019) 640 final of 11.12.2019, p. 1 ff. (p. 2).

The Communication on a "Pathway to a healthy planet for all" with the sub-title "EU action Plan: 'Towards zero pollution for air, water and soil'" of 12 May 2021 will be adopted by the European Commission in 2021[43] and intends to address the interlinked challenges of these environmental media. Within the soil protection context, it means the main ambition of this measure is to prevent pollutants from various pathways into the soil and to reduce soil contamination. Reference point of this Communication of the Commission is the definition of pollution in Art. 3 para. 2 of Directive 2010/75/EU referring, among other aspects, to whether the introduction of substances into land may be harmful to human health or to the quality of the environment. The Green Deal Communication already mentions the prevention of pollutant discharges from industrial installations, pharmaceuticals and other chemicals, packaging material, plastics and other wastes etc.[44] According to the 2030 biodiversity strategy, the Commission will also promote the goal of zero pollution from nitrogen and phosphorus flows from fertilisers through reducing nutrient losses by at least 50% and the use of fertilisers by at least 20%.[45] The risk and use of chemical pesticides are to be reduced by 50% and the use of more hazardous pesticides by 50%. In this regard, the farm-to-fork strategy will further address the reduction in the use and risk of pesticides.[46]

Although soil pollution and biodiversity loss interdependence are well-known (other main biodiversity loss drivers are also related to soil protection: changes in land use and overexploitation of natural resources), the Commission does not consistently include soil protection in its prevention considerations. Furthermore, while the Commission advocates for the inclusion of "pollution prevention in all relevant EU policies", zero pollution targets have been set up for air pollution, noise pollution, pesticides, plastics, waste etc.,[47] representing different soil endangerment paths, but not for soil per se or its related threats. This inconsistent and near-sighted vision gives the impression the EU only legislates reactively where the (lobby) pressure is particularly high, whereas a future-oriented, responsible and sustainable policy and legislation based on scientific findings starts earlier and also gives a voice to less lobby-supported concerns, such as environmental and even soil protection.

As part of the EDG zero pollution ambition, the European Commission has also published a new "chemicals strategy for sustainability 'Towards a toxic-free environment'" Communication[48] on 14 October 2020. Under this new strategy—requiring a targeted and hopefully precisely coordinated revision of the CLP Regulation (regulation on the classification, labelling and packaging of chemical substances and mixtures) and the REACH Regulation (Regulation on registration, evaluation,

[43]COM (2019) 640 final of 11.12.2019, p. 1 ff. (p. 14); COM(2021) 400 final of 12.05.2021, p. 1 ff.

[44]COM (2019) 640 final of 11.12.2019, p. 1 ff. (p. 14 f.); COM (2020) 380 final of 20.05.2020, p. 1. (p. 14).

[45]COM (2020) 380 final of 20.05.2020, p. 1 ff. (p. 13).

[46]COM (2020) 380 final of 20.05.2020, p. 1 ff. (p. 13 f.).

[47]COM (2021) 400 final of 12.05.2021, p. 1 ff. (p. 3).

[48]COM (2020) 667 final of 14.10.2020, p. 1 ff.

authorisation and restriction of chemicals)—the focus falls on banning the most harmful chemicals in consumer products. This will allow for further EU's resilience regarding critical chemicals supply and sustainability while taking into account they can have various combined effects when assessing chemical-related risks.[49] In order to fully address environmental toxicity, persistency, mobility and bioaccumulation, setting minimum requirements for horizontal criteria and new hazard classes is important. In this perspective, a stricter regulation will hopefully also consider the accumulation of chemical substances as well as their effects in the soil; the revisions are planned till the last quarter of 2021 and the first quarter of 2022.[50] The Covid-19 pandemic might as well have "added to the urgency to protect human and planetary health"[51] and the Commission will reinforce the regulation for food chemical contaminants,[52] a matter directly related to soils. The "chemicals strategy for sustainability" Communication also acknowledged the long-term and large-scale effects hazardous chemicals can have on terrestrial ecosystems such as impacting their resilience as well as human health, through the presence of contaminants in the food chain.[53] In addition, the Communication highlighted 2.8 million sites were potentially contaminated in the EU and are currently "posing a significant environmental hazard for terrestrial and aquatic ecosystems and affecting the productivity of soils".[54] From the European Commission's point of view, "the current regulatory and policy framework struggles to take this into account and needs to be strengthened". Nevertheless, significant steps have not yet been proposed with regard to the entry of harmful substances from chemicals into soils. Admittedly, the Commission wants to further examine and regulate per- and polyfluoralkyl substances (PFAS) since these industrially produced chemicals are broadly used in consumer products and treatment techniques are currently being tested for PFAS-contaminated soil hotspots. However, the focus falls once again on certain substances whereas a comprehensive risk assessment would be necessary.

On the one hand, the Commission stresses the necessity to implement the existing EU regulatory frameworks protecting air and several types of water faster and better and on the other hand the "Zero pollution action plan Communication" only pleads for "urgently working towards a framework to regularly address the status of EU soils and take action on all levels to address soil pollution and degradation".[55] In this regard, soft law language such as "working towards a framework" or "as much as possible" can be found throughout the zero-pollution action plan communication and is indicative of the "poor cousin" status still conferred to soils, being more obvious for some sectors than for others. Then, this communication text only referred to the

[49] European Commission (2020), Chemicals Strategy, para. 6.

[50] European Commission (2020), Chemicals Strategy.

[51] COM (2020) 667 final of 14.10.2020, p. 1 ff. (p. 3).

[52] COM (2020) 667 final of 14.10.2020, p. 1 ff. (p. 13).

[53] COM (2020) 667 final of 14.10.2020, p. 1 ff. (p. 13).

[54] COM (2020) 667 final of 14.10.2020, p. 1 ff. (p. 13).

[55] COM (2021) 400 final of 12.05.2021, p. 1 ff. (p. 8).

upcoming EU soil strategy which "will develop measures to significantly increase efforts to identify, investigate, assess and remediate contaminated sites" which have already been mentioned in the "chemicals strategy for sustainability" communication.[56] Overall, the action plan merely announces the proposal for legally binding EU nature restoration targets "will consider addressing" the restoration of degraded soil ecosystems; in this respect, a focus on soil protection aspects is not even certain in this strategy[57] which will be presented in 2021.

While the prevention of other environmental media is stricter and legally sound, new soil contamination should be prevented "as much as possible"—which says less than nothing in terms of legally binding targets. Ultimately, the Commission will just continue supporting the Global Soil Partnership's efforts.[58] Moving "towards" zero pollution for soil will therefore mean that the legally non-binding and contraproductive situation will continue to be maintained. Indeed, the announced "vision of a pollution-free world"[59] is not nearly as visionary as it suggests and is based on an outdated mindset which repeatedly does not adequately consider the ecological context for soil resources. Therefore, the Action plan does not live up to its ambitious name in being limited to rethoric and lacking a fresh impetus so far. In the long term, the reluctance of the Commission to put soil in the same legal position as air and water endangers the important basis of our living and survival.

4.2 *Biodiversity Strategy*

The EU biodiversity strategy for 2030 "Bringing nature back into our lives"[60] is a restoration and protection plan setting legally binding targets to address several environmental dimensions, including soil biodiversity and other soil threats; it acknowledges the role of soil organisms to "regulate and control key ecosystem services such as soil fertility, nutrient cycling and climate regulation".[61] The strategy aims at maintaining at least 30% of European lands under the protected areas regime, including EU primary and old-growth forests, and ensuring the effective management of these protected areas by defining clear conservation objectives and monitoring indicators. It encourages the maintenance of at least 10% of agricultural areas containing highly diverse landscape elements such as buffer zones (crop rotation, fallow lands, hedges, non-productive trees)—as they have a high potential for soil organic carbon storage and sequestration—as well as soil erosion and loss

[56] 2.8 million sites across the EU are potentially contaminated and 390.000 are expected to require remediation, see COM (2021) 400 final of 12.05.2021, p. 1 ff. (p. 9).

[57] COM (2021) 400 final of 12.05.2021, p. 1 ff. (p. 9).

[58] COM (2021) 400 final of 12.05.2021, p. 1 ff. (p. 10).

[59] COM (2021) 400 final of 12.05.2021, p. 1 ff. (p. 21).

[60] COM (2020) 380 final of 20.05.2020, p. 1.

[61] COM (2020) 380 final of 20.05.2020, p. 1 (p. 8).

prevention, air and water filtering and resilience to climate change.[62] The strategy ultimately aims at incentivizing a transition to fully sustainable agricultural practices, considering that biodiversity conservation is crucial for food security and it has potential direct economic benefits for many sectors.[63]

The biodiversity strategy aims to ensure the entire world's ecosystems restoration, resilience and adequate protection by 2050 by setting out very specific targets (mostly by 2030):

- Protecting a minimum of 30% of the EU's land area and strictly protecting at least a third of the EU's protected areas (as part of the Trans-European Nature Network),
- Legally binding EU nature restoration targets to restore degraded ecosystems in order to achieve their good condition, including soil health,
- Reaching favourable conservation status for at least 30% of species and habitats,
- Halting and reversing the decline of pollinators, habitats and species,
- Planting three billion trees,
- Restoring at least 25,000 km of EU rivers to a free-flowing state (e.g. through restoration of floodplains and wetlands),
- Increasing organic farming to at least 25% of the agricultural area and significantly increasing the uptake of agro-ecological practices,
- Reducing the use and risk of chemical pesticides by 50%, likewise from hazardous pesticides, and no use of chemical pesticides at all in sensitive areas,
- Reducing the use of fertilizers by at least 20% and the loss of nutrients from fertilizers by 50%,
- Significant progress in remediating contaminated soil sites,
- Restoring degraded carbon-rich ecosystems[64]

Nevertheless, substantial progress is still hampered by considerable gaps toward regulation and implementation. The first gap lies in the absence of an obligation for Member States to generate biodiversity restoration plans, to monitor or map ecosystem services and restoration efforts in a holistic manner. There are also still no clearly defined binding targets along with a set time frame, nor definitions or indicators for ecosystem restoration or sustainable use.[65] There is currently no EU comprehensive governance framework to foster the implementation of biodiversity commitments at the national, regional or international levels.[66] To bridge this gap, the Commission will submit a proposal in 2021 to establish a new biodiversity governance framework to set a commonly agreed and clear set of indicators within a

[62] Prandecki and Wrzaszcz (2020), p. 164.

[63] COM (2020) 380 final of 20.05.2020, p. 1 (p. 1 f.; concerning the cost of inaction from land degradation: p. 2).

[64] Maréchal and Hart (2020), pp. 13–15.

[65] COM (2020) 380 final of 20.05.2020, p. 1 ff. (p. 1 ff.).

[66] COM (2020) 380 final of 20.05.2020, p. 1 (p. 15).

monitoring and review mechanism targeting degraded ecosystems restoration.[67] This proposal will target, among other things, the increasing land artificialisation issue, the management of areas with the greatest carbon sequestration potential as well as poor land management in general, such as deforestation, overgrazing, unsustainable agricultural and forestry practices, construction activities or soil sealing.[68]

Next to this new governance framework, the stepping up of implementation and enforcement of relevant environmental legislation will contribute to a change in the "thinking in boxes" way and will hopefully provide comprehensive consideration of all natural components of a habitat, including soil.[69] In this regard, the Habitats Directive ensures substantial aspects of soil protection, in particular with respect to its physical and biological protection.

Nevertheless, compared to the great importance the EU Environment Commissioner Sinkevičius attributed to soil protection on The Global Symposium on soil biodiversity on 19 April 2021,[70] it must be stated the biodiversity strategy does not substantiate this significance. Indeed, the Commission did not propose any legally binding targets for the relevant soil aspects in the biodiversity context for the European Union, while focusing on a new global framework for the post-2020 at the 15th Conference of the Parties to the Convention on Biological Diversity[71] (CBD). Such a failure is very unfortunate as this Conference was to represent a "turning point similar to the Paris Agreement for climate" and include a "substantial and ambitious dimension for soil" in the corresponding agreement.[72] This ambition was further stressed by the Commissioner who explicitly emphasised "Europe [was] determined to lead by example [with] the commitments [..] made for soil in [the] strategy for biodiversity."[73] In the light of the absence of any binding specific target for the protection against biological or physical soil threats, this statement appears to be merely lip service, especially since both the Commission and the Council have reaffirmed their commitment to reaching land-degradation neutrality on other occasions.[74]

[67] COM (2020) 380 final of 20.05.2020, p. 1 (p. 15).

[68] EEA, Land and soil in Europe, p. 10.

[69] Heuser (2006), JEEPL, p. 190 ff. (p. 198).

[70] European Commission (2021), Commissioner Sinkevičius speech at The Global Symposium on Soil Biodiversity.

[71] COM (2020) 380 final of 20.05.2020, p. 1. (p. 19).

[72] European Commission (2021), Commissioner Sinkevičius speech at The Global Symposium on Soil Biodiversity, p. 3. See also the statement of the President of the European Commission, Ursula von der Leyen, Biodiversity needs its on Paris Agreement, Euractiv of 21.03.2021, p. 2.

[73] European Commission (2021), Commissioner Sinkevičius speech at The Global Symposium on Soil Biodiversity, p. 3.

[74] Council of the European Union (2020), Conclusion 24, p. 10.

4.3 European Climate Law

Due to its capacity to regulate the climate by storing carbon, soils, after the oceans, represent the largest pool of carbon on earth; the soil organic carbon pool is about twice the size of the atmospheric carbon pool and about three times the size of the biotic carbon pool.[75] EU forests store about 2.5 times more carbon in soils than in tree biomass.[76] Sustainable soil management and other soil protection measures can help to rapidly sequester excess atmospheric carbon into the soil. Therefore, soils are of great importance in the context of tackling and reversing climate change. The IPCC's special report on climate change and land[77] identified increased soil organic carbon (SOC) as one of the most cost-effective options for climate change adaptation and mitigation, as well as for combating desertification, land degradation and food insecurity. Given the important role of soils in achieving the climate goals, it is to be hoped that this will also be valued in the context of the Green Deal legislation. In line with the EU climate roadmap, a greater emphasis therefore needs to be put on the role healthy soils and soil sustainable management (SSM) practices play in the climate mitigation strategy since healthy soils are key to tackle climate change adverse impacts and increase resilience toward increasing extreme events.

Since the adoption of the Paris Agreement, climate change has also been a legal focus, most recently because of decisions by various constitutional courts,[78] at least partly based on the European Convention on Human Rights, recognizing also the principles by the Charter of Fundamental Rights of the European Union. The European Parliament, in its session at the first reading of the regulation on the "European Climate Law" on 24 August 2021 also emphasised the recognition of the principles of the Charter of Fundamental Rights of the European Union.[79]

In March 2020, the European Commission proposed the first European Climate Law, enshrining the 2050 climate-neutrality objective into legal terms. This means a legally binding target of net zero greenhouse gas emissions by 2050 for the EU as a whole; all EU policies will have to contribute to this objective by cutting emissions, investing in green technologies and protecting the natural environment. On 21 April 2021, the European Commission turned the European Green Deal targets for a European Climate Law into a provisional agreement and agreed on legal obligations for reducing net greenhouse gas emissions by at least 55% by 2030 as compared to 1990 levels (the European Parliament had voted for 60%) and reaching climate neutrality by 2050.[80] The "2030 EU Climate Target Plan" is seen as a necessary step to get to the 2050 target and the proposed European Climate Law will have to be

---

[75] European Commission (2016), p. 1 ff. (p. 55).

[76] European Parliament (2021), Motion for a resolution for soil protection, p. 17, no. 27.

[77] IPCC (2019).

[78] E.g. decision of the German Federal Constitutional Court of 24.3.2021, see press release: https://www.bundesverfassungsgericht.de/SharedDocs/Pressemitteilungen/DE/2021/bvg21-031.html.

[79] European Parliament (2021), European Climate Law, Recital No. 5.

[80] European Commission (2021), Provisional agreement on the European Climate Change Law.

approved by Parliament and Council. In the first reading of the European Climate Law on 24 August 2021 in the European Parliament, the 55% reduction target in Art. 4 para. 1 was maintained (the European Parliament had previously voted for 60%) and it was confirmed that the 2030 climate target is also lowered to de facto only −52.8% compared to 1990 due to a calculation trick, because it is a "net" target, in which for the first time CO_2 sinks, e.g. forests or soils, will also be included in the calculation (even if the contribution of net removals to the Union 2030 climate target shall be limited to 225 million tonnes of CO_2 equivalent according to the EP proposal in Art. 4 para 1). In addition, the European Parliament further stressed the essential role of carbon sinks in the transition to climate neutrality for the Union, in particular the contribution of the agriculture, forestry and land use sectors,[81] but in the regulation itself only specified very vaguely in Art. 4 para. 1 that "in order to enhance the Union's carbon sink in line with the objective of achieving climate neutrality by 2050, the Union shall aim to achieve a higher volume of its net carbon sink in 2030". The 55% net reduction target is already problematic because this level of climate ambition is far from sufficient to limit global warming to 1.5 degrees (we are even aiming for global warming of well over 2 °C). Moreover, CO_2 storage by more forests or soils is based on optimistic assumptions and ignores the fact that climate change is already causing droughts or forest fires. It seems unclear where these CO_2 sinks will be created and how safely the carbon will be stored in them. In order to take the responsibility for the planet and the young generation seriously, other targets would be necessary.

The decision on a European Climate Law also included a process for setting a 2040 climate target, the establishment of the European Scientific Advisory Board on Climate Change and a EU strategy on Climate Change Adaptation, to make Europe a climate-resilient society by 2050:[82] In order to accompany this process of adjusting to current or expected climate change and its unavoidable impacts, the Commission proposed a Communication with the title "Forging a climate-resilient Europe - the new EU strategy on Adaptation to Climate Change" on 24 February 2021.[83] The strategy pursues the objectives to make adaptation smarter, swifter and more systemic, and to step up international action on adaptation. A revision of all climate-related policy instruments to achieve the additional emissions reductions for 2030 is scheduled for July 2021.[84] Concerning the Emissions Trading System (ETS), the plan is to allocate fewer free certificates to airlines and to extent the ETS to new sectors (especially the maritime transport) or to reduce emissions in sectors outside the ETS.

When proposing the Union 2040 climate target, the Commission shall also consider according to the new Art. 4 para. 5 (j) proposed by the European Parliament "the need to maintain, manage and enhance natural sinks in the long term and protect

[81] European Parliament (2021), European Climate Law, Recital No. 22.

[82] European Commission (2021), EU Climate Action and European Green Deal.

[83] COM (2021) 82 final of 24.02.2021, p. 1 ff.

[84] European Commission, EU Climate Action and European Green Deal.

and restore biodiversity". The Parliament has also made an ammendment in Art. 5 para. 4 regarding the national adaptation strategies, that "Member States shall take into account the particular vulnerability of the relevant sectors, inter alia agriculture, and of water and food systems, as well as food security"; surprisingly, the overall context of environmental media is not mentioned here either and soil is not explicitly included as a decisive factor also for adaptation.

Nevertheless, the Commission also recognized the need for emission reductions through soils as carbon sinks and will propose a more ambitious Regulation on Emissions from Land Use, Land Use Change and Forestry (LULUCF) in July 2021 as part of the Commissions "Fit for 55 Package", which will cover wide-ranging policy areas to ensure emission reductions of at least 55% by 2030 (as explained before). As an Annex to the European Parliament legislative resolution of 24 August 2021 on the proposal for a European Climate Law, the Commission acknowledged that the LULUCF sector both emits greenhouse gases and absorbs CO_2 in its soil and biomass and that restoring and growing our land carbon sink is crucial to our climate goals.[85] In order to illustrate this, the Commission drew on a study of the interrelations between soil and climate change and a conference in August 2008[86] on this topic. It came to the conclusion that potential losses arising from land use changes, especially on highly organic peat soils, are a major risk; therefore, the most effective strategy to prevent soil carbon loss would be to halt land conversion to cropland, although this may conflict with the growing global food demand.[87] Dealing with carbon emissions is of particular importance if we consider only 0.1% of carbon emitted into the atmosphere from European soils would be equivalent to generated CO_2 emissions from 100 million extra cars on our roads.[88]

Under the current "Regulation (EU) 2018/841 of 30 May 2018 on the inclusion of greenhouse gases and removals from land use, land use change and forestry in the 2030 climate and energy framework, and amending Regulation (EU) No 525/2013 and Decision No 529/2013/EU" (on the basis of the conclusions of the European Council in October 2014 with an emission reduction target of at least 40%), Member States have to ensure that accounted emissions from land use are entirely compensated by an equivalent removal of CO_2 from the atmosphere through action in the sector for the 2021 to 2030 period. In the LULUCF sector, besides the need to take the role of forest soils into account in the preparation of a new EU forest strategy,[89] sustainable land management practices can contribute to climate change mitigation in several ways, in particular by reducing emissions, maintaining, enhancing

[85] European Parliament (2021), European Climate Law, Annex.

[86] European Commission (2009), Press release Climate Change, p. 1.

[87] See the key messages of the project CLIMSOIL: European Commission (2008), p. 16.

[88] European Commission (2009), Press release Climate Change, p. 1.

[89] European Parliament (2021), Motion for a resolution for soil protection, p. 17, no. 27.

terrestrial sinks and carbon stocks[90] through nature-based solutions.[91] This policy particularly focuses on CO_2 emissions reduction in the land sector, dealing with all land use forms.[92] Of course, restoring land and soil or even whole landscapes does of have the ability to solve the climate crisis alone, but the potential of restoration for climate change mitigation and adaptation is huge.[93] The LULUCF regulation (next to biomass or wood) mainly deals with soil organic carbon as carbon pools (Annex I, B. (e)), i.e. a biogeochemical feature or system within which carbon (or any greenhouse gas containing carbon) is stored. According to Article 4 of the current LULUCF Regulation, each Member State shall ensure that emissions do not exceed removals, calculated as the sum of total emissions and total removals on its territory in all of the land accounting categories referred to in Article 2. The expected proposal for a revision of this regulation will have to make clear how the additional CO_2 sinks will be created and it will hopefully appreciate the environmental value of soils as carbon sinks in land use.

4.4 Conclusion

When the president of the European Commission, Ursula von der Leyen, compared the Green Deal with "Europe's man on the Moon moment",[94] she ignored the fact that the astronauts also had soil under their feet during this moon landing, albeit with a different gravitational force than on earth. Achieving soil protection (or sustainable soil use or management) across Europe is conditional to the effective implementation of several of the EGD targets as it is a practice at crossroads addressing global challenges related to food security, biodiversity and climate change. In the fight against climate change, the plan to make Europe the first climate-neutral continent can only succeed if the capacity of soils to regulate the climate by storing carbon is valued and the need to protect them from chemical, physical and biological threats is

[90] Regulation (EU) 2018/841 of the European Parliament and of the Council of 30 May 3018, Official Journal of 19 August 2021, L 156, p. 1 ff., Recital 7.

[91] COM (2021) 82 final of 24.02.2021, p. 1 ff. (p. 11: 2.2.4).

[92] The six land-use categories according to the 2006 IPCC Guidelines are: forest land, cropland, grassland, wetlands, settlements and other land. See 2019 Refinement to the 2006 IPCC Guidelines for National Greenhouse Gas Inventories, https://www.ipcc-nggip.iges.or.jp/public/2019rf/pdf/4_Volume4/19R_V4_Ch01_Introduction.pdf, p. 1.10 (accessed 15 August 2021).

[93] This was explored in many forums and workshops, especially of the Global Landscapes Forum, e.g. on 29 April 2021, in a digital forum on restoration, hosted jointly by the Collaborative Partnership on Forests (CPF) and the Global Partnership on Forest and Landscape Restoration (GPFLR) where the synergies generated between mitigation and adaptation were discussed. There seems to be a need to connect on-the-ground monitoring with global commitments and that the role of community members in restoration initiatives, including gender dynamics and changing social and power relations, is important. See https://events.globallandscapesforum.org/bringing-evidence-to-bear-for-scaling-landscape-restoration/.

[94] European Commission (2019), Press remarks by President von der Leyen, 8th paragraph.

underpinned by effective legal instruments. Currently, the EGD aims at enshrining the goal of climate neutrality by 2050, at strengthening adaptation efforts, of at providing a process for setting and updating the trajectory to 2050, for regular assessments, and for a procedure to be followed in case of insufficient progress or inconsistencies, in the EU legislation. Although the protection of soils is somewhat included in the EGD areas of pollution prevention and control, biodiversity and climate protection as a component of nature, its decisive role and essential regulatory functions are misunderstood. The European Green Deal is Europe's response to the climate and biodiversity crisis, but mainly seems to keep out one very important ally: our soils. Therefore, given the current challenges of steadily increasing soil degradation in the European Union, it is very astonishing that the European Green Deal does not include an explicit recommendation for a specific soil protection (framework) directive in the near future nor a well-defined legal protection status so far.

5 Agricultural Provisions of the European Green Deal

European soils have been damaged with intensive monocrop farming and heavy reliance on pesticides for more than a century. So far, EU policies targeting the agricultural sector, embodied in the common agricultural policy (CAP), have been unable to not only exploit the soil as a basis for production, but also to sufficiently ensure its protection. Following the Council's general agreement on the post-2020 CAP reform package reached in October 2020, when the ministers agreed on strong commitments for higher environmental ambition by introducing instruments like mandatory eco-schemes and an enhanced conditionality, negotiations with the European Parliament are currently underway. The objective is to reach an overall agreement to adopt three new regulations shaping the reformed CAP for the period 2023–2027. The main discussion points at the Agriculture and Fisheries Council on 26–27 May 2021 included social conditionality, targeting of payments, and green architecture.[95] For the time being, many aspects of the CAP are not yet sustainable from an over-the long-term and against-planetary-boundaries perspective. Since 95% of our food is directly or indirectly produced from our soil,[96] the concretization of the European Green Deal with regard to agriculture through the farm to fork strategy is most relevant for achieving the goals of the whole concept.

The "farm to fork strategy" (F2F)[97] promotes a transition towards fair, healthy and environmentally-friendly European food systems and is at the heart of the European Green Deal with regard to soil protection by targeting, more specifically, food insecurity, climate resilience, biodiversity and environmental protection in

[95]European Council, Agriculture and Fisheries Council, 26–27 May 2021, see https://www.consilium.europa.eu/en/meetings/agrifish/2021/05/26-27/.

[96]See https://ec.europa.eu/environment/soil/index_en.htm.

[97]COM (2020) 381 final of 20.05.2020, p. 1 ff.

general. With the F2F Strategy, the European Union addresses food sustainability in a comprehensive manner, from primary production to food processing and the reduction of food loss and waste to consumption by consumers,[98] and includes ambitious and concrete targets to be achieved by 2030: a reduction of chemical and hazardous pesticides by 50%, a reduction of fertilizer use by 20%, a quota of 25% of EU arable land dedicated to organic farming and a reduction of sales of antimicrobials by 50%.[99] The proposal for a revision of the Sustainable Use of Pesticides Directive, to significantly reduce the risk and dependency on pesticides and enhance Integrated Pest Management, will be presented at the beginning of 2022;[100] it will contribute to the reduction of soil pollution,[101] soil biodiversity loss and other soil threats. Regarding biodiversity, the farm to fork strategie's targets are very similar to the ones laid out in the 2030 biodiversity strategy.[102]

On 25 March 2021, the Commission presented an Action Plan for the development of organic farming which aims at boosting the production and consumption of organic products, to reach at least 25% of agricultural land under organic farming by 2030, as well as increasing organic aquaculture. However, ensuring the achievement of the "25% of agricultural land under organic farming by 2030" target will be challenging since the share of organic farming in the total agricultural area varies considerably across the EU. This will be particularly challenging in those countries, like Poland, Ireland, the Netherlands, Romania or Luxembourg, where the total area under organic farming is decreasing.[103]

Organic farming contributes to increased soil protection due to its less intensive use of soil and low use of pollutants. The organic farming objective is an essential step toward the transition to sustainable agriculture, however it needs to incorporate additional socio-economic specific measures to boost supply such as updating the related regulations to ease organic certification for small farmers. Discussions must also be conducted in parallel to come to a common definition of sustainable agriculture, along with the development of measurable indicators to verify and monitor organic farming sustainability.[104] Unfortunately, the entry into force of a new legislation on organic farming, complementary to the objective set out in the F2F strategy, has been postponed, which further delays the chances for compliance with organic farming standards within the EGD framework. Nevertheless, this delay

---

[98] COM (2020) 381 final of 20.05.2020, p. 1 ff. (pp. 5–14).

[99] COM (2020) 381 final of 20.05.2020, p. 1 ff. (pp. 6–8).

[100] Annex to COM (2020) 381 final of 20.05.2020, p. 1 ff. (p. 2: Farm to Fork Draft Action Plan, No. 4).

[101] COM (2020) 381 final of 20.05.2020, p. 1 ff. (p. 6) concerning soil pollution.

[102] Prandecki and Wrzaszcz (2020), p. 163.

[103] Prandecki and Wrzaszcz (2020), p. 164.

[104] El Chami (2020), p. 2.

can give the time needed for Third countries to adapt to such a framework to ensure they will be able to continue exporting to European countries.[105]

We may wonder whether the future common agricultural policy (CAP) is flexible enough to support and reinforce the sustainability and climate dimensions of agricultural policies while the framework still needs considerable adjustments to be more coherent with the EDG objectives.[106] Currently, the 2018 Commission's proposal for the future CAP is not green enough and not fit for achieving the EGD objectives. Indeed, the future 2023–2027 CAP content will enhance the green architecture based on the incentivization of environmental and climate actions under the second pillar and via eco-schemes under the first pillar. However, the CAP financial support for the implementation of the Green Deal and its related initiatives has been unbalanced as 80% of payments go to 20% of farmers.[107] Furthermore, while the future CAP will integrate additional requirements in the cross-compliance mechanism under pillar 1 targeting soil protection, some practices, like crop rotation, wouldn't be included for farms smaller than 10 hectares (arable lands). This would mean that only 24% of farmers would be concerned with direct support[108] and therefore incentivized to apply soil protection measures. Furthermore, these additional green actions under both pillars are still to be voluntary in nature.[109] Indeed, on the one hand, each Member State (MS) will ultimately decide how much from the Pillar 1 budget will go to eco schemes and which sectors will be targeted. On the other hand, farmers will decide whether they align with the eco schemes requirements and to what extent.[110] It is therefore very likely a large part of the payments to farmers will continue to be linked to the size of the area, but not sufficiently to ecological and not at all to soil protection criteria. Such a regime could dramatically weaken the effectiveness of the EGD-related-strategies within the new CAP framework. Thus, forthcoming negotiations must focus on ensuring the reformed CAP's objectives are to be aligned with the EGD's ones, especially with the farm to fork and biodiversity strategies since their objectives fall within the soil protection scope and are not yet reflected in the currently ongoing trilogy negotiations on the future 2023–2027 CAP. It is up to MS to generate national CAP strategic plans aligning with the EGD ambition and to the EU Commission to set up adapted monitoring systems to ensure objectives are leveled up across countries.[111]

Nevertheless, substantial and effective changes needed in agri-food systems cannot be brought about by the CAP and the Green Deal alone,[112] but would require

[105] EU Commission (2020), Role of the organic sector in the Green Deal (Questions and Answers), para. 15.

[106] Jongeneel et al. (2020), p. 21; Pe'er et al. (2019), p. 451.

[107] Metta (2020), para. 8.

[108] European Commission (2020), The Green Deal Factsheet, p. 8.

[109] European Commission (2019), The environmental objectives of the future CAP, para. 12, 15.

[110] Jongeneel et al. (2020), p. 23.

[111] Matthews (2020), p. 5.

[112] Guyomard and Bureau (2020), pp. 14, 61–83.

the development of a comprehensive performance-based system[113] along with a set of coherent and well-coordinated environmental policies directly addressing soil-related issues. In this regard, sustainable soil management practices could be integrated in the various relevant existing instruments to foster better coordination and coherence across European agri-environmental policies. Presently, some of the sustainable soil management practices are already being promoted by a set of EU standards as part of the cross-compliance mechanism, namely the Good Agricultural and Environmental Conditions (GAEC), aiming at achieving sustainable agriculture by conditioning direct payment to compliance.[114] GAECs mainly address soil protection issues such as prevention measures to reduce soil erosion with minimum soil cover and land management reflecting soil specific conditions, reduce the use of heavy machinery to maintain soil structure and soil organic matter content through the implementation of adapted practices such as crop rotation.[115] This strategy needs to be pursued and further promoted within the CAP and the EGD framework strategies.

In this regard, the April 2021 Resolution of the European Parliament on soil acknowledged sustainable soil management was a "key component of farming and food policy in long term".[116] In the light of these challenging aspects, ambition therefore needs to be scaled up to ensure the EGD effective enforceability across the European Union and beyond. Indeed, EU agricultural imports' sustainability has not really been considered so far, while standards should have been established.[117] Indeed, promoting sustainable regional agriculture without taking adequate measures for international trade could result in an incoherent green transition by outsourcing environmental damages.[118]

In a nutshell, it is important to advocate at the EU level for the CAP to become a sustainable land use and food policy. By 2028, flat-rate area payments must be replaced by an ecologically effective and fair remuneration system for concrete environmental services provided by land users, keeping in mind that healthy soils mean healthy life. We need to use all options for a gradual phase-out of direct payments. In the short term, at least one billion euros should be made available annually for targeted (soil) biodiversity measures through a shift from the first to the second pillar of the CAP.

[113] European Court of Auditors (2018), Opinion No 7/2018.

[114] European Parliament and Council (2013), Regulation No 1306/2013 Annex II.

[115] Ecologic Institute (2003), p. 52.

[116] 2021/2548(RSP) Resolution on soil protection, para. 33.

[117] Fuchs et al. (2020), p. 672.

[118] Fuchs et al. (2020), p. 671

6 Other Policy Areas of the European Green Deal

The European Green Deal intends to put the EU on the track with a "just transition" towards a carbon-neutral and sustainable economy by 2050, ensuring that no one is left behind.[119] In addition to environmental and agricultural policies, other sectoral policies will need to be reconsidered, in particular energy supply across the economy, industry, production and consumption, infrastructure, transport, construction, taxation and social benefits.[120] The EU has already started to modernise and transform its economy and reduced greenhouse gas emissions between 1990 and 2018 by 23%, while the economy grew by 61%. However, reaching the 2050 climate neutrality goal will require further action from the economic and societal spheres as well as many more transformative changes. The "Fit for 55 Package" of the Commission will hopefully cover wide-ranging policy areas to ensure emission reductions of at least 55% by 2030. To overcome the old-established prejudice that environmental protection and economic growth are mutually exclusive, the sustainable use of soil should also be promoted in all areas of EU law. Given the multiple functions soil performs, it is no surprise that soil is "the glue that binds the different strategies of the Green Deal".[121]

An EU industrial strategy, presented by the Commission in March 2020,[122] is to address both the green and the digital transformation challenges and to support the transformation towards greenhouse gas neutrality. The European Commission has also presented another action plan for the circular economy, which focuses on a comprehensive product policy that is already intended to make the production stage more sustainable. According to the "circular economy" concept, the European Commission considers that we are taking more out of our planet than it can afford to give us. In order to tackle the effects of this overuse, we must urgently reduce the environmental and carbon footprint of the goods we consume—and this includes also of the soil we use. The current EU waste management law contributes to the reduction of pollutant entries from waste into the soils and the reuse and recycling of waste. Provisions of classification and treatment of contaminated land are still missing[123] and will hopefully be established in the new soil strategy.

In the context of soil protection, it is relevant that the European Commission is planning the use and renovation of buildings (called "renovation wave") for private and public buildings to promote energy efficiency and affordability. The Construction Products Regulation will be revised so that the design of new and renovated buildings meet the needs of the circular economy at all stages. While the construction of new buildings is mostly associated with soil sealing, the renovation wave,

---

[119] European Commission (2020), The Just Transition Mechanism.

[120] COM (2019) 640 final of 11.12.2019, p. 4.; see the Fig. 1 on page 6 of this entry.

[121] European Commission, Homepage: https://ec.europa.eu/jrc/en/eu-soil-observatory (accessed 15 August 2021).

[122] COM (2020) 102 final of 10.03.2020, p. 1 ff.

[123] Heuser (2006), JEEPL, p. 190 ff. (p. 196).

although requiring some amounts of energy and mineral resources, helps to reduce especially the physical threats to the soils.

In the field of transport, the Commission intends to adopt a "sustainable and smart mobility strategy".[124] It aims at promoting multimodality and, among other things, at shifting 75% of inland freight traffic to rail and inland waterways. Even if the transportation sector is only addressed for climate protection reasons, because transport accounts for a quarter of the EU's greenhouse gas emissions, a shift to sustainable and smart mobility may also contribute to the prevention of further soil sealing, if the focus is on public transportation.

Research and innovation have been at the heart of the EGD under Horizon 2020 as well. However, although they will probably be indirectly targeted, soils are not mentioned once in the thematic areas and topics to be targeted in the "Horizon 2020 European Green Deal Call fact sheet" published in September 2020, setting out the first steps of the road map for autumn 2021.[125] Hopefully, the "Horizon Europe" (2021–2027) will be able to complement what the Horizon 2020 missed out for the 2014–2020 period, with respect to soil protection in Europe. Horizon 2020 for soils is part of a long-term strategy to understand soil functions in order to propose solutions to optimise their biological, chemical and physical properties and the European Commission has been published since 2014 a set of documents under the Horizon 2020 challenge, stressing the importance of supporting research and innovation for soils. The new Horizon Europe will be in line with the EGD objectives as well[126] and with regard to soil protection, one of the Horizon Europe missions areas will be targeting "Soil health and food". Accordingly, the soil mission support—launched in November 2020—will help to "identify, compile and review knowledge gaps and priority areas for soil and land management-related Research & Innovation across land uses and soil functions".[127] The "Caring for soil is caring for life" Horizon 2020 Europe mission advocates towards the achievement of the "75% of healthy soils by 2030" objective.[128] The 2020 report of the mission Commission acknowledges soils health has been a neglected issue while "life on Earth depends on healthy soils".[129] The mission will also deal with the international dimension of soil protection by promoting ways to ensure European "soil degradation problems" will not be exported to other countries.[130] Hopefully, the research and innovation sector for soils will continue to gain visibility at the European level in the coming years, countering the main argument that justifies or associates the lack of a comprehensive framework for soil protection with a lack of knowledge. Since the Horizon 2020

[124] European Commission (2020), COM 789 final, p. 1 ff.

[125] European Commission (2021), European Green Deal Call raises high number of proposals under Horizon 2020, para. 3.

[126] European Commission (2019), Research and innovation for the European Green Deal, para. 5.

[127] Ecologic (2020), paras. 2,6.

[128] European Commission (2020), Caring for soil is caring for life, p. 1 ff.

[129] European Commission (2020), Caring for soil is caring for life, p. 3.

[130] European Commission (2020), Caring for soil is caring for life, p. 4.

initiative has been laid out, soil has been well integrated in the European research and innovation framework. However, the focus is still too much oriented toward the improvement of agricultural systems and soil ecosystems services. This limitation in the research scope is indicative of a specific understanding of soils in policy making, still limited to soil use which struggles to obtain recognition of its per se value.

In a broader, more socio-political sense, the European Green Deal intends to support the transition of the EU to a fair and prosperous society that responds to the challenges posed by climate change and environmental degradation, improving the quality of life of current and future generations.[131] The protection and sustainable use of soils is a key aspect for the future sustainability of our society.

7 Preliminary Conclusion on the Relevance of the EGD for Soil Protection Law

Soils play a major role in climate change mitigation and adaptation, agricultural production and food security, nature and biodiversity preservation, and are the foundation of our health and our wealth. In order to address and protect the multifunctionality of soils and especially its ecological functions, specific strategies need to be developed taking into account the different terminological dimensions the concept of soil protection encompasses.

European soils have been damaged with intensive monocrop farming and heavy reliance on pesticides for more than a century. So far, EU policies targeting the agricultural sector, embodied in the common agricultural policy (CAP), have been unable to not only exploit the soil as a basis for production, but also to sufficiently ensure its protection. Following the Council's general agreement on the post-2020 CAP reform package reached in October 2020, when the ministers agreed on strong commitments for higher environmental ambition. The scope of legislation predetermines its effectiveness and many legal texts are shaped by their definitions. It is thus essential to differentiate the various soil protection dimensions as a first step in building common definitions and thus common objectives, but also in view of an equal level of ambition for implementation across EU Member States. As soil resources continue to dramatically degrade in Europe, urgent political and legislative action is more than ever needed to address the eight soil threats identified by the European Commission in the 2006 EU soil strategy, to protect Europe's great variety of soil types and the services they provide. In this regard, the European Green Deal's set of measures has emerged as an opportunity to foster ambition for a greener Europe and for the inclusion of more soil protection measures across the targeted policy areas. Indeed, the EGD targets different key policy areas being all soil related to a certain extent. However, a closer look at the detailed single EGD legal proposals and references to soil protection reveals the EU Commission long-term strategy for

[131] COM (2019) 640 final of 11.12.2019, p. 1 ff. (p. 23).

environmental, and especially soil protection, lacks ambition. Amongst the targeted EGD key aspects across policy sectors, only the farm to fork, biodiversity and climate strategies as well as the zero-pollution action plan, have been identified as being the most relevant policy areas for soil protection.

Figure 2 summarises the extent to which these EGD strategies address the eight soil threats identified in the 2006 EU soil strategy and incorporate relevant solutions to improve soil health. First, soil threats are mainly indirectly addressed by the different EGD strategies in the F2F and biodiversity 2030 strategies. Then, the zero-pollution action plan is the only strategy directly targeting one soil threat, namely soil contamination. Finally, the climate strategy is the only sector directly promoting soil-related solutions, such as soil organic carbon sequestration, land use planning, soil and land degradation and restoration prevention and the implementation of sustainable soil management practices.

This analysis reveals the European Green Deal lacks consistency in the distribution and level of ambition for soil protection across key policy areas. Thus, translating the EGD guidelines into concrete actions will require shifting the focus to soil protection even more as the significance of soil within the different EGD strategies is still low.

Therefore, the emergence of the EGD with its high ambition has set Europe on a promising path towards greater soil protection. However, ensuring an adequate sustainable green transition for Europe would entail higher soil protection standards. Indeed, soil is still regarded as the "poor cousin" within the European environmental policy framework, compared to air and water resources, which have been protected by EU law since many years. In contrast, strategies for soil protection are not adequately developed at European level and there is currently no comprehensive EU legal instrument dealing with soil protection. As a reaction, the European Commission keeps stressing the need for a new soil strategy since soils are regarded as being essential for achieving the EGD objectives. However, at the legal level, it emphasises that the integration of soil protection concerns into all other policies still makes sense—an assumption that falls short. In this context, the existing European legal instruments fail in combating all relevant soil threats and mainly achieve soil protection or sustainable use of soil as side-effects of other goals. The reluctance of the Commission to put soil on the same level as air and water can be seen in the almost systematic use of soft law language in the various strategies and plans of the EGD but also of the CAP, whose green aspects have mainly been rhetoric so far. Presently, substantial progress is still hampered by considerable gaps toward regulation and implementation at the national level, since legally binding measures are still lacking.

The future CAP would in theory have the capacity to support and reinforce the sustainability and climate dimensions of agricultural policies, especially for soil protection via Pillar 1 and the eco schemes. However, the framework still needs considerable adjustments to be more coherent with the EDG objectives and many aspects of the CAP are not yet sustainable from an over-the long-term and against-planetary-boundaries perspective. So far, EU policies targeting the agricultural sector have been unable to provide a sufficient framework to ensure the

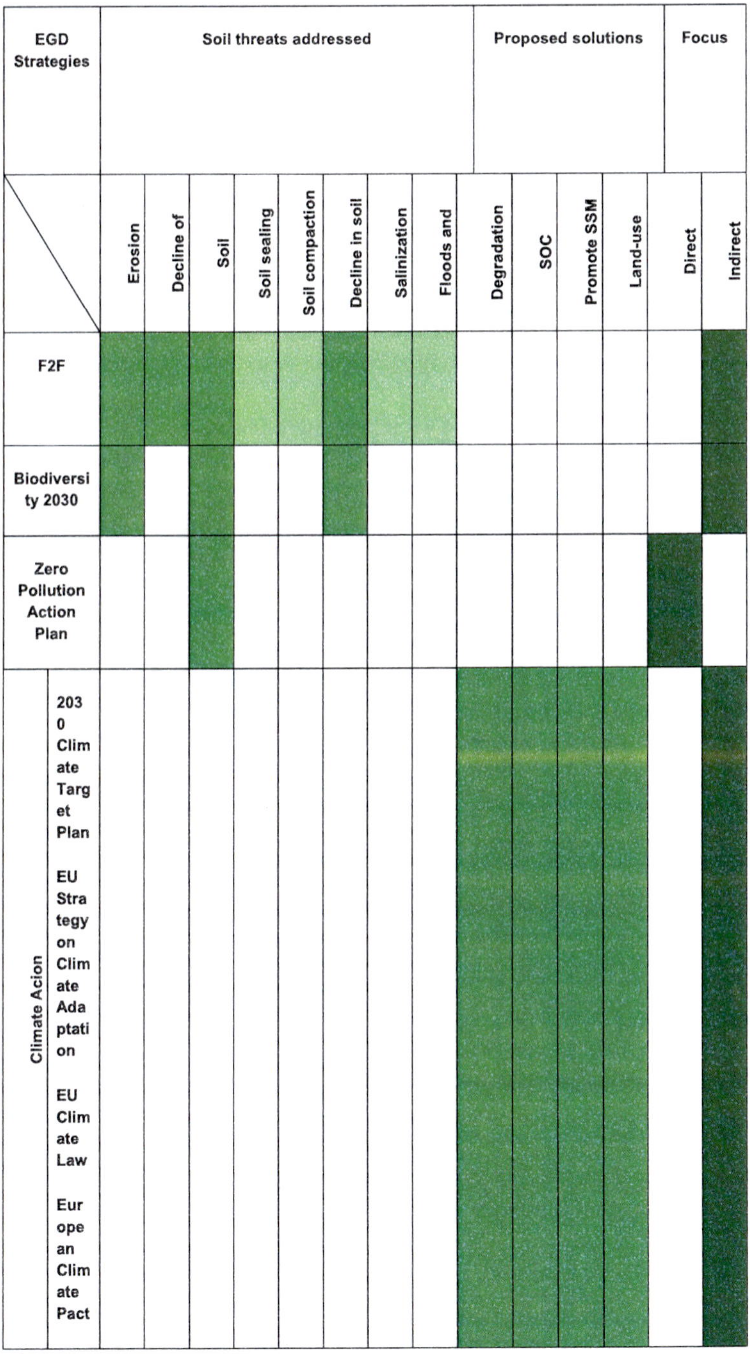

Fig. 2 Integration of soil protection aspects in the EGD related strategies (Maréchal and Hart 2020; Cleary Gottlieb Steen & Hamilton 2020; COM (2020) 562 final; COM (2021) 82 final; European Commission 2021)

homogeneous adoption and implementation of soil protection measures and sustainable soil management practices. Forthcoming negotiations must therefore focus on ensuring the reformed CAP's objectives are to be aligned with the EGD's ones, especially with the farm to fork and biodiversity strategies since their objectives fall within the soil protection scope and are not yet reflected in the currently ongoing trilogue negotiations on the future 2023–2027 CAP. The CAP eco-schemes could also further leverage ambition and coordination regarding SSM practices between the new green structure of the future CAP and the EDG. Furthermore, the EU cannot continue promoting sustainable regional agriculture without taking adequate measures for international trade. If the CAP negotiations do not strengthen the coherence and coordination with EGD instruments, it is very likely that the CAP will not be able to sufficiently contribute to the implementation of the EGD strategies and thus to the achievements of its objectives.

The EU research and innovation sector has highlighted the importance of soils with the Horizon 2020 programme (2014–2020), which has led to the development of numerous projects promoting better soil management and protection based on the latest available knowledge, even though soil is not one of the directly targeted thematic areas. Hopefully, the Horizon Europe (2021–2027) will be able to complement what the Horizon 2020 missed for the 2014–2020 period with respect to soil protection in Europe and beyond.

Overall, the EGD is providing, in theory, great inputs with regard to the general green transition, however, the concept of soil protection has not been given the attention it deserves, especially with respect to the eight soil threats. The EGD strategies along with the CAP instruments have not yet succeeded in providing the necessary tools to shift EU legislation away from the utilitarian vision tied to soil.

8 Further Development of EU Soil Protection Law

8.1 Proposal for a Soil Framework Directive of 2006 and Reasons for Its Failure: Lessons Learned?

Although the Single European Act of 1987 already established the theoretical possibility of enacting regulations about soil protection, this field has remained almost neglected until today. At the moment, there is no comprehensive legal instrument dealing with soil protection in the European Union. As shown in the previous sections, the European Green Deal seems to continue to ignore the immense social, economic and ecological importance of this environmental medium and to leave out the legal protection of soil as well.

What prevents the EU Commission from submitting a new proposal for a directive on soil protection? Perhaps it is the negative experiences from the first attempt for soil protection legislation in the EU during the years 2001 to 2007: from the first official Communication "Towards a Thematic Strategy for Soil Protection" in 2002 and a second Communication "Thematic Strategy for Soil Protection"[132] with a "Proposal for a Directive establishing a framework for the protection of soil and amending Directive 2004/35/EC"[133] being put on the table[134] in 2006 till this proposal failed in 2007 for the first time. The decision on a Soil Protection Framework Directive (SFD) required majority vote and was rejected in the Environment Council on the 20 December 2007 due to a blockage by Germany, France, the United Kingdom, the Netherlands, Austria and Italy, although the European Parliament (and other institutions) had issued a positive resolution with agreement on the need for a soil framework directive shortly before.[135] The proposal was finally withdrawn by the Juncker Commission in 2014; the soil thematic strategy didn't require any further decision as a communication from the Commission.

The EU environmental action programmes (EAP) have repeatedly demanded the adoption of a directive on soil protection to ensure an adequate level of protection in the European Union. The draft Soil Framework Directive would require Member States to follow a systematic approach to identifying and combating soil degradation and essentially consisted of designating risk areas to address the major soil threats. Nevertheless, it contained far-reaching regulations only for a few soil endangerment paths and excluded many important issues, such as contaminated sites. From an environmental point of view, the proposed regulations were not sufficient, although at least a start. In contrast, the industry and the agroindustry lobby put forward considerable arguments against the draft directive and influenced the views of national governments.

What were the main reasons for the rejection? The Member States having caused the blockage in the Environment Council didn't all have specific national soil protection acts (Germany and the Netherlands did, Italy nominally also, Austria in most of its states, France did not). Most national laws had or still have a limited scope: for example, the German Federal Soil Protection Act ("Bundes-Bodenschutzgesetz") contains a reference to agricultural soil use in its § 17,[136] but just refers to "good farming practice",[137] a term which is still not sufficiently defined. When evaluating the process in the Environment Council, it must be taken into account that only a soil framework directive should have been adopted, so that it could have been transposed into the respective national legal system later on. The more honest reason for this failure is that the lobby influences (agricultural and industrial lobbies were in agreement) were very strong at that time, while soil protection awareness was not yet as developed.

---

[132] COM (2006) 231 final of 22.09.2006, p. 1 ff.

[133] COM (2006) 232 final of 22.09.2006, p. 1 ff.; Directive No. 2004/35/EC is the Environmental Liability Directive.

[134] COM (2002) 9179 final of 16.04.2002, p. 11 ff.

[135] European Parliament (2007), no. 10 and 11.

[136] § 17 para 1 sentence 1: "In cases of agricultural soil use, the obligation to take precautions pursuant to Article 7 shall be fulfilled by good agricultural practice."

[137] Federal Soil Protection Act of Germany, published in Federal Law Gazette I p. 502.

Another reason is that soil has been privately owned for centuries and, unlike other environmental goods, it was assumed that the owner would not only use, but also protect the soil. For a long time, soil was regarded as private property only and seemed to be removed from any regulatory competence. The need for soil conservation and its legal framework did not develop until the early years of the last century in Europe, when soil degradation increased so significantly that agriculture became difficult in many parts of the world. Iceland was the first country in the world with soil specific legislation, the "Act on Forestry and Protection against Soil Erosion" of 1907.[138] In the U.S., the Dust Bowl removed entire landscape areas from use and also established a Soil Conservation Service (today Natural Resources Conservation Service) in 1935. Since the increase of environmental awareness and the creation of an environmental competence on the basis of EU primary law, including the protection of soils, there has been a growing insight in the need for soil protection in the European Union.

In the proposed 2006 draft framework directive or on an amended version, the Member States were unable to agree on an EU-wide setting of minimum standards for soil protection and rejected it based on costs, subsidiarity, technical and administrative aspects. Officially, the failure to adopt a soil framework directive was mainly due to concerns about a violation of the principle of subsidiarity, with some Member States insisting that soil was a matter that should essentially be regulated at the national level; a need for regulation at EU level was partly not seen at that time. With regard to EU law, this argument is not convincing, essentially pretextual, and legally incorrect, as the Commission's Legal Service already expressed in its opinion of March 2007.[139] But, nevertheless, it is now being repeated (e.g. by the Federation of German Businesses), as was recently shown during the European Commission's EGD consultation on soil protection. In general, many soil threats (like wind or water erosion, floods and landslides or soil biodiversity loss) can have transboundary effects. Furthermore the competitiveness is affected because of the high costs of soil remediation and further prevention of soil degradation.[140] It would therefore be good if the Member States agreed on common legal instruments and standards, such as "sustainable soil management", so that the European Green Deal does not remain just lip service. However, the former experiences could be seen as a dress rehearsal for a new start in soil protection legislation.

[138] Crofts (2011), p. 11.

[139] Rat der Europäischen Union, Gutachten des Juristischen Dienstes, Vorschlag für eine Richtlinie des Europäischen Parlaments und des Rates zur Schaffung eines Ordnungsrahmens für den Bodenschutz und zur Änderung der Richtlinie 2004/35/EG, Document No. 7381 of 14.03.2007, p. 6 ff.

[140] Homepage EU COM, DG ENV: https://ec.europa.eu/environment/soil/index_en.htm.

8.2 Towards a New Soil Protection (Framework) Directive

In the almost 15 years that have passed since the proposal of a Soil Framework Directive, a great number of new evidences on soil degradation and its negative consequences has proven the necessity of a common European soil protection directive. The Seventh EU-Environment Action Plan,[141] which entered into force in 2014 (the year of the withdrawal of the Draft Soil Framework Directive), recognised the severe challenges of soil degradation. It provided that by 2020, soil should be managed sustainably and be protected adequately in the EU, and the remediation of contaminated sites should be well underway. Furthermore, it commits the EU and its Member States to increasing efforts to reduce soil erosion, to increase soil organic matter and to remediate contaminated sites. The Environmental Action Programmes (EAP) of the EU generally reflect the current long-term vision and therefore serve as a barometer for the development of environmental—and here especially soil—awareness in the EU.

Over the last years, the European Union has been increasingly emphasising the strategic role soil plays in many different sectors due to its cross-cutting nature. As such, various soil-related issues are being indirectly addressed across policy areas, but an EU soil specific framework is still lacking.[142] Nevertheless, such a policy framework recognising the environmental importance of soil and taking into account the problems arising from the competition among its concurrent uses (ecological and socio-economical), and aiming at maintaining its multiple functions, is urgently needed. The European Environment Agency outlined in its Environment State and Outlook 2020 that the lack of a coherent and comprehensive land and soil legislative framework was hampering the effective implementation of existing initiatives and of the EU to achieve its green transition.[143] As such, a new policy framework dedicated to soil and a revision of the 2006 thematic soil strategy is very much needed.[144] The EU Commission held a consultation on a new EU strategy for soil protection in February 2021;[145] the results have not yet been published.[146]

The European Parliament adopted a quite exhaustive resolution on soil protection on 28 April 2021.[147] It includes the call on the Commission "to design an EU-wide

[141] European Parliament and Council (2013), Decision No. 1386/2013/EU, pp. 171–200 (p. 180, No. 23-25).

[142] Paleari (2017), p. 164.

[143] European Environment Agency (2010), The European Environment State and Outlook, p. 3.

[144] EEA (2020), SOER, p. 131.

[145] European Commission (2021), New EU soil strategy, p. 12.

[146] The consultation inputs were initially published transparently on the COM's homepage, but after the end of the consultation they can no longer be found there (unlike other consultations, see e.g. the summary of the public consultation of the LULUCF Regulation: https://ec.europa.eu/info/law/better-regulation/have-your-say/initiatives/12657-Land-use-land-use-change-and-forestry-review-of-EU-rules/public-consultation_de).

[147] European Parliament (2021), Motion for a resolution on soil protection, p. 1 ff.

common legal framework for the protection and sustainable use of soil, addressing all major soil threats".[148] Hopes for a new directive are therefore high as failure to come to an agreement at the regional level could have dramatic consequences for the EGD (and the EU's) future.

The update of the 2006 EU soil thematic strategy to address soil and land degradation processes in a comprehensive way has been announced and this could be an opportunity to integrate complementary policy instruments for soil protection within the existing European policies and structures to avoid the development of a stand-alone instrument for soil protection. Indeed, a mix of overlapping policies is often considered more effective than a single instrument in fostering synergies across relevant areas. This statement is generally true but must be put into perspective for soil protection. One thing seems to be clear: The mere continuation of the focus on integrating soil protection and sustainable use concerns into other policy areas has not proven successful, as the increase in soil degradation in the EU has shown, and this does not sufficiently address the protection needs.

Although a coherent set of rules for soil protection has still not been implemented, EU policies are inscribed within a soil-related international commitments framework and some EU institutions are calling for further actions, at least at regional level. Since 2018, the EU has developed a strong monitoring system for SDGs based on 100 indicators.[149] In line with the Sustainable Development Goal (SDG) 15.3 to "combat desertification, restore degraded land and soil, including land affected by desertification, drought and floods, and strive to achieve a land degradation-neutral world by 2030",[150] the European Court of Auditors instructed the Commission to focus on land degradation and desertification in accordance with the need to step up efforts for soil and biodiversity protection within the EU soil legal framework.[151] Even EU Environment Commissioner Sinkevičius emphasises in the launch of the respective consultation that "the aim of the new EU Soil Strategy will be to address soil- and land-related issues in a comprehensive way and to help achieve land degradation neutrality by 2030, one of the key targets of the Sustainable Development Goals".[152] This could potentially strengthen the EGD ambition with a new SFD toward this specific goal as the land degradation neutrality (LDN) objective does not currently receive the attention it deserves.[153]

An updated SFD could build on the 2006 draft directive for the soil threats classification and scale up ambition for biodiversity protection, soil degradation processes, land artificialisation toward the achievement of the no-net-land-take goal by 2050, soil contamination, desertification and land degradation neutrality.

[148] European Parliament (2021), Motion for a resolution on soil protection, p. 1 ff. (p. 14, no. 10).

[149] European Commission (2018), Sustainable development in the European Union; Keesstra et al. (2016), pp. 111–128.

[150] UNEP (2015).

[151] European Court of Auditors (2018), Combating Desertification in the EU, Annex III pp. 2–10.

[152] European Commission (2021), New EU soil strategy, para. 4.

[153] Eisemann (2021), p. 40.

Indeed, a new directive taking past failures into account could step-up efforts for the widespread adoption of soil protection measures, including SSM practices, especially with regard to soil organic carbon sequestration and storage, mandate the improvement of soil information systems, encourage the development of research and innovation. A new SFD could also incorporate the EGD soil-related objectives to foster synergies and coordination by complementing and steering actions across the different EU policy areas. Accordingly, the new soil strategy could as well include the nature restoration objectives the Commission should submit in 2021 and that will supposedly be legally binding.

Hopes for a new and ambitious soil directive are high as it will be supported by a multitude of stakeholders—from a MS soil protection working group to farm advisors, businesses, agro-food industries, research and academic, non-governmental organisations and general public—, but also based on data compiled at the European and international levels[154] to ensure constant soil monitoring, reporting and verification. Looking even further ahead, it should be noted that the adoption of an EU (framework) directive on soil protection is legally possible and would be in line with the subsidiarity principle, depending on the specific focus of the legal instrument.

Nevertheless, some currently critical aspects will need to be considered as part of a new SFD to live up expectations, but as part of an overall European soil protection framework. Presently, such a framework consists of a policies patchwork scattered across sectors as no EU directive does require Member States to implement commonly agreed targets yet. As a result, EU and national regulations are fragmented and do not offer an overarching guiding and legally binding framework with enforceable policies for soil protection. On the EU level itself, the new European Parliament resolution lists all of the relevant policy modules that relate to soil protection;[155] this extensive list shows that the previous policy of integrating soil protection into the individual policy areas and thematic fields provided a patchwork quilt without seams.

From the beginning, this policy patchwork lacked a common definition[156] for soil protection in Europe. This is all the more problematic since the EU is integrating more soil related objectives within the agricultural and environmental policies which are still at risk to be ineffectively structured and inconsistently implemented.[157] Another consequence lies in the lack of comprehensiveness in the way soil threats are addressed across EU that may lead to the implementation of inadequate responses at the national level.[158] Naturally, soil types and their levels of protection vary across Europe. However, these issues need more attention from European bodies as a status quo could further widen the gap between the level of ambition

---

[154] European Commission (2020). Ares (2020)6391319, p. 1.

[155] European Parliament (2021), Motion for a resolution on soil protection, p. 1 ff. (pp. 2–6).

[156] Castelo-Grande (2018), pp. 74, 88.

[157] Paleari (2017), p. 170.

[158] Boschetti (2015), pp. 1–44.

proposed in the EU legislation and the fragmented implementation across countries, hampering overall effectiveness and making monitoring of progress difficult.[159]

Therefore, setting common thresholds and ensuring harmonisation of practices across countries, the monitoring of targets and indicators, the development of sharing knowledge platforms and defining soil-use, need to be set as priorities in the EU soil framework agenda. Indeed, these mechanisms are believed to benefit soil restoration, protection and conservation targets on the long term.[160] So far, a list of possible legal instruments for the protection and sustainable use of soils has not been substantiated in detail under EU law, although in the international arena such considerations are taking place for the national level due to the lack of an international legal instrument on soil protection as well.[161]

Ultimately, the European Commission could further include soil protection issues in the external policy and scale up ambition and start addressing transboundary issues such as conditioning the access of other countries to the EU market on meeting the EU standards for all stages of food products, from product quality and safety to production practices.

In the climate change context, EU represents less than 10% of global emissions, meaning European action alone will not be enough to slow down global warming. To keep the increase in global temperature as close to 1.5 °C as possible, the EU must also support decarbonisation efforts beyond our borders. Within the Emissions Trading Scheme of the EGD framework, for example, a global perspective and an outer dimension have been incorporated and the proposal for a border carbon tax could be extended to the agri-environment-climate sectors.[162] Similarly, the farmers could be incentivised to sequester carbon via regional or nationally adapted carbon markets, but the EU Commission would need to address the inherent technical issues associated to it such as measurement methods, monitoring, additionality, reversibility, transaction costs, accountability and transparency.[163] Ensuring coherence between the upcoming EU forest strategy and a legal framework for the protection and sustainable use of soils might also contribute to storing carbon in forest soils and trees.

However, the discussed ambition regarding soil protection within the existing and future instruments/strategies should be meaningful, as some objectives may not be realistic, such as the current measurement methods proposed by the Commission for biodiversity, climate, pesticide harm and on farm sustainable practices.[164] In order to

[159] Ekardt (2020), pp. 255–280; Vrebos (2017), p. 14.

[160] Glæsner (2014), pp. 9539–9540.

[161] Hannam and Boer (2004), p. 5 ff.

[162] Matthews (2020), p. 4.

[163] Matthews (2020), p. 2.

[164] In this regard, the EGD proposal to transform the Farm Accountancy Data Network into a Farm Sustainability Data Network for 2022, could be a milestone toward the development of an effective MRV systems.

be serious about climate protection, we must not leave out the strongest ally in this context: the soil, which acts as a CO_2 sink. Soil must be put on an equal footing with the other environmental media at EU level, also in legal terms.

The EGD also has another dimension as an integral part of the strategy to implement the United Nation's 2030 Agenda and the sustainable development goals (SDGs) and by advocating for the objectives of the Green Deal at the international level with regard to trade policy, development policy, security policy and other foreign policy strategies.[165] Soil degradation can result in hunger, famine, migration and sometimes even in war. In the context of the implementation of SDG 2 to "end hunger, achieve food security and improved nutrition and sustainable agriculture", a profound change in the global food and agriculture system is needed if we are to nourish the more than 690 million people who are hungry today and the additional 2 billion who will be by 2050.[166] Human population is projected to increase to nearly 9.8 billion people by 2050 and 11.2 billion by 2100 (United Nations 2018).[167] In order to feed this rapidly increasing population, coherent and immediate action against land and soil degradation is essential. Without fertile soils, food security, poverty alleviation and climate change mitigation and adaptation will not be achieved. In the context of the European Union's diplomatic cooperation with the African Union, for example, the issue of preventing soil and land degradation is even more important, due to the effects of drought, desertification and land degradation for the food production and the fast growing population, especially in Sub-Saharan Africa,[168] and also with regard to the role played by the UNCCD. As the EU is the world's largest single market, it can set standards that are in line with environmental ambitions and that apply across global value chains.[169]

All these dimensions show that legal instruments to protect and sustainably use the soil are desperately needed at EU level and it is to be hoped that the call of the European Parliament on the Commission to "design an EU-wide common legal framework, with full respect to the subsidiarity principle, for the protection and sustainable use of soil, addressing all major soil threats", will not fade away.[170]

9 Outlook on the Green Deal in Transition Towards a Soil Framework

Soil protection has commonly been considered in a scattered way across EU and international legislation, as a tool to achieve secondary goals in related environmental policies. Currently, EU and international Environmental Law are still

[165] COM (2019) 640 final of 11.12.2019, p. 1 ff. (p. 20).

[166] United Nations (2015), SDG 2, see https://www.un.org/sutainabledevelopment/hunger/.

[167] IPCC (2019), p. 88.

[168] Tetteh (2019), p. 3.

[169] COM (2019) 640 final of 11.12.2019, p. 1 ff. (p. 22).

[170] European Parliament (2021), Motion for a resolution on soil protection, p. 1 ff. (p. 14, no. 10).

apprehending soil and related environmental concerns in a sectoral and compartmentalized manner rather than a set of interconnected problems.[171] As a result, both international and European environmental governance are still lacking a soil specific instrument along with soil specific policies incorporated within the existing and relevant legislation, while soils play a central role in achieving the many set environmental goals. For the European green transition, this means soils have to represent a key element of the existing environmental legislation framework and especially of the new European Green Deal. The European Green Deal first policy document emphasized the EU's intention to integrate the United Nation's Sustainable Development Goals, to put sustainability and the well-being of citizens at the heart of economic policy and EU's policymaking and action.[172]

The EGD has proposed a comprehensive strategy to tackle environmental and climate related challenges with four central and soil related strategies: the zero-pollution action plan, the farm to fork and the biodiversity 2030 strategies and the Climate Law framework. However, in addition to these strategies' individual approaches, there is an urgent need for EU legislation to scale up ambition towards the protection and sustainable use of soil because this void is reducing the existing initiatives effectiveness and may limit Europe's ability to achieve its green objectives. Additional action must therefore be taken to avoid missing out the long-awaited green transition. As compared with air and water for which legal and policy frameworks are already in place, strategies for soil protection and sustainable use of soil are neither adequately developed nor legally "embedded" at EU level. A comparable policy framework recognising the environmental importance of soil and taking into account the problems arising from the competition among its concurrent uses (ecological and socio-economical), and aiming at maintaining its multiple functions, is urgently needed—a view shared by the European Environment Agency.[173]

Many challenges associated with the EGD's consistent implementation across Europe remain and need to be addressed. At the moment, the fields of biological, chemical and physical soil protection are still regarded as the "poor cousins" compared to other environmental media legislation in the EU. A major challenge in this regard would be to guarantee the coherent use of all available planning tools within the European Green Deal framework while verifying the adequacy and effective implementation of the Member States proposed strategic national plans with measurable outcomes.[174] Indeed, the heterogeneous structural realities faced by Member States make the EGD effective implementation a challenge. Several questions remain to statute on the future EGD performance, especially whether it will remain a priority in the EU agenda and will receive adapted financial endowment.[175]

---

[171] See explanations in: Stevens and Kanie (2016), p. 1 ff.

[172] COM (2019) 640 final of 11.12.2019, p. 3.

[173] EEA (2020), SOER, p. 1 ff.

[174] Prandecki and Wrzaszcz (2020), p. 173.

[175] Siddi (2020). The European Green Deal: Assessing its current state and future implementation.

Therefore, some steps need to be undertaken in order to make a coherent environmental framework—incorporating the objectives of a revised EU soil thematic strategy—out of the Green Deal. A new and comprehensive soil directive might fill the existing vacuum.

The European Green Deal bears an inherent transboundary dimension towards the successful completion of the green transition. Its ambition cannot be achieved if Europe acts alone as climate change and biodiversity loss drivers as well as soil degradation are global challenges that do not stop at national borders. Therefore, the EU can use its influence, expertise and financial resources to encourage its neighbours and partners to join it on its sustainable path.[176]

Given their multiple functions and the role that soils play as a critical resource for achieving many of the 17 UN Sustainable Development Goals, including food security, clean water and sanitation, and health and well-being, our solutions must take into account the future sustainability of our soil resources and translate this into concrete policy and legislation.

Ultimately, the EU Commission needs to ensure the EGD and soil protection will be maintained at the top of on the European political agenda—despite the Covid-19 pandemic and its consequences which should not be used as an excuse for a lack of political action. The following months will be decisive to avoid economic boost to be prioritised over some of the EGD components as well as for scaling up ambition for the 2030 and 2050 roadmaps. Hopefully, soil degradation will receive the same political attention and legal consideration as climate change and biodiversity.

References

Boschetti B (2015) The demand for land: rethinking land use management and regulation. In: Www.Icon-Society.Org, pp 1–46. https://publires.unicatt.it/en/publications/the-demand-for-land-rethinking-land-use-management-and-regulation-16. Accessed 7 Aug 2021

Bundesverband Boden (German Soil Association), Roadmap "New Soil Strategy – healthy soil for a healthy life", Statement of 8 December 2020, p 1

Castelo-Grande T, Augusto P, Fiúza A, Barbosa D (2018) Strengths and weaknesses of European soil legislations: the case study of Portugal. Environ Sci Policy 79:66–93. https://doi.org/10.1016/j.envsci.2017.10.010

Cleary Gottlieb (2020) A sustainable recovery for Europe: the EUs Green Deal. https://www.clearygottlieb.com/news-and-insights/publication-listing/a-sustainable-recovery-for-europe-the-eus-green-deal. Accessed 8 Aug 2021

Commission of the European Communities (2006) Communication from the Commission to the Council, the European Parliament, the European Economic and Social Committee and the Committee of the Regions, Thematic Strategy for Soil Protection, COM (2006) 231 final of 22.09.2006, p 1 ff

Commission of the European Communities (2002) Communication from the Commission to the Council, the European Parliament, the Economic and Social Committee and the Committee of the Regions, Towards a Thematic Strategy for Soil Protection, COM (2002) 179 final

[176] Maréchal and Hart (2020), p. 7.

Commission of the European Communities (2006) Proposal for a Directive of the European Parliament and of the Council establishing a framework for the protection of soil and amending Directive 2004/35/EC, COM (2006) 232 final
European Commission, Cordis (2020) EU research results. Towards a European Research and Innovation Roadmap on Soils and Land Management. https://cordis.europa.eu/programme/id/H2020_FNR-04-2020. Accessed 7 Aug 2021
Council of the European Union (2020) Note from the General Secretariat of the Council, Conclusions on Biodiversity, Doc. No. 11829/20 of 16 October 2020
Crofts R (2011) Healing the land. Soil Conservation Service of Iceland, Gunnarsholt, Iceland
Deutscher Naturschutzring, Feedback on the Initiative "New Soil Strategy – healthy soil for a healthy life", Statement of 3 December 2020, p 1
El Chami D (2020) Towards sustainable organic farming systems. Sustainability 12(23):9832. https://doi.org/10.3390/su12239832. Accessed 6 Aug 2021
European Commission (2020) Directorate-General for Research and Innovation, elaborated by: Grand, A et al. (2020) Caring for Soil Is Caring for Life: Ensure 75% of Soils Are Healthy by 2030 for Food, People, Nature and Climate, Report of the Mission Board for Soil Health and Food. https://op.europa.eu/en/publication-detail/-/publication/4ebd2586-fc85-11ea-b44f-01aa75ed71a1/language-en. Accessed 7 Aug 2021
Ecologic Institute (2003) EU Soil Protection Policy: Current Status and the Way Forward. https://www.ecologic.eu/sites/default/files/publication/2015/1965_soil_protection_background_paper.pdf. Accessed 7 Aug 2021
Ecologic (2020) Soil Mission Support: Towards a European Research and Innovation Roadmap on Soils and Land Management, November 1, 2020. https://www.ecologic.eu/17696. Accessed 7 Aug 2021
Eisemann L (2021) Widening the European Green Deal's Perspective towards a Sustainable Europe, Umweltbundesamt. https://www.umweltbundesamt.de/en/publikationen/widening-the-european-green-deals-perspective. Accessed 6 Aug 2021
Ekardt F (2020) Sustainability: transformation, governance, ethics, law. Environmental humanities: transformation, governance, ethics, law. Springer International Publishing, Cham. https://doi.org/10.1007/978-3-030-19277-8
European Commission (2008) DG environment, CLIMSOIL, Technical Report 2008, Review of existing information on the interrelations between soil and climate change. https://ec.europa.eu/environment/archives/soil/pdf/climsoil_report_dec_2008.pdf. Accessed 26 Aug 2021
European Commission (2009) Joint Research Centre, Final Report on the Project 'Sustainable Agriculture and Soil Conservation' (SoCo), 2009, publication. Available at: https://data.europa.eu/doi/10.2791/10052. Accessed 7 Aug 2021
European Commission (2009) Press release of 5 March 2009: Climate Change: Commission dishes the dirt on the importance of soil
European Commission (2016) Joint Research Centre (JRC) Technical Reports: Soil Threats in Europe, Luxemburg
European Commission (2019) Communication from the Commission to the European Parliament, the Council, the European Economic and Social Committee and the Committee of the Regions, The European Green Deal, COM(2019) 640 final
European Commission (2020) Communication from the Commission to the European Parliament, the European Council, the European Economic and Social Committee and the Committee of the Regions, A New Industrial Strategy for Europe, COM (2020) 102 final
European Commission (2020) Communication from the Commission to the European Parliament, the Council, the European Economic and Social Committee and the Committee of the Regions, A Farm to Fork Strategy for a fair, healthy and environmentally-friendly food system, COM (2020) 381 final
European Commission (2020) Communication from the Commission to the European Parliament, the Council, the European Economic and Social Committee and the Committee of the Regions, Chemicals strategy for sustainability 'Towards a toxic-free environment', COM (2020) 667 final

European Commission (2020) Communication from the Commission to the European Parliament, the Council, the European Economic and Social Committee and the Committee of the Regions, EU Biodiversity Strategy for 2030, Bringing nature back into our lives, COM (2020) 380 final

European Commission (2020) Communication from the Commission to the European Parliament, the Council, the European Economic and Social Committee and the Committee of the Regions, Stepping up Europe's 2030 Climate Ambition. Investing in a Climate-Neutral Future for the Benefit of Our People, COM (2020) 562 final

European Commission (2020) Communication from the Commission to the European Parliament, the Council, the European Economic and Social Committee and the Committee of the Regions, Sustainable and Smart Mobility Strategy – putting European transport on track for the future, COM (2020) 789 final

European Commission (2021) Communication from the Commission to the European Parliament, the Council, the European Economic and Social Committee and the Committee of the Regions, Forging a climate-resilient Europe – the new EU Strategy on Adaptation to Climate Change, COM (2021) 82 final

European Commission (2021) Communication from the Commission to the European Parliament, the Council, the European Economic and Social Committee and the Committee of the Regions, "Pathway to a healthy planet for all" with the sub-title "EU action Plan: 'Towards zero pollution for air, water and soil'", COM (2021) 400 final

European Commission (2010) Joint Research Centre (JRC)/European Environment Agency, JRC Reference Reports. The State of Soil in Europe, A contribution of the JRC to the European Environment Agency's Environment State and Outlook Report — SOER. https://publications.jrc.ec.europa.eu/repository/handle/JRC68418. Accessed 15 Aug 2021

European Commission (2018) Sustainable Development in the European Union — Monitoring Report on Progress towards the SDGs in an EU Context. https://ec.europa.eu/eurostat/web/products-statistical-books/-/KS-02-20-202. Accessed 9 Aug 2021

European Commission (2020) Contribution to the Green Deal and the Just Transition Scheme | InvestEU. https://europa.eu/investeu/contribution-green-deal-and-just-transition-scheme_en. Accessed 18 Aug 2021

European Commission (2019) Speech of 11 December 2019, Press remarks by President von der Leyen on the occasion of the adoption of the European Green Deal Communication. https://ec.europa.eu/commission/presscorner/detail/EN/SPEECH_19_6749. Accessed 18 Aug 2021

European Commission (2019) AgriResearch Factsheet Soils. https://ec.europa.eu/info/sites/default/files/food-farming-fisheries/farming/documents/factsheet-agri-soils_en.pdf. Accessed 7 Aug 2021

European Commission (2020) Directorate-General Environment, Chemicals Strategy. https://ec.europa.eu/environment/strategy/chemicals-strategy_en. Accessed 7 Aug 2021

European Commission (2021) EU Climate Action and European Green Deal. https://ec.europa.eu/clima/policies/eu-climate-action_en. Accessed 20 Aug 2021

European Commission (2021) European Green Deal Call Raises High Number of Proposals under Horizon 2020.Text. European Commission - European Commission. https://ec.europa.eu/info/news/european-green-deal-call-raises-high-number-proposals-under-horizon2020-2021-feb-01_en. Accessed 7 Aug 2021

European Commission (2018) Soil erosion costs European farmers €1.25 billion a year. EU Science Hub. https://ec.europa.eu/jrc/en/news/soil-erosion-costs-european-farmers-125-billion-year. Accessed 8 Aug 2021

European Commission (2010) Les Sols: Une Ressource Essentielle Pour l'Union Européenne. Factsheet, September 2010. https://ec.europa.eu/environment/archives/soil/pdf/factsheet_2010_fr.pdf (document only available in French and German). Accessed 7 Aug 2021

European Commission (2021) Press release of 21 April 2021: Commission welcomes provisional agreement on the European Climate Change Law. https://ec.europa.eu/commission/presscorner/detail/en/ip_21_1828. Accessed 18 Aug 2021

European Commission (2019) Research and Innovation for the European Green Deal. Text. European Commission - European Commission, 2019. https://ec.europa.eu/info/research-and-innovation/strategy/strategy-2020-2024/environment-and-climate/european-green-deal_en. Accessed 7 Aug 2021

European Commission (2021) Research and Innovation to Drive the Green Deal. https://op.europa.eu/en/web/eu-law-and-publications/publication-detail/-/publication/93de16a0-821d-11eb-9ac9-01aa75ed71a1. Accessed 7 Aug 2021

European Commission (2020) Role of the Organic Sector in the Green Deal. Questions and Answers. European Commission, April 9, 2020. https://ec.europa.eu/commission/presscorner/detail/en/QANDA_20_1539. Accessed 6 Aug 2021

European Commission (2019) The Environmental Objectives of the Future CAP, Website, January 25, 2019. https://ec.europa.eu/info/news/environmental-care-and-climate-change-objectives-future-cap-2019-jan-25_en. Accessed 7 Aug 2021

European Commission (2020) Ares (2020)6391319. Healthy Soils – New EU Soil Strategy. https://ec.europa.eu/info/law/better-regulation/. Accessed 7 Aug 2021

European Commission (2006) Soil protection: the story behind the strategy. https://ec.europa.eu/environment/archives/soil/pdf/soillight.pdf. Accessed 2 Aug 2021

European Commission (2013) What Is Horizon 2020? Text. Horizon 2020 - European Commission, October 23, 2013. https://ec.europa.eu/programmes/horizon2020/en/what-horizon-2020. Accessed 7 Aug 2021

European Commission (2020) Working with Parliament and Council to Make the CAP Reform Fit for the European Green Deal. Factsheet, November 2020. https://ec.europa.eu/info/sites/default/files/food-farming-fisheries/key_policies/documents/factsheet-cap-reform-to-fit-european-green-deal_en.pdf. Accessed 6 Aug 2021

European Commission (2021) Directorate-General for Environment, Commission consults on new EU Soil Strategy, 2 February 2021. https://ec.europa.eu/environment/news/commission-consults-new-eu-soil-strategy-2021-02-02_de. Accessed 27 Aug 2021

European Commission (2020) Horizon 2020: European Green Deal Call. https://data.europa.eu/doi/10.2777/200559. Accessed 7 Aug 2021

European Commission (2021) Commissioner Sinkevičius speech at The Global Symposium on Soil Biodiversity April 19, 2021. https://ec.europa.eu/commission/commissioners/2019-2024/sinkevicius/announcements/commissioner-sinkevicius-speech-global-symposium-soil-biodiversity-april-19-2021_en. Accessed 17 Aug 2021

European Council (2019) A New Strategic Agenda: 2019-2024, Brussels. https://www.consilium.europa.eu/media/39914/a-new-strategic-agenda-2019-2024.pdf. Accessed 25 Aug 2021

European Court of Auditors (2018) Opinion No 7/2018: Concerning Commission Proposals for Regulations Relating to the common agricultural policy for the Post-2020 Period. https://www.eca.europa.eu/en/Pages/DocItem.aspx?did=47751. Accessed 25 Aug 2021

European Court of Auditors (2018) Combating Desertification in the EU: a Growing Threat in Need of More Action, Special Report No 33, Luxembourg, Annex III pp 2-10. https://www.eca.europa.eu/Lists/ECADocuments/SR18_33/SR_DESERTIFICATION_EN.pdf. Accessed 11 Aug 2021

European Environment Agency, The European environment – state and outlook 2020. https://www.eea.europa.eu/soer/2020. Accessed 9 Aug 2021

European Environment Agency (2019) Land and Soil in Europe, EEA Signals. https://www.eea.europa.eu/publications/eea-signals-2019-land. Accessed 11 Aug 2021

European Environment Agency (2021) Soil Organic Carbon. Indicator Assessment. https://www.eea.europa.eu/data-and-maps/indicators/soil-organic-carbon-1/assessment. Accessed 4 Aug 2021

European Environment Agency/United Nations Environment Programme (2000), Down to earth, Copenhagen

European Parliament and Council (2013) Decision No. 1386/2013/EU of 20 November 2013 on a General Union Environment Action Programme to 2020 ‚Living well, within the limits of our planet', Official Journal L 354 of 28.12.2013, pp 171-200

European Parliament and Council (2013) Regulation (EU) No 1306/2013 of the European Parliament and of the Council of 17 December 2013 on the Financing, Management and Monitoring of the common agricultural policy and Repealing Council Regulations EEC) No 352/78, (EC) No 165/94, (EC) No 2799/98, (EC) No 814/2000, (EC) No 1290/2005 and (EC) No 485/2008. Annex II., 59

European Parliament (2021) Text adopted, P9_TA (2021)0309, European Climate Law, European Parliament legislative resolution of 24 August 2021 on the proposal for a regulation of the European Parliament and the Council establishing the framework for achieving climate neutrality an damending Regulation (EU) 2018/1999 (European Climate Law) (COM(2020)0080 – COM(2020)0563 – C9-0077/2020 – 2020/0036(COD))

European Parliament (2021) Motion for a resolution on soil protection (2021/2548(RSP)), Doc. No. B9-0221/2021

European Parliament (2007) Resolution of 13 November 2007 on the Thematic Strategy for Soil Protection (2006/2293(INI), Official Journal C 282E of 8 November 2008, pp 138-144

European Parliament. "2021/2548(RSP)- Resolution on Soil Protection.," April 28, 2021. https://www.europarl.europa.eu/doceo/document/TA-9-2021-0143_EN.html. Accessed 4 Aug 2021

Eurostats (2019) Organic Farming Statistics. https://ec.europa.eu/eurostat/statistics-explained/index.php?title=Organic_farming_statistics. Accessed 7 Aug 2021

Fetting C (2020) The European Green Deal, ESDN Report, Vienna, ESDN Office. https://www.esdn.eu/fileadmin/ESDN_Reports/ESDN_Report_2_2020.pdf. Accessed 2 Aug 2021

Federation of German Industries, Feedback on the Initiative "Healthy soils – new EU soil strategy", Statement of 9 December 2020, p 1

Food and Agriculture Organisation of the United Nations (FAO), Intergovernmental Panel on Soils/ Global Soil Partnership (2017) Voluntary Guidelines on Sustainable Soil Management, Rome. https://www.fao.org/3/bl813e/bl813e.pdf. Accessed 23 Aug 2021

Food and Agriculture Organisation, Global Soil Partnership (2017) GSP Guidelines for sharing national data/information to compile a Global Soil Organic Carbon (GSOC) map, Pillar 4 Working Group of 01 February 2017. https://www.fao.org/3/bp164e/bp164e.pdf. Accessed 23 Aug 2021

Fuchs R, Brown C, Rounsevel M (2020) Europe's Green Deal offshores environmental damage to other nations. Nature 586(7831):671–673. https://doi.org/10.1038/d41586-020-02991-1

Glæsner N, Katharina H et al (2014) Do current European policies prevent soil threats and support soil functions? Sustainability 6(12):9538–9563

Gottlieb C, A sustainable recovery for Europe: the EUs Green Deal. https://www.clearygottlieb.com/news-and-insights/publication-listing/a-sustainable-recovery-for-europe-the-eus-green-deal. Accessed 4 Aug 2021

Guyomard H, Bureau J-C (2020) Research for the AGRI Committee - The Green Deal and the CAP: policy implications to adapt farming practices and to preserve the EU's natural resources. European Parliament, November 23, 2020. https://www.europarl.europa.eu/thinktank/en/document.html?reference=IPOL_STU%282020%29629214. Accessed 6 Aug 2021

Hannam I, Boer B (2004) Drafting legislation for sustainable soils: a guide, IUCN Environmental Policy and Law Paper No. 52, Gland/Cambridge

Hannam I, Boer B (2002) Legal and institutional frameworks for sustainable soils: a preliminary report, IUCN Environmental Policy and Law Paper No. 45, Gland/Cambridge

Heuser I (2006) Milestones of soil protection in EU environmental law. JEEPL (Journal for European Environmental & Planning Law) 2006, p 190 ff

Heuser I, Montanarella L (2016) Protection: policy of the European Union. In: Rattan L (ed) Encyclopedia of soil science, 3rd edn

Intergovernmental Panel on Climate Change (2019) Climate change and land: an IPCC special report on climate change, desertification, land degradation, sustainable land management, food

security, and greenhouse gas fluxes in terrestrial ecosystems. https://www.ipcc.ch/site/assets/uploads/2019/11/SRCCL-Full-Report-Compiled-191128.pdf. Accessed 22 Aug 2021
Institute for European Environmental Policy, Elements for soil quality and health, 27 April 2021, p 1
Jongeneel R, Hoes A-C, Berkhout P (2020) Climate-smart policy, the need for a coherent approach. Wageningen University and Research. https://research.wur.nl/en/publications/climate-smart-policy-the-need-for-a-coherent-approach-at-food-sys. Accessed 9 Aug 2021
Keesstra S, Bouma J, Wallinga J et al (2016) The significance of soils and soil science towards realization of the United Nations Sustainable Development Goals. SOIL 2(2):111–128. https://doi.org/10.5194/soil-2-111-2016. Accessed 9 Aug 2021
Maréchal A, Hart K (2020) Aligning the Post-2020 common agricultural policy with the European Green Deal. https://ieep.eu/publications/aligning-the-post-2020-common-agricultural-policy-with-the-european-green-deal. Accessed 7 Aug 2021
Matthews A (2020) Agriculture in the European Green Deal: From Ambition to Action, CAP Reform, 21 October 2020. https://capreform.eu/agriculture-in-the-european-green-deal-from-ambition-to-action/. Accessed 6 Aug 2021
Metta M (2020) The EU's common agricultural policy after 2020: High Ambition, Low Reform!, Heinrich Böll Stiftung, Brussels. https://eu.boell.org/en/2020/11/03/cap-post-2020-high-ambition-low-reform. Accessed 7 Aug 2021
Munta M (2020) The European Green Deal: A Game Changer or Simply a Buzzword?, Friedrich Ebert Stiftun. https://www.bib.irb.hr/1064244. Accessed 7 Aug 2021
Paleari (2017) Is the European Union protecting soil? A critical analysis of community environmental policy and law. Land Use Policy 64:163–173. https://doi.org/10.1016/j.landusepol.2017.02.007
Pe'er G, Zinngrebe Y, Moreira F, Sirami C, Schindler S (2019) A greener path for the EU common agricultural policy. Science 365(6452):449–451. https://doi.org/10.1126/science.aax3146
Prandecki K, Wrzaszcz W (2020) Agriculture and The European Green Deal, Rochester. https://papers.ssrn.com/abstract=3821462. Accessed 7 Aug 2021
Rat der Europäischen Union, Gutachten des Juristischen Dienstes, Vorschlag für eine Richtlinie des Europäischen Parlaments und des Rates zur Schaffung eines Ordnungsrahmens für den Bodenschutz und zur Änderung der Richtlinie 2004/35/EG, Dokument No. 7381 of 14.03.2007, p 6 ff. (document only available in German)
Secretariat of the Intergovernmental Science-Policy Platform on Biodiversity and Ecosystem Services, Montanarella L, Scholes R, Brainich A (eds) (2018) The IPBES assessment report on land degradation and restoration, Bonn, Germany. https://www.ipbes.net/sites/default/files/2018_ldr_full_report_book_v4_pages.pdf. Accessed 23 Aug 2021
Siddi M (2020) The European Green Deal: assessing its current state and future implementation. FIIA Working Paper 114. https://www.fiia.fi/wp-content/uploads/2020/05/wp114_european-green-deal.pdf. Accessed 23 Aug 2021
Stevens C, Kanie N (2016) The transformative potential of the Sustainable Development Goals (SDGs). Int Environ Agreements: Polit Law Econ 16(3):393–396. https://doi.org/10.1007/s10784-016-9324-y
Tetteh F (2019) Challenges to soil protection and sustainable management in Africa. International Yearbook on Soil Law and Policy (IYSLP) 2019, p 3 ff
United Nations Environment Programme (2015) GOAL 15: life on land. https://www.unep.org/explore-topics/sustainable-development-goals/why-do-sustainable-development-goals-matter/goal-15. Accessed 7 Aug 2021
United States Department of Agriculture (USDA) (2021) Natural Resources Conservation Service, Soil Health Glossary: term "soil health or soil quality". https://www.nrcs.usda.gov/wps/portal/nrcs/detailfull/soils/health/?cid=nrcs142p2_053848. Accessed 22 Aug 2021
Vrebos D, Bampa F, Creamer R, Gardi C, Ghaley B, Jones A, Rutgers M, Sandén T, Staes J, Meire P (2017) The impact of policy instruments on soil multifunctionality in the European Union. Sustainability 9(3):407. https://doi.org/10.3390/su9030407

Soil Protection Legislation and Policy in South Africa: An Overview

Oliver C. Ruppel, Keeley M. Knutton, and Rirhandzu A. Marivate

Each time one of us touches the soil of this land, we feel a sense of personal renewal. The national mood changes as the seasons change. We are moved by a sense of joy and exhilaration when the grass turns green and the flowers bloom.
Nelson Mandela at his Inauguration as President of the Republic of South Africa, Union Buildings, Pretoria, South Africa, 10 May 1994

Abstract The maintenance and protection of soil in South Africa is vital in order to allow a continued reliance upon it. This maintenance and protection needs to be governed by legislation and policy, along with prescribed and practiced enforcement measures. This chapter analyses the current position of South Africa regarding soil protection legislation and policy, whilst further providing recommendations for future legislation and policy needed to ensure that South Africa's soil is both protected and used sustainably. The also chapter considers the importance of healthy soil and land, the negative impacts of soil degradation and the relevant legal position in relation to such in the South African context. The main drivers of soil degradation, as well as the causes of soil threats found in South Africa are examined, while dealing with the relevant international law and the law of South Africa governing soil.

O. C. Ruppel (✉)
Faculty of Law, Stellenbosch University, Stellenbosch, South Africa

Research Center for Climate Law, University of Graz, Graz, Austria
e-mail: ruppel@sun.ac.za

K. M. Knutton
Stellenbosch University, Stellenbosch, South Africa

R. A. Marivate
Living Soils Community Learning Farm, a Regenerative Agriculture and Food Security Project Based at the Sustainability Institute, Stellenbosch University, Stellenbosch, South Africa

H. Ginzky et al. (eds.), *International Yearbook of Soil Law and Policy 2020/2021*, International Yearbook of Soil Law and Policy 2020,
https://doi.org/10.1007/978-3-030-96347-7_13

1 Introduction

Soil is made up of a layer of the earth, that can range from a few centimetres to several metres deep. Within this layer lies the entire capacity of earth to produce any life, and therefore it is the single most important resource for all biological, including, human existence, on the planet. Soil has been responsible for the development of all civilisations and is connected to climate and biogeochemical systems that support biodiversity on earth.[1] The interaction of these various systems with soils play an important role in the capability of ecosystems to provide a variety of ecosystem services that are beneficial to human well-being.[2]

There are a number of reasons why soil forms an important part of our ecosystem. Soil is used for a number of purposes, including often being used for building materials, mining, urban development, together with storage of water and a purifying effluent. Soil can be seen as a key element to the productive capacity of land, with good structured and healthy soils being able to clean air and water, process organic waste, cycle nutrients, capture carbon dioxide form the atmosphere, and one of the most important properties is that it is a medium for which plants grow and humans are most reliant in order to grow their own food.[3] Soils, regardless of characteristic, must not be neglected or degraded for the benefit of development on any scale.[4]

The maintenance and protection of soil in South Africa is vital in order to allow a continued reliance upon it.[5] This maintenance and protection needs to be governed by legislation and policy, along with prescribed and practiced enforcement measures. This chapter analyses the current position of South Africa regarding soil protection legislation and policy, whilst further providing recommendations for future legislation and policy needed to ensure that South Africa's soil is both protected and used sustainably.

This chapter considers the importance of healthy soil and land, the negative impacts of soil degradation and the relevant legal position in relation to such in the South African context. The main drivers of soil degradation, as well as the causes of soil threats found in South Africa are examined. The chapter then deals with the relevant international law and the law of South Africa governing soil and (to a lesser extent) land.

---

[1] Hillel (1992), p. 511; Yaalon (2000), p. 407; Braimoh and Vlek (2008), p. 1.

[2] Young and Crawford (2004).

[3] Meyer et al. (2018), p. 283; UNEP (2017).

[4] Braimoh and Vlek (2008), p. 1.

[5] As in other African counties, cf. Ruppel and Ginzky (2021).

2 Country Information

South Africa is located at the southern tip of the continent of Africa. It is bordered by Botswana, Mozambique, Namibia,[6] Lesotho, Swaziland and Zimbabwe. The size of South Africa is 1,219,090 km^2 in total, with 1,214,470 km^2 being land and 4620 km^2 being water. South Africa is the 26th largest country in the world. The population approximation for South Africa is 56,463,617, which is the estimation for July 2020. This estimate takes into account the effects of excess mortality due to AIDS, which lowers life expectancy, increases infant mortality and death rates, lowers population growth rates and changes the distribution of the population by age and sex. South Africa has the 26th largest population in the world and has a 0.97% population growth rate.[7]

When considering a country's information, its' geostrategic relevance is often considered. There are a number of factors that contribute to the geostrategic relevance of South Africa. One of these factors being that South Africa is arguably the most developed country situated in Africa with very important ports and a well-functioning infrastructure.[8] Further, South Africa is also one of the most important political actors on the continent, being one of the few African countries ranked as an upper-middle income country and the only African country with a G20 seat.[9] As a result of this, the actions taken by South Africa allow recognition to be given to the continent of Africa as a whole. A further example of this is the way in which Africa is given a 'voice' in BRICS as a result of South Africa being member of BRICS (Brazil, Russia, India, China and South Africa). Further factors that contribute to the geostrategic relevance are the natural resources that are found in South Africa, which links to the agricultural and mining industries, the tourism that is offered as well as the trade industry.

South Africa is a middle-income developing market with an abundant supply of natural resources.[10] The economy of South Africa is considered to have well developed financial, legal, communications, energy and transport sectors. Further, South Africa has a stock exchange that is Africa's largest and is also among the top 20 in the world.[11]

Traditionally South Africa's economy has been rooted in the primary sectors as a result of a wealth of mineral resources and favourable agricultural conditions. However, in recent decades this has changed due to a structural shift in output resulting in economic growth mainly being driven by the tertiary sector, which includes wholesale and retail trade, tourism and communications. Further, South Africa is now moving towards becoming a knowledge-based economy, with

[6] Cf. also Ruppel and von Finckenstein (2016), p. 309ff.

[7] Central Intelligence Agency (2020).

[8] Mutsindikwa (2018), p. 2.

[9] Ruppel and Borgmeyer (2018), p. 284.

[10] Central Intelligence Agency (2020).

[11] Central Intelligence Agency (2020).

a greater focus on technology, e-commerce, financial and other services. As the economy currently stands, key sectors that contribute to the gross domestic product are manufacturing, wholesale and retail trade, financial services, transport, mining, agriculture and tourism.[12]

The Gross National Product, which is now known as Gross National Income (GNI), is the sum of value added by all resident producers plus any product taxes, less subsidies, not included in the valuation of output plus net receipts of primary income from abroad.[13] The Gross National Product was reported at 84.190 billion United States dollars in September 2019, a decrease from June of the same year.[14] South Africa's GNP is updated quarterly, averaging 24.494 billion United States dollars from March 1960 to September 2019.[15]

The external debt of South Africa compares the total public and private debt owed to non-residents repayable in foreign currency, goods or services. South Africa's external debt, as recorded in 2017, was $156.3 billion.[16] This figure obviously stems from the time before COVID-19 came into play.

The public debt of South Africa compares the cumulative total of all government borrowings less repayments that are denominated in the country's currency.[17] This amount was considered in 2017 and came to 53% of the GDP.[18] Like the GNP, direct foreign investments into South Africa are calculated quarterly.[19] Over the last 10 years there has been constant fluctuation in direct foreign investments, with the highest being 52712.00 billion ZAR and the lowest −13910.00 billion ZAR.[20] When considering foreign investments in South Africa, it can be found that as it currently stands in the current quarter, there is 17024.00 billion ZAR worth of direct foreign investments in South Africa, which is a decrease from the 26301.00 billion ZAR in the previous quarter.[21]

South Africa offers primary, secondary and tertiary education; however, these are not accessible to all due to the fees involved. There is constant strain placed on education in South Africa as the government attempts to achieve equal opportunities for all citizens.[22] In South Africa, education is compulsory until the end of grade nine, which would encompass primary school and 2 years of high school.[23] Following this would be grades 10 to 12, which would see the completion of secondary

---

[12] Brand South Africa (2018).

[13] Macrotrends (2020).

[14] CEIC Data (2020).

[15] CEIC Data (2020).

[16] Central Intelligence Agency (2020).

[17] Central Intelligence Agency (2020).

[18] Central Intelligence Agency (2020).

[19] Trading Economics (2020).

[20] Trading Economics (2020).

[21] Trading Economics (2020).

[22] Scholaro Pro (2020).

[23] Scholaro Pro (2020).

education with a Matric certificate. The number of South Africans having attained secondary education has increased by 9% from 2008 to 2018.[24] If a certain level of grades is obtained in secondary school, application to tertiary institutions can follow. Tertiary education is provided by a number of private and state managed institutions located all over the country. Tertiary education qualifications are less common in South Africa, with only 13% of adults having had access to this level of education.[25] It is clear that importance education in South Africa needs to be considered and acted upon in order to improve the current standing of education.

The climate conditions in South Africa are mostly semiarid. The east coast of South Africa is considered to be subtropical. The country experiences sunny days and cool evenings.[26] The impacts of climate change are already being felt across various sectors.[27] In South Africa, climate change poses multiple challenges to economic growth and sustainable development and to the various facets of human security. Climate change amplifies existing risks and create new risks for natural and human systems.[28] Risks are unevenly distributed and are generally greater for disadvantaged people and communities in countries at all levels of development. Ever since the end of Apartheid, South Africa has been struggling to find and implement a roadmap to address distributive injustices of the past. In this light, anthropogenic climate change also contributes an urgent threat to human societies and the environment, and requires an effective, progressive and well-coordinated response. South African responses aim to highlight that anticipated domestic climate change impacts have the potential to undermine the country's development goals.[29]

Moreover, South Africa has embarked on the implementation of the Sustainable Development Goals (SDGs), within the context of existing regional and national strategic plans—such as the African Union's Agenda 2063 and the country's own National Development Plan (NDP) 2030. Not only does Agenda 2063 articulate a Pan-African vision of integration, solidarity and unity on a continental level, but it also calls for coordination and cooperation in mutually beneficial partnerships between regions and continents to enable the realisation of this African vision. By adopting the 2030 Agenda, South Africa is committed to leave no one behind in the implementation of the SDGs. This means that the specific vulnerability needs must

---

[24] OECD (2020).

[25] OECD (2020).

[26] Central Intelligence Agency (2020).

[27] "Impacts of climate change are the "effects on natural and human systems of extreme weather and climate events and of climate change. Impacts generally refer to effects on lives, livelihoods, health, ecosystems, economies, societies, cultures, services, and infrastructure due to the interaction of climate changes or hazardous climate events occurring within a specific time period and the vulnerability of an exposed society or system. Impacts are also referred to as consequences and outcomes. The impacts of climate change on geophysical systems, including floods, droughts, and sea level rise, are a subset of impacts called physical impacts." IPCC (2014), p. 5.

[28] IPCC (2014), p. 13.

[29] Cf. https://businesstech.co.za/news/energy/250747/south-africa-has-a-new-climate-change-bill-heres-what-you-need-to-know, accessed 05 May 2020.

be addressed for sustained, inclusive and sustainable economic growth and social progress.[30] In this light, South Africa has aligned, as far as possible, the implementation of the SDGs and Agenda 2063 with that of its NDP. This alignment means that efforts to achieve the SDGs have the potential to significantly impact the achievement of the country's national priorities as expressed in the NDP, and vice versa.[31]

The NDP sets out the road map addressing South Africa's priorities for the years to come. Its overarching aim is to eliminate poverty and reduce inequality by 2030 by ensuring, amongst other priorities, a transition to an environmentally sustainable, climate change resilient, low carbon economy and just society. The NDP identifies climate change as a major factor that will influence the context in which South Africa operates. Notwithstanding the aforementioned, the South African state of the environment is in general decline, with agriculture, human settlements and mining playing a large role in the decline. The country has faced an enormous pressure on land resources since its democracy. Economic growth and the need for social improvements have cost the country its natural resources and the loss of ecosystems. The South African government is critical with ensuring that it takes action in the improvement of land and natural resource management in order to prevent further land degradation and the loss of ecosystems. The main macro trends in land use are mainly in urbanisation from rural to urban migration, mining and agriculture that place pressure on land resources and productivity. These activities place have large land requirements and place significant pressure on the environment as they encroach onto important conservation areas, water resources and arable areas that have resulted in the loss of productive agricultural land and loss of ecosystem services.[32] The 2016 South African Environmental Outlook estimates that around 18% of the country's natural land cover has been transformed, mainly through cultivation followed by natural resource degradation, and then urban land use and forestry.

3 Main Drivers of Soil Degradation

Soil degradation is when there is a decline in soil function or its capacity to provide economic goods or ecosystems services.[33] The degradation of soil can take place in the form of soil erosion, loss of fertility, compaction and or the loss of soil carbon,

[30] Cf. https://www.un.org/development/desa/en/news/sustainable/leaving-no-one-behind.html, accessed 20 May 2020.

[31] Cf. PWC (Price Waterhouse Coopers). 2016. Make it your business: Engaging with the Sustainable Development Goals. Available at: https://www.pwc.co.za/en/assets/pdf/make-it-your-business.pdf, accessed 20 May 2020.

[32] Department of Environmental Affairs (2012).

[33] Lal (2010), p. 709.

these are accelerated by increased and excessive land clearing and using inappropriate agricultural practices.[34]

South Africa has experienced land degradation since the early colonial period. The ghost of history is evident in the South African landscape. Former homelands occupied by black South Africans are generally more degraded than former white commercial farming areas.[35]

The largest contributor to soil degradation across the world is the reduction of its productive capacity because of poor land management.[36] Even before human history, land resources on the planet were under pressure, but are even more so today because of the mounting demands from increasing land resource consumption by humans; increase pressure on land also means increases in its degradation.[37] Much attention is now being paid to land and soil degradation because the rate of degradation exceeds that of soil formation, and it has become not only an environmental, but a development issue globally.[38] The continued long-term land exploitation that has led to its degradation without replacement or restoration is not sustainable.

Agriculture is one of the main sectors of the economy in South Africa, at the same time it is diverse, and ranges from intensive, large-scale commercial agricultural sector to low-intensity, small scale, and subsistence farming sector. South Africa's land area is 121.9 million hectares (ha), and approximately 100 million ha (over 80%), is used for agriculture, 11% has arable potential, while the 69% is used for grazing.[39] The cultivated areas are utilised mainly for commercial purposes (horticulture, vegetable production and horticulture), making up 82% of the arable land. However, larger areas of low or marginal agricultural potential are being cultivated across the country, which has led to crop failure, abandonment and consequently land and soil degradation. Unsustainable agricultural practices and monoculture has resulted in soil degradation, biodiversity loss and poor water management.

The mining industry is under the regulation of government, which is under capacitated and requires funding and support in order to manage, rehabilitate, and reduce their environmental impacts.[40] South Africa is facing a mushrooming of new mining ventures particularly in coal, platinum group metals and other precious metals that are placing a heavy burden on land and significantly impact soil. The mining site has a small footprint compared to its associated network of industrial infrastructure that is extensive and disruptive to the land. Mining overall creates many environmental problems, physically and chemically. Mining activities have

[34] Van Lynden and Oldeman (1997), p. 10.

[35] Cf. https://knowledge.unccd.int/sites/default/files/ldn_targets/South%20Africa%20LDN%20TSP%20Country%20Report.pdf, accessed 20 May 2020.

[36] Zika and Erb (2009), pp. 310–318.

[37] UNCCD (2017).

[38] Scherr (1999), p. 2.

[39] Department of Environmental Affairs (2012).

[40] Department of Environmental Affairs (2012).

significant impacts on air and water quality, biodiversity and productivity of soil, by causing erosion, sedimentation, subsidence and landslides. From the removal of topsoil in the initial preparation of mining sites, to chemical soil contamination from toxic or harsh substances used to separate valuable minerals from waste products. Mineral processing also contributes to soil contaminations as contaminants build up in the soil near smelter, and can find their way into water, soil and vegetation, and cause harm. Other soil degradation in the form of weathering and erosion can be caused by mine closures with bad rehabilitation management. For example, strip mining destroys landscapes, forests and wildlife habitats at the site of the mine when trees, plants, and topsoil are cleared from the mining area, which in turn leads to soil erosion and destruction of agricultural land.[41]

South Africa has a variety of industrial sites that are related to a range of sectors such as agriculture, forestry, food and drink, tanneries, mining, pharmaceuticals, etc. There are a number of ways that industrial pollution can impact soils. Industrial air pollution can cause an acidifying effect, with chemicals such as sulphate and nitrate aerosols causing acid rain that especially impact the pH of sandy or loamy soils. Industrial effluent is thought to be difficult to treat because of their complexity; waste water from effluents from food and drink industries, and industries that use animal and vegetable materials are biodegradable but can have an effect on soils degradation as they increase the hardness of the surface layer and more pronounced angular blocky structures at depth. On the other hand, effluent from metal and some chemical industries are not biodegradable and many times toxic. The latter two industries pollute the soils and can inhibit plant growth.[42]

There currently is a scarcity in South Africa of literature on brownfields, which are sites that have been affected by the former use of the site and surrounding land, are derelict and underused, may have real or perceived contamination problems, are mainly in developed urban areas and require intervention to bring them back to beneficial use.[43] However, there is growing awareness for the need for brownfield sites in urban areas as South African cities mature with continually increasing population growth, urbanisation and development pressure. Brownfields are seen as an important means of limiting urban sprawl, preventing urban decay, making infrastructure more efficient and promote sustainable development of cities in South Africa. The number of brownfields sites in South Africa has not been established, and there is still uncertainty on the level of risk and understanding of remediation measures needed for these sites.[44] The impact of brownfields on soil degradation in South Africa, is therefore uncertain. However, in a study conducted in the UK, it was found that the impact from brownfield development on soil properties is wide-ranging and in the extreme case where brownfield development results in a

[41] Pretty and Odeku (2017), p. 17.

[42] Schoeman and van Deventer (2004), pp. 369–373.

[43] Mathey et al. (2018), p. 384.

[44] Potts and Cloete (2012), pp. 390–391; Goosen and Fitchett (2020), pp. 41–59.

nongreen after- use, the soil is removed permanently and hence all soil functions will be lost.[45]

Population growth and urban migration in South Africa has come with significant challenges for the government, as they aim to ensure that human settlements are adequate and sustainable. With the growth of the population, urbanisation and consumption also increases, and so does the impact on the natural environment. According to the Department of Environmental Affairs, the most significant impacts on the environment are urban and rural sprawl, housing demand, transportation and basic service infrastructure. The legacy of apartheid has played a large part and proves to be a challenge as the government has to work with distorted, and concentrated forms of human settlement, particularly informal housing, safety and land tenure that have magnified the unsustainable use of land resources, with increase in consumption and demand.[46]

Pre-1994 land policies have led to the crowding of people in rural 'homelands'. Rural 'homelands' were the areas where 'non-white South Africans' were made to live during the Apartheid era and have continued to be used as settlements. Until today, these areas continue to be sites of retreat, recuperation and retirement from the urban and industrial economy.[47] These 'homelands' also constitute shared access to land resources commonly referred to communal areas. Communal areas are used for producing crops and livestock, mainly for subsistence and livelihood security. Communal areas have long been neglected and this has resulted in overgrazing, over harvesting and soil erosion.[48]

Forestry is one of the economic sectors in South Africa mainly located in the north eastern parts of the country. Some varieties of plants have a positive effect on the soils, where the organic matter increase the pH leading to higher soil fertility. However, the organic matter from a number of plantations in timber-growing areas in South Africa have been linked to increases the water repellence in soils; when the canopy and ground cover are removed during clear-felling, regardless of how well the organic matter stores water, the soil is exposed to drying and is at risk of surface flow from rain water that leads to soil erosion and reduced soil water replenishment.[49]

---

[45]Gregory (2015), p. 4.

[46]Department of Environmental Affairs (2012).

[47]Lugogo (2018), p. 11.

[48]Department of Environmental Affairs (2012).

[49]Schoeman and van Deventer (2004), pp. 369–373.

4 What Is Causing Soil Threats?

Soil is a resource of vital importance in South Africa, resulting in any threat towards it needing to be taken seriously. South Africa's soil is under threat from a variety of sources, including soil erosion, topsoil degradation, compaction and surface crusting, acidification, salinization and infestation by weeds and pathogens.[50]

Soil erosion is one of the largest forms of degradation and is a major environmental challenge globally. Croplands are also more susceptible due to the practices of tillage and continuous exposure of soil. Erosion occurs when energy is transmitted from rain water or wind that results in the dislodging of soils particles.[51] In South Africa, soil erosion is largely caused through water erosion that affects more than 70% of the surface area. Soil erosion is also considered as one of the most important phenomenon in South Africa. Farming practices such a tilling and over time surface exposure when there are no crops in the field increase the erodibility of soil. The loss of soil through erosion reduces on site nutrient availability and soil productivity, while water erosion causes sedimentation of water channels downstream. Soil contamination is as a result of solid waste and effluents, heavy metals, pesticides, herbicides and other pollutants. Chemical pollutants, found mainly in industrial and agricultural waste products, can be toxic to plant growth.

Another form of soil contamination is through acidification that results from leaching chemicals through rainwater.[52] The most prevalent soil contamination in South Africa is in the forms of land pollution and acidification because of mining activities, which is most common in the Gauteng and Mpumalanga provinces of the country.[53] This chemical contamination increases soil acidity that is seen as a great problem for crop production. For example, acid mine drainage from mining operations and mine tailings dams and dumps contain a variety of toxic metals, some of which are found in high concentrations in soils surrounding mining areas in South Africa.[54]

Soil compaction is a result of intensive farming practices, and it is estimated that 68 million hectares of land are compacted for vehicle traffic. Soil compaction leads to increased bulk density due to the reduced gaps between the soil particles.[55] The compaction of soil can be worsened by the lack of organic matter or the use of heavy machinery when the soil has high moisture content. Soil compaction leads to reduced plant growth, reduced incorporation of fresh organic matter, a reduced

[50] Kidd (2011), p. 132.

[51] Pimentel et al. (1995), p. 179.

[52] Garland et al. (2000), pp. 69–107.

[53] Garland et al. (2000), pp. 69–107.

[54] Utembe et al. (2015), p. 34.

[55] Soil Science Society of America (1996).

nutrient cycle and water erosion.[56] According to Garland et al. (2000), soil compaction is not an extensive phenomenon in South Africa, but still plays and important part in overall soil degradation.

Salinization is when the accumulation of salt occurs under poor drainage conditions, and it usually is seen in cropland soil. Naturally, the accumulation of salts occurs when the soil solution takes up soluble salts during wet periods, which is then brought to the surface in the dry periods when the moisture evaporates, and the salts gradually accumulate and are precipitated out of the soil. Human activities through mainly irrigation and water supply, have increase salinization in the soil. It is thought that salinization as a form of soil degradation is insignificant in South Africa.[57]

Loss of soil organic matter or carbon is a result of soil disturbance from various land use activities. Soil organic matter is important for healthy soil, as it has the ability to store water, carbon and nutrients, and provides soil structure that is important for aeration, drainage and also reduces the susceptibility of topsoil to erosion. In the larger context soil is important in the carbon capturing, and good farming practices that reduce the degradation of soil can promote carbon sequestration.[58]

5 Who Are the Most Relevant Actors with Regard to Soil Degradation?

Human activities are a large contributor to the environmental impacts on soil and the decline is soil quality.[59] Identified sources of human activity that have contributed to soil pollution in South Africa include those of agricultural activities, mining, roadside emissions, auto-mechanic workshops, refuse dumps and e-waste.[60] The agricultural sector is estimated to have annual economic loss in the billions of rands bracket from the loss of soil nitrogen, phosphorus and potassium; nutrients important for plant growth. In addition, the mining industry, particularly in Mpumalanga, through soil contamination and acidification as a result of acid rain is seen to cost the region multiple millions of rands annually.[61]

Furthermore, the use of manure and wastewater are increasingly prevalent with South African farmers due to the low availability and high cost of inorganic fertilizers.[62] However, these practices could also be a source of metals, which pollutes the soil and contaminates the food chain. Additionally, poor farm

[56]Hamza and Anderson (2005), pp. 122–124.

[57]Garland et al. (2000), pp. 69–107.

[58]UNEP (2017).

[59]Meyer et al. (2018), p. 309.

[60]Fayiga et al. (2018), p. 41.

[61]Garland et al. (2000), pp. 69–107.

[62]Fayiga et al. (2018), p. 43.

management practices, which include the excessive use of agrochemicals, irrigation with contaminated water and the addition of untreated biosolids also contribute to the pollution of farmlands, which results in the pollution of the soil.[63]

Changes in available water and climate change have also challenged farmers to look toward alternate methods of irrigation. For example, several boreholes in the Western Cape have been found retrospectively to be unacceptably saline for agricultural use. Furthermore, there are a number of behavioural practices that have contributed to the deterioration of soil resource. These practices include a general lack of public awareness, inadequate enforcement of existing legislation (which is a universal problem with regard to environmental protection), and a major shortage of skilled labourers and insufficient allocation of funds to promote soil conservation. Along with these practices has been a tendency to view repossession of land and resettlement as solutions to land degradation, which they have not been found to be, these are politically rather than socially oriented methods and generate extreme change as opposed to managed change.[64]

The Intergovernmental Panel on Climate Change (IPCC) has in its findings in Global Warming of 1.5 °C ("2018 Report"),[65] and Climate Change and Land ("2019 Report")[66] looked closely at the relationship between soil and the impact of climate change. It acknowledges that human changes in land use result in atmospheric carbon being stored in plants and soil and this is differentiated from natural CO2 stored in the land.[67] It further contends that climate change hinders various processes of natural ecosystems, leading to reduced supply and quality of fresh water, loss of species, decline in biodiversity and degradation of landscapes. Some key soil properties are highly sensitive to climate change and climate variability.[68]

Agricultural soils are commonly affected by acidification through a series of chemical processes that are influenced by interacting effects of management (fertilizer, biomass removal) and climate and changes in precipitation under climate change may influence the extent of soil acidification. Out of many soil environmental factors, soil moisture plays a key role in regulating soil temperature and biological activity, particularly, high soil moisture maintains an even temperature while reducing the thermal conductivity of soil and activity of different biological processes, such as soil C and nutrient mineralisation and root growth.[69] Drought and dryness due to global warming, an increase in global temperatures of 1.5–2 °C, is said to impact soil moisture—particularly in areas of the southern African region.[70]

[63] Fayiga et al. (2018), p. 56.
[64] Cooper (1996), p. 14.
[65] IPCC (2018).
[66] IPCC (2019).
[67] IPCC (2018), p. 114.
[68] Reicosky (2018), p. 42.
[69] Reicosky (2018), p. 45.
[70] IPCC (2018), p. 196.

6 The Organisational Structure of the State and the Legal System

The progression towards South Africa's current constitutional dispensation was a road that was filled with challenges based on a mixture of legal regimes. Colonisation, settlement and Apartheid are major influences on the current multi-faceted nature of South Africa's legal system. However, the introduction of the 1996 Constitution has meant extensive change to South Africa's previous legal and political system. South Africa has not neglected its legal past but rather has found ways to adapt it in the post-Apartheid constitutional dispensation.

Today, South African law consists of the Constitution, legislation, judicial precedent, the common law (rules developed from Roman-Dutch and British authorities), (African) customary law and international law. The "common and customary law embodies official legal pluralism, whilst those two 'official' legal systems, together with all other 'unofficial' legal systems (e.g. Hindu law, Jewish law and Muslim law) embody 'deep' legal pluralism."[71] This pluralistic nature is an indication of vast and expansive legal potential that our democracy incorporates. South Africa is made up of a melting pot of various cultures; religions and community practices that make up the complex and all-encompassing nature of the legal system. This system is mirrored by the inclusion of African customary law, Roman Dutch Law and British common law in pluralistic practice.[72]

South Africa's legal journey in terms of constitutionalism[73] and democratisation started in 1990 with the unbanning of prohibited political parties and the release or return of anti-apartheid leaders. During 1993, multi-party negotiations continued, and discussions took place around the establishment of an Interim Constitution; arrangements to ensure free and fair elections; and the election of a committee to oversee the implementation of the Interim Constitution of 1994. Between 1994 and 1996, the Constitutional Assembly then concentrated on the drafting of the new Constitution which was approved by 1996 and took effect as of 1997.[74]

The Constitution is considered the direct and ultimate source of law. Section 1 and 2 of the 1996 Constitution, provide that the Constitution is the supreme law of the land and that "law or conduct inconsistent with it is invalid, and the obligations imposed by it must be fulfilled". Section 8(1) of the Constitution provides that "[the] Bill of Rights applies to all law, and binds the legislature, the executive, the judiciary and all organs of state" Section 8(2) states that "a provision of the Bill of Rights binds a natural or a juristic person if, and to the extent that, it is applicable, taking into account the nature of the right and the nature of any duty imposed by the right."

[71] Rautenbach (2010), p. 145.

[72] Du Plessis (2019), p. 15.

[73] Ruppel (2017), p. 51.

[74] Rautenbach (2012), p. 17.

These provisions of the Constitution read together indicate that the Constitution is applicable in all contexts and to all levels of the Government system.

The South African legal system follows the rule of precedent, whereby a court is bound by its own previous decisions or the decision of a higher court, which had been established prior to British occupation but remains a crucial part of South African's modern legal system. Judicial precedent is established by decisions of court, more formally known as the judiciary, and provides that previous court decisions are seen as authoritative rather than merely persuasive. The structure of courts is therefore important in determining precedent. The Latin maxim *stare decisis et non quieta movere* (henceforth known simply as "*stare decisis*") means to "stand by previous decisions" or "to stand by decisions and not to disturb settled law". Judicial precedent emerges when new rules or legal principles are established in particular judgements, this does not happen in every case. Often judicial precedent is applied as is from previous judgements where the courts deems appropriate. If, in a given case, there is no precedent that sufficiently address the issue in question the court may then take to establish precedent to be used by future courts in similar instances. "The doctrine of precedent not only binds lower courts but also binds courts of final jurisdiction to their own decisions. These courts can depart from a previous decision of their own only when satisfied that that decision is clearly wrong. Stare decisis is therefore not simply a matter of respect for courts of higher authority. It is a manifestation of the rule of law itself, which in turn is a founding value of our Constitution. To deviate from this rule is to invite legal chaos".[75]

Moreover, "a court is bound by precedent unless the facts of a matter are not materially the same as those in a previous matter or where the decision of the previous court is manifestly incorrect. In other words, where the facts of a particular matter are distinguished from those of a previous matter for which a precedent exists, the existing precedent will not be applicable to the distinguished facts"[76] Strict adherence to the doctrine of *stare decisis* would mean that courts are obliged to follow earlier decisions regardless of whether an earlier decision still makes sense. Therefore, and for greater fairness and legal certainty, courts are bound by their own decisions unless and until they are overruled by a superior court. It is, however, conceivable that circumstances arise that would render it possible for a court to override its own legal opinion.[77]

The South African government has a legislative branch, which is in charge of delivering legal acts. This legislative branch is a bicameral Parliament made up of the National Council of Provinces and the National Assembly.[78] The main functions of the national legislature are to pass new laws, amend existing laws and to repeal or abolish old laws.[79]

---

[75] Humby et al. (2012), p. 152.

[76] Humby et al. (2012), p. 218.

[77] Havenga et al. (2002), p. 8ff.

[78] Central Intelligence Agency (2020).

[79] Parliament of the Republic of South Africa (2020).

Further, traditional authorities play an important role in South Africa although not everyone is prepared to recognise them as role players.[80] Traditional entities have been established over time by the multiple cultures that South Africa holds. Customary law is that which established and recognised traditional leaders originally. The existence of customary law is not denied by the Constitution to the extent the it is consistent with it.[81] Traditional leaders are recognised by the 1996 Constitution, with chapter 12 providing for recognition and the role of traditional leaders.[82] The Constitution states that the institution, status and role of traditional leadership, according to customary law, are recognised in South Africa, subject to the Constitution.[83] It further provides that national legislation may provide for a role for traditional leadership as an institution at local level on matters that affect the local communities.[84] The 2003 White Paper on Traditional Leadership and Governance sets out the national framework, norms and standards that define the role and place of the institution of traditional leadership in South Africa.[85] It is clear that through the constitution, statutory law and customary law there is still a role to be played by traditional entities in South Africa.

In the past, only English and Afrikaans were official languages used in governmental affairs, which excluded large parts of population who had little or no languages proficiency in those languages. The 1996 South African Constitution therefore gave official status to all the major languages of South Africa. In terms of section 6(1) (T)he official languages of the Republic are Sepedi, Sesotho, Setswana, siSwati, Tshivenda, Xitsonga, Afrikaans, English, isiNdebele, isiXhosa and isiZulu. Section 6(2) provides that (R)ecognising the historically diminished use and status of the indigenous languages of our people, the state must take practical and positive measures to elevate the status and advance the use of these languages. Interestingly in section 6(5) a Pan South African Language Board has been established by national legislation. Such organ is tasked to promote, and create conditions for, the development and use of all official languages; the Khoi, Nama and San languages; and sign language; and promote and ensure respect for all languages commonly used by communities in South Africa, including German, Greek, Gujarati, Hindi, Portuguese, Tamil, Telegu and Urdu; and Arabic, Hebrew, Sanskrit and other languages used for religious purposes in South Africa.

[80] Du Plessis W & the Traditional Authorities Research Group (1999), p. 295.

[81] Constitution of the Republic of South Africa (1996).

[82] Constitution of the Republic of South Africa (1996).

[83] Constitution of the Republic of South Africa (1996).

[84] Constitution of the Republic of South Africa (1996).

[85] South African Government (2020).

7 International Law

7.1 Relevant International Agreements

The term international agreement in the South African Constitution is synonymous with the term treaty as defined in article 2(1) of the 1969 Vienna Convention on the Law of Treaties.[86] Numerous international agreements—bilateral, regional or multilateral—have been concluded by South Africa and international customary rules, as evidence of a general practice accepted as law, have been well established. International agreements are binding upon states if the consent to be a party to them is expressed by a signature followed by ratification, or by accession, where the state is not a signatory to a treaty, or by declaration of succession to a treaty concluded before such a state existed. The sources of international law in general are listed in Article 38 of the Statute of the International Court of Justice (ICJ), the principal judicial organ of the United Nations. However, considering that Article 38 of the Statute of the ICJ was first drafted in 1920, these provisions no longer reflect all the sources of today's international law.[87]

South Africa has often been at the forefront of international law efforts since the end of Apartheid. Such has South Africa signed the United Nations Framework Convention on Climate Change (UNFCCC) in 1993 and ratified it in 1997. The UNFCCC is a treaty in terms of international law and Article 2.1(a) of the Vienna Convention on the Law of Treaties. Moreover, South Africa acceded to the Kyoto Protocol in 2002 and ratified the Paris Agreement in 2016 respectively. The UNFCCC recognises the role of healthy ecosystems in providing valuable services such as food, clean water, flood and erosion control, while at the same time building resilience against climate change impacts. It provides for implementing adaptation through national action plans (NAPs) and national adaptation programmes of action (NAPA).[88]

The Paris Agreement entered into force on 4 November 2016 and contains the following cornerstones: As a long-term goal it is envisaged to keep global warming well below two degrees and to pursue efforts to limit the temperature increase even further to 1.5 degrees. The Paris Agreement provides for a system of Nationally Determined Contributions (NDCs), which equally has much significance for the protection of soil. The Paris Agreement raises the importance of adaptation as a response to climate change—requiring Parties to engage in adaptation planning and implementation that takes into account "vulnerable people, places and ecosystems" and builds "the resilience of socio-economic and ecological systems, including through economic diversification and sustainable management of natural resources". It specifically establishes a global goal on adaptation of "enhancing adaptive

[86]De Wet (2015), p. 27.

[87]Ruppel (2013), p. 49.

[88]Cf. https://www.sanbi.org/wp-content/uploads/2018/09/Strategic-Framework-and-Overarching-Implementation-Plan-for-EbA-in-SA.pdf, accessed 20 May 2020.

capacity, strengthening resilience and reducing vulnerability to climate change". The agreement also recognises that adaptation efforts of developing countries are a key part of their contribution to fighting climate change.[89]

South Africa ratified the United Nations Convention to Combat Desertification (UNCCD) in 1997. The objective of the Convention is to combat desertification and mitigate the effects of drought in countries experiencing serious drought and/or desertification, particularly in Africa. This is done through effective action at all levels, supported by international cooperation and partnership arrangements, in the framework of an integrated approach which is consistent with the SDGs, with a view to contributing to the achievement of sustainable development in affected areas.[90] Since its inception, the UNCCD has sought to enhance the adaptive capacities of dryland populations to highly variable environmental conditions. As vulnerability varies across sectors, regions and social groups, adaptation measures range from reducing vulnerability to enhancing the long-term sustainability of the poorest and most vulnerable populations in dryland areas. The provisions of the UNCCD reflect a strong link between the adaptation of biodiversity to climate change and desertification, land degradation and drought.[91]

Land is a vital resource for producing food and other ecosystem goods and services including conserving biodiversity, regulating hydrological regimes, soil nutrients cycling, and storing carbon, among others. Indeed, the most significant natural capital asset is productive land and fertile soils. For those communities that rely heavily on land as their main source, especially the rural poor, human well-being and sustainable livelihoods are completely dependent upon and intricately linked to the health and productivity of the land.[92] SDG Goal 15 (Life on Land) stipulates as follows:

> Protect, restore and promote sustainable use of terrestrial ecosystems, sustainably manage forests, combat desertification, and halt and reverse land degradation and halt biodiversity loss.

Target 15.3 of the same goal envisages that:

> By 2030, combat desertification, restore degraded land and soil, including land affected by desertification, drought and floods, and strive to achieve a land degradation-neutral world.

As of 2019, 123 countries had committed to set their voluntary targets for achieving Land Degradation Neutrality (LDN) and in 60 countries, governments had already

[89] Cf. https://www.sanbi.org/wp-content/uploads/2018/09/Strategic-Framework-and-Overarching-Implementation-Plan-for-EbA-in-SA.pdf, accessed 20 May 2020.

[90] Deputy Minister of Environment, Forestry and Fisheries to address the African Regional Meeting for the United Nations Convention to combat desertification in Pretoria, 6 August 2019, available at https://www.gov.za/speeches/combat-desertification-6-aug-2019-0000.

[91] Cf. https://www.sanbi.org/wp-content/uploads/2018/09/Strategic-Framework-and-Overarching-Implementation-Plan-for-EbA-in-SA.pdf.

[92] South Africa: Final country report of the LDN Target Setting Programme (2018), available at https://knowledge.unccd.int/sites/default/files/ldn_targets/South%20Africa%20LDN%20TSP%20Country%20Report.pdf, accessed 20 May 2020.

officially endorsed these targets.[93] South Africa has set national targets that are supplemented with sub-national and specific targets. At the national scale, South Africa aims to achieve LDN by 2030 as compared to 2015 and an additional 5% of the national territory has improved. At the sub-national scale South Africa aims to achieve LDN in the grassland biome, thicket biome and renosterveld.[94]

7.2 *Application of International Law*

Chapter 4 in sections 231–233 of the South African Constitution regulates details regarding international agreements and the application of (customary) international law. Regarding the distinction between the monist and dualist approaches regarding the status of international law in a domestic legal order, the South African Constitution makes use of both approaches.[95] In case of the application of international customary law, section 232 of the Constitution states that customary international is law in the Republic unless it is inconsistent with the Constitution itself or an Act of Parliament and thus, there is no further implementing statute required for international customary law to become legally binding. In contrast hereto, section 231(4) of the Constitution determines that an international agreement becomes law in the republic only when it is enacted into law by national legislation. As international economic law is mostly regulated by international treaties, this part focuses more on the implementation and the direct effect of those treaties in the South African legal order.

The Constitution distinguishes between the obligations of the Republic on an international level that derive from international agreements to which South Africa is a party and their applicability in the domestic legal system.[96] Section 231(2) of the Constitution requires the ratification of an international treaty by both the National Assembly and the National Council of Provinces except for those agreements that can be classified as technical, administrative or executive.[97] The Constitution does not contain any definitions or indication when these criteria are met. It is, however, important to emphasise that ratification by parliament does not replace the need for implementing legislation to guarantee the international treaty's domestic applicability.[98] Therefore, section 231(4) of the Constitution provides for the enactment of

[93] Progress towards the Sustainable Development Goals Report of the Secretary-General (2020), online available at https://sustainabledevelopment.un.org/content/documents/26158Final_SG_SDG_Progress_Report_14052020.pdf, accessed 20 May 2020.

[94] Cf. https://knowledge.unccd.int/sites/default/files/ldn_targets/South%20Africa%20LDN%20TSP%20Country%20Report.pdf, accessed 20 May 2020.

[95] Ferreira and Ferreira-Snyman (2014), p. 1471.

[96] Sections 231 (2) & (4) Constitution of the Republic of South Africa (1996).

[97] Section 231(3) Constitution of the Republic of South Africa (1996).

[98] Sucker (2014), p. 426.

additional national legislation to transform the international treaty provisions into municipal law, unless the particular agreement contains a self-executing provision. The legislature seems to use three different methods to incorporate international agreements into South African national law.[99] Firstly, the provisions of the particular agreement may be embodied in the text of an act; secondly, the agreement may be included as a schedule to a statute and thus be incorporated by reference; and lastly, the legislation may authorise the executive to bring the agreement into effect as domestic law by publishing it in the Government Gazette. These principles were also mentioned in *Glenister v President of the Republic of South Africa* (*'Glenister'*).[100] Once an international agreement is incorporated into national legislation its provisions enjoy the same legal status as the implementing legislation itself.[101]

Ultimately, Section 39(1)(b) of the Constitution provides that international law must be considered when a court interprets the Bill of Rights. Although Section 39 provides that international law must be considered it does not require that international law must be applied.[102] However, as stipulated in section 233, "when interpreting any legislation, every court must prefer any reasonable interpretation of the legislation that is consistent with international law over any alternative interpretation that is inconsistent with international law".

8 Statutory Soil Legislation

Soil degradation has continuously been seen as a long-term problem in South Africa, and the government has historically attempted to address the issue in one way or another.

> Many colonial laws produced impacts that are visible today. The laws favoured white commercial farmers and discriminated against black farmers. The extensive degradation evident today may be attributed to the creation of black homelands. Any evaluation of drivers of land degradation that ignores the historical perspective will lead to a simplistic understanding of a complex problem. Proximate (direct) drivers of land degradation include over- abstraction of water, mining, urbanisation and infrastructure development, over-harvesting of natural resources, bush encroachment, alien invasive species and unsustainable land management. Underlying (indirect) drivers include high population density, market access, land tenure, poverty, decentralisation and international policies.[103]

The country's government has previously promulgated legislation that was aimed at controlling or reducing soil loss. These legislations included the Forest and Veld Conservation Act 13 of 1941, the Soil Conservation Act 76 of 1969, and the

[99] Dugard (2018), p. 61.

[100] *Glenister v President of the Republic of South Africa* 2011 (3) SA 347 (CC).

[101] *Glenister*, para 100.

[102] Rautenbach (2018), p. 75.

[103] Cf. https://knowledge.unccd.int/sites/default/files/ldn_targets/South%20Africa%20LDN%20TSP%20Country%20Report.pdf, accessed on 20 May 2020.

Conservation of Agricultural Resources Act 43 of 1983 and the Environment Conservation Act 73 of 1989. The above mentioned legislation however, were only ever applied to white-owned areas, while soils protection legislation in black areas fell under the Native Administration Act 38 of 1927, the Bantu Homelands Constitution Act 21 of 1971, legislation formulated by individual homelands governments, and proclamations by the State President, the such as Proclamation 116 of 1949, and the Betterment Areas Proclamation R196 of 1967.[104] The last, Environment Conservation Act 73 of 1989, was subsequently replaced by the National Environmental Management Act 107 of 1998 ("NEMA") together with the primary legislation aimed at addressing soil conservation, the Conservation of Agricultural Resources Act 43 of 1983 ("CARA"), which are complementary frameworks for the safeguarding soil.[105]

CARA promotes the conservation of soil, water resources and vegetation, with the objects of the Act being to provide for the conservation of the natural agricultural resources of the Republic through the maintenance of the production potential of land, by combating and preventing erosion and weakening or destruction of the water resources, and by the protection of the vegetation and the combating of weeds and invader plants.[106] The act provides management plans that guide to eradicate weeds, alien invasive vegetation and promote and protect the integrity of indigenous vegetation, as well as guidelines to the management and protection of soil. CARA provides the Minister of Agriculture with the power to prescribe control measures that need to be complied with and applied by land users which include control measures for the cultivation of virgin soil, the utilisation and protection of land, which is cultivated.[107]

CARA makes provision for the Minister of Agriculture to establish schemes in terms of which financial assistance can be granted to land users, with which the land user may be subsidised in order for the reparation of damage to soil conservation works that have been caused by natural disasters such as flooding; the restoration of eroded, disturbed, denuded or damaged land; and the planting and cultivation of particular crops for the improvement of soil fertility or that counteract soil erosion.[108]

The National Environmental Management Act 107 of 1998 ("NEMA") promotes the cooperative environmental governance through the establishment of principles of decision making in an integrated and co-operative manner for any matters that affect the environment. NEMA further stipulates the procedures and administration and enforcement of environmental management laws, through various management instruments. Furthermore, NEMA Section 24 management principles, requirements, and procedures for environmental authorisation are outlined in order to reduce the

[104] Garland et al. (2000), pp. 69–107.

[105] Kidd (2011), p. 132.

[106] Ibid.

[107] Kidd (2011), p. 133.

[108] Ibid.

negative impacts and degradation on the environment. Section 24 further makes stipulations for actions to be taken in the case of pollution or degradation of the environment by the persons responsible.

Chapter 3 of NEMA provides stipulations for the procedures for cooperative governance. These procedures stipulate the need for the national and provincial government departments to exercise functions that included the development and enforcement of environmental implementation plans and management plans in cases that certain activities may affect the environment. The purpose of these environmental plans is to coordinate and harmonise environmental policies, plans, programmes and decisions of the various national departments that exercise the functions that may affect the environment or are entrusted with powers and duties aimed at the achievement, promotion and protection of a sustainable environment. These plans must be complied to in accordance to section 2 and section 146(2)(b)(i) of the Constitution and as set out by the relevant Minister.

The Implementation Framework for the LandCare Programme (1999) by the Department of Agriculture, Forestry and Fisheries[109] is a discussion document that considers the overall objectives of the LandCare programme in relation to the conservation of natural resources. LandCare South Africa will be a community-based programme supported by both the public and private sector through a series of partnerships. It is a process focused on the conservation of the natural resources (soil, water and vegetation) through sustainable utilisation and the creation of a conservation ethic through education and awareness. In addition, it seeks to address rural poverty by means of sustainable job creation.

The National Action Programme Combating Land Degradation to Alleviate Rural Poverty (2004), set by the Department of Environmental Affairs and Tourism[110] aims to combat desertification and mitigate the effects of drought in line with the framework of Agenda 21, in order to achieve sustainable development. The achievement of the objective will involve long term integrated strategy that focus on the improvement of land productivity, and the rehabilitation, conservation and sustainable management of land and water resources that will lead to the improved living conditions of people at community level.

The Soils Protection Strategy (2005) (yet to be completed)[111] was prompted by the need to identified priority areas for the implementation of integrated soil rehabilitation programmes that have been identified through modelling and mapping land capability and predicted soil erosion. The strategy recommended the establishment of a National Advisory Board (NAB) on Soil Protection, which would be supported by three thematic areas, namely erosion, decline in organic matter/nutrient depletion and soil contamination; and two technical working groups on monitoring and research. Through the NAB the strategy would be implemented through the

[109] https://www.nda.agric.za/docs/landcare/landcare.htm, accessed 9 Mar 2022.

[110] Department of Environment and Tourism (2004).

[111] Department of Agriculture (2009).

coordination of all the stipulated activities of the working groups in order to provide recommendations for the development of a Soil Protection Policy.

South Africa has been relatively active dealing with regulatory aspects of climate change.[112] The National Climate Change Response White Paper (2011) reflects the government's vision on how to effectively response to climate change and the long-term, just transition to a low carbon economy, climate resilient society.[113] The climate change response has two objectives; firstly to effectively manage the impacts of climate change by developing interventions and appropriate response capacity that build the social, economic and environmental resilience of South Africa. Secondly, to make effective contribution to reducing and stabilising greenhouse gases to levels in order to avoid dangerous anthropogenic impacts on the climate system in order to continue sustainable development. The response paper particularly highlights the need for transformation of the agricultural sector, particularly in the appropriate use of small-scale labour-intensive models and techniques to increase agricultural jobs, promote food security and conservation of soil, and biodiversity. Additionally, in the investment of research, technology and techniques for water, nutrient and soil conservation.

The Ecosystem-based Adaptation Strategic Framework & Overarching Implementation Plan (2016–2021) for the Department of Agriculture, Forestry and Fisheries highlights EbA as one of its foremost interventions particularly through restoration efforts and sustainable land use management.[114] EbA is the use of biodiversity and ecosystems to support communities to adapt to life in the clanging climate, and focuses on the management, restoration and rehabilitation do ecosystem service by improving water and soils quality, and contributing to livelihood opportunities.

The 2018 Draft National Climate Change Adaptation (NCCAS) Strategy aims to articulate South Africa's approach to adaptation in order to ensure a coordinated effort by the relevant key role players involved within civil society, private sector, and government for the effective implementation of adaptation interventions.[115] The NCCAS focusses on area such as water, early warning systems, and ecosystem based adaptation for building climate resilience in particularly in vulnerable communities. Adapting to build a strong South Africa has been conceptualised as the key message of the NCCAS underpinned by the following:

- South Africa needs to adapt to climate change;
- Climate change threatens development;
- The poor are most vulnerable to climate change impacts;
- The NCCAS will give South Africa an advantage in the future;

[112] Ruppel et al. (2020), p. 274.

[113] Department of Environmental Affairs (2011).

[114] Department of Environmental Affairs and South African National Biodiversity Institute (2016).

[115] Department of Environmental Affairs (2018).

- Transformational, systemic change is required to address the challenges presented by climate change;
- Linkages between adaptation and mitigation are increasingly vital.[116]

In June 2018, the South African Minister of Environmental Affairs, also published the Draft Climate Change Bill ("Climate Change Bill") in the Government Gazette, thus opening the public to submit comments.[117] The Climate Change Bill seeks to "build the Republic's effective climate change response and the long term, Just transition to a climate resilient and lower carbon economy and society in the context of an environmentally sustainable development framework; and to provide for matters connected therewith."[118] The Bill addresses policy alignment and institutional arrangements; climate change response for provinces and municipalities; national adaptation to impacts of climate change; greenhouse gas emissions and removals; and general matters and transitional arrangements i.e. public participations, delegation, offences and penalties etc.[119]

The overall objectives of the Climate Change Bill are also relevant to soil protection in that it pursues to promote a "constitutional right of everyone to dignity, life and an environment that is not harmful to their health and well-being, and to have the environment protected for the benefit of present and future generations while allowing justifiable environmentally sustainable economic and social development. To ensure that this constitutional right is met the government has committed to introducing measures that will respond to the recognition that 'anthropogenic climate change represents an urgent threat to human societies and the planetary environment and requires an effective, progressive and incremental response."[120]

Prior to the Climate Change Bill, climate change concerns in South African had been addressed through the 2011 White Paper on National Climate Change Response Plan and the Nationally Determined Contribution submitted to the Combating Climate Change secretariat in terms of the Paris Agreement. The inability of current policy, legislative and regulatory frameworks to effectively meet the unique challenges of the trans-sectoral impacts of climate change is recognised by the Climate Change Bill, which "seeks to introduce a nationally driven, coordinated and cooperative legal and administrative response that acknowledges the centrality of the provincial and municipal spheres in achieving its objectives."[121]

Since 2018, no clear communication from the Government or Parliament concerning the Climate Change Bill has been released. The most recent communication from the office of the President regarding climate change was at the 2020 State of the Nation Address, where he stated that "the Presidential Commission on Climate

---

[116] Cf. Annex 1: https://cer.org.za/wp-content/uploads/2019/05/DEA-Draft-climate-change-adaptation-strategy.pdf, accessed 20 May 2020.

[117] Government Gazette 41689 of 8 June 2018.

[118] Climate Change Bill (draft) No. 580, Government Gazette 41689 of 8 June 2018.

[119] Ibid.

[120] Grist (2018), p. 53.

[121] Grist (2018), p. 54.

Change will ensure that as we move towards a low carbon growth trajectory that we leave no one behind. We will finalise the Climate Change Bill, which provides a regulatory framework for the effective management of inevitable climate change impacts by enhancing adaptive capacity, strengthening resilience and reducing vulnerability to climate change – and identifying new industrial opportunities in the green economy."[122]

9 Environmental Provisions in the Constitution

The Constitution of South Africa holds a number provisions that either directly address or may pertain to the environment. Most evidently is that of section 24 of the 1996 Constitution, which stipulates that everyone has a right to an environment that is not harmful to their health and wellbeing; and to have the environment protected for the benefit of present and future generations, through reasonable legislative and other measures that prevent pollution and ecological degradations, promote conservation and secure ecologically sustainable development and use of natural resources while promoting justifiable economic and social development.[123] Further, a number of clauses found in the Bill of Rights may also pertain to the environment. These clauses are the locus standi clause, the just administrative action clause, the access to information clause, the property clause, the limitations clause and the interpretation of the Bill of Rights clause.[124] Although these to not directly relate to the environment, they can be used or addressed when considering a matter pertaining to the environment.

The Constitution is the foundation of South African legislation, and although not explicitly stipulated, soil is considered part of the environment and is broadly covered under the provision of the protection of the environment. CARA is a legislation that more specifically makes provision for the promotion of conservation of soil. As stated in previous section, CARA provides the control over the use of natural agricultural resources of the Republic of South Africa in order to promote the conversation of soil, water and vegetation and the combating of weeds and invasive vegetation. CARA makes further stipulations that any directions that have been declared applicable to land under the now repealed Soil Conservation Act 76 of 1969, need to be enforced on the land user.

The Constitution mandates the government to protect the environment, and the framework used to enforce this is the National Environmental Management Act

[122] South African Government "State of the Nation Address by President Cyril Ramaphosa, Parliament, Cape Town" (13 February 2020) *President Cyril Ramaphosa: 2020 State of the Nation Address*, https://www.gov.za/speeches/president-cyril-ramaphosa-2020-state-nation-address-13-feb-2020-0000, accessed 9 May 2020.

[123] Section 24 of the Constitution of the Republic of South Africa (1996).

[124] Sections 38, 33, 32, 25, 36, and 39 of the Constitution of the Republic of South Africa (1996).

107 of 1998. NEMA provides an environmental management framework for decision making across sectors, from individuals, institution and government. NEMA aims to promote cooperative governance and ensure that people's rights are upheld, while also recognising the importance of economic development. The act has a broad definition of the environment, and states that the environment in which humans exist includes land, water and the atmosphere, and the interrelationships, properties and conditions of all organisms that exist within the surrounds. NEMA, thus promotes integrated environmental management in recognition of these interrelationships.

Under section 152 of the Constitution, it further stipulates the obligatory requirements to ensure that the right to a clean and healthy environment is fulfilled; section 152(1)(b) and (d) on the part of local government, that is translated in sections 4(2)(d) and 4(2)(i), 73(1) and (2) of the Municipal Systems Act 32 of 2000.

10 Legislation on Land Property/Land Ownership and on Land Rights to Uses/Access

The right to property is found in section 25 of the 1996 Constitution, stating that no person may be deprived of property except in terms of the law of general application and that no law may permit arbitrary deprivation of property.[125] Further, section 25 provides that property may be expropriated only in terms of the law of general application for the public purpose or in the public interest and subject to compensation.[126]

The Constitution allows for the protection of property, but also for the reform of property relations. Common law, precedent and statutory law were traditionally regarded as sources of South African property law but are no long regarded as exhaustive under the new constitutional dispensation.[127] Public law dealing with land ownership and rights before 1996, can be found to be discriminatory. As a result of this, a number of instruments have been introduced in order to right the wrongs of the past. One of the most applicable measures is the 1997 White Paper on Land Policy, which was set up with the purpose if redressing the Apartheid era, nurturing national reconciliation and sustainability to support economic growth, to improve welfare and to relieve poverty.[128]

In South Africa, as a result of the post 1996 constitutional dispensation, there is a demand for equal consideration be given to (African) customary law as is to public law.[129] Customary law is both written and unwritten, having developed from the

[125] Constitution of the Republic of South Africa (1996).

[126] Ibid.

[127] Mostert and Pope (2010).

[128] Grain (2015).

[129] Mostert and Pope (2010).

customs and traditions of communities over time.[130] In order for the customs and traditions to be considered as law, they must be known to the community, followed by the community and be enforceable. When there is a conflict between public and customary law, both laws are looked at together in order to determine the differences and ensure that both are in line with the Constitution. The choice of law rule will then be considered. The purpose of this rule to select the law that will do justice in the given case, not for one party but for all. Further the system of law that is chosen must be in line with the Constitution.

Planning law is also relevant when considering land ownership and rights to use and access of land. Planning law can be defined as the area of law that provides for the creation of a sustainable land management planning framework as well as for the management of land development with the purpose of ensuring health, safety and welfare of society as a whole, while accounting for overarching interests such as the environment.[131] This is an important division of the law in South Africa and is predominantly governed by the Spatial Planning and Land Use Management Act (16 of 2013) ("SPLUMA").

A large number of environmental issues stem from the way land is used and managed.[132] The environment needs to be taken into consideration when exercising planning law due to potential negative effect that it can have. Environmental conservation and land use planning are intricately connected, with land use planning being the basis and necessary condition of potentially successful pollution control and conservation of natural resources.[133] The form of tenure on specific pieces of land has environmental consequences and is particularly relevant to communal land throughout South Africa and the land previously known as the homelands.[134] Wise land use planning has the potential to assist in the conservation of natural resources.[135] The manner in which land is used clearly has an impact on the environment, and therefore is an important component of environmental management.[136] Urban planning in South Africa is aimed at the development of sustainable living and working environments to address spatial imbalances such as people living long distances from their workplaces, which inevitably adds to their living costs. The government has initiated a policy for the design and development of more productive and inclusive urban areas that are properly governed. The South African urban landscapes are still very much reflective of inequalities and although steps have been taken to remedy this, there is still a need for further development. Transformation is necessary and entails changing the urban landscape to address obvious imbalances. Urban planning in South Africa needs to include both public and private

---

[130] Ruppel (2008), p. 23.
[131] Van Wyk (2018), p. 1131.
[132] Glazewski (2000), p. 229.
[133] Rabie (1976), p. 6.
[134] Glazewski (2000), p. 230.
[135] Rabie (1976), p. 6.
[136] Kidd (2011), p. 209.

sectors in order to bring the necessary changes towards more sustainable cities and land use in the future. Strategic spatial planning, as is relevant to South Africa, entails methods used and approaches followed by government and private sector stakeholders in managing the distribution of activities and people in spaces. This is a coordinated effort that includes policies and practices that affect spatial organisation of people and activities that they are involved in. The process of spatial planning entails the creation of a plan that is relevant to a specific geographic area and must be implemented effectively. Spatial planning in South Africa is done with consideration given to financial, regulatory and institutional environments. One of the aims of modern spatial planning is to create a balance between economic development, spatial equality and sustainable environmental practices and management. Although spatial planning in South Africa is mainly focused on long term development goals, it is still important to connect these goals with immediate developmental needs. An Act that assists in this type of planning is the Physical Planning Act 125 of 1991.[137] This Act, as amended by the Regional and Land Affairs General Amendment Act 89 of 1993, was designed

> (t)o promote the orderly physical development of the Republic, and for that purpose to provide for the division of the Republic into regions, for the preparation of national development plans, regional development plans, regional structure plans and urban structure plans by the various authorities responsible for physical planning, and for matters connected therewith.

While South African planning law remains substantially patchy until today, the number of sectoral legislation that operates parallel to the aforementioned planning laws has increased in terms of newer legislation and (in parts) more effective administration.

11 Public Environmental Law

The Conservation of Agricultural Resources Act 43 of 1983 ("CARA") promotes the conservation of soil, water resources and vegetation. The act provides management plans that guide to eradicate weeds, alien invasive vegetation and promote and protect the integrity of indigenous vegetation, as well as guidelines to the management and protection of soil. CARA specifically states that the degradation of the agricultural potential of soil is illegal. Under the Section 6 (2)(a) of the Act, control measures may be prescribed by the relevant Minister for the cultivation of virgin soil, Section 6 (2)(o), the construction, maintenance, altercation or removal of soil conservation works or other structures on land. Section 8 and 12 (1) of the Act further stipulate schemes or appropriation of funds by Parliament and management measured and funds by other land users as deemed fit by the relevant Minister(s) for the construction of soil conservation works, the reparation of damage to the natural

[137] Kidd (2011), p. 212.

agricultural resources or soil conservation works due to various specified forces/impacts, and the planting and cultivation of particular crops which improve soil fertility or counteract the vulnerability of soil erosion.

The National Environmental Management Act's ("NEMA") Environmental Impact Assessment ("EIA") Regulation 326 of December 2014 (as amended on 7 April 2017), makes provision for the protection of soil or land prior to the commencement of any development. The provisions are made through the specific listed activities that may be triggered by a specified development and an environmental impact assessment may need to be conducted in order to minimise and manage those impacts on the environment. These listed activities that concern soils are as follows: Government Notice Regulation ("GNR") 327 Activity 27 that states that the clearance of 1 ha or more, but less than 20 ha of indigenous vegetation; GNR 325 Activity 24 makes provision for the extraction or removal of peat or peat soils, including the disturbance of vegetation or soils in anticipation of the extraction or removal of peat or peat soils. GNR 324 Activity 12 speaks on the clearance of an area of 300 m^2 or more of indigenous vegetation within critically endangered or endangered ecosystems, biodiversity areas, or land that was zoned as conservation areas.

National Environmental Management: Biodiversity Act 10 of 2004, supports the conservation of plant and animal biodiversity, and the soil and water that it depends on. The act promotes the management and conservation of South Africa's biodiversity within the framework of NEMA in order to protect species and ecosystems, and equitable and sustainable use of natural resources.

National Environmental Management: Protected Areas Act 57 of 2003, makes provision for the protection and conservation of ecologically viable areas that represent South Africa's biological diversity and natural land and seascapes, and calls for the establishment of national, provincial and local protected areas and their management in accordance with national norms and standards. Amendment 31 of 2004 of the Act, supports the conservation of soil, water and biodiversity.

The National Environmental Management: Waste Act 59 of 2008, ("NEMWA") stems from NEMA, and is the act is primarily concerned for waste management in order to protect health and the environment. NEMWA provides reasonable measures for the prevention of pollution and ecological degradation and for securing ecologically sustainable development. Under section 73(1)(a) of NEMWA, the national norms and standards for 'The Remediation of Contaminated Land and Soil Quality' in the Republic of South Africa are stipulated and took effect in 2013. These norms and standards focus on the contamination of through various land use activities that have an impact on soil quality, human health and water resources; from this contamination the norms and standards them outline remediation measures that assist with the management of the contaminated site to ' prevent, minimise, or mitigate' the damage to human health or the environment.

Further, when addressing public environmental law, the agricultural sector must be addressed as a result of its relation to and impact on the environment. Agriculture plays an important role in the economy of South Africa; therefore, it is necessary to have legislation with provisions dealing with agriculture. The Conservation of

Agricultural Resources Act 43 of 1983 ("CARA"), aims to provide for the conservation of the natural agricultural resources of the South Africa by maintaining the production potential of land, by combating and preventing erosion and weakening or destruction of the water sources, and by protecting the vegetation and combating weeds and invader plants.[138]

The Subdivision of Agricultural Land Act 70 of 1970 is another act contributing to the provisions concerning agriculture. This act provides that agricultural land may not be subdivided unless the Minister of Agriculture, Forestry and Fisheries has consented to the subdivision. This act was supposed to be replaced with the Subdivision of Agricultural Land Repeal Act 64 of 1998, however, this has not been put into operation yet. Further, the act has not yet been brought into alignment with the 1996 constitutional framework, leaving much to still be determined. New draft legislation has been developed to replace the Act; however, the finalisation and promulgation is yet to have happened.[139]

Along with legislation, there are a number of draft bills that address agriculture, both published and unpublished. Two unpublished draft bills dealing with the protection of agricultural resources and agricultural land, being the Draft Conservation of Agricultural Resources Amendment Bill of 2013 and the Draft Sustainable Utilisation of Agricultural Resources Bill of 2003, both aimed to provide for the shortcomings of CARA. Additionally, there are two published draft bills dealing with the protection of agricultural land. These are the Draft Preservation and Development of Agricultural land Bill and the Draft Regulation of Agricultural Land Holdings Bill.[140]

Other than relevant legislation and draft bills, the Constitution also deals with agriculture in Schedule 4 (Part A), where agriculture is defined and placed as a functional area of concurrent national and provincial legislative competence.

12 Cross-Cutting Issues

The NEMA EIA Regulations stipulate that the need for an EIA in terms of sections 24(2) and 24D of NEMA arises in the listed activities under Government Notice Regulation 326 of December 2014 (as amended on 7 April 2017). Any listed activity that is triggered required the implementation of an environmental impact assessment prior to the commencement of the proposed development.

As per section 24 NEMA, all activities listed or specified by the Minister of Environmental Affairs must be granted an environmental authorisation before they are permitted to commence. Such for instance is the construction of a coal-fired power station one of these listed activities. Therefore, a company that for example

[138] Meyer et al. (2018), p. 335.

[139] Ibid.

[140] Meyer et al. (2018), pp. 338–339.

intends to build a new coal-fired power station in South Africa requires an environmental authorisation, which must be issued by the Chief Director of the DEA. Section 24(1) of NEMA stipulates that the environmental impact of a listed activity must be considered, investigated, assessed, and reported to the competent authority responsible for deciding on an environmental authorisation. Pursuant to section 24O (1) of NEMA, the competent authorities are required, when deciding on an application for an environmental authorisation, to take into account all relevant factors, including any pollution, environmental impact or environmental degradation likely to be caused by the approval or rejection of the application. This was subject to a case decided on 8 March 2017 by the Gauteng High Court,[141] the so-called Thabametsi-case.[142]

In the Thabametsi-case, the court in its decision came to the conclusion that it would suspend a grant of the environmental authorisation until a full investigation and consideration of the climate change impacts assessment report of the proposed coal-fired power station had been conducted. Although the decision was later on appealed, the case of first instance was a good example for the question of whether the authorities should take account of the climate objectives of the UNFCCC and the Paris Agreement in their domestic decision-making processes and what relevance they have within the national legal systems. One key question of the case was whether national administrative regulations must be interpreted in such a way that they are consistent with the climate objectives of international law.[143] After the High Court remitted the matter back to the Minister, the Minister decided to confirm the environmental authorisation. The Minister stated that she had noted that the Thabametsi Power Station will result in significant GHG emissions and will therefore have climate change impacts. However, she pointed out that, although the environmental and social costs associated with the proposed power plant are high, this is not necessarily a critical deficiency, as the additional energy production capacity will bring benefits to the country. She argued that she has carefully considered all relevant factors, including climate change, but came to the result that the environmental authorisation should nevertheless be granted.[144]

Notwithstanding the outcome of the aforementioned case, the EIA generally provides detailed information and the degree of severity of the potential impacts of the development and management interventions, and or decisions on whether the development proceed. It is ultimately the decision of the competent environmental authority, prescribed by government under NEMA, to give the final decision on whether the proposed development is granted environmental authorisation. In accordance with the NEMA EIA Regulations of December 2014, (as amended on 7 April 2017), a public participation process must be carried out as part of the EIA process. An Environmental Assessment Practitioner must ensure that the public participation

[141] [2017] ZAGPPHC 58 (2017) 65662/16.

[142] This was arguably the first climate litigation case on the African continent; Ruppel (2018).

[143] Saiger (2020), p. 49.

[144] See Minister of Environmental Affairs, Appeal Decision, Reference LSA 142346.

process for the EIA is adequate with provision made for meetings and consultations if necessary. Public concerns that arise are to be addressed and made available to the competent authority prior to the issuing of the environmental authorisation. As part of the environmental assessment process the, reports and information regarding the process are to be made available to the public. Provisions are also to be made available to ensure that the information is accessible through various mediums, as well as by request. The case of *Fuel Retailers Association of Southern Africa v Director-General Environmental Management, Department of Agriculture, Conservation and Environment, Mpumalanga Province* found that development must take into account the needs of society as expressed by members of the public, who must be given the opportunity to participate in the decision-making process.[145]

Chapter 2 of the National Appeal Regulations published in GN R993 on 8 December 2014 in terms of Section 44(1)(a) of NEMA, and as amended in GN R205 on 12 March 2015, which prescribes the appeal procedure to be followed. The regulation stipulates that should a person wish to appeal any aspect of an environmental authorisation decision, an appeal must be submitted to the Appeal Administrator by the Appellant, and a copy thereof must be submitted to the Applicant, any registered Interested and Affected Parties ("I&AP") and any Organ of State that has interest in the matter, within 20 days from the date that the notification of the decision was sent.

A further cross-cutting issue is that of foreign investments. Foreign investments establish an important tool for the generating of capital inflow, as well as economic growth and development, particularly in developing countries.[146] Due to the impact that foreign investment can have on a developing country, it is imperative that such investments are regulated properly. If these investments are not regulated properly, there will be a reluctance by foreign investors to invest where their interests are not protected and promoted. However, the regulation of foreign investments should ideally be such that they promote and protect the interests of both the investor and the country in which they are investing.[147]

When considering the legal rules that pertain to foreign investors when they invest in South Africa, it is important consider the legislation that deals directly with investments. This legislation will give clarity to position of the investor and the country in which they wish to invest. While the respective investment will be subject to the particular agreement related to that investment, compliance with the legislation is also required. The relevant legislation dealing with the regulation of investments in South Africa is the Protection of Investment Act 22 of 2015. This Act defines what an investment is for the purpose of its application. The purpose of the Act is to protect investments in accordance with and subject to the Constitution, in a manner that balances public interest and the rights and obligations of the investor, to affirm South Africa's sovereign right to regulate investments in the public interest

[145] 2007 6 SA 4 (CC) paras 60 and 70.

[146] Picker (2018), p. 17.

[147] Mhlongo (2019), p. 2.

and to confirm the Bill of Rights and the laws that apply to all investors and their investment in the country (section 4 of the Protection of Investment Act 22 of 2015). The Protection of Investment Act, in section 7(1), stipulates that all investments must be established in accordance with the laws of the Republic. This indicates that there is a requirement that investments made by foreign investors, as well as the entity that they are investing into in South Africa, must comply with the relevant public law applicable to their investment and relationship. The legal protection of property under the Protection of Investments Act is subject to section 10, which stipulates that investors have the right to property in terms of section 25 of the Constitution, just as South African citizens do.[148]

In terms of unfair or illegal land acquisition, there are two sides. The first side is unfair or illegal land acquisition by a foreign investor in South Africa and the second is unfair or illegal land acquisition from the foreign investor. On the side of acquisition by the investor, due to foreign investors having to comply with South African public law, any land acquisition that they may be involved in would have to legally attained in order for it to be legally valid. On the side of acquisition from the investor, the position is slightly different. The Protection of Investment Act allows government to take any measures, in accordance with the Constitution and relevant legislation, to redress historical inequalities, to uphold the values and principles of the Constitution, to promote cultural heritage, to foster economic development and to protect the environment.[149] This land acquisition from the investor by the state would amount to expropriation, which is allowed as stated in section 25(2) of the Constitution. This expropriation may only be done if it is for public purpose or in the public interest and is subject to compensation (section 25(2)(b) of the Constitution). If the land acquisition from the foreign investor is done in compliance with section 25 of the Constitution, then it will not amount to unfair or illegal land acquisition.

However, if the land acquisition either by or from the foreign investor is done is such a way that it does not comply with the relevant legislation and/or is not in line with the Constitution, then this will amount to unfair or illegal land acquisition. In such a case, a dispute will more than likely arise, and this will then need to be settled. The parties can go about such a settlement by either following the dispute procedure set out in their investment contract, if any, or they can follow the rules set out in section 13 of the Protection of Investment Act or take the matter to international arbitration.[150]

There are a number of different forms of tax that need to be paid in terms of South African law. These forms of tax are not exclusive to South African citizens but must also be paid by foreign investors. The way in which South African tax law is structured sees foreign investors being fairly taxed in that they are taxed similarly to South African citizens. Tax on the income of non-South African residents is

[148] Picker (2018), p. 28.
[149] Mhlongo (2019), p. 12.
[150] Schlemmer (2009), p. 743.

source-based, meaning that any income from a source within South Africa is taxed. The amount as which they are taxed is a set amount, ensuring fairness for all foreign investors in terms of much they are taxed. In terms of Value Added Tax (VAT), if the foreign investments is one into a company that sells goods or provides services, it must register as a VAT vendor and pay the set amount of 15% VAT to the South African Revenue Service, unless the supplies are exempt or zero-rated.[151] Capital gains tax will be levied upon a foreign investor in the case of them disposing of their assets situated in South Africa. Further, South Africa has double taxation agreements with most of its trading partners, which prevents the investors from being taxed twice.[152] This analysis indicates that foreign investors who invest in South Africa are fairly taxed on their investment.

South Africa's Low Emission Development Strategy (SA-LEDS) dated February 2020, sets out a direction in meeting a low carbon emission development pathway to meet our commitments to the international community and address our developmental agenda/priorities and needs.[153] This strategy is supposedly 'a living document, the beginning of South Africa's journey towards ultimately reaching a net zero carbon economy by 2050'. Therein it is stipulated as follows regarding the Carbon Tax:

> The Carbon Tax Act (15 of 2019) was brought into effect from 1 June 2019. The Act gives effect to the polluter pays principle and aims to price carbon by internalising the negative costs of emitting GHGs. The tax rate is set at R120 per tonne of CO2-eq produced. To allow businesses time for transition, a basic tax-free allowance of 60% will initially apply to all emissions, with a number of further allowances depending on the activities. These provide for maximum tax-free allowances of between 60 and 95%. The allowances include those on fugitive emissions; for trade exposure; for performance above the industry benchmark; for purchasing offsets; and for voluntarily being allocated a (non-enforceable) carbon budget. The allowances bring the effective tax rate to between ZAR 6 and ZAR 48 per tonne.

13 Legal Requirements on Environmental Monitoring

Environmental monitoring is done through the process of integrated environmental assessment and management. Environmental assessment was formalised in South Africa between 1996 and 2004, with formal environmental impact assessment (EIA) legislation emerging in terms of the 1997 EIA Regulations.[154] The EIA regulations in terms of Chapter 5 (Integrated Environmental Management) of NEMA only appeared in April 2006. These regulations deal with the powers and responsibilities of competent authorities, and the requirements of environmental

[151] Brand South Africa (2013).

[152] Ibid.

[153] Cf. https://www.environment.gov.za/sites/default/files/docs/2020lowemission_developmentstrategy.pdf, accessed 26 August 2021.

[154] Kidd et al. (2018), p. 1220.

assessment practitioners.[155] The EIA involves an assessment of potential impacts of the activity on the environment, whether and to what extent the impacts on the environment can be mitigated and whether or not there are significant concerns and impacts that require further investigation into the activity in concern. The regulations further provide time requirements and list the kinds activities that require EIA.[156] In 2010, new EIA regulations, along with three lists of activities for which EIA is required, were published and came into effect on 2 August 2010, along with the Environmental Management Framework Regulations GN R665.[157] The EIA regulations were expanded in order to avoid any detrimental impacts on the environment, or where it cannot be avoided, to ensure the impacts are mitigated and managed to acceptable levels, and to optimise positive environmental impacts.[158]

In terms of the South African context, the lack of environmental monitoring and performance evaluation amidst the wealth of practice can be considered to be one of the main deficiencies and missed opportunities for environmental assessment. There is very little EIA follow up, enforcement and compliance monitoring. Efficiency problems mainly lie with the inability of the government to implement the legal mandate prescribed by legislation. Enforcement entities can take action in terms of different sections under Chapters 5 and 7 of NEMA in response to detected non-compliances, including punishment of the violator and actions taken to ensure that harm done to the environment is halted or rehabilitated. The agency responsible for the enforcement of South Africa's environmental laws is the Environmental Management Inspectorate,[159] also known as the "Green Scorpions".[160] The competences required of the enforcement entities are a knowledge of environmental law, the ability to ensure compliance with the law and the ability to ensure rehabilitation takes place if necessary. As a result of the lack of experienced staff, as discuss below, the competencies of the enforcement entities often lack in conjunction.

When dealing with disputes or investigations into matters that concern the environment, it is important to be able to obtain the necessary information needed in order to resolve such issues. However, not all of the needed information may be readily ascertainable and may even be withheld by an opposing party. The National

---

[155] Kidd (2011), p. 248.

[156] Kidd (2011), p. 249.

[157] Government Gazette 33411 of 30 June 2010.

[158] Kidd (2011), p. 251.

[159] EMI must see to it that environmental legislation is followed and enforced. The EMIs have the powers to: investigate: question witnesses, inspect and remove articles, take photographs and audiovisual recordings, take samples and remove waste; inspect: enter premises to ascertain whether legislation is being followed and seize evidence of criminal activity; enforce: search premises, containers, vessels, vehicles, aircraft and pack animals; seize evidence and contraband; establish road blocks and make arrests; administrate: issue compliance notices and admission of guilt fines. The EMIs are not empowered to prosecute cases in court. All cases continue to be handed over to the National Prosecuting Authority (NPA) for prosecution. The EMIs work closely with prosecutor's country wide to ensure the successful prosecution of offenders.

[160] Kidd et al. (2018), p. 1272.

Environmental Management Act originally contained provisions in section 31 relating to the access to environmental information, however, the majority of this section was deleted by section 14 of the National Environmental Laws Amendment Act 14 of 2009. As a result of this deletion, NEMA only considers access to information in section 2(4)(k), which requires that access to information must be provided for in accordance with the law. Due to the minimal consideration given to importance of access to environmental information, the most effective way to ensure access to environmental information is through the use of the Promotion of Access to Information Act 2 of 2000 ("PAIA").

There has been argument for data regarding environmental management to be openly accessible. This argument notes the centrality of access to information as a precondition to public participatory governance and finds that it could be argued that open data could enhance and support public participatory governance structures.[161] This idea of 'open data' would promote the right of access to information and assist in the acquisition of environmental data needed in any given case. There exists a desperate need for improved government environmental assessment capacity both in terms of the number of personnel and their levels of experience. Although there are indications that the levels of training are improving, this cannot substitute the need for experience in this field of execution of the law. Further, due to environmental assessment still being fairly 'new' in South Africa, there is insufficient technical equipment and a lack of personnel who are able to operate such equipment as well as a lack of personnel who have experience in enforcing environmental monitoring. The future of environmental assessment requires a sufficient number of well trained and experiences officials as well as an increase in availability of technical equipment needed in order to conduct environmental monitoring.[162]

14 Recommendations and Conclusion

Sustainable land management describes "the stewardship and use of land resources, including soils, water, animals and plants, to meet changing human needs while simultaneously assuring the long-term productive potential of these resources and the maintenance of their environmental functions."[163] By introducing soil management practices on an extensive scale, South African soils can be used as a carbon sink to help achieve a net removal, depletion or mitigation of overall CO2 levels within the atmosphere.[164] Exercising sustainable soil management practices can help to tackle food security and climate change challenges faced by agricultural

[161] Adams and Adeleke (2016), p. 82.

[162] Kidd et al. (2018), p. 1276.

[163] IPCC (2019), p. 100.

[164] Paustian et al. (2019), p. 568.

systems.[165] A close collaboration between the private and public sectors will be essential to achieve this.

While South Africa is one of the most unequal societies in the world due to past apartheid policies that impoverished the majority black population, the major current national development priorities are to reduce poverty, unemployment and inequality.[166] This seems much in line with the most recent 2020 SDG Progress Report, which finds that "continued unevenness of progress" still requires significant improvement. And while soil protection wasn't explicitly mentioned in this report, "progress was stalled or reversed on the number of people suffering from hunger, the rate of climate change, and increasing inequalities".[167] The report also enumerates impacts of the COVID-19 pandemic on groups of people and operations around the world, and how those effects are "imperilling progress" towards the SDGs. It further reports that countries stand to be hit hardest due to fragile health systems, limited social protection coverage, limited financial and other resources, vulnerability to shocks, and dependence on international trade. The UN Secretary-General therefore urged that recent gains need to be protected as much as possible and a truly transformative recovery from COVID-19 must be pursued, one that reduces risk to future crises and brings much closer the inclusive and sustainable development required to meet the goals of the 2030 Agenda and the Paris Agreement on Climate Change.[168]

South Africa generated its first Low Emission Development Strategy (SA-LEDS) 2050, generated after the adoption of the Paris Agreement. Through submitting this document to the UNFCCC South Africa reiterated its commitment to achieving the Paris Goals and the Sustainable Development Goals (SDGs). When it comes to Agriculture, Forestry and Land Use (AFOLU) in South Africa, the existing policy framework already makes provision for a selection of actions that will support emissions reductions and expansion of the carbon sink. These include promotion of land afforestation, conservation agriculture farming methods and agroforestry. Further interventions in this sector are currently being explored by government.

South African national development imperatives have resulted in calls for radical economic transformation and accelerated land reform. It is therefore clear that SDG 15 that focusses on life on land and target 15.3 that aims at achieving land

---

[165] IPCC (2019), p. 500.

[166] Cf. https://knowledge.unccd.int/sites/default/files/ldn_targets/South%20Africa%20LDN%20TSP%20Country%20Report.pdf, accessed 20 May 2020.

[167] Progress towards the Sustainable Development Goals Report of the Secretary-General (2020), online available at https://sustainabledevelopment.un.org/content/documents/26158Final_SG_SDG_Progress_Report_14052020.pdf, accessed 20 May 2020.

[168] Cf. http://sdg.iisd.org/news/un-secretary-general-releases-2020-sdg-progress-report/?utm_medium=email&utm_campaign=SDG%20Update%20-%2020%20May%202020&utm_content=SDG%20Update%20-%2020%20May%202020+CID_033f37b52fdb4016cbee1f4db6346058&utm_source=cm&utm_term=Read, accessed 20 May 2020.

degradation neutrality by 2030 will need to come even more to the fore in national development programmes.[169]

> Desertification, land degradation and drought are intricately linked to food and water insecurity, poverty, urbanisation, climate change, and biodiversity loss. Desertification, land degradation and drought is therefore amongst the most critical sustainable development challenges (...) as billions of tons of fertile top soil are lost every year with far reaching implications for poor and marginalised populations who depend on land for livelihood. Land degradation has the potential to undermine government efforts of poverty eradication, fighting unemployment and enhancement of economic opportunities especially in rural areas.[170]

Therefore, legal measures in relation to soil are conditional, such as legislation and policies that are necessary in order to enforce and control the protection of soil and to minimise the environmental impact with the accompaniment of social education and sensitisation. Although South Africa has applicable legislation, like that of CARA, there are improvements that need to be made in order to have a legal system that promotes and achieves soil protection. For example, CARA prescribes the creation of soil conservation committees, however, unfortunately a number of these committees have yet to start functioning, which leaves them redundant until they begin to do so. This highlights that action needs to be taken in order for the legislations to be fully functioning. Legislation without action, enforcement and regulation is of little value. A further setback with this Act is that it manifests persuasion rather than coercion and there is a lack of manpower that is necessary to enforce the Act. With only a small number of inspectors throughout the country, it is not possible to effectively enforce the Act in all areas of South Africa. Progressing from this is the fact that once a soil conservation issue has been discovered, there are a lack of technicians that can assist farmers in remedying the situation.[171] As a result of this it is difficult to enforce the legislation promoting the environmental conservation of soil. In order for the legislation and policies that pertain to soil protection to be effective, there needs to be implementation of what is envisaged and prescribed within the legislation and policies. Ultimately, it would be desirable to have legislation that solely dealt with the protection and management of soil in South Africa. Although South Africa's soil protection legislation and policies are not currently at a satisfactory level of environmental conservation that is necessary, it is a positive measure towards effecting change.

When considering improvements that need to be made to legislation, it can also be said that the legislation relating to foreign investors in South Africa could also be developed. There is a need for legislation that clearly sets out how the foreign investor needs to comply with South African public law, as well as a need for

[169] Cf. https://knowledge.unccd.int/sites/default/files/ldn_targets/South%20Africa%20LDN%20TSP%20Country%20Report.pdf, accessed 20 May 2020.

[170] Department Environment, Forestry and Fisheries (2016) South Africa marks the World Day to combat desertification, available at https://www.environment.gov.za/mediarelease/southafrica_markswdcd2016, accessed 20 May 2020.

[171] Meyer et al. (2018), p. 335.

legislation that prescribes how foreign investors can acquire land in South Africa. These improvements should also highlight and clarify the social and environmental responsibilities of the foreign investor, along with the consequences that will apply should they not comply with these responsibilities or should they cause damage to the environment.

Through examining the relevant legislation and policies that pertain to soil protection, it is clear that improvements with regard to enforcement and monitoring of such legislation and policies, as well as access to environmental information are needed. The implementation and enforcement of legislation effectively in the context of limited resources is a challenge which is a concern for South African environmental law in general.[172] As discussed above, there are an inadequate number of designated enforcers of the provisions. Further, a number of measures that are set out by the legislation, like that of the soil conservation committees, have yet to be established. In order for any sort of legislation and policy system to be successful, their provisions need to be enforced and monitored, which requires manpower with the required expertise. There is a clear need for an increase in the number of enforcement officers who have been educated with the relevant information needed in order to carry out their enforcement duties. Further, an increase in the number of people with the expertise to monitor, as well as the number of technicians that have the expertise to resolve potential environmental issues with regards to soil protection, is needed. Additionally, in order to ensure that the information needed in order to enforce and monitor the adherence of legal provisions, access to relevant environmental information needs to be more easily and readily ascertainable.

Progression in terms of land tenure rights is also necessary in order to enable soil protection as a result of the effect that land tenure can have on the soil. Human influence on the land and natural resources, such as soil, is accelerating as a result of the growth in population of South Africa, as well as the increase of food requirements.[173] Food security and food availability is highly dependent on access to and the productivity of land, however, population increases, along with soil degradation, and changes in climatic conditions suggest that productivity increases and not an increase in cultivated areas are the only viable option to sustained future food production.[174] It has been widely recognised that mankind plays a significant role in soil degradation through their interactions with the natural environment, especially with regard to land-use practices like agriculture.[175] Land-use change is displayed in the change of land-cover and this change is a key component of the global environmental change that is being experience currently, which is affecting the climate, biodiversity, and ecosystem, resulting in land-use decisions being impacted.[176] For example, increased hardened areas associated with urban

[172] Kidd (2011), p. 141.

[173] Kanianska (2016), p. 4.

[174] Sandrey (2019/20), p. 134.

[175] Cooper (1996), p. 1.

[176] Kanianska (2016), p. 6.

development increase runoff during rainfall, which accelerates erosion and runoff downstream of the urban catchment.

The way land is used plays a great role on the quality of soil that is present. Land tenure, therefore, plays a role in soil conservation. This indicates that during the process of land tenure, and when considering land tenure rights, it is necessary to consider soil protection. It is important here to consider land tenure rights and their interaction with soil protection. Currently, official information in respect of communal land rights is insufficient and unreliable due to the present registration system of land rights not providing for the registration of communal land rights, but only providing for the registration of common law land ownership rights. This then has an impact on the general land administration system of South Africa. This administration system is an integrated process of determining, recording and disseminating information on the tenure, value and usage of land, with the aim of developing a suitable land management and development system and policies. It is imperative that the land rights administration system deals with not only formal land rights, but also informal land rights in order to have complete information regard the tenure, value and usage of all land in South Africa. It is not possible to have reliable information with regards to land usage and tenure with the aim of successful administration and sustainable development if only the formal land tenure rights are considered. Communal land rights have been exercised for centuries and they encompass a substantial amount of land, to not recognise them as registerable land tenure rights would see inadequate information amount the usage of all South African land.[177]

The recognition of individual land tenure rights and not those of communal land tenure rights with result in a lack of clarity and control over how the land is used. This would find sizeable areas of land being used in a way that can lead to further soil degradation, but would not be under administration, which is necessary in order to implement soil protection measures that are prescribed by relevant legislation and policies. Furthermore, it has been shown that an effective land registration system is one which wards off land disputes. Equal protection for both land tenure systems is important in order to ensure that they are both administered, which would see a land administration system that could assist in the protection of soil from incorrect or environmentally harmful land use. There is a need for the formalisation of land tenure in respect of both individual and communal land rights, which would see the integration of the two land tenure systems, resulting in a unitary system that can be used to ensure all land in South Africa is administered and regulated, ensuring land protection and in particular soil protection.[178]

The current lack of suitable land policy has negative consequences on people living in the previous homelands, which are characterised by tenure insecurity and a lack of administrative support for agricultural activities, which it directly related to the protection and sustainability of South Africa's soil.[179] Although legislation

[177] Pienaar (2013), p. 21.

[178] Tlale (2018), p. 267.

[179] Pienaar (2013), p. 27.

cannot independently secure land tenure, it goes without saying that a legal framework is necessary for any registration programme to function and bring about the desired outcomes, like the administration of land usage and protection of natural resources.[180]

Although there are many improvements that can be made to the legal system in order to enhance the protection of soil in South Africa, education in soil law is also important in order to stress the importance of soil protection measures within the legal system. The training of lawyers and law students in the subject of soil protection law can promote the need for a legal system that has secure and effective soil protection measures. This training would see an increase in legal personnel who have the relevant expertise and knowledge to ensure that soil protection law is complied with and improved upon, as well as enforced and monitored. An increase of lawyers and law students who have been trained in soil protection law, would see an increase in the number of advocates for the growth, improvement and primacy of soil protection law and ultimately an increase in related litigation cases.

In this light, the example of the Thabametsi-case is encouraging in that it emphasises potential environmental consequences or impacts of a planned operation, which needed to be assessed before creating further damage. This assessment required the court a quo to look into the future and interpret NEMA in accordance with the constitution and international law regarding the effects of climate change. Although not immediately related to soil protection, the case not only demonstrates that the enforcement of international law (such as the Paris Agreement) most effectively lies within national law and is dependent on its interpretation in line with the constitution. It also reflects the constant obligation of the state to take into account future and potential effects of climate change when granting operating authorisations under national legislation, which—with the persuasive power of analogy—could perhaps similarly apply to the context of soil protection in the interest of future generations.

Interestingly, in the recent 2021 case of *Maxrae Estates (Pty) Ltd v The Minister of Agriculture, Forestry and Fisheries & Another*[181] the South African Supreme Court of Appeal (SCA) upheld an appeal against an order of the Gauteng High Court, Pretoria, in terms of which an application brought by the appellant, Maxrae Estates, for review of a decision by the Minister of Agriculture, Forestry and Fisheries, refusing subdivision of a Farm, was dismissed. The application for permission to subdivide was rejected by the delegate of the Minister mainly, on the ground that approval thereof would encourage establishment of sub- economic farming units thus endangering food security. An appeal to the Minister against that refusal failed for the same reason. The Minister was also of the view that the smaller subdivided unit would 'not be resistant in the long run considering the impact of climate change'. The appellant brought an application in the high court challenging

[180] Tlale (2018), p. 266.

[181] *Maxrae Estates (Pty) Ltd v Minister of Agriculture, Forestry and Fisheries & Another* (407/2020) [2021] ZASCA 73 (9 June 2021).

the decision of the Minister on the grounds that the decision was irrational, unreasonable and arbitrary because the Minister took into account irrelevant factors, ignored relevant factors, and reached conclusions which were not supported by evidence. The high court dismissed the review application. It found that the Minister had exercised his wide discretion properly, in line with the purpose of the relevant laws and had considered all the evidence placed before him. In upholding the appeal against the decision of the high court, the SCA held that the high court erred in its understanding of the Minister's wide discretionary powers. It held that wide Ministerial discretion was no licence for disregard of factors that were relevant to the decision he was required to make. The SCA found that the Minister ignored the evidence and conclusions set out in the expert reports submitted by Maxrae and instead made vague conclusions which included matters in respect of which there was no evidence before him. Also, rather than making an independent assessment of the information placed before him, the Minister extracted conclusions made in a memorandum prepared for him to sign by the Deputy Director of the department. The SCA ordered that the matter revert to the Minister for fresh consideration of Maxrae's appeal.[182] In the judgement, Dambuza JA explicitly discusses the Minister's reliance on climate change to achieve his conclusions; whether the Minister failed to apply his mind to relevant factors; and whether the decision was irrational:

> [...] his decision comprised vague conclusions which included matters in respect of which there was no evidence before him. His conclusion that 'the creation of a small portion will not be sustainable viable and will not be resistant in the long run considering the impact of climate change' was one such example.[183]

References

Adams R, Adeleke J (2016) Assessing the potential role of open data in south African environmental management. Afr J Inf Commun 19:79–99

Braimoh AK, Vlek PLG (2008) Impact of land use on soil resources. In: Braimoh AK, Vlek PLG (eds) Land use and soil resources. Springer, Dordrecht

Brand South Africa (2013) South African Tax for Investors https://www.brandsouthafrica.com/investments-immigration/business/investing/regulations/tax-investors. Accessed 10 Mar 2020

Brand South Africa (2018) SA's key economic sectors https://www.brandsouthafrica.com/investments-immigration/business/investing/economic-sectors-agricultural. Accessed 16 Feb 2020

CEIC Data (2020) South Africa Gross National Product https://www.ceicdata.com/en/indicator/south-africa/gross-national-product. Accessed 16 Feb 2020

Central Intelligence Agency Database (2020) The World Fact Book: South Africa. https://www.cia.gov/library/publications/the-world-factbook/geos/sf.html. Accessed 12 Feb 2020

---

[182] Cf. Media Statement from the SCA, http://www.supremecourtofappeal.org.za/images/judgment_2021/sca2021-73ms.pdf, accessed 26 August 2021.

[183] *Maxrae Estates (Pty) Ltd v Minister of Agriculture, Forestry and Fisheries & Another* (407/2020) [2021] ZASCA 73 (9 June 2021), para 15.

Cooper A (1996) Soil conservation policy in South Africa, 1910–1992: the 'human dimension" DPhil thesis. University of Natal, South Africa
De Wet E (2015) Reception of international law in the SA legal order: an introduction. In: De Wet E, Hestermeyer H, Wolfrum R (eds) The implementation of international law in Germany and South Africa. Pretoria University Press, pp 23–50
Department of Agriculture (2009) Implementation framework for the LandCare Programme. Department of Agriculture, Pretoria
Department of Environment and Tourism (2004) National Action Programme Combating Land Degradation to Alleviate Rural Poverty. Department of Environment and Tourism, Pretoria, pp 1–84
Department of Environmental Affairs (2012) National Climate Change Response, White Paper. Department of Environmental Affairs, Pretoria, p 51
Department of Environmental Affairs (2011) 2nd South Africa Environment Outlook. A report on the state of the environment. Executive Summary. Department of Environmental Affairs, Pretoria, p 60
Department of Environmental Affairs (2018) Draft National Climate Change Adaptation Strategy. Department of Environmental Affair, Pretoria, South Africa
Department of Environmental Affairs and the South African National Biodiversity Institute (2016) Strategic Framework and Overarching Implementation Plan for Ecosystem-Based Adaptation (EbA) in South Africa; 2016-2021. Department of Environmental Affairs, Pretoria, South Africa
Du Plessis H (2019) Legal Pluralism, uBuntu and the Use of Open Norms in the South African Common Law of Contract. Potchefstroom Electr Law J 22:1–37
Du Plessis W & the Traditional Authorities Research Group (1999) The role and future of traditional leaders in South Africa. Koers J 64:295–324
Dugard J (2018) Dugard's international law: a South African perspective, 4 edn. JUTA
Fayiga AO, Ipinmoroti MO, Chirenje T (2018) Environmental pollution in Africa. Springer 20:41–73
Ferreira G, Ferreira-Snyman A (2014) The incorporation of public international law into municipal law and regional law against the background of the dichotomy between monism and dualism. PER 17:1471–1496
Garland GG, Paterson DG, Newby TS, Laker MC, Hoffman MT (2000) Soil degradation. In: Hoffman MT, Todd S, Ntshona Z, Turner S (eds) A national review of land degradation in South Africa. South African National Biodiversity Institute, Pretoria, pp 69–107
Glazewski J (2000) Environmental law in South Africa. Durban, South Africa
Goosen J, Fitchett A (2020) Success factors for Urban Brownfield developments in Johannesburg, South Africa. Urban Forum, 65
Grain SA (2015) Unpacking the various forms of land ownership. https://www.grainsa.co.za/unpacking-the-various-forms-of-land-ownership. Accessed 21 Feb 2020
Gregory AS, Ritz K, McGrath SP, Quinton JN, Goulding KWT, Jones RJA, Harris JA, Bol R, Wallace P, Pilgrim ES, Whitmore AP (2015) A review of the impacts of degradation threats on soil properties in the UK. Soil Use Manag 31:1–15
Grist J (2018) The Climate Change Bill. Accountancy SA, 53–55
Hamza MA, Anderson WK (2005) Soil compaction in cropping systems. A review of the nature, causes and possible solutions. Soil Tillage Res 82:121–145
Havenga P, Garbers C, Havenga M, Schulze WG, van der Linde K, van der Merwe T (eds) (2002) General principles of commercial law, 4th edn. Juta Law, Lansdowne
Hillel D (1992) Out of the earth: civilization and the life of the soil. University of California Press, Berkeley, CA
Humby T, Kotze LJ, du Plessis A (eds) (2012) Introduction to law and legal skills in South Africa. Jurisprudence Oxford University Press
IPCC/Intergovernmental Panel on Climate Change (2014) Climate change 2014: synthesis report – contribution of working groups I, II and III to the Fifth Assessment Report of the Intergovernmental Panel on Climate Change. In: Pachauri RK, Meyer LA (eds) IPCC, Geneva

IPCC/Intergovernmental Panel on Climate Change (2018) Special Report on the impacts of global warming of 1.5°C above pre-industrial levels and related global greenhouse gas emission pathways, in the context of strengthening the global response to the threat of climate change, sustainable development, and efforts to eradicate poverty. In: Masson-Delmotte V, Zhai P, Pörtner H-O, Roberts D, Skea J, Shukla PR, Pirani A, Moufouma-Okia W, Péan C, Pidcock R, Connors S, Matthews JBR, Chen Y, Zhou X, Gomis MI, Lonnoy E, Maycock T, Tignor M, Waterfield T (eds) IPCC, Geneva

IPCC/Intergovernmental Panel on Climate Change (2019) Special Report on climate change, desertification, land degradation, sustainable land management, food security, and greenhouse gas fluxes in terrestrial ecosystems. Shukla PR, Skea J, Calvo Buendia E, Masson-Delmotte V, Pörtner H-O, Roberts DC, Zhai P, Slade R, Connors S, van Diemen R, Ferrat M, Haughey E, Luz S, Neogi S, Pathak M, Petzold J, Portugal Pereira J, Vyas P, Huntley E, Kissick K, Belkacemi M, Malley J (eds) IPCC, Geneva

Kanianska R (2016) Agriculture and its impact on land-use, environmental, and ecosystem services. Matej Bel University, Slovakia

Kidd M (2011) Environmental law. Cape Town, South Africa

Kidd M et al (2018) Integrated environmental assessment and management. In: King ND, Strydom HA, Retief FP (eds) Fuggle & Rabie's environmental management in South Africa, 3rd edn. Cape Town, South Africa, pp 1213–1277

Lal R (2010) Soil Erosion and land degradation: the global risks. In: Lal R, Stewart BA (eds) Advances in soil science. Advances in soil science, vol 11. Springer, New York

Lugogo S (2018) The practice and politics of state-funded rural development in the former homeland of Transkei, Eastern Cape. Dissertation, University of the Western Cape

Macrotrends (2020) South Africa GNI 1962-2020. https://www.macrotrends.net/countries/ZAF/south-africa/gni-gross-national-income. Accessed 16 Feb 2020

Mathey J, Arndt T, Banse J, Rink D (2018) Public perception of spontaneous vegetation on brownfields in urban areas – Results from surveys in Dresden and Leipzig (Germany). Urban For Urban Green 9:384–392

Meyer TC et al (2018) Soil, land and agriculture. In: King ND, Strydom HA, Retief FP (eds) Fuggle & Rabie's environmental management in South Africa, 3rd edn. Cape Town, South Africa, pp 283–350

Mhlongo L (2019) A critical analysis of the protection of investment act 22 of 2015, vol. 34. Unisa Press, pp 2–22

Mostert H, Pope A (eds) (2010) The principles of the law of property in South Africa. Cape Town, South Africa

Mutsindikwa C (2018) South Africa in African geopolitics: the imagined and real role of South Africa in Africa from case analysis. Dissertation, University of South Africa

OECD (2020) Education at a Glance 2019 https://www.oecd.org/education/education-at-a-glance/EAG2019_CN_ZAF.pdf. Accessed 16 Feb 2020

Parliament of the Republic of South Africa (2020) How a law is made; https://www.parliament.gov.za/how-law-made. Accessed 12 Feb 2020

Paustian K, S Collier, J Baldock, R Burgess, J Creque, M DeLonge, J Dungait, B Ellert, S Frank, T Goddard, B Govaerts, M Grundy, M Henning, RC Izaurralde, M Madaras, B McConkey, E Porzig, C Rice, R Searle, N Seavy, R Skalsky, W Mulhern & M Jahn (2019). Quantifying carbon for agricultural soil management: from the current status toward a global soil information system 10(6). Carbon Management

Picker CS (2018) Legal protection of property under the protection of investment act 22 of 2015. Acta Juridica 17:17–42

Pienaar G (2013) Land tenure security: the need for reliable land information. J Helen Suzman Foundation 70:20–27

Pimentel D, Harvey C, Resosudarmo P, Sinclair K, Kunz D, McNair M, Crist S, Shpritz L, Fitton L, Saffouri R, Blair R (1995) Environmental and economic costs soil erosion and conservation benefits. Science 267:1117–1123

Potts L, Cloete CE (2012) Developing guidelines for brownfield development in South Africa. Environ Impact 162:389–399
Pretty MM, Odeku KO (2017) Harmful mining activities, environmental impacts and effects in the mining communities in South Africa: a critical perspective. Environ Econ 8:14–24
Rabie A (1976) South African environmental legislation. Pretoria, South Africa
Rautenbach C (2010) Deep legal pluralism in South Africa: judicial accommodation of non-state law. J Legal Pluralism Unofficial Law 60:143
Rautenbach C (2018) Introduction to legal pluralism in South Africa, 5th edn. LexisNexis, South Africa
Rautenbach IM (2012) Rautenbach-Malherbe constitutional law, 6th edn. LexisNexis, Butterworths
Reicosky D (ed) (2018) Managing soil health for sustainable agriculture Volume 1: Fundamentals
Ruppel OC (ed) (2008) Women and custom in Namibia: cultural practice versus gender equality. MacMillan Education Namibia, Windhoek
Ruppel OC (2013) Intersections of law and cooperative global climate governance – challenges in the Anthropocene. In: Ruppel OC, Roschmann C, Ruppel-Schlichting K (eds) Climate change: international law and global governance volume i: legal responses and global responsibility. Nomos, Baden-Baden, pp 29–93
Ruppel OC (2017) Constitutionalism and constitutional reform: selected aspects from a regional perspective. In: Ruppel OC, Scherr KM, Berndt AD (eds) Assessing progress in the implementation of Zimbabwe's new constitution. National, regional and global perspectives. Law and constitution in Africa, Vol. 32. Nomos, Baden-Baden, pp 51–83
Ruppel OC (2018) Conference presentation (unpublished) on the "Thabametsi Case" at the international conference on climate change, responsibility and liability, Faculty of law. University of Graz, Austria. on 8 November 2018
Ruppel OC, Borgmeyer T (2018) The BRICS partnership from a South African perspective: sustainable development space in a new global governance. In: Ndulo M, Kayizzi-Mugerwa S (eds) Financing innovation and sustainable development in Africa. Newcastle, pp 282–306
Ruppel OC & H Ginzky (2021 forthcoming) African Soil Protection Law: Mapping out options for a model legislation for improved sustainable soil management in Africa - a comparative legal analysis from Kenya, Cameroon and Zambia. Law and Constitution in Africa. Nomos, Baden-Baden
Ruppel OC, von Finckenstein A (2016) The protection of soil under Namibian law. In: Ginzky H, Heuser I, Qin T, Ruppel OC, Wegerdt P (eds) International yearbook of soil law and policy. Springer, pp 309–346
Ruppel OC, Junger GW, Knutton KM (2020) Der Klimawandel in der Governance, Gesetzgebung und Rechtsprechung Südafrikas: Ein Überblick über die jüngsten Entwicklungen. Vol. 5. Zeitschrift für Umweltrecht (ZUR), pp 274–280
Saiger AJ (2020) Domestic Courts and the Paris Agreement's climate goals: the need for a comparative approach. Transnatl Environ Law 9(1):37–54
Sandrey R (2019/20) African agricultural land: past, present and future. In: Chidede T, Mbakhwa F, Moobi M, Phaleng L, Sandrey R, Wagner N, Viljoen W, Zosuliwe K (eds) Agricultural and food security in Africa. Trade Law Centre, South Africa, pp 133–150
Scherr SJ (1999) Soil degradation: a threat to developing-country food security by 2020. Intl Food Policy Res Inst 27:1–63
Schlemmer EC (2009) Protection of investors and investments. South Afr Mercantile Law J 21: 734–752
Schoeman JL, van Deventer PW (2004) Soils and the environment: the past 25 years. South Afr J Plant Soils 21(5):369–387
Scholaro Pro (2020) Education System in South Africa. https://www.scholaro.com/pro/countries/South-Africa/Education-System. Accessed 16 Feb 2020
Soil Science Society of America (1996) Glossary of soil science terms. Madison, WI
South African Government (2020) Traditional Affairs https://www.gov.za/about-government/government-system/traditional-leadership. Accessed 16 Feb 2020

Sucker F (2014) Approval of an international treaty in parliament: how does Section 231(2) 'bind the republic'? Constitutional Court Rev V:417–434
Tlale MT (2018) A critical evaluation of the South African land tenure policy: a comparison with selected aspects of the Kenyan and Tanzanian law. Doctoral Thesis, North-West University
Trading Economics (2020) South Africa Foreign Direct Investment. https://tradingeconomics.com/south-africa/foreign-direct-investment. Accessed 16 Feb 2020
United Nations Convention to Combat Desertification (UNCCD) (2017) The global land outlook, 1st edn. Bonn, Germany
United Nations Environment Programme (UNEP) (2017) The emissions gap report 2017. United National Environmental Programme, Nairobi, Kenya
Utembe W, Faustman EM, Matatiele P, Gulumian M (2015) Hazards identified and the need for health risk assessment in the South African mining industry. Hum Exp Toxicol 34:1212–1221
Van Lynden GWJ, Oldeman LR (1997) Soil degradation in South and Southeast Asia. ISRIC, Wageningen
Van Wyk J (2018) The law on planning and the environment. In: King ND, Strydom HA, Retief FP (eds) Fuggle & Rabie's environmental management in South Africa, 3rd edn. Cape Town, South Africa, pp 1131–1176
Yaalon DH (2000) Down to earth: why soil and soil science matters. Nature 407:301
Young IM, Crawford JW (2004) Interactions and self-organization in the soil-microbe complex. Science 304(5677):1634–1637
Zika M, Erb K-H (2009) The global loss of net primary production resulting from human-induced soil degradation in drylands. Ecol Econ 69(2):310–318

Soil Protection Governance in Iran

Masoud Faryadi

Abstract Iran is a vulnerable country to desertification and soil degradation and to respond to this problem, some laws and regulations such as Act on Soil Protection Act, as well as institutional arrangements for environmental protection and soil preservation, have been initiated. However, the efficiency of these instruments in building an effective soil governance system is disputable. This chapter intends to exercise the legal challenges of Iran in soil protection governance. To do so it firstly considers the state of soil degradation and protection in Iran and then examines the development of soil protection governance in its legal system. This chapter, emphasizing on the contribution of institutional efficiency and legal enforcement in soil governance, concludes that Iran's legal system faces with a body of legal, institutional and social obstacles in development and improvement of soil protection governance pillars, *i.e.* integrated instruments and cross-sectoral collaboration in multilevel protection.

1 Introduction

After a decade of drafting and scrutinizing, finally, Iran's parliament passed the Act on Soil Protection (ASP) on 25 May 2019. Prevention of soil pollution had previously been stipulated either clearly or implicitly in some laws and by-laws, but the new Act specifically aims to protect the soil of Iran. A high rate of erosion and the dramatic reduction in quality of actively used and arable soils urged the Iranian Parliament to pass a special Act for their protection after decades of delay.

What makes an examination of Iran's steps in the legal protection of soil interesting is its efforts to overcome at least three conundrums: natural barriers such as desertification and a thin layer of topsoil of the country and its scarcity for

M. Faryadi (✉)
Department of Law, Faculty of Law and Political Sciences, University of Mazandaran, Babolsar, Iran
e-mail: m.faryadi@umz.ac.ir

H. Ginzky et al. (eds.), *International Yearbook of Soil Law and Policy 2020/2021*, International Yearbook of Soil Law and Policy 2020,
https://doi.org/10.1007/978-3-030-96347-7_14

productive cultivation; increasing demands of a growing population; and the institutional and legal efficiency to respond to these problems. Iran is located in the west of Asia in a region known as the arid belt. Approximately 90 per cent of the country consists of arid and semi-arid land,[1] and therefore, experiences 250 mm average annual rainfall in about 70% of the country, which fluctuates from less than 50 mm in the central plateau to more than 1600 mm on the Caspian Sea coastal region (north), Iran is overall an arid land.[2] One of the most noticeable natural characteristics of Iran is its high climatic variability; such that the country is divided into a number of climatological categories: wet and semi-wet in the north, west and north-west, and dry and semi-dry in the center, south and east. Two huge saline deserts, Dasht-e-Kavir and Dasht-e-Lut, are located in the arid areas of the center and south-east.[3] Given its location in the arid and semi-arid zone of the Earth, the area of Iran is exposed to low precipitation, high evaporation and a thin layer of surface soil (which limits the formation of topsoil). In addition to the natural shortage of active and arable soils, natural and human induced soil erosion is the other threat. According to the FAO in 2004, Iran loses 1–2 billion tonnes of soil per year because of erosion and 76% of its total area is exposed to erosion threat.[4]

Since numerous natural and anthropogenic factors contribute to soil degradation, soil protection is a complex task that requires a multidimensional body of legal mechanisms to regulate all of the contributing factors. To protect soil resources effectively, rather than isolated instruments, a holistic protection approach in soil governance is needed to integrate policies, plans, laws and by-laws, institutions, economic and social initiations for preservation and sustainable use of soils. Success in soil protection to a great degree depends on a comprehensive and well-functioning soil governance system.

Within the Iranian legal system, some mechanisms have been devised for soil protection in response to alarming threats of soil degradation, however, can these instruments build an efficient soil governance system or not? With this in mind, the main questions to be asked are what is the legal approach of Iran with respect to soil protection and whether its legal system could develop integrated soil protection governance or not?

This chapter examines the legal initiatives of Iran in soil protection and its approach in soil governance development. Inspired by the definition[5] of soil protection governance, in this chapter it means a collection of integrated legal and socioeconomic instruments and collaborative public and private institutions for soil protection at all decision-making, implementation and supervision levels. Thus, it will be examined whether the two pillars of soil protection governance, *i.*

[1]Najafi Nejad (2013), p. 379.

[2]Qadir et al. (2008), pp. 214–215. See also Mesgaran et al. (2017), p. 2 and also Gall (2004), p. 219.

[3]Harris (2003), pp. 90–91

[4]FAO & ITPS (2015), p. 411.

[5]Juerges and Hansjürgens (2018), p. 1629.

e. integrated instruments and cross-sectoral collaboration in multilevel protection, have been developed or not.

In doing so, firstly, the causes of soil degradation in Iran are discussed in order to understand better the necessities for legislative intervention and the level of urgency for the development of soil governance in its legal system. Then, the legal mechanisms of environmental protection and soil protection are discussed, specifically the ASP. Pointing out several challenges in development of soil governance, the chapter concludes that in spite of considerable efforts to control desertification and soil erosion in Iran, there has been insufficient development of an integrated system of soil protection governance and the probability of success in curbing desertification and soil erosion in the next years does not appear to be achievable.

2 Soil Degradation in Iran

To initiate an efficient soil protection governance mechanism, there should a correlation between the chosen legal instruments and causes and effects of soil degradation. In other words, it is necessary to sketch out the process of soil degradation and its causes succinctly. Each year, Iran loses considerable amounts of soil because of natural and human-induced causes, such as soil erosion, desertification, salinization, and destruction of plant cover by deforestation and overgrazing. Degradation of soil by construction and urbanization, a high rate of exploitation of alluvial soils by industry, soil contamination,[6] and dumping of waste are among the other causes of soil degradation in Iran. Since there is no accurate scientific estimate of soil erosion in Iran,[7] it is difficult to state an exact estimation of its rate in the country. According to the report of the Forest, Range and Watershed Management Organization of Iran (FRWMO) in 2004, approximately 75 million hectares (Mha) of land area of the country was exposed to water erosion, 20 Mha to wind erosion, 5 Mha to other types of chemical and physical degradation processes, about 2 Mha were exposed to infertility, and 2 Mha to salinization.[8] The expansion of agricultural areas, together with a rapid and incompatible population growth (from 19 million in 1956 to 82 million people in 2018),[9] industrialization and development, and urbanization, aside from natural factors, have increased the pressure intensely on the soils of Iran. The main causes of soil degradation include the following:

---

[6] FAO & ITPS (2015), pp. 420, 423.

[7] Najafi Nejad (2013), p. 378.

[8] Forest, Range and Watershed Management Organization (2004), p. 13.

[9] Emadodin et al. (2019), p. 8. For more information, see the Statistical Yearbook of Iran at: <https://www.amar.org.ir/english/Iran-Statistical-Yearbook>.

2.1 Desertification

Located in a dry region, Iran is one of the most vulnerable and prone countries to desertification in the world. Geographically, a huge part of Iran is occupied by deserts especially in its central and southeast regions, which have been increasing in area over the last few decades. Two massive salt deserts in the center and southeast are gradually being extended. Massive parts of Iran have suffered from extensive dust storms in the current decade.[10] Besides the dominant dry and semi-dry climate conditions, droughts and low precipitation further intensify the natural processes of desertification.[11] The effects of climate change are causing fluctuations in precipitation and its impacts on water resources is especially telling in dry regions.[12] According to predictions, warming weather and reduction in precipitation instigated by climate change will increase the vulnerability to desertification and change some semi-arid lands into arid land.[13] The International Panel on Climate Change expects that the deserts of Iran will become warmer and drier at faster rates,[14] which can compound the loss of soil moisture and drought and prevent steady soil formation or vegetation growth in the desert lands. Moreover, anthropogenic factors such as water overconsumption and poor irrigation systems, soil mismanagement, decline in the land productivity and degradation of rangelands through overgrazing and cultivation, further increase the potential for desertification.[15]

2.2 Salinization

Due to inefficient irrigation and lack of drainage facilities, Iran has experienced an increasing rate of soil salinization where the area of salt-affected soils have increased from 15.5 Mha in 1960 to 18 Mha in 1980, to 23 Mha in 1990, and to more than 25 Mha today.[16] The FAO estimated the total area affected by salinity at about 15.5 million ha or 9.4% of the total area of Iran.[17] According to another assessment, the area of saline soil cover 25% of the total area of Iran.[18] Increasing salinization of soils, especially in cultivated lands, is an additional risk in Iran[19]

[10] See for instance Akhzari et al. (2014), p. 636.

[11] Emadodin et al. (2019), p. 4.

[12] Abbaspour et al. (2009), pp. 2, 5.

[13] Mahmoudi et al. (2011), p. 65.

[14] IPCC (2014), p. 312. See also IPCC (2018), pp. 215, 247.

[15] Amiraslani and Dragovich (2011), pp. 2–3. See also Emadodin et al. (2019), p. 2.

[16] Qadir et al. (2008), p. 216.

[17] FAO (2008), p. 13.

[18] Mesgaran et al. (2017), p. 3.

[19] Cherlet et al. (2018), p. 104.

Since irrigation is the usual practice in Iran's agriculture,[20] unsustainable water management and heavy inefficient irrigation with groundwater subject the farm soils to salinization and erosion.[21]

2.3 *Unsustainable Agriculture*

Unsustainable agriculture is the main human induced driving force of soil degradation in Iran,[22] which has not been taken seriously in soil protection policy because of various economic and social issues and a shortage of technical information. It contributes to soil degradation in many ways:

2.3.1 Overextension and Misallocation of Farmland

Any agricultural development policy must be based on scientific assessment and soil quality monitoring of the land. In this regard, agricultural zones should be located according to the climatic features, spatial zoning plans and long-term agricultural policies, which is something that is missing in Iran's approach to agricultural policy and this absence has contributed to serious soil degradation. The total area of agriculturally usable soils in Iran is limited due to its particular geological and pedological qualities. The area of arable lands was estimated in 2018 at more than 15.5 Mha, of which about 9.3 million hectares are rainfed arable lands.[23]

In spite of the limited arable and active soils, land use has changed rapidly in the past 50 years, during which the amount of cultivated land has risen by more than five times.[24] There has been an insistent political and social emphasis on spreading farmlands in the country, in particular, bringing non-irrigated arable lands under cultivation.[25] Since principles 3.13 and 43.9 of the Constitution of Iran obliges the government to provide self-sufficiency in the economic and agriculture sectors, and in order to attain self-sufficiency in staples and to meet the needs of the growing population, the government has persistently supported further development of cultivation and farmland. This policy has resulted in the expansion of land use as well as extravagance in use of water resources, fertilizers and pesticides, in contrast to the contribution of the agricultural sector to Iran's gross domestic product, which has decreased over time.[26]

---

[20] Madani et al. (2016), p. 1000.

[21] FAO, p. 416. And also FAO (2008), p. 11.

[22] Amiraslani and Dragovich (2011), p. 4.

[23] Statistical Center of Iran (2019). See also Statistical Centre of Iran (2018).

[24] Bahrami et al. (2010), p. 601.

[25] Government of Iran (2012a), p. 7.

[26] Mirzaei et al. (2019), p. 3.

Cultivation farms have been developed usually in the absence of spatial zoning plans. In some regions, farmland allocation and crop patterns have been heterogeneous with climatic, hydrological, edaphic and pedological features.[27] Consequently, farmland has expanded unsustainably, which has been a significant factor in the increase in soil degradation.[28] An assessment of the total amount of agriculturally suitable soils of Iran reveal the irrationality of this extreme interest in cultivation at any price: while the FRWMO had estimated approximately 37 million hectares of soils suitable for cultivation in 2004, a scientific study has recently classified Iran's land suitability for cropping at (million ha): very good 0.4% (0.6), good 2.2% (3.6), medium 7.9% (12.8), poor 11.4% (18.5), very poor 6.3% (10.2), unsuitable 60.0% (97.4), and excluded areas 11.9%.[29] It seems that the ambitious policies to expand cultivation have been based on overestimations of the natural capacity of the land or underestimation of its natural limits. For instance, the Third Five-Years Development Plan Act (FYDPA) (1999) authorized the State to issue land ownership titles and accommodate the farmers with large scale public lands being reclaimed and allocated for cultivation without stipulation of any specific proviso for appropriate farming practices (Article 108.1). Its objective was to develop the agricultural sector and empower the farmers through land ownership and reclamation. However, this policy led to an anomalous pattern of allocation of the lands for cultivation and as a result some cultivated lands were not pedologically or climatically compatible with the chosen cultivation practices or patterns.

2.3.2 Inefficient Cultivation

Inefficient cultivation practices or patterns, to a great extent, precipitate soil degradation through hyper cultivation of lands, traditional patterns of farming, low efficiency in agricultural production, poor on-farm water management, soil depletion and diminution of soil fertility and nutrients, as well as soil compaction by conventional tillage practices.[30] Soil fertility in many areas of Iran has decreased because of inefficient irrigation and agrochemical usage. It is estimated that about 70% of agricultural soils contain less than 0/5% organic matter,[31] which requires many years of good land husbandry to restore.

Inefficient irrigation systems, in addition to soil salinization, drive land subsidence. Dwindling groundwater resources through thousands of authorized and unauthorized water wells for artificial irrigation in various flat areas of Iran contributes to the subsidence of agricultural land[32] and exposes the surface soils to other

[27] Madani et al. (2016), p. 1000.
[28] Emadodin et al. (2012), p. 618.
[29] Mesgaran et al. (2017), p. 2.
[30] FAO & ITPS (2015), p. 409.
[31] Government of Iran (2012b), p. 15.
[32] FAO (2011), pp. 119, 120.

processes of degradation. In 2008, groundwater depletion was estimated at 3.8 km^3/year,[33] and recently more than 55% of the total water demand is satisfied through groundwater extraction.[34]

Traditional technics, inefficient technologies and insufficient agricultural knowledge and expertise[35] are additional factors in the cause or exacerbation of soil degradation. Cultivation is mostly practiced by small-scale private farming units[36] where the application of modern cultivation technologies is not economically accessible.

2.3.3 Soil Pollution

The accumulation of heavy metals in soil due to industrial activities,[37] overuse of pesticides, irrigation with untreated wastewater in the farmlands, especially around the industrial urban areas such as Tehran,[38] and waste concentration, are the main causes of soil pollution. Although the Waste Management Act (2004) regulates waste treatment, however, many types of wastes are not treated separately but dumped into landfills.

2.3.4 Deforestation and Destruction of Vegetation Cover

Overgrazing and deforestation destroy the surface plant cover and lead to soil degradation. Although the forests of Iran are legally protected, every year, hectares of forests and rangelands are degraded when converted into farmland or settlement facilities.[39] Land grabbing and conversion of rangeland and farmland into residential land use is a cause of soil degradation and many land areas have been degraded through building and construction activity.

2.3.5 Construction

Iran has experienced a large urban sprawl during the last decades.[40] According to the Statistical Center of Iran, 74% of Iran's 79 million people lived in urban areas and

[33]FAO (2008), p. 5.

[34]Madani et al. (2016), p. 1002.

[35]FRWMO (2004), p. 18.

[36]FAO (2008), p. 3.

[37]See for example Sayadi and Sayyed (2011), pp. 1181–1187.

[38]Hani and Pazira (2011), p. 678.

[39]Makhdoum (2008), p. 573. See also Sotoudeh Foumani et al. (2017), pp. 187–188.

[40]FAO & ITPS (2015), p. 409.

cities in 2016, compared to 54% in 1986.[41] This rapid urban sprawl over the past six decades[42] accelerated construction activities, such as housing, which have permanently degraded urban soils and green zones and sealed large surface layers of land. Although urban and rural green zones are legally protected, there is no effectual legal requirement to control urban sprawl according to the physical capacity and resilience of the land, whereupon irregular urbanization devours the urban and suburban virgin soils more and more.

2.4 *Impacts of Soil Degradation*

Soil degradation creates a chain of interrelating severe social, economic and environmental problems and threatens the agriculture sector and food security of Iran. With regard to climate change, according to an assessment in 2000, the degradation of agricultural soil contributed to 55% of total emissions of the greenhouse gases in this sector.[43] Due to the reduction in quality of agricultural soils, many rural famers have been forced to leave their lands and migrate to urban areas.[44] As a result, employment in the agricultural sector has fallen from 24.7% in 2001 to 18% in 2016.[45] Soil degradation and desertification also impose long-term economic direct and indirect costs and harms on biodiversity that have not been calculated properly yet. The variety, longitude, range and severity of the current and future process of soil degradation in Iran provide a satisfactory justification to establish special policies for soil protection governance.

3 Legal Mechanisms of Soil Protection

Natural factors such as climate variability, topography, dryness, desertification and water shortage have restricted the availability of agriculturally suitable soils in Iran, which has been exacerbated by the misadministration of the soil. This has led to many negative effects, as described above. Therefore, it is necessary to establish an efficient legal and administrative system for soil protection. A legal system for environmental protection should be based on the climatic and geographical features of the country and responsive to the main threats to environmental stability. Likewise, instruments of soil protection should be adjusted with site-specific

[41] Statistical Center of Iran, Statistical Yearbook of Iran, 2017, (in Persian), at: <https://www.mpoes.ir//Dorsapax/userfiles/Sub1/g_sarshomari95.pdf>.

[42] Emadodin et al. (2019), p. 8.

[43] Government of Iran (2010), p. 4.

[44] Madani et al. (2016), pp. 999–1000.

[45] Tehran Chamber of Commerce, Industries, Mines and Agriculture (2017).

geographical and socioeconomic conditions of a country.[46] Iran has special natural features and runs the gamut of pressing environmental problems[47] which both should have been addressed in its legal system already in the past. Threats such as drought, desertification, dust storms, water scarcity, flooding, soil loss and population growth are critical enough to urge the legislator to enact appropriate responsive and adaptive legislation. Nonetheless, environmental laws and regulations that have been developed are more or less incompatible with these natural features and are not responsive to the severity of these perils. At the moment, a collection of conventions, laws and regulations constitutes Iran's legal environmental protection system which *inter alia* addresses soil conservation that will be explained in this part of the chapter.

3.1 International Conventions and Cooperation

Ratified international conventions are valued as ordinary laws in Iran's legal system. The Parliament has ratified a number of international environmental conventions such as the Biodiversity Convention, CITES, UNCCD, UNFCCC, as well as the Kyoto and Montreal Protocols. The UNCCD (Art. 5 & 10) calls for the member states to construct their own policies, strategies and priorities in the format of a national action programme to fulfill the objectives of the Convention. In doing so, the FRWMO (an agency of Ministry of Agriculture) as the national focal point of the Convention prepared The National Action Programme to Combat Desertification and Mitigate the Effects of Drought (2004), where one of its main targets was to promote soil conservation via integrated and sustainable management, education, development of vegetation cover and erosion control.[48] In fulfilment of its obligations, the Ministry of Agriculture has implemented considerable plans to combat desertification and stabilize shifting sand dunes through man-made plantations and mulching.[49] Iran has committed to set its National Land Degradation Neutrality target but has not submitted its report yet. Iran is a member of FAO and has implemented many joint projects with this organization since 1977, especially in water and soil preservation.[50] Also, under the UNFCCC (3.3 & 4.1.d), Iran is required to conserve its natural greenhouse gas sinks such as soil and vegetation cover.

Nowadays, soil protection requires extensive collaboration between international and national bodies, cooperation among governments and international organizations, and the stipulation of internationally accepted conservation responsibilities

[46] Helming et al. (2018), pp. 6, 10.

[47] Government of Iran (2010), p. 2.

[48] Forest, Range and Watershed Management Organization (2004), pp. 21, 26, 38.

[49] For instance, see Najafi Nejad (2013), pp. 388–389.

[50] Government of Iran (2012a), pp. 70–82.

and standards in national laws.[51] However, collaboration between Iranian governmental and non-governmental bodies and international bodies in this field has not been effective and there is a divergence between international law and its domestic law in the protection of soil.

3.2 *National Legislation*

Soil protection is a function of environmental governance in general and cannot be examined separately from the general environmental matters. Due to the diversity of environmental problems in Iran, a variety of laws and regulations have been passed in various environmental related fields. Iran began its legal protection of the environment with forest preservation. The Act on Forests (1942) was its first experience in natural resource preservation and over the following decades, further laws and regulations came into force for forest preservation and prevention of illegal logging. The Act on Environmental Protection and Improvement (AEPI) (1974) was the main legal initiative for environmental protection and prevention of pollution. The AEPI established the Department of Environment (DOE) as the agency of the president for the administration of environmental protection, research and education, and drafting of legal standards. In the following decades, with the intensification of environmental problems, additional regulatory instruments were progressively introduced to protect natural ecosystems, including the Prevention of Water Pollution By-law (1994), the Prevention of Air Pollution Act (1995) and the Clean Air Act (2017). Recently, the environmental regulation process has tended towards comprehensive and multidimensional environmental protection through the application of environmental considerations in economic, industrial and agricultural activities. Even though air and water had been protected earlier by special Acts, pursuing soil protection through a separate legal instrument had been neglected for decades until the recent enactment of the ASP. Provisions in favor of soil protection can be examined in two set of general regulations and a special Act, *i.e.* ASP.

3.2.1 General Provisions

Because soil degradation in Iran stems from various economic and social forces, such as institutional, agricultural, technical, legal and financial failures, an integrated system of soil protection governance is required. Nonetheless, provisions of soil protection are spread across various regulations and the legal system has not taken a systematic and holistic approach in this regard. Meanwhile, the lack of a comprehensive law for agriculture, land administration or land planning or natural resource

[51] Boer et al. (2017), p. 69.

use, has prevented the development of an integrated approach to legal soil protection. However, the existing regulations do address various aspects of soil protection:

3.2.1.1 Prevention of Pollution

The environmental law of Iran relevant to soil protection has mainly concentrated on pollution prevention rather than degradation in general. Environmental pollution has been forbidden in the AEPI (Art. 9) and the Islamic Penal Act (1996) (Art. 688), and technical standards of soil pollution are approved by the DOE as required under the AEPI.

3.2.1.2 Soil Preservation

In order to regulate its development process, Iran's government started to enact five-year development plans that bind the State to address environmental considerations in its development plans. The Second FYDPA (1994) was the first of these Acts that obliged the State to preserve the soil resources of the country. The Third FYDPA (Art. 104.1), for the first time stipulated that natural resources, including soil, should be utilized according to their natural bearable capacity. The Forth FYDPA (2004) required the State to analyze and integrate the economic values of natural resources in the national accounts (Art. 59), to develop soil and water infrastructures and enhance irrigation and drainage systems (Art. 18.1), to restrict grazing rights in the forests and pastures (Art. 68.2), and to implement plantations and afforestation in order to halt desertification (Art. 68). Water management and irrigation reform, afforestation and desertification control were the other mechanisms laid down in the Fifth FYDPA (2010) aiming at soil conservation (Art. 148).

3.2.1.3 Cultivation Reform

The main priority in soil protection in Iran is the reform and adaptation of the agriculture sector, regarded as the main source of soil degradation, which is now demanded by the legal system. The Sixth FYDPA (2017) decreed more mechanisms for soil preservation in cultivation and urged the State to enrich and conserve farmland soils (Art. 32), and carry out irrigation improvement (Art. 35) and watershed preservation (Art. 38). More recently, conservation agriculture has been recommended as an appropriate method for soil preservation, including minimum soil disturbance and conservation of soil cover,[52] where the Act charges the State to develop this model of agriculture in order to enhance soil productivity and food security and promote conservation agriculture (Art. 31).

---

[52]Farooq and Siddique (2015), p. 4.

The Act on Improvement of Efficiency of Agriculture Sector and Natural Resources (2010), which promotes efficient cultivation, commits the State to decrease the soil erosion rate and to halt desertification (Art. 11), to reform cultivation patterns (Art. 6), and to reduce livestock grazing in the pastures (Art. 14).

3.2.1.4 Vegetation Cover Protection

Several laws have been enacted for forest protection during the last century, and at the moment, due to tragic decline of forests areas, logging or any kind of taking of trees have been restricted. To control illegal farmland grabbing, the Act on Conservation of Farmlands and Gardens (1995) requires an administrative commission in the Ministry of Agriculture to authorize the use of farmland. Nonetheless, mechanisms of the Act are not adequate, as every year, a noticeable area of the farmlands are legally or illegally being grabbed and degraded. To protect pastoral lands for its key role in soil preservation, a collection of harmonized legal and administrative procedures and established judicial arrangements for dispute resolution are required,[53] which are currently neglected in Iran's legal system. Therefore, the conversion of the public pastoral lands into farmlands or settlement facilities is not banned definitely. The Clean Air Act, which aims to expand the area of natural sinks, obliges the municipalities and Ministry of Road and Urbanization and Ministry of Agriculture to expand the green zones in and around the urban areas and industrial complexes (Arts. 15, 21, 27).

3.2.2 Act on Soil Protection (ASP)

The abovementioned legal mechanisms regulate just some aspects of soil protection. However, in the absence of a specific legal system for soil protection, substantial areas of Iran have experienced soil degradation. Thus, a framework legal document was seen as essential to put the building-blocks of soil protection governance in place. The enactment of the ASP is anticipated as an important step to build an integrated approach for soil protection and to devise appropriate legal instruments to overcome all of these constraints. The Act, with 26 articles, presents various mechanisms for soil protection and can be assessed in terms of the following criteria.

3.2.2.1 Definitions and Targets

For more clarity, Article 1 defines the important terms in this Act. It defines the soil as "a natural and dynamic body consisted of chemical, physical and bio reactions which affected by water, climate and living entities is shaped on the land surface".

---

[53] Hannam (2018), p. 102.

So, it adopts a wide definition to cover all kinds of soils. But neither a strategy to implement the Act, nor a timetable to achieve its targets, have been determined.

3.2.2.2 Instruments

The Act has introduced some preventive and reactive instruments such as: valuing soil for national accounts, compensation of soil degradation, regulation of agro-chemical production and its use in agriculture, pollution control, prohibition of waste dumping on soil, banning soil export and allocation of special court divisions for infringement of soil related provisions. Due to a shortage of scientific and technical data about soils in Iran, the Act stipulates requirements for soil mapping, monitoring and regular survey, developing soil scientific information and a national database for soils.

3.2.2.3 Implementation

Institutionally, the Ministry of Agriculture (which is also responsible for forests, rangelands and watersheds preservation) has assumed the most responsibilities in the implementation of this Act. Soil pollution control, systematic monitoring and warning to polluter entities are assigned to DOE, which in doing so, may cease the polluter's activity without pre-warning in hazardous emergency circumstances. Nonetheless, the Act does not devise an institutional mechanism for cross-sectoral cooperation in soil protection. The following Table 1 lists the organizations in charge of the Act and their related responsibilities.

The controlling system of the Act is primarily centered on self-manifestation and reporting of polluter entities in case of pollution, and soil monitoring by the DOE. The polluters are obligated to inform the DOE if soil pollution occurs and obey its instructions, otherwise, they will be liable to be fined and ordered to pay compensation. Due to the vastness of the country and the rate of spread of industrial entities, exact and inclusive soil monitoring in each part of the country is practically difficult for the officials. The other critical point of this Act is that it does not provide effective and deterrent penalties for contravention.

3.2.2.4 Restoration and Rehabilitation

The ASP obligates soil polluters to restore and compensate for polluting the soil or else the offender will be fined. The Act also obliges the DOE and Ministry of Agriculture to initiate and exercise plans for the restoration of soils degraded by natural disasters such as flood. However, it poses no stipulation about restoration of the degraded soils in the rest soils which certainly mounts up to millions of hectares.

Table 1 Participant organizations in ASP

Organization	Responsibility
Ministry of Agriculture	Drafting policies of the sustainable management, protection and usage of soils, Mapping and classification of soils and lands, Providing national database for soils, Preparing manuals for soil sciences, Quality and quantity monitoring of soils, Regulation of agrochemicals in agriculture in cooperation with DOE and Ministry of Health, Control of production, import and application of agrochemicals.
Department of Environment	Cooperation with Ministry of Agriculture in drafting policies for sustainable management and protection of soils, Monitoring polluted soils, warning and declaring the emergency circumstances, Warning to the soil polluter entities, Judicial prosecution of soil polluters.
Organization of Planning and Budget	Valuing soil in national accounts.
All public institutions	Financing soil protection and compensation in their administration.
National Organization of Standard	Standardization of agrochemicals in cooperation with Ministry of Agriculture, DOE and Ministry of Health.
Judiciary	Judicial prosecution of offenders in special court divisions.
Governmental media	Public awareness making and education.

3.2.2.5 Education

Although public and professional education on soil protection awareness is urgently needed, the ASP does not provide considerable requirements for proper education and training, but merely urges the Organization of Radio and Television to broadcast programs in this regard.

3.2.2.6 Legal Innovations

The ASP, as the newest environmental Act, was expected to provide innovative legal approaches and mechanisms prevalent in modern environmental law (such as green economy, economic incentives, technology development, environmental co-regulation and self-regulation, retroactive restoration, adaptation to the climate change, proactive instruments for the future soil problems and so on). Instead, with the exception of requirements regarding soil survey, soil mapping and control of fertilizers, it has relied on command and control instruments and provides almost no innovative legal mechanism to improve the protection of soil.

4 Challenges of Soil Protection Governance

As the final debate, the question remains whether the ASP and the other Acts are efficient in responding to the aforementioned vast-range of problems and requirements for soil protection or not? And whether the current legal system can develop into an integrated system of soil protection governance or not? The legal challenges of soil protection in Iran cannot be viewed in isolation of environmental protection and the legal system as a whole. Thus, some challenges for soil protection stem from the general environmental governance framework and basic legal system that cannot be expounded here by detail. So in this part just the significant legal challenges for soil protection governance will be outlined.

4.1 Significant Threats, Minor Reactions

Given that Iran experiences almost every form of soil degradation, and that to at a large scale, this reciprocally requires equivalent legal responses. It was expected that the ASP, as the first specific regulatory initiative to provide diverse innovative mechanisms to combat several drivers of soil deterioration, would create an exhaustive legal framework to eliminate or mitigate some legal shortcomings and build up a primary legal system for soil protection. However, the Act is not a comprehensive framework law and overarching document, and it is missing in some important regulatory tools in soil protection as will be mentioned below. After an elongated legislative delay in soil protection, the Act is currently faced with so many barriers to adequately protect the soil and to restore existing degraded lands. In a comprehensive system of soil protection governance, all of the soil threats should be addressed. However, the ASP mainly controls pollution and does not address all causes of soil degradation, such as poor cultivation techniques, mining and extraction of gravel and sand, logging, flooding, construction, overgrazing, desertification and salinization. For a country severely exposed to desertification, the ASP and other regulations should have provided strict legal instruments to reduce the high rate of desertification and oblige the government to apply more effective techniques in this way rather than implementation of disputed mulching technique.

To give proper effect to the objectives of soil regulations, according to the first pillar of soil protection governance, soil resources should be protected by diverse and appropriate instruments.[54] When existing authority structures or legal instruments fail in their objects in a society, new and innovative instruments of governance are required[55] to make the change towards better governance. The sensitive status of the ecosystems and institutional failures necessitate a shift towards environmental

[54]Juerges and Hansjürgens (2018), pp. 1631–1632.

[55]Fiorino (2006), p. 162.

governance based on diversification of instruments[56] particularly in soil governance. As in order to protect soil as a dynamic ecosystem it is not sufficient to focus on the explicit soil regulation alone, it is also necessary to strengthen all instruments for addressing all of the soil-related aspects[57] such as land use amelioration, spatial planning, sustainable urbanisation and market based instruments. Nonetheless, almost all of the legal instruments concerning soil protection have focused purely on soil and overlooked its economic, social and technical aspects.

Apart from that, co-regulation, self-regulation and voluntary programs[58] are among the other alternative affordable instruments for soil protection governance, which are applied in the broader aspects of environmental law of Iran. Environmental protection in Iran is based on the command and control approach, which is exercised just through the state organizations (Public Services Management Act, 2007, Art. 8). This constraint in instrument choice limits public participation in soil protection and forces the legal system to take a "one size fits all" approach in soil protection. Furthermore, while special sensitive soils require specific or more restrictive protection rules, the soil protection regulations have not adapted the legal mechanisms with functions, location, quality and sensitivity for all soil types and conditions.

4.2 Land Planning

One of the main triggers in anthropogenic soil degradation is the trivialization of spatial planning and absence of a national strategy for land use planning.[59] Although some regulations oblige the government to provide spatial plans for all regions of the country, these plans have not yet been prepared to regulate the land use. The lack of such plans and complexity of land use regulations have halted better land use and resulted in the misallocation of land for cultivation and urbanization. At the moment, the implementation of spatial plans is highly urgent to fully address the ecological capacity of the regions in urban sprawl and agricultural development, to avoid more pressure on regions that are environmentally vulnerable or sensitive such as arid zones, and to prevent the allocation of farmland for inconsistent cropping patterns. Misuse of land is the other result of negligence in land planning. The State, in pursuit of its social and economic objects (such as entrepreneurship for farmers), has prioritized the expansion of farmland and land use as far as possible, to the extent that together with inefficient cultivation and urban sprawl, has jeopardized the soil security of the country. There is no special regulation for conservation of virgin soils

[56] *Ibid*, p. 3. See also Evans (2012), p. 34.

[57] Hurni and Meyer (2002), p. 26.

[58] Potoski and Prakash (2004), p. 158.

[59] Makhdoum (2008), pp. 568–569.

and soil conservation reserves to ensure minimal soil disturbance and to achieve a land degradation neutrality.

4.3 Cultivation Reform

Undoubtedly, a principal solution for soil protection in Iran is the improvement of cultivation techniques. Cultivation reform requires a combination of legal, social, technical and economical instruments, but in the absence of an active free market, the State has to do it with regulation. Although the Sixth FYDPA obliged the State to develop sustainable agriculture, nevertheless, more comprehensive legal and economic mechanisms are required for pursuing sustainable agriculture and soil protection in cultivation. In spite of the growing number of agricultural regulations, many elements of an efficient cultivation system, such as productivity, rational irrigation, soil preservation, and sustainable agriculture, adaptive and compatible patterns of cultivation, are not appropriately backed by the legal and economic system. Harmful practices to soil, such as poor tillage practices, over irrigation and hyper cultivation, have not been yet regulated effectively.

The other constraint is the limited soil protection on private farms, in which the legal system cannot effectually guarantee soil protection. Firstly, as the social protection of farmers has always been given top priority in development plans, the State regulates the pricing of staple products and provides subsidies to farmers regardless of their cultivation efficiency, which does not motivate them to change their inefficient cultivation patterns. Secondly, due to private law dominance over private land, the State is not legally authorized to control all of these farms and to obligate the farmers to follow the best available practices and efficient patterns in cultivation. Legally, the State does not apply any supervision or control over private land use and cropping, irrigation patterns and use of agrochemicals. Based on a private law rule, the private landowner has the legal power to exercise complete control over his property and take any advantage of his ownership rights freely unless prohibited by law (Civil Act, art. 30). According to this rule, the usage of private farmlands cannot be intervened by public authority even if the farmer applies inefficient methods in cultivation. The Ministry of Agriculture issues some recommendations for efficient cultivation, however, these recommendations are not legally binding for private farmers, and in effect, the ministry applies no supervision to check the cultivations methods in private farmland. Of course, the ministry is authorized to exercise control over the state owned farmland. But direct intervention and imperative regulations to change the attitudes of the farmers are socially problematic. Thus, the legal system needs to find a solution to apply environmental rules on private farmland. An effective way to urge the farmers to follow efficient cultivation methods is to train and empower farmers through self-regulation, land retirement programs,[60] product liability rules and providing economic incentives such as a free market for efficient agricultural products, technical assistance and cost sharing.[61]

4.4 *Institutional Barriers*

An efficient institutional environment is vital for good environmental governance[62] since it can pave the way for more effective implementation of its goals. Soil protection governance, given its multidimensional and cross-sectoral aspects, requires an efficient and cooperative institutional environment. However, many environmental regulations and soil protection initiatives are not appropriately implemented in practice because of some institutional shortcomings, the most significant of which are as follows.

4.4.1 Administrative Coordination

One of the significant institutional obstacles in environmental protection in Iran is administrative disharmony among the related organizations. Basically, the main organization responsible for environmental protection is DOE, however, many sectors are administrated by other ministries such as the Ministry of Agriculture, which manages the soils and forests. But the main responsibility of this ministry is the economic exploitation of natural resources rather than preservation, whereas the dramatic decline in quality and quantity of the country's forests and soils reveal its inefficiency in this regard. In practice, the division of administrative competences restrains the DOE from effectively supervising the protection of these vital resources. To improve coordination in environmental management, the High Council of Environmental Protection was established as a reconciliation body for environmental management by the AEPI, consisting of the DOE and some related ministries. But due to its undeveloped administrative structure and irregular meetings, it is ineffective in achieving administrative collaboration among the cross-sectoral environment-related organizations. So, the establishment of a multi-sectoral administrative council specialized for soil protection may be an alternative to the High Council of Environmental Protection.

---

[60] Angelo et al. (2013), p. 22.

[61] McEowen (2017), pp. 557–558.

[62] Potoski and Prakash (2004), p. 157.

4.4.2 Implementation and Enforcement

The question addressed in this subsection is how legal provisions are implemented in practice? A variety of legal, political, social and economic reasons contribute to this weak enforcement of environmental law, from fragile rule of law to cumbersome bureaucracy, and from inadequate monitoring to ineffective punishments of environmental crimes. For instance, due to the historical and social role of the farmers and nomadic communities, and in favor of social protection of the vulnerable groups, the State takes lenient and permissive approaches in controlling overgrazing, overconsumption of water and unsustainable patterns of agriculture and does not tackle seriously the problems of trespass, land grabbing, logging and misuse of agrochemicals. This lenient form of social protectionism renders the regulatory instruments in protection of natural resources and amelioration of cultivation ineffective. Other factors include the misdistribution of administrative powers among the respective organizations, shortage of technical tools, lack of legal knowledge and ambiguous and complicated regulations with no detailed by-laws and directives. Financial shortcomings have also played a deterrent role in the full implementation of many soil protection plans, which mainly derives from misallocation of funds, improper budgeting and funding. For instance, the State (Ministry of Agriculture) is engaged to tackle desertification through inclusive research and executive projects over the last decades and these projects can contribute indirectly to about 3.75% of the country's annual GDP.[63] Nevertheless, because of the continuance of desertification and the increasing costs of control, the State is constrained by insufficient funding for anti-desertification plans.[64]

4.4.3 Cross-Sectoral Participation

The second pillar of soil governance is the meaningful participation of stakeholders in the overall process of soil protection, from decision-making to implementation and supervision. In spite of some developments in the promotion of cross-sectoral cooperation and public participation in land management in Iran, an efficient enabling institutional environment for this purpose is absent.[65] Due to the way the government generally governs environmental protection, non-governmental bodies are left with a minor role in this process, especially in decision-making and supervision, which remain completely under the realm of the government. Likewise, legal documents concerning soil protection, specially the ASP, empower only the state organizations and gives no role for non-governmental public and private entities such as municipalities, farmers' unions and NGOs. Bearing this in mind, the Sixth FYDPA has urged the State to pave the way for local communities and NGOs to play

---

[63] Amiraslani et al. (2018), p. 147.

[64] *Ibid.*, p. 149.

[65] Amiraslani and Caiserman (2018), pp. 5, 10, 13.

a greater role in environmental protection. However, there is still a long path forward to include the non-governmental sectors in environmental governance. The farmers' unions have not been developed independently and have had little role in agricultural administration. This explains why the farmer community is underrepresented in soil protection activities. Moreover, academic institutions, professional associations, business activists and media are not systematically involved in decision-making and consultation about soil protection. In spite of the establishment of many environmental NGOs, there is not yet any form of serious collaboration between these organizations and the governmental ones.[66]

4.5 *Protection of Soil in Climate Change*

Over the coming years, the agricultural sector in Iran will be more vulnerable to the impacts of climate change,[67] in particular, its impacts on temperature, precipitation, soil moisture, crop yield and water resources. Climate change will result in some irregularities in crop and plant production over the coming years in Iran, which will contribute to soil degradation.[68] In spite of affirmation of vulnerability to a high rate of soil erosion and desertification in its INDC,[69] Iran has not prioritized soil protection in its national climate change programs and its legal system is poorly prepared for climate change impacts on soil protection, vegetation cover, agricultural production and flood control. The Executive By-law of the UNFCCC and Kyoto Protocol (Council of State, 2009) obliges the FRWMO to planning, conservation and restoration of the pastures so they can adapt to climate change (Art. 8.2), Moreover, there are financial and technological shortcomings in climate change adaptation,[70] compounded with the absence of a forceful comprehensive legal mechanism in climate change management. The soil protection system in Iran is reactive rather than proactive. In the coming years, soil degradation will leave a trail of problems behind with uncountable economic, social and ecological impacts. Considering this, in order to be adaptive to effects of degradation in the future, institutional resilience in soil governance is an absolute necessity.[71] However, no mechanisms for mitigation of soil vulnerability to climate change and adaptation to its impacts have been designed in Iran's legal system. Also, there is so far no programme for the promotion of preparedness of the public in adaptation to soil degradation and desertification impacts in the future.

[66] Hashemi et al. (2019), pp. 5, 6.

[67] Alefi and Ghahremanzadeh (2019), p. 58. See also Karimi et al. (2018), pp. 2, 9.

[68] Amiri and Eslamian (2010), p. 212. See also van de Steeg and Tibbo (2012), p. 32.

[69] DOE (2015), p. 8.

[70] Karimi et al. (2018), p. 9.

[71] Juerges et al. (2018), p. 2.

4.6 Research and Education

A noticeable number of research institutes are active in soil and land matters.[72] However, there is a shortfall of adequate information about the various aspects of soil conditions in Iran, such as accurate area of the deserts and degraded soils, the processes of land degradation and its causes and effects in the future. There is no systematic interconnection between the scientific agricultural community and farmers, as well as between the regulatory and executive bodies. The Ministry of Agriculture, in spite of its considerable efforts in the promotion of scientific methods in cultivation, should take more action for education on appropriate land use practices and sustainable cultivation methods for farmers.

4.7 Restoration

The reigning body of legal documents do not initiate a firm framework for the amelioration and restoration of degraded soils. Although the State has been obligated to the restoration of degraded forests, however, restoration and reclamation of abandoned farmlands and degraded lands throughout the country are not regulated. There are no special mechanisms to reduce the impervious coverage of the soil and preservation of organic matter and soil nutrients to enhance its fertility. Lastly, the enormous cost of soil degradation to Iran is not calculated in the national accounts and surely will be transferred to the future generations.

5 Conclusion

Iran is one of the most vulnerable countries in the world to desertification and soil degradation. To respond to these severe threats, a considerable body of laws and regulations, as well as institutional arrangements for environmental protection and soil preservation, have been initiated. Soil protection and cultivation amelioration have been addressed recently in the five-years development plan Acts. The new Act on Soil Protection of Iran provides minimal instrumental control over the high rate of soil degradation in the country, mainly through pollution prevention and soil survey. Iran, as a member of UNCD and UNFCCC processes, has prepared its national program to control desertification and mitigate the effects of climate change. Numerous research centers and university faculties are active in soil research. By and large, these regulatory instruments are legally and technically well-drafted, aiming highly in land preservation, and the Ministry of Agriculture has undertaken considerable effort to restore forests and combat soil erosion.

---

[72] Najafi Nejad (2013), p. 393.

However, practically, these legislative initiatives have not enhanced the conservation of natural ecosystems successfully and there are still a plenty of barriers that reduce the effectiveness of these legal instruments. The inefficient institutional environment, unstable rule of law, legal shortfalls and ambiguities, procrastination in enforcement and failure in reconciliation of economic development with environmental protection are among the significant barriers.

Any regulatory policy for soil protection must be integrated with the policies of water and vegetation cover protection, land use planning and spatial zoning, and agricultural strategies. To intertwine these policies and strategies, an integrated system of soil governance is required. This has been an important area of neglect in the soil protection system of Iran as it has not integrated soil protection concerns in the core of environmental governance and the sustainable development process. The current legal mechanisms do not take a comprehensive approach for soil protection, in which the best available techniques and diverse regulatory tools should be applied. Moreover, as environmental law of the country is based on a state command and control system, there is no real possibility for the participation of private and non-governmental stakeholders in soil governance. Therefore, the lack of diverse instruments and cross-sectoral participation undermine the two pillars of soil governance. Enactment of the ASP was a unique opportunity for taking such a comprehensive approach in soil protection, but in reality, it is an inchoate and incomplete Act that should be enhanced by other related legal instruments. Perhaps the steady rate of soil degradation in Iran and interactions of its several causes may persuade the legislator to amend this Act to make it complete and more responsive.

Legal instruments, on their own, are not able to protect soils of a given country alone. Iran's experience in soil protection confirms this point, whereas a noticeable number of legal instruments cannot be applied appropriately in practice and the legislature continues to fail to achieve its targets. In practice, a variety of obstacles such as institutional barriers, financial deficits, inadequate professional and legal training, misadministration, inappropriate judicial and political supervision constrain the legal instruments. This experience reveals that the obvious need to develop an effective soil protection governance framework, given that dependence on just governmental driven legal instruments does not necessarily always work. Rather, a collection of economic, social, political, technical and legal instruments, combined with the collaboration of cross-sectoral stakeholders, are needed to develop a responsive soil governance. With this in mind, in answer to the main question of this chapter, one can say that soil protection governance, in spite of the existing initiatives, has not been developed as an integrated coherent system in Iran, its two pillars are not consolidated and there is an urgent need to integrate the soil considerations in the core of its economic and social development plans as well as agricultural and land use policies. To be geologically young with thin layer of soil and to be located in an arid region has not been Iran's choice, however, she can choose how to adapt to these natural traits and take a sustainable approach in soil use.

References

Abbaspour KC, Faramarzi M, Ghasemi SS, Yang H (2009) Assessing the impact of climate change on water resources in Iran. Water Resource Res 45, W10434:2, 5. https://doi.org/10.1029/2008WR007615

Akhzari D et al (2014) Impact of source area desertification on dust storm emission in the western parts of Iran. Curr World Environ 9(3):636

Alefi K, Ghahremanzadeh M (2019) Assessing climate change impacts on land use in Iran: the spatial fractional multinomial logit modeling approach. In: Rashidghalam M (ed) Sustainable agriculture and agribusiness in Iran. Springer Nature, Singapore

Amiraslani F, Caiserman A (2018) Multi-stakeholder and multi-level interventions to tackle climate change and land degradation: the case of Iran. Sustainability 10(6). https://doi.org/10.3390/su10062000

Amiraslani F, Dragovich D (2011) Combating desertification in Iran over the last 50 years: an overview of changing approaches. J Environ Manag 92:1–13. https://doi.org/10.1016/j.jenvman.2010.08.012

Amiraslani F, Dragovich D, Caiserman A (2018) A long-term cost-benefit analysis of national anti-desertification plans in Iran. Desert 23(1):141–151

Amiri MJ, Eslamian SS (2010) Investigation of climate change in Iran. J Environ Sci Technol 3(4): 208–216

Angelo MJ, Czarnezki JJ, Eubanks II WS (2013) Food, agriculture, and environmental law. Environmental Law Institute, USA

Bahrami A, Emadodin I, Atashi MR, Bork HR (2010) Land-use change and soil degradation: a case study, North of Iran. Agric Biol J N Am 1(4):600–605

Boer BW, Ginzky H, Heuser IL (2017) International soil protection law: history, concepts and latest developments. In: Ginzky H, Heuser IL, Qin T, Ruppel OC, Wegerdt P (eds) International yearbook of soil law and policy 2016. Springer, Switzerland

Cherlet M, Hutchinson C, Reynolds J, Hill J, Sommer S, von Maltitz G (eds) (2018) World Atlas of desertification. Publication Office of the European Union, Luxembourg

DOE (2015) Intended Nationally Determined Contribution (INDC) National Climate Change Committee, Department of Environment of Iran, 19 November 2015

Emadodin I, Narita D, Bork HR (2012) Soil degradation and agricultural sustainability: an overview from Iran. Environ Dev Sustain 14(5):611–625. https://doi.org/10.1007/s10668-012-9351-y

Emadodin I, Reinsch T, Taube F (2019) Drought and desertification in Iran. Hydrology 6:66. https://doi.org/10.3390/hydrology6030066

Evans JP (2012) Environmental governance. Routledge, USA

FAO (2008) AQUASTAT country profile – Iran. Food and Agriculture Organization of the United Nations (FAO). Rome, Italy, at: http://www.fao.org/3/ca0339en/CA0339EN.pdf. Accessed 1 July 2021

FAO (2011) The State of the World's Land and Water Resources for Food and Agriculture (SOLAW) – managing systems at risk. Food and Agriculture Organization of the United Nations, Rome and Earthscan, London

FAO & ITPS (2015) Status of the World's Soil Resources: Main Report. Rome, Italy

Farooq M, Siddique KHM (2015) Conservation agriculture: concepts, brief history, and impacts on agricultural systems. In: Farooq M, Siddique KHM (eds) Conservation agriculture. Springer, Switzerland

Fiorino DJ (2006) The new environmental regulation. MIT Press, USA

Forest, Range and Watershed Management Organization (FRWMO) (2004) The National Action Programme to combat desertification and mitigate the effects of drought of Islamic Republic of Iran. Ministry of Agriculture, Iran, at: https://knowledge.unccd.int/sites/default/files/naps/2017-08/iran-eng2004.pdf. Accessed 1 July 2021

Gall TL (Editor in Chief) (2004) Worldmark encyclopedia of world nations, vol 4. Gall Publication, Asia and Oceania, USA

Government of Iran (2010) Iran Second National Communication to UNFCCC, Department of Environment

Government of Iran (2012a) Country Programming Framework (CPF) 2012-2016 for Iran's Agriculture Sector, prepared by Government of Iran and FAO, at: http://www.fao.org/3/a-bp603e.pdf. Accessed 2 July 2021

Government of Iran (2012b) Country report on history and status of soil survey in Iran, GSP regional workshop, Jordan, Amman, 1-5 April 2012, at: http://www.fao.org/fileadmin/user_upload/GSP/docs/Presentation_NEMA_Inception/The_Islamic_Republic_of_Iran_2.pdf. Accessed 25 July 2021

Hani A, Pazira E (2011) Heavy metals assessment and identification of their sources in agricultural soils of Southern Tehran, Iran. Environ Monit Assess 176:677–691. https://doi.org/10.1007/s10661-010-1612-3

Hannam I (2018) Governance of pastoral lands. In: Ginzky H, Dooley E, Heuser IL, Kasimbazi E, Markus T, Qin T (eds) International yearbook of soil law and policy 2017. Springer, Switzerland

Harris N (2003) Atlas of the world's deserts. Fitzroy Dearborn, USA

Hashemi F, Sadighi H, Chizari M, Abbasi E (2019) The relationship between ENGOs and Government in Iran. Heliyon 5:e02844. https://doi.org/10.1016/j.heliyon.2019.e02844

Helming K, Daedlow K, Hansjürgens B, Koellner T (2018) Assessment and governance of sustainable soil management. Sustainability 10:4432. https://doi.org/10.3390/su10124432

Hurni H, Meyer K (eds) (2002) A world soils agenda: discussing international actions for the sustainable use of soils, prepared with the support of an international group of specialists of the IASUS Working Group of the International Union of Soil Sciences (IUSS). Centre for Development and Environment, Switzerland

IPCC (2014) Climate Change 2014 impacts, adaptation, and vulnerability, Part A: global and sectoral aspects. In: Field CB et al (eds) Working Group II Contribution to the Fifth Assessment Report of the Intergovernmental Panel on Climate Change. Cambridge University Press, USA & UK

IPCC (2018) Global warming of 1.5°C. An IPCC Special Report on the impacts of global warming of 1.5°C above pre-industrial levels and related global greenhouse gas emission pathways, in the context of strengthening the global response to the threat of climate change, sustainable development, and efforts to eradicate poverty. Masson-Delmotte, Valérie et al., in Press, at: https://www.ipcc.ch/site/assets/uploads/sites/2/2019/06/SR15_Full_Report_High_Res.pdf. Accessed 23 July 2021

Juerges N, Hansjürgens B (2018) Soil governance in the transition towards a sustainable bioeconomy – a review. J Clean Prod 170:1628–1639. https://doi.org/10.1016/j.jclepro.2016.10.143

Juerges N, Hagemann N, Bartke S (2018) A tool to analyse instruments for soil governance: the REEL-framework. J Environ Policy Plan. https://doi.org/10.1080/1523908X.2018.1474731

Karimi V, Karami E, Keshavarz M (2018) Climate change and agriculture: impacts and adaptive responses in Iran. J Integr Agric 17:1–15. https://doi.org/10.1016/S2095-3119(17)61794-5

Madani K, AghaKouchak A, Mirchi A (2016) Iran's socio-economic drought: challenges of a water-bankrupt nation. Iran Stud 49(6):997–1016. https://doi.org/10.1080/00210862.2016.1259286

Mahmoudi P, Kalim D, Amirmoradi MR (2011) Investigation of Iran vulnerability trend to desertification with approach of climate change. 2011 2nd International Conference on Environmental Science and Development IPCBEE, vol 4. IACSIT Press, Singapore

Makhdoum MF (2008) Management of protected areas and conservation of biodiversity in Iran. Int J Environ Stud 65(4):563–585. https://doi.org/10.1080/00207230802245898

McEowen RA (2017) Agricultural law in a nutshell. West Academic Publishing, USA

Mesgaran MB, Madani K, Hashemi H, Azadi P (2017) Iran's land suitability for agriculture. Sci Rep 7:7670. https://doi.org/10.1038/s41598-017-08066-y

Mirzaei A, Saghafian B, Mirchi A, Madani K (2019) The groundwater-energy-food nexus in Iran's agricultural sector: implications for water security. Water 11:1835. https://doi.org/10.3390/w11091835

Najafi Nejad A (2013) Soil and water conservation for desertification control in Iran. In: Ali Heshmati G, Squires VR (eds) Combating desertification in Asia, Africa and the Middle East: proven practices. Springer, Netherlands

Potoski M, Prakash A (2004) The regulation dilemma: cooperation and conflict in environmental governance. Public Adm Rev 64(2):152–163

Qadir M, Qureshi AS, Cheraghi SAM (2008) Extent and characterization of salt-affected soils in Iran and strategies for their amelioration and management. Land Degrad Dev 19(2):214–227. https://doi.org/10.1002/ldr.818

Sayadi MH, Sayyed MRG (2011) Comparative assessment of baseline concentration of the heavy metals in the soils of Tehran (Iran) with the comprisable reference data. Environ Earth Sci 63: 1181–1187. https://doi.org/10.1007/s12665-010-0792-z

Sotoudeh Foumani B, Rostami Shahraji T, Mohammadi Limaei S (2017) Role of political power in forest administration policy of Iran. Caspian J Environ Sci 15(2):181–199

Statistical Center of Iran (2019) Iran's agriculture statics results in abstract for 2017-2018, Presidency of Iran, (in Persian) at: https://www.amar.org.ir/Portals/0/News/1398/1_chntaz%201397.pdf. Accessed 21 July 2021

Statistical Center of Iran, Statistical Yearbook of Iran, 2017, (in Persian), at: https://www.mpoes.ir//Dorsapax/userfiles/Sub1/g_sarshomari95.pdf. Accessed 30 July 2021

Statistical Centre of Iran (2018) Selected results of crop survey, the year 1397. Presidency of Iran, at: https://www.amar.org.ir/Portals/1/releases/agriculture/Selected_Results_of_Crop_Survey_the_Year_1397.pdf. Accessed 20 July 2021

Tehran Chamber of Commerce, Industries, Mines and Agriculture (2017) "Agricultural Sector and Necessity of Sustainable Development in Iran", expert report, (in Persian)

Van de Steeg J, Tibbo M (2012) Livestock and climate change in the near east region measures to adapt to and mitigate climate change. FAO Regional Office for the Near east, Cairo, Egypt

Part IV
Cross-cutting Issues

Sustainable Soil Management Threats Resulting from the COVID-19 Pandemic

Pamela Towela Sambo

Abstract This chapter adopts an exploratory approach towards analyzing the sustainable soil threats arising from the COVID-19 pandemic. The pandemic arises from a range of coronaviruses (CoVs) which are believed to be transferable from animals to humans, and are likely to generate interaction with almost all environmental media—the land, soils on it, water and air. Soils are undisputedly the foundation for all environmental considerations, being the basic support for all forms of life. Soils are found in the environment where we live, work, play, learn and worship. Therefore, the outbreak of the COVID-19 pandemic is both an environmental matter of grave concern as well as a public health emergency which is threatening sustainable soil management (SSM) and the entire global environment. The pandemic is not only threatening environmental wellbeing, but also the very existence of human life. Both the COVID-19 pandemic and environmental challenges such as climate change and loss of biological diversity have no boundaries and can negatively affect global SSM. This brings the enforcement of a wide range of environmental laws into perspective—waste management, natural resources conservation, pollution control, water and sanitation, including wildlife and other animal-related legislation.

In most transition and developing countries, the global pandemic is manifesting alongside inadequate socio-economic bases, poor healthcare facilities, severe environmental degradation, and poor enforcement of laws and policies in the governance and environmental sectors which are closely aligned to public health. This chapter uses desktop research to explore likely threats arising from the outbreak of the COVID-19 pandemic and whether the existing law and policy frameworks can provide effective response to the resultant SSM threats. In the wake of the COVID-19 global health challenge, any lapses in enforcing environmental laws and policies may result in worsened challenges. For instance, the global food security threat is likely to escalate owing to soil being polluted and thereby returning poor yields. In the final analysis, if the anticipated threats on SSM are not well

P. T. Sambo (✉)
Private Law Department, School of Law, University of Zambia, Lusaka, Zambia
e-mail: Pamela.sambo@unza.zm

H. Ginzky et al. (eds.), *International Yearbook of Soil Law and Policy 2020/2021*,
International Yearbook of Soil Law and Policy 2020,
https://doi.org/10.1007/978-3-030-96347-7_15

managed, there might be significant disturbances in ecosystem support services, leading to further loss of biological diversity among other challenges. In the absence of conclusive information on the full extent of threats to SSM during and post the COVID-19 pandemic, the purpose of this exploratory chapter is to call attention to more research.

1 Introduction

The first confirmed case of the Novel Coronavirus (COVID-19) was reported in Wuhan, Hubei Province of China in December, 2019, which then caused a rapid outbreak in the rest of China. Within a few months, COVID-19 had emerged in almost all countries across all the continents, presently infecting more than 170 million people and causing more than 3 million deaths. Although the pandemic and research on its full implications are still evolving, it is clear that the links between environmental law, science and technology must be clearly understood in relation to global sustainable soil management (SSM). In addition, the results of such research will need to be disseminated in record time for assimilation into law and policy. This chapter is exploratory, but provides insightful 'food for thought' by focusing on soil sustainability and waste management in the wake of the COVID-19 pandemic particularly in countries with transition and developing economies.

This chapter is comprised of six sections. Section 2 outlines the background and establishes the scientific link between COVID-19 and SSM, while Section. 3 reviews the literature review relating to possible threats presented by the COVID-19 pandemic to the global environment. Section 4 analyzes the socio-economic impacts of the pandemic and also evaluates how this might affect soil health. Further, the section briefly discusses the approach towards managing the effects of the COVID-19 pandemic in some southern African countries with economies in transition. This discussion is then linked to the exercise of social, cultural and economic rights, which include environmental protection. It is clear in this section that the COVID-19 pandemic presents an array of threats to SSM, most of which are yet to be uncovered. Against the numerous threats and uncertainties of the pandemic, Sect. 5 explores preliminary research, law and policy considerations in the SSM discourse and how it will be shaped in the light of the COVID-19 outbreak. Section 6 concludes the chapter by re-emphasizing more research and the implementation of the One Health approach towards addressing present and future threats to SSM and the environment.

2 Background and Scientific Link: COVID-19 and Sustainable Soil Management (SSM)

The 'environment' is defined in a very broad way to include the health of all living organisms including human beings, whether approached from the global or jurisdictional perspectives. The term 'environment' is usually given a very wide definitional ambit,[1] and it might appear in a different light and more diverse context in legal scholarship, to include both anthropocentric and ecocentric approaches.[2] This ethical approach shows the interrelatedness of the environment and brings to the fore matters to do with preservation of biological diversity, soil security and climate change amongst several other global concerns.[3]

In the wake of the COVID-19 pandemic, these issues will consequently raise several scientific questions for investigation. Given this background, the definition of the 'environment' is arguably on equal footing with the threats arising from the COVID-19 pandemic, which have also been described in many ways since it broke out in December, 2019.[4] In general, research has shown that Emerging Infectious Diseases (EIDs) have over the years been given impetus by growing human populations which have led to rapid disruption of natural ecosystems; globalization that has sent EIDs around the world in record time, and "changing climates that are affecting drivers such as increasing demand for animal protein, unsustainable agricultural intensification, and destructive harvest of natural resources."[5]

2.1 *The Background to the Outbreak of COVID-19*

COVID-19 is generally thought to have originated from bats or pangolins, and available medical evidence explains this further as being a consequence of natural evolution and not genetic engineering.[6] According to this research, there are two possible scenarios on how natural evolution may have produced COVID-19. In the first scenario, it is plausible that the virus evolved to its current pathogenic state through natural selection in a non-human host and then jumped to humans; while the second scenario postulates that a non-pathogenic version of the virus possibly

[1] https://www.soas.ac.uk/cedep-demos/000_P500_ESM_K3736-Demo/unit1/page_08.htm (Accessed on 31.08.2021); Larsson (1999), pp. 156–157.

[2] Fracchia (2005), pp. 2–7.

[3] McBratney et al. (2014), p. 203.

[4] Rume and Islam (2020), pp. 2–5.

[5] McNeely (2021), p. 768.

[6] Scripps Research Institute "COVID-19 coronavirus epidemic has a natural origin." ScienceDaily https://www.sciencedaily.com/releases/2020/03/200317175442.htm (Accessed on 31.08.2021); Andersen et al. (2020), p. 450.

jumped from an animal host into humans and then evolved to its current pathogenic state within the human population.[7]

Other research findings validate the finding that the COVID-19 pandemic may be a product of natural evolution, owing to the global presence of enabling factors and underlying vulnerabilities, which include ecosystem and land use change, species endangerment or extinction, illegal and poorly regulated wildlife trade and climate change.[8]

Further, recent research reveals that zoonotic diseases appear to emerge most commonly when humans create situations in which the probability of transmission of disease-causing microbes increases,[9] and these situations involve elevated contact between humans and some wild species due to three major factors. The first factor is increased conversion of natural land cover to human-dominated land uses with resulting loss of habitat, increased direct and indirect contact between humans and their livestock and a widening range of wildlife species and their viromes, while habitat disturbance perturbs host-disease equilibria and in some cases increases disease agent shedding. The second factor is agricultural intensification, often combined with production of domesticated animals in more intensive settings, which creates ideal conditions for disease organisms to thrive, evolve, and make the jump to human hosts. The third factor is wildlife exploitation—the production, sale and consumption of a subset of wildlife with an active trade moving animals to urban centers, while mixing species of wild animals and their microbes with each other and domestic animals and of course increasing the exposure to humans.[10]

The analysis in this section underscores the 'One Health' concept which has, for almost a century, advocated that "human health and animal health are interdependent and are bound to the health of the ecosystems in which they exist."[11] Human health is closely linked to animal and environmental health; and as such, health issues are integrated between and amongst human beings, animals, and the environment, including other living beings, such as plants. 'One Health' has become a mainstream approach for recognizing the intimate connections among humans, animals, ecosystems, and economies,[12] by building collaboration in policies, legislation, and research. All these factors rest on the direct and indirect consideration of soil health,

[7]Scripps Research Institute "COVID-19 coronavirus has a natural origin." ScienceDaily https://www.sciencedaily.com/releases/2020/03/200317175442.htm 451 (Accessed on 31.08.2021).

[8]Claudia S. de Windt, Environmental Rule of Law: Pandemics Now and the Next Time in the Americas (2020) 1–2. Available at https://www.eli.org/events/breaking-news-pandemics-now-next-time (Accessed on 01.09.2021).

[9]Ginzky et al. (2020), p. 10. Available at https://www.umweltbundesamt.de/sites/default/files/medien/1410/publikationen/pp_neueleitmotive_final_bf.pdf (Accessed on 09.09.2021).

[10]https://ipbes.net/sites/default/files/2020-11/20201028%20IPBES%20Pandemics%20Workshop%20Report%20Plain%20Text%20Final.pdf (Accessed on 30.08.2021).

[11]Bonilla-Aldana et al. (2020), p. 234.

[12]Ginzky et al. (2020), p. 6. Available at https://www.umweltbundesamt.de/sites/default/files/medien/1410/publikationen/pp_neueleitmotive_final_bf.pdf (Accessed 09.09.2021).

through deforestation, the growth of crop and livestock production, and urbanization.

This background shows that the COVID-19 out-break is closely linked to human and animal health, which are both consequently interrelated to the environment in which they exist. The following section goes further to make a scientific nexus between human and animal health on one hand and environmental and soil sustainability on the other.

2.2 *The Scientific Link: COVID-19 Pandemic and Sustainable Soil Management*

To start with, soil is a key component of the Earth and the environment, providing "essential services for agricultural production, plant growth, animal habitation, biodiversity, carbon sequestration and environmental quality."[13] The COVID-19 pandemic is currently presenting environmental threats on soil, water and air, wildlife and biological diversity in general. Further scientific research will in time, reveal more linkages between different sources of environmental pollution and effects of the pandemic. It will also be important to understand the projected management scenario of freshwater and particularly, soil pollution in the post-COVID-19 pandemic era.

Healthy soils are important for both recovery from the COVID-19 pandemic and long-term environmental sustainability. Soil health is connected to many environmental benefits, including sustained good yields and many other ecosystem functions for present and future generations. In order to hasten global recovery from the COVID-19 pandemic, soil quality and functionality must be restored through the application of "modern innovations that strengthen the resilience of the local food production system while improving environmental quality."[14] This argument further underscores the importance of the link between the COVID-19 pandemic and human wellbeing on the one hand and that of soil health and functionality on the other.

Soil science addresses several interrelated environmental challenges such as climate change, global food production, biodiversity loss, water quality and quantity, and biodiversity. The global COVID-19 crisis has forced the soil science discipline to address short-term but profound issues such as changes in the food supply chain and its possible relation to soil management as well as human health implications. According to Poch et al., the resilience of global populations facing the COVID-19 pandemic can be increased by ensuring more heavy reliance on agriculture and local food production,[15] using soils that are managed sustainably. To reach this goal, the Intergovernmental Technical Panel on Soils (ITPS) of the Food and Agriculture

[13] Hou et al. (2020), p. 729.

[14] Lal et al. (2020), p. 2.

[15] Poch et al. (2020), pp. 542–543.

Organization of the United Nations (FAO), has recommended five active strategies: improved access to land, sound land use planning, SSM, enhanced research, and investments in education and extension.[16]

Forecasting beyond the pandemic, further research will be required to understand the effect of COVID-19 on food production and public health issues. The pandemic has spread across the globe, including "rural regions where subsistence farming and smallholders are dominant,"[17] and thereby possibly compromising soil quality. While a lot is known about viruses in the soil, the potential presence and behavior of COVID-19 in the soil is not fully known. Suffice that the effects of the pandemic on soil quality may occur through several routes: interrupted food chain, irregular food production, disturbance to agricultural commodities, and opportunistic soil management behavior.[18]

Another challenge that forestalls development and enforcement of principles around soil during environmental challenges is that soil is not effectively marketed as an important resource. According to Lal et al, "soil science must also be connected with the general public—as the food consumer and beneficiary of other ecosystem services provided through soil and its management."[19] This will ensure that all players in the soil cycle are rewarded through appropriate incentives. Imbedded in the suggestion to connect soil science to the general public is the issue of increasing awareness about the importance of soil in addressing existing and emerging global issues such as the COVID-19 pandemic.

The outbreak of the pandemic and ensuing responses needs to bring soil science, experts, society, policy making, and politics together. There is urgent need for both international scientific collaboration and the creation of open, accessible, and reliable sources of information on the full impact of the pandemic. Given this opportunity, it is important that national and international efforts be channeled towards developing law and policy reform to address effects of such pandemics now and in the future.

3 COVID-19 Pandemic Threats on the Global Environment

The COVID-19 pandemic continues to threaten almost all the aspects of the global environment. As seen in the previous section, human, animal and environmental health are very closely intertwined and transcend national boundaries. This section examines how the pandemic might affect international relations and the global environment, and in turn compromise soil sustainability.

---

[16] Poch et al. (2020), pp. 544–545.

[17] Lal et al. (2020), p. 9.

[18] Nicola et al. (2020), pp. 185–193.

[19] Lal et al. (2020), p. 10.

3.1 Threat to Global Peace and Security

The United Nations has previously declared matters of international public health as threats to global peace and security.[20] In line with this precedent, it has been argued that the effects of the COVID-19 pandemic surpass those of earlier break outs of diseases such as the Ebola virus and must therefore be declared as one of the biggest threats that the world has ever had to resolve.[21] The UN Secretary General outlined eight likely ways of the pandemic undermining global peace and security; noting that humanitarian workers were being hampered in their responses to war torn areas and those ravaged by environmental and natural calamities, making their work all the more insurmountable in the wake of the COVID-19 pandemic.[22] Given the numerous ramifications of a breach in global peace and security, it is not far-fetched that there might be some adverse impacts on SSM.

3.2 Threat to Global Food Security

A healthy soil "with optimal ranges of physical, chemical and biological properties is a prerequisite for the production of healthy and safe food, feed, fiber, and bioenergy crops."[23] Regrettably, the outbreak of COVID-19 is likely to have a negative effect on the optimal ranges of soil composition in various ways, and in turn pose a huge threat to global food security. From the vantage viewpoint of the World Food programme, the pandemic is ravaging global food security from the source; COVID-19 poses a great threat to food security in countries already battling pervasive poverty and poor healthcare infrastructure.[24]

Additionally, it is arguable that the outbreak of COVID-19 has shown the flaws and weaknesses that prevail in the existing food supply system that depends on "high volumes of external inputs, long distance and international supply chains" and thrives on "disregard for soil health and the health of those who consume the food, overlooking global health as being intrinsically linked to the health of our soils."[25] While it is acknowledged that significant progress has been made towards addressing global hunger, one major flaw of the food supply system is that a significant proportion of the global population still faces extreme hunger; and at

[20] Brundtland (2003), pp. 417–420.

[21] Svicevic (2020). Available at: https://www.ejiltalk.org/covid-19-as-a-threat-to-international-peace-and-security-what-place-for-the-un-security-council/ (Accessed on 31.08.2021).

[22] https://www.un.org/press/en/2020/sc14241.doc.htm (Accessed 02.09.2021).

[23] Altier and Abreo (2020), p. 9.

[24] https://www.wfp.org/stories/coronavirus-and-5-major-threats-it-poses-global-food-security (Accessed 01.09.2021).

[25] Moyer (2020), p. 581.

least two billion people face 'hidden hunger' owing to "lack of key micronutrients, particularly iron, zinc, vitamin A and iodine."[26]

3.3 Turning Point for Globalisation

The COVID-19 pandemic has been identified as a barrier to globalisation especially in terms of negatively impacting international trade and travel,[27] which in turn have yielded both positive and negative environmental and soil related impacts. Goffman argues generally that "as horrible as the coronavirus is, it might be an opportunity to correct a global economy that ravages the environment.[28] Analyzed in terms of air quality and public health, this might be plausible; but the tonnage of waste occasioned on the soil, land, and environment in general tends to tilt this argument. Therefore, the conclusion is that the pandemic has mostly generated negative consequences for soil sustainability. On the contrary, environmental protection and soil sustainability, as global concerns, both require the cooperation of both national and international players. These are some of the possible threats arising from the pandemic that need to be followed through with research in order to ascertain whether they actually exist and how they can be mitigated.

3.4 COVID-19 Pandemic and Air Pollution

The scaling down of adverse human-made factors in number and intensity has reportedly led to improved environmental health in some situations. For instance, there are several reports highlighting the positive role of the lockdowns which have been implemented in the wake of COVID-19; some reportedly helping the environment, habitats and associated biodiversity in many countries.[29] The most frequently reported positive aspect is in the improvement of air quality. Some of the positive environmental out-comes of the COVID-19 pandemic have largely focused on reduced industrial activity and mechanized transport, which have consequently lowered emissions and air pollution worldwide.[30]

According to a recent research focusing on the Gulf of Mannar between the west coast of Sri Lanka and the south-eastern tip of India, it was found that there has been a reduction in the impact of human-induced factors that are capable of enhancing

[26] https://www.futureoffood.ox.ac.uk/food-system-challenges (Accessed 01.09.2021).

[27] Shrestha et al. (2020), pp. 3–4.

[28] Goffman (2020), pp. 48–52.

[29] Henriques (2020), Mani (2020) and Stone (2020).

[30] Airborne Nitrogen Dioxide Plummets over China (NASA Earth Observatory, 2020); https://go.nature.com/397mtEl (Accessed 29.08.2021).

environmental health.[31] The research compared pre- and post-lockdown data relating to water parameters such as turbidity, nutrient concentration and microbial and dissolved oxygen levels, phytoplankton and fish densities. The lockdowns imposed in India and many other countries in the world as a result of the COVID-19 pandemic have restricted and halted human activity, thereby reducing anthropogenic stress to nature.

3.5 COVID-19 Pandemic and Water Pollution

Remarkable improvement in the water quality during the COVID-19 pandemic has been observed in certain countries, owing to heightened measures for purification as numerous studies show.[32] Healthier water has generally led to healthier aquatic and non-aquatic forms of life, including the soil, because water and soil are the basic natural resources that sustain ecosystems. Further research has however established substantial increase in evidence showing the presence of pathogenic viruses in the wastewater and/or treatment plants, including the novel coronavirus.[33] The research reviewed the occurrence and fate of viruses in wastewater, associated risks of disease outbreaks and the possible methods for virus disinfection during wastewater treatment. Some other research has since concluded that there is presently no additional risk of infection from wastewater. For example, in a research conducted in different cities of the Federal State of North Rhine-Westphalia (Germany) on the same day in April 2020, around the time of the first peak of the pandemic, the analysis of evidence suggested that wastewater is presently not a major route for transmission of COVID-19 to human beings.[34]

Moreover, the bibliometric tracking of global publications relating to all the aspects of COVID-19 has identified over 200,000 publications. Among these, 262 papers were published by authors from 67 countries regarding the relationship between COVID-19 and water science and technology. Although the transmission routes of COVID-19 in the water cycle have not been proved in any published research, water and wastewater play an important role in the control of the COVID-19 pandemic. Further, whereas drinking water is generally safe, most of the research publications show that wastewater may only be a potential risk during the pandemic.[35] Given this potential risk, further detailed research is important to determine the possibility of eventual soil contamination arising from the COVID-19 pandemic.

While it is widely acknowledged that lock down periods have since brought about an appreciable scaling down of some anthropogenic activities that threaten aspects of

[31] Edward et al. (2021), pp. 16–17.

[32] Link (2020), Selvam et al. (2020) and Yunus et al. (2020).

[33] Bhatt et al. (2020), p. 8.

[34] Westhaus et al. (2021), p. 5.

[35] Bin et al. (2021), pp. 1–2, 8.

environmental health, the effect of enhanced hygienic practices has had quite the opposite effect. According to research undertaken in Poland, the increased use of water, various soaps, cleaning detergents, hand sanitisers, surgical masks and other Personal Protective Equipment (PPE) which eventually end up deposited as solid waste on the environment, posing varying degrees of negative impacts on soil sustainability such as....[36] Further, this potential for increased water and soil pollution has arisen from the use of soaps and detergents in large quantities during the pandemic as washing hands and cleaning surfaces using soap and other detergents have been entrenched as new and unavoidable habits in the local communities.[37] This shows that frequent handwashing with detergents may be releasing more toxic substances into the water and soil profile and eventually the general environment.

3.6 *COVID-19 Pandemic and Land/Soil Pollution*

The discovery and emergence of the Covid-19, and its declaration as a public health emergency of international concern (PHEIC) by the World Health Organization (WHO) have influenced decisions around the generation, management, and disposal of solid waste. This waste is primarily deposited on the soil and land, presenting contamination prospects and problems of waste management. According to research, COVID-19 remains viable on the surface of objects for a few hours and up to several days,[38] thereby retaining the potential to contaminate the soils.

Owing to the COVID-19 pandemic, there has emerged a huge increase in the amount of used personal protective equipment (PPE) such as facemasks, gloves, and other protective stuffs, and wide distribution of infectious wastes from hospitals, health care facilities, and quarantined households.[39] Therefore, the huge amount of PPE disposed of on a daily basis is infectious waste, and it must, in line with most global environmental laws and policies, be treated as medical or hazardous waste in order to forestall negative environmental and health impacts. In addition, the amount of food and plastic waste is also likely to have increased during the pandemic when most food outlets are offering services on take-away basis. These factors have undoubtedly overwhelmed waste management facilities in many developing and transitional countries. This is especially so because the insurmountable tonnage of infectious wastes being produced during the COVID-19 pandemic is "not only generated from the health facilities such as hospitals and temporal medical facilities, but also households and quarantine centre facilities."[40]

---

[36] Nowakowski et al. (2020), p. 7.

[37] Daverey and Dutta (2021), p. 104754.

[38] Kampf et al. (2020), pp. 246–251.

[39] Sharma et al. (2020), p. 105052.

[40] Hantoko et al. (2021), p. 8.

These activities are likely going against established laws and policies for waste management and are injurious to environmental wellbeing. SSM comes under serious threat owing to the increased likelihood of soil contamination. In the specific case of China, the COVID-19 pandemic has greatly increased reliance on single use plastics despite them having been banned prior to the outbreak,[41] owing to the real threat they pose to the environment. Single use plastics are being banned across the globe, due to their existential threat to soil, air, and aquatic ecosystems, wildlife and human health.[42]

Although global best practices have emerged on disposal of COVID-19 hazardous wastes, such as delaying collection for up to 72 h to avoid the life span of COVID-19 in the environment,[43] it must be pointed out that most of the developing and transitional countries are unlikely to implement such a rigorous and meticulous plan needing infrastructural support. To augment this point, Nzediegwu and Chang, 2020 argue that in developing countries, solid wastes are discharged in open or unsanitary landfills where scavengers collect recyclable materials without any personal protective tools.[44] Further, poorly managed landfills in most of the developing world are recognised as a "food source" for domesticated animals, which practice may arguably increase the spread of zoonotic diseases including COVID-19. Land use protection mechanisms are not being observed, yet the interaction between animals and human beings is what probably led to the COVID 19 outbreak. Ultimately, these high levels of contamination eventually end up in the soil and on other components of the environment such as water and then filter through ecosystems. It is clear that the application of environmental law needs to be enhanced in the post-COVID 19 era.

The impact of 'mask waste' on the environment is immense and has increased across the world as people are not following the appropriate disposal methods for used masks.[45] This is likely to create a new environmental challenge which ultimately may have a negative impact on soil sustainability due to increased waste volume. Further, there are no appropriate mask or plastic waste collecting method specified in most countries, such as parts of Sri Lanka, India, Pakistan and China.[46] In a study that was conducted in the UK, to analyze the possibility of each individual using one disposable surgical mask every day for a year, it was found that this would translate into several hundred thousands of tons of unrecyclable plastic waste; several thousands of tons of contaminated waste and another several thousand tons of plastic packaging.[47] However, there is currently no specific waste stream for these

[41] Liu et al. (2021), pp. 237–238.

[42] Barboza et al. (2018), pp. 344–345.

[43] Hantoko et al. (2021), pp. 1–5; Parashar and Hait (2021), pp. 1–2; 5–13.

[44] Nzediegwu et al. (2020), p. 519.

[45] Selvaranjan et al. (2021), pp. 10–12.

[46] Sangkham (2020), p. 11.

[47] Ayse et al. (2020), p. 9.

highly contaminated products. Such waste is most likely thrown recklessly in the streets or collected and treated as mixed, normal domestic waste.

The used masks which are eventually littered on the surface of the soil can negatively impact the fauna. For instance, it has been reported in Columbia that a bird was tangled in a discarded coronavirus facemask in a tree.[48] Further, where a used mask is mistaken for food by an animal, the plastic parts can fill stomachs, decrease food intake, and cause animals to starve and die. Furthermore, such dead animals can potentially contaminate the soil and land on which their bodies eventually decompose.

The increase in waste generation during the COVID-19 pandemic notwithstanding, it must be mentioned that there may be a disparity between amounts accumulated in urban and rural areas. This assumption rests on the fact that rural communities especially in poorer parts of most communities the world over, generally bear the brunt of social ills; and the COVID-19 crisis has "massively accelerated some pre-existing trends . . . setting in motion waves of change with a wide range of possible trajectories."[49] In other words, the pandemic has merely exacerbated existing differences in wealth, income, and social status.

A huge volume of medical waste is now being produced in response to the COVID-19 outbreak, including used personal protective equipment, tissue papers, plastic bags, and empty bottles of sanitizers and hand soaps.[50] Effective management of medical waste through incineration or decontamination processes should be carefully conducted before any exposure to the environment occurs (e.g., soil, water, and ocean). The daily usage of face masks and gloves by millions of people may create high risks to the community and environment.[51] Soil pollution that emanates from increased medical and other organic and inorganic wastes should be closely monitored and controlled as it is a novel occurrence, yet threatening the already threatened existence of biological diversity.[52]

This section has shown that almost all environmental media are under threat of contamination from the COVID-19 pandemic. As further shown in this section, these possibilities of contamination transcend national boundaries in the same way the pandemic has touched every part of the globe. This is a call for concerted action in addressing and reversing COVID-19 related environmental impacts.

[48] Boyle (2020), p. 1; Selvaranjan (2021), p. 6.

[49] https://www.oecd.org/coronavirus/policy-responses/the-territorial-impact-of-covid-19-managing-the-crisis-across-levels-of-government-d3e314e1/ (Accessed 03.09.2021).

[50] Zambrano-Monserrate et al. (2020), p. 728.

[51] Saadat et al. (2020), p. 56.

[52] Lal et al. (2020), p. 9.

4 Soil Related Socio-economic Impacts of COVID-19

The discussion in this chapter has so far revealed that there is an interactive link between COVID-19 and the environment, through the public environmental health limb. This section examines the COVID-19 pandemic responses in Zambia and other emerging economies in southern Africa.

4.1 *Zambia Country Specific Measures for Managing the COVID-19 Pandemic*

According to the OECD, Zambia is a lower middle-income country in the group of least developed countries (LDCs), with high levels of economic and social vulnerabilities.[53] Given these socio-economic vulnerabilities, most LDCs usually the divert financial resources from the social and economic sectors, to public health crises such as the COVID-19 pandemic.[54] It will be important to discuss how such possible diversion of financial resources would affect the sustainability agenda in Zambia.

To start with, the Zambian government, declared COVID-19 as a public health emergency pursuant to two statutory instruments (SIs), under the Public Health Act, and not via the constitutional emergency provisions.[55] The effect of adopting this legal route is that the "observance of human rights, constitutionalism and accountability have largely been glossed over, thereby undermining the success of the [Zambian] government's response to fight the COVID-19 pandemic."[56] This notwithstanding, some directives aimed at containing the pandemic, such as restriction of public gatherings and mandatory wearing of masks in public places, amongst others were implemented. The effect of these measures on soil sustainability is discussed below in relation to availability of financial resources for the provision of socio-economic services.

By March, 2020, the Government of the Republic of Zambia, had prepared the COVID-19 National multi Sectoral Contingency and Response Plan (NMRCP). The NMCRP outlines potential health and non-health impact of the COVID-19 pandemic in the country, of which two (2) non-health impacts are related to the enforcement of environmental laws and policies and soil sustainability:

---

[53] https://www.oecd.org/dac/financing-sustainable-development/development-finance-topics/Zambia-Transition-Finance-country-pilot.pdf (accessed 03.09.2021).

[54] https://www.oecd.org/coronavirus/policy-responses/covid-19-and-africa-socio-economic-implications-and-policy-responses-96e1b282/ (Accessed 03.09.2021).

[55] Kalunga (2020) https://ancl-radc.org.za/node/632 (Accessed 05.09.2021).

[56] Kalunga (2020) https://ancl-radc.org.za/node/632 (Accessed on 02.09.2021).

(a) Diversion of resources from routine services to response to COVID-19;
(b) Reduced tourist visits to national parks in the affected areas due to suspended flights;
(c) Reduced forex due to low earnings from non-traditional exports;
(d) Disruption in the provision of socio-economic services.[57]

It is notable that the diversion of resources from routine services and the disruption in the provision of socio-economic services in Zambia have in turn inconvenienced the enjoyment of economic, social, and cultural rights. In relation to the discussion in this chapter, the guarantee of the rights to an adequate standard of living, food, a clean, safe and healthy environment are relevant. With increased waste generation due to the pandemic and disruption in the provision of routine services such as proper waste management, the enjoyment of the right to a clean, safe and healthy environment has been compromised.

It has been argued in the context of Zambia and other countries with economies in transition that some of the government measures and approaches towards combating the COVID-19 pandemic have impacted negatively on social and economic rights; the argument being premised on the doctrine of the indivisibility and interdependence of human rights.[58] The exercise of the right to a clean, safe and healthy environment cannot be guaranteed in a physical space devoid of other indivisible rights. To extend this argument, the suppression of any rights compromises the existence and quality of human life in a wholesome environment, with functional ecosystems that benefit from SSM. Overall, a new climate of uncertainty has been unleashed and is "fueling protectionism and playing into nationalist narratives."[59] This view shows that there is a newly emerging argument pointing to the possibility of the COVID-19 pandemic endangering the rule of law in democratic societies of countries in transition.

4.2 *COVID-19 and Social, Cultural and Economic Rights in Southern Africa*

As seen from research conducted in some countries in Southern Africa like South Africa, Zambia and Zimbabwe, the fear of harsh socio-economic conditions might even be worse than that of the pandemic. For instance, the majority of the African population works in the informal sector and live from hand to mouth; and the COVID-19 pandemic has now disrupted their major sources of livelihood.[60]

---

[57] Non-Health Impacts 2.2.2 COVID-19 13.
[58] Kaaba (2020), p. 108.
[59] Yaya et al. (2020), pp. 1–4.
[60] Metta et al. (2021), p. 628.

Banda et al. interrogate the public health systems and socio-economic implications of the COVID-19 pandemic on both South Africa and Zimbabwe and the authors conclude that the level and effectiveness of the response strategies to this alarming health crisis are affected by the fact that South Africa assumes a middle-income status and Zimbabwe a low-income one.[61] According to the authors, these economic disparities and differential income status, notwithstanding, the pandemic has revealed that both countries have similar deep-rooted problems in their already overburdened public health systems.[62] These problems are in turn exacerbated by underlying socio-economic factors such as poverty, unemployment, inequality, slow economic development, inadequate water and sanitation, and food insecurity.[63] Given this scenario, the conclusion is that the citizens in most southern African countries, are not guaranteed their social, economic and cultural rights, let alone, the right to life. The right to a clean, safe and healthy environment, embedded in the right to life, appears to be a luxury. In this case, efforts to ensure SSM are less likely to be given due attention, unless, perhaps presented as a food security issue.

As already discussed in this chapter, most of Africa is already overwhelmed by many issues such as severe effects of climate change poverty, high unemployment rates, wildlife poaching, desertification, deforestation, water pollution, waste management and land degradation.[64] These challenges are already placing immense pressure on natural resources especially water and soil, without the separate threats occasioned by the COVID-19 pandemic.

5 Law and Policy Research: COVID-19 Pandemic Threats and Sustainable Soil Management

How can the concomitant threats on both environmental and soil sustainability be addressed? The full effects of the diverse threats induced by the COVID-19 pandemic on governance of natural resources in general can only be well-understood and mitigated by more concerted research. There is need for further research on the threats posed by the COVID-19 pandemic on SSM in order to inform corresponding law and policy reform.

[61] Banda et al. (2021), p. 2.

[62] Banda et al. (2021), p. 3.

[63] Banda et al. (2021), pp. 17–19.

[64] Chirisa et al. (2020), pp. 2–5.

5.1 Research Prospects: COVID-19 Pandemic Threats and SSM

This chapter has already alluded to the necessity of further research on the threats posed by the COVID-19 pandemic on soil sustainability. Numerous studies have shown that the COVID-19 pandemic lockdowns have yielded both negative and positive outcomes for the environment in general. This has also presented a rare opportunity to compare the health of entire ecosystems before and after lockdowns in order to understand the magnitude of ecological recovery without human intervention. This outcome is important for present and future law and policy purposes in the public health and environmental sphere as it would confirm the full extent of any negative human involvement and how to minimize it.

Lindsey et al. argue that although the COVID-19 pandemic associated environmental changes may benefit biodiversity conservation in some ways, in the context of Africa, these "positive environmental outcomes are likely temporary and prone to reversal when travel restrictions ease. . ."[65] Ultimately, the totality of these negative outcomes is likely to have a snowball effect on soil sustainability.

In addition, it might be worth contemplating what would happen to the environment generally and soil sustainability specifically, once the pandemic is over. For instance, given that the pandemic has brought about significant variations in air, land and water pollution, would these suffer reversal once the pandemic ends? Would there be rebound effects? Further, researchers could benefit from this unprecedented time by commissioning studies into the impact of human activity on the natural world. Arguably, such research would elicit information that could possibly help improve conservation and biodiversity efforts.[66]

5.2 Need to Legislate for COVID-19 Pandemic Threats on SSM

Efforts have been made at the global level towards addressing environmental sustainability in general and SSM in particular. It merits discussion whether these efforts have been scaled up during the emergence of the COVID-19 pandemic. Are these emerging efforts to develop law and policy on global soil management and sustainability likely to bear fruit?

[65] Lindsey et al. (2020), p. 1302.

[66] https://www.medicalnewstoday.com/articles/how-covid-19-has-changed-the-face-of-the-natural-world#COVID-19-and-air-and-water-pollution (Accessed on 05.09.2021).

5.2.1 Laws and Policies Addressing Public Health Disasters

Most countries have general legal provisions addressing emergencies and disasters of the COVID-19 magnitude. For instance, a research conducted in 2021 reveals that three main approaches have been employed. First, some countries rely on constitutional provisions permitting emergency approaches. Secondly, other countries interpret existing legislation to provide the basis for emergency measures or adopt new legislation; and yet another group of countries adopt a third, but more ambiguous approach that has a less explicit legal basis, such as executive measures that lack (or appear to lack) a specific legal grounding.[67] What is not certain, in most instances, however, is how to address the impact of these disasters on specific aspects of the environment. Drawing from scientific input on the effects of the COVID-19 pandemic on natural resources such as air, water and soil, it should be possible to craft guidelines for ensuring SSM.

5.2.2 UN Sustainable Development Goals (SDGs)

Although the United Nations adopted seventeen sustainability goals in 2015, to date, the effects of the UN Sustainable Development Goals (SDGs) global goals are not yet reflected in people's lifestyles. This is so, despite research showing that sustainable management of soil health is important for achieving several SDGs—SDGs 1 (End Poverty), 2 (Zero Hunger), 3 (Good Health and Wellbeing), 5 (Gender Equality), 6 (Clean Water and Sanitation), 7 (Affordable and Clean Energy), 9 (Industry, Innovation and Infrastructure), 11 (Sustainable Cities and Communities), 12 (Responsible Consumption and Production), 13 (Climate Action) and 15 (Life on Land), particularly SGD 15 (15.3), which directly concerns soils. Most of these SDGs rely considerably on plant production while the others depend on soil processes such as "water movement, heat transfer, sorption and physical filtration, ion exchange, and biochemical and biophysical transformations."[68]

According to de Windt, the present pandemic is just another symptom of how humankind has resisted changing behavior that negatively affects human and nature systems; for instance, "our interactions-land use and ecosystem change, change, species endangerment and extinction, human induced climate change on top of underlying vulnerabilities" still shape our "diseased ecology and no-one seems to have the capacity to absorb or cope with the negative effects and guarantee our fundamental rights."[69]

The COVID-19 era has increasingly highlighted the need for a broader understanding of sustainability, as many of the current ways of working are changing

---

[67] https://www.idea.int/sites/default/files/publications/emergency-law-responses-to-covid19.pdf (Accessed on 09.09.2021).

[68] Rattan et al. (2021), p. 23.

[69] De Windt (2020), p. 12.

completely. At the forefront is the holistic concept of sustainability, which refers to the social, cultural, economic, environmental, technological, and ethical fields and touches not only the global society, but also derives from the values of the individual.[70] The achievement of the SDGs can therefore be mirrored in efforts aimed at combatting the COVID-19 pandemic; both the COVID-19 pandemic and the achievement of the SDGs can be accentuated by focusing on soil sustainability. This reality therefore begs the question for the development of guiding principles for soil sustainability.

5.2.3 Towards Global Legislation for Soil

The COVID-19 crisis has shown the need for immediate and effective implementation of the 'One Health' strategy where soil health, plants, animals, people, and environment are a whole indivisible resource. The 'One Health' approach leverages animal health, human health and environmental health; and can therefore be used as a guiding principle for pandemic prevention policies.[71] According to research, there are many explanations for the connectedness of soil to the 'One Health' strategy. Two explanations are relevant here; first, soil health is responsible for the nutrients that eventually filter to the crops grown in it and used as food for both animal and human sustenance.[72] Secondly, soil is important for the sequestration of excess carbon that can amplify the effect of global warming.[73] These two explanations are important because they relate to addressing food security and well as climate change, both of which are current global challenges of huge impact. This underscores the urgency for concerted efforts towards legislating in the interest of soil sustainability.

In addition to the age old "One Health' approach, there have been international initiatives especially under the auspices of the UN Food and Agriculture Organization's (FAO) and the UN Convention on Biological Diversity (CBD) on the status of global soil biodiversity. On December 4, 2020 on the occasion of World Soil Day, the report "*State of Knowledge on Soil Biodiversity: Status, Challenges, and Potentialities*" was released. This report focuses on the campaign "*Keep soil alive, protect soil biodiversity*" and primarily aims at raising awareness of the importance of maintaining healthy ecosystems and human wellbeing, by addressing the growing challenges in soil management. Further, the report increases awareness about soils,

---

[70] Vovk Korze et al. (2021), p. 157.

[71] Ginzky et al. (2020), pp. 16–17; Keith et al. (2016), pp. 40–42.

[72] Brevik and Sauer (2015), pp. 35–46.

[73] Hannam (2019), pp. 400–402; 420–422.

fighting soil biodiversity loss and encouraging governments and communities around the world to commit to proactively improving soil health.[74]

In a first comprehensive meta-analysis of medical waste management practices, including medical waste generation rates, types, and composition conducted in 78 countries, it was concluded that immediate, collective, voluntary, and effective measures must be initiated for environmentally sustainable management of medical waste.[75] This is particularly required in countries with transitional economies, where the medical waste generation rate has increased, courtesy of the COVID-19 pandemic.

Evidence shows that conservation of intact ecosystems and their characteristic biodiversity can reduce[76] the emergence of infectious diseases. To echo Poch et al., "soils are a finite, non-renewable, multi-systemic source of life, and still they are easily overlooked in decision-making acts and policies"[77] This renews earlier global calls and initiatives for the development of soil protection legislation, where soil healthy should be focus across all sectors.

In a comprehensive study of the history, concepts, milestones and developments in relation to the international law and policy aspects of soil conservation in 2016, Boer et al. established that soil, as "the substantive constitutive element of land, should be regarded as a common concern of human-kind."[78] The authors then conclude that international regulation of soil is important if, first, soil degradation would result in transboundary effects; secondly, if the control of soil degradation and achievement of LDN could work best through international cooperation and thirdly, if soil degradation could result in political or military conflict between states.[79] These views are further echoed by Ginzky, who presents some conceptual thoughts which survey effective international soil governance for the future which include possibilities of a soil treaty.[80] Given the present circumstances of the COVID-19 crisis which is presenting new threats to soil sustainability, good governance for the future of soil sustainability must be effected now.

At international law, the legal basis for soil protection is provided by the UNCCD, UNFCC and CBD. In a 2021 article, Heuser argues that these international legal instruments are inadequate for purposes of advancing "a comprehensive approach to protect and sustainably use the soils. Soils demand attention at the same level as biological diversity and climate change."[81] Consequently, in order to

[74]FAO, ITPS, GSBI, SCBD, and EC (2020). State of knowledge of soil biodiversity - Status, challenges and potentialities, Report 2020. Rome, FAO. Accessed from the following link on 01.09.2021, http://www.fao.org/3/cb1928en/CB1928EN.pdf.

[75]Singh et al. (2021), pp. 2–3; 18.

[76]Diaz et al. (2015), pp. 1–16.

[77]Poch et al. (2020), p. 545.

[78]Boer et al. (2017), p. 50.

[79]Boer et al. (2017), p. 67.

[80]Ginzky (2020), pp. 44–50.

[81]Heuser (2021) Abstract.

provide adequate protection to soils, what is needed is "a specific binding legal instrument is needed on the international level, for example a protocol on land degradation neutrality and the protection and sustainable use of soil, as proposed in a draft by the World Commission on Environmental Law of the International Union for the Conservation of Nature."[82]

In the wake of the COVID-19 pandemic and the threats it brings to SSM, this chapter extends these well-thought arguments for the development of soil protection legislation. The recommendation for global soil regulation, amongst others is very important in relation to addressing threats from the pandemic. A new post-pandemic reality should ensure that soil is recognized in law and policy as the great connector and service provider that links our lives to all human needs of food, health and security.[83] Caring for soils is imperative to reduce the impacts of global disturbances, such as the current COVID-19 crisis. According to McNeely, the COVID-19 pandemic can be a "catalyst to set the global society on a new path to a sustainable relationship between people and the rest of nature."[84] The COVID-19 crisis should be the turning point for ensuring that similar disasters do not cause widespread negative threats on soil sustainability.

6 Conclusion and Future Policy Considerations

In spite of its exploratory nature, this chapter proposes some measures for addressing SSM threats from the COVID-19 pandemic. It is important to start by exploiting the available and workable mechanisms, institutions and procedures under the existing laws. These might relate to how both SSM and the COVID-19 pandemic are being addressed in the current law and policy. In addition, it is suggested that innovative ways of addressing both SSM and the pandemic be explored, with the inclusion of soils at all levels of education curricula if awareness on the importance of soils in our lives is to be increased.

Although there are numerous uncertainties surrounding the actual threats that the COVID-19 pandemic presents to global SSM, it must be considered that at the national level, enhancing and coordinating the existing national policies on land tenure, land use planning, waste management and other sectors closely aligned to soil management and public health can be the starting point. In relation to waste management, it has been shown that in most countries with emerging and transition economies, hazardous waste generation has increased and there are no corresponding disposal facilities. Further, few measures have been developed for ensuring that contaminated waste does not come into contact with land, water and soil. Furthermore, the prospect of contamination from waste water, detergents and

[82] Heuser (2021) Abstract.

[83] Moyer (2020), pp. 581–582.

[84] McNeely (2021), p. 778.

other chemical substances being used during the pandemic also needs to be investigated. At the international level, it will also be important to capitalise on the existing initiatives towards legislating for sustainable soil governance. The severity of the threats on soil health that are projected from the pandemic should give impetus to the international soil law initiatives to come to fruition at the earliest.

In line with the 'One Health' approach, there is need to develop sustainable and resilient soil systems for soil sustainability and food production. This will require concerted research efforts with renewed approaches towards interdisciplinarity, if the linkages between and among soil, plant, animal, human and environmental health are to be exhausted. The have been threats to SSM before the outbreak of the pandemic, however, new research dealing with the preservation of soil quality for agriculture will be even more important now that the COVID-19 crisis presents novel threats of soil contamination which could in turn compromise global food production and security.

In addition to all the proposed initiatives and future policy suggestions, this chapter restates recommendations relating to the need for resilience in the global populations facing the COVID-19 pandemic. The Intergovernmental Technical Panel on Soils (ITPS) of the Food and Agriculture Organization of the United Nations (FAO), has advanced five active strategies for ensuring the availability of adequate productive soil that can be managed sustainably in order to feed national populations, namely: improved access to land, sound land use planning, SSM, enhanced research and investments in education.[85] Although the Global Soil Partnership of FAO has been addressing these five active strategies through its five pillars of action—soil management, awareness raising, promoting research, information and data harmonization, it is important that these initiatives be adopted at national and international law levels.[86]

References

Altier N, Abreo E (2020) One health: considerations in the International Year of Plant Health. https://doi.org/10.31285/AGRO.24.422

Andersen KG et al (2020) The proximal origin of SARS-CoV-2. Nat Med. https://doi.org/10.1038/s41591-020-0820-9

Ayse LA et al (2020) The environmental dangers of employing single-use face masks as part of a COVID-19 exit strategy. Available at https://ucl.scienceopen.com/document_file/97551 6df-aa0e-4b79-b950-c54e5608a3e0/ScienceOpenPreprint/Covid19-Masks-Journal-paper.submitted.pdf. Accessed 3 Mar 2022

Banda BC et al (2021) 'A tale of two paradoxes in response to COVID-19': public health system and socio-economic implications of the pandemic in South Africa and Zimbabwe. Cogent Soc Sci 7:1. https://doi.org/10.1080/23311886.2020.1869368

[85]Poch et al. (2020), p. 545.

[86]Poch et al. (2020), p. 547.

Barboza LGA et al (2018) Marine microplastic debris: an emerging issue for food security, food safety and human health. Mar Pollut Bull 133:336–348. https://doi.org/10.1016/j.marpolbul.2018.05.047

Bhatt A et al (2020) Occurrence, fates and potential treatment approaches for removal of viruses from wastewater: a review with emphasis on SARS-CoV-2. J Environ Chem Eng 8(5). https://doi.org/10.1016/j.jece.2020.104429

Bin J et al (2021) Water science under the global epidemic of COVID-19: bibliometric tracking on COVID-19 publication and further research needs. J Environ Chem Eng 9(4). https://doi.org/10.1016/j.jece.2021.105357

Boer BW, Ginzky H, Heuser IL (2017) International soil protection law: history, concepts and latest developments. In: Ginzky H, Heuser I, Qin T, Ruppel O, Wegerdt P (eds) International yearbook of soil law and policy 2016. International Yearbook of Soil Law and Policy, vol 2016. Springer, Cham. https://doi.org/10.1007/978-3-319-42508-5_7

Bonilla-Aldana DK, Dhama K, Rodriguez-Morales AJ (2020) Revisiting the one health approach in the context of COVID-19: a look into the ecology of this emerging disease. Adv Anim Vet Sci 8(3):234–237. https://doi.org/10.17582/journal.aavs/2020/8.3.234.237

Boyle L (2020) Bird dies after being tangled in coronavirus mask. Independent New York

Brevik EC, Sauer TJ (2015) The past, present, and future of soils and human health studies. Soil 1:35–46

Brundtland G (2003) Global health and international security. Glob Gov 9(4):417–423. Retrieved June 17, 2021, from http://www.jstor.org/stable/27800493

Chirisa I et al (2020) The urban penalty of COVID-19 lockdowns across the globe: manifestations and lessons for Anglophone sub-Saharan Africa. GeoJournal. https://doi.org/10.1007/s10708-020-10281-6

Daverey A, Dutta K (2021) COVID-19: eco-friendly hand hygiene for human and environmental safety. J Environ Chem Eng 9(2):104754. https://doi.org/10.1016/j.jece.2020.104754. Accessed 25 Mar 2021

de Windt C (2020) Environmental rule of law: pandemics now and the next time in the Americas. Available at: https://www.iucn.org/sites/dev/files/content/documents/2020/pandemics_now_and_next_time_final_paper.pdf

Diaz et al (2015) The IPBES conceptual framework. Connecting nature and people. Curr Opin Environ Sustain. https://doi.org/10.1016/j.cosust.2014.11.002

Edward JKP et al (2021) COVID-19 lockdown improved the health of coastal environment and enhanced the population of reef-fish. Mar Pollut Bull. https://doi.org/10.1016/j.marpolbul.2021.112124

FAO (2020) FAO, CBD and GSBI: state of knowledge of soil biodiversity. Status, challenges and potentialities. Food and Agriculture Organization of the United Nations, Rome, Italy

Fracchia F (2005) The legal definition of environment: from rights to duties, Bocconi Legal Studies Research Paper No. 06-09. Available at SSRN: https://ssrn.com/abstract=850488 or https://doi.org/10.2139/ssrn.850488

Ginzky H (2020) Good governance for "sustainable management of soil" on national and international level: how to do it? In: Yahyah H et al (eds) Legal instruments for sustainable soil management in Africa. International yearbook of soil law and policy, regional perspectives, pp 35–54

Ginzky H et al (2020) Lessons from the corona crisis: new guiding principles required for environmental and sustainability policy? – a discussion paper. www.umweltbundesamt.de/publikationen

Goffman E (2020) In the wake of COVID-19, is glocalization our sustainability future? Sustain: Sci Pract Policy 16(1):48–52

Hannam I (2019) Aspects of a legislative and policy framework to manage soil carbon sequestration. In: Ginzky H et al (eds) International yearbook of soil law and policy, pp 399–433

Hantoko D et al (2021) Challenges and practices on waste management and disposal during COVID-19 pandemic. J Environ Manag. https://doi.org/10.1016/j.jenvman.2021.112140

Henriques M (2020) Will COVID-19 have a lasting impact on the environment? Available at: https://www.bbc.com/future/article/20200326-covid-19-the-im-pact-of-coronavirus-on-the-environment. Accessed 20 Feb 2021

Heuser LI (2021) The legal dimension of soil protection and sustainable use of soils. In: Mauerhofer V (ed) The role of law in governing sustainability, 1st edn. Routledge. https://doi.org/10.4324/9781003158837

Hou D et al (2020) Sustainable soil use and management: an interdisciplinary and systematic approach. Sci Total Environ 729

Kaaba O (2020) Deliberate disregard: socio-economic rights, constitutionalism and the Covid-19 pandemic in Zambia. In: Justice Mavedzenge (ed) COVID-19 pandemic and socio-economic rights in selected east and southern African countries. Juta, Cape Town, pp 107–125

Kalunga FK (2020) Executive disorder, constitutionalism ad COVID – 19 in Zambia. Available at: https://ancl-radc.org.za/node/632. Accessed 10 Jan 2021

Kampf G et al (2020) Persistence of coronaviruses on inanimate surfaces and their inactivation with biocidal agents. J Hosp Infect 104

Keith AM, Schmidt O, McMahon BJ (2016) Soil stewardship as a nexus between ecosystem services and one health. Ecosyst Serv 17:40–42

Lal R et al (2020) Managing soils for recovering from the COVID-19 pandemic. Soil Syst 4(3):46. https://doi.org/10.3390/soilsystems4030046

Larsson M (1999) Legal definitions of the environment and of environmental damage. Scand Stud Law:155–176

Lindsey P et al (2020) Conserving Africa's wildlife and wildlands through the COVID-19 crisis and beyond. Nat Ecol Evol 4:1300–1310. https://doi.org/10.1038/s41559-020-1275-6

Link D (2020) Fact check: COVID-19 crisis has not created decreased long-term human environmental impact. USA Today. Mar 25. https://www.usatoday.com/story/news/factcheck/2020/03/25/fact-check-coronavirus-crisis-benefiting-environment/2908300001/. Accessed 3 Mar 2022

Liu J, Vethaak AD, An L et al (2021) An environmental dilemma for China during the COVID-19 pandemic: the explosion of disposable plastic wastes. Bull Environ Contam Toxicol. https://doi.org/10.1007/s00128-021-03121-x

Mani KS (2020) The lockdown cleaned the Ganga more than 'Namami Gange' ever did. Available at: https://science.thewire.in/environment/ganga-river-lock-down-cleaner-namami-gange-sewage-treatment-ecological-flow/. Accessed 20 Feb 2021

McBratney A et al (2014) The dimensions of soil security. Geoderma 213:203–213. https://doi.org/10.1016/j.geoderma.2013.08.013

McNeely JA (2021) Nature and COVID-19: the pandemic, the environment, and the way ahead. Ambio 50:767–781. https://doi.org/10.1007/s13280-020-01447-0

Metta E et al (2021) Response of the social systems to COVID – 19 in Mozambique, Tanzania and Zambia: a synthesis of the challenges and opportunities. Int J Innov Sci Res Rev 03(01): 625–631. Available online at http://www.journalijisr.com

Moyer J (2020) A time of reflection: a time for change. Agric Hum Values 1(37):581–582. https://doi.org/10.1007/s10460-020-10075

Nicola M et al (2020) The socio-economic implications of the coronavirus and COVID-19 pandemic: a review. Int J Surg

Nowakowski et al (2020) Disposal of personal protective equipment during the COVID-19 pandemic is a challenge for waste collection companies and society: a case study in Poland. Resources 9(10):116. https://doi.org/10.3390/resources9100116

Nzediegwu C et al (2020) Improper solid waste management increases potential for COVID-19 spread in developing countries. Resour Conserv Recycl 161(519):104947

Parashar N, Hait S (2021) Plastics in the time of COVID-19 pandemic: protector or polluter? Sci Total Environ 759. https://doi.org/10.1016/j.scitotenv.2020.144274

Poch RM et al (2020) Soil: The Great Connector of our lives now and beyond COVID – 19. SOIL 6: 541–547. https://doi.org/10.5194/soil-6-541-2020

Rattan L et al (2021) Soils and sustainable development goals of the United Nations (New York, USA): an IUSS perspective. Geoderma Reg. https://doi.org/10.1016/j.geodrs.2021.e00398
Rume T, Islam S (2020) Environmental effects of COVID-19 pandemic and potential strategies of sustainability. Heliyon 6(9):e04965. https://doi.org/10.1016/j.heliyon.2020.e04965
Saadat S et al (2020) Environmental perspective of COVID-19. Sci Total Environ 728:138870
Sangkham S (2020) Face mask and medical waste disposal during the novel COVID-19 pandemic in Asia. Case Stud Chem Environ Eng 2:100052
Scripps Research Institute. COVID-19 coronavirus epidemic has a natural origin. ScienceDaily. www.sciencedaily.com/releases/2020/03/200317175442.htm. Accessed 24 Feb 2021
Selvam S et al (2020) Imprints of pandemic lockdown on subsurface water quality in the coastal industrial city of Tuticorin, south India: a revival perspective. Sci Total Environ 738:139848
Selvaranjan K et al (2021) Environmental challenges induced by extensive use of face masks during COVID-19: a review and potential solutions. Environ Challenges. https://doi.org/10.1016/j.envc.2021.100039
Sharma et al (2020) Challenges, opportunities, and innovations for effective solid waste management during and post COVID-19 pandemic. Resour Conserv Recycl 162:105052
Shrestha N, Yousaf Shad M et al (2020) The impact of COVID-19 on globalization. One Health 11: 100180. https://doi.org/10.1016/j.onehlt.2020.100180
Singh N et al (2021) Medical waste: current challenges and future opportunities for sustainable management. Crit Rev Environ Sci Technol. https://doi.org/10.1080/10643389.2021.1885325
Stone M (2020) Carbon emissions are falling sharply due to coronavirus. But not for long. Available at: https://www.nationalgeographic.com/science/2020/04/co-ronavirus-causing-carbon-emissions-to-fall-but-not-for-long/. Accessed 17 Jan 2021
Svicevic M (2020) COVID-19 as a threat to international peace and security: what place for the UN Security Council? Available at: https://www.ejiltalk.org/covid-19-as-a-threat-to-international-peace-and-security-what-place-for-the-un-security-council/. Accessed 10 Feb 2021
Vovk Korze A et al (2021) Re-interpretation of "sustainability" concept, in post-COVID-19 period. Int J Manag (IJM) 12(2). Available at: http://www.iaeme.com/IJM/issues.asp?JType=IJM&VType=12&IType
Westhaus S et al (2021) Detection of SARS-CoV-2 in raw and treated wastewater in Germany – suitability for COVID-19 surveillance and potential transmission risks. Sci Total Environ 751. https://doi.org/10.1016/j.scitotenv.2020.141750
Yaya S, Otu A, Labonté R (2020) Globalisation in the time of COVID-19: repositioning Africa to meet the immediate and remote challenges. Global Health 16:51. https://doi.org/10.1186/s12992-020-0058
Yunus AP et al (2020) COVID-19 and surface water quality: improved lake water quality during the lockdown. Sci Total Environ 731. https://doi.org/10.1016/j.scitotenv.2020.139012. Accessed 3 Mar 2022
Zambrano-Monserrate MA et al (2020) Indirect effects of COVID-19 on the environment. Sci Total Environ 728:138813

Soil and Land Stewardship: An Action Perspective for More Soil Care and Protection?

Bernard Vanheusden and Sarah Jacobs

Abstract Soil and land deliver essential services to humankind and our ecosystems. However, they are under increasing pressure and have become vulnerable resources. The principle of "Soil and Land Stewardship" could be a good instrument to increase soil and land awareness. The goal of this contribution is to give a general introduction to the Soil and Land Stewardship concept. We start with our interpretation of the Soil and Land Stewardship concept. We examine the legal points and the more controversial points of departure regarding Soil and Land Stewardship (e.g. soil as a "common good", ecosystems services and granting rights to nature). We also pay attention to the most important constraints to implementing this concept. This is necessary to work out practical, workable (legal) translations of Soil and Land Stewardship. The contribution ends with some suggestions for next steps that we think would be very valuable in further implementing the Soil and Land Stewardship principle.

1 Introduction

> A nation that destroys its soils destroys itself.—Franklin D. Roosevelt – Letter to all State Governors on a Uniform Soil Conservation Law, 1937

This quote is a powerful reminder that soil and land are the basis for life on our planet. Soil and land provide 90% of total food, feed, fibre and fuel production, supply the necessary raw material for activities ranging from horticulture to construction and are essential for clean water and healthy ecosystems.[1]

The awareness that mankind needs soil and land in order to survive is growing. To protect all life on land, the United Nations developed, in 2015, the 15th

[1] European Environment Agency, https://www.eea.europa.eu/themes/soil/intro (consulted on September 27, 2021).

B. Vanheusden (✉) · S. Jacobs
Faculty of Law, Hasselt University, Hasselt, Belgium
e-mail: Bernard.vanheusden@uhasselt.be; Sarah.jacobs@monardlaw.be

H. Ginzky et al. (eds.), *International Yearbook of Soil Law and Policy 2020/2021*, International Yearbook of Soil Law and Policy 2020,
https://doi.org/10.1007/978-3-030-96347-7_16

Sustainable Development Goal (SDG 15).[2] With its sub target 15.3, the United Nations wishes to "combat desertification, restore degraded land and soil, including land affected by desertification, drought and floods, and strive to achieve a land degradation-neutral world by 2030". In addition to this specific Development Goal, land and soil play a crucial part in achieving several other SDGs,[3] for instance:

- SDG 1: no poverty;
- SDG 2: zero hunger;
- SDG 3: good health and well-being;
- SDG 6: clean water and sanitation;
- SDG 7: affordable and clean energy;
- SDG 11: sustainable cities and communities;
- SDG 13: climate action;
- SDG 16: peace, justice and strong institutions;
- SDG 17: partnerships for the goals.

This again emphasises the importance of soil and land for a sustainable world.

Nonetheless, soil and land are exposed to the constant, often contradictory demands of our society, putting them under increasing pressure.[4] In 2015 the European Commission's Joint Research Centre (JRC) published the *Soil Threats in Europe* report, in which the JRC[5] identifies the main soil and land threats within the European Union: soil erosion, decline in soil organic matter, soil compaction, soil sealing, soil contamination, soil salinisation, desertification, flooding and landslides and a decline in soil biodiversity.[6]

The vital role of soil and land is often insufficiently visible to society, which is also reflected in the limited legal framework for soil care and protection. The existing international regulations concerning soil are fragmented. In fact these regulations are not specifically soil-focused. In addition, there are no legally binding international instruments available which specifically address all the soil threats. Also, at the European level specific regulations on soil and land are lacking. Following opposition from several EU Member States, a proposal for an EU Soil Framework Directive was withdrawn in 2014. National regulations mainly focus on soil pollution and the restoration of already degraded soil, rather than on the prevention and striving for a world that is neutral in terms of land degradation. Another discussion is whether

---

[2] Sustainable Development Goals Knowledge Platform, https://sdgs.un.org/goals/goal15 (consulted on September 27, 2021).

[3] European Environment Agency, https://www.eea.europa.eu/signals/signals-2019-content-list/infographics/soil-and-united-nations-sustainable/view (consulted on September 27, 2021).

[4] Ibid.

[5] The Joint Research Centre (JRC) is the European Commission's science and knowledge service, which employs scientists to carry out research in order to provide independent scientific advice and support to EU policy.

[6] Stolte et al. (2016), pp. 9–12.

current soil legislation is adapted to new soil challenges, such as diffuse soil contamination[7] or the new emerging contaminants or emerging pollutants (EPs).[8]

It is within this context that, in recent years, more attention has been given to new concepts such as *Soil and Land Stewardship,* soil as a common good, thinking in terms of ecosystems and ecosystem services, rights for nature, circular economy, etc. as levers for more soil care. These principles could be good instruments to increase soil and land awareness in order to promote societal care for one of the most threatened assets of natural capital worldwide.

In this contribution, which should be seen as a general introduction to the Soil and Land Stewardship (SLS) concept, we examine successively what SLS is (Sect. 2), whether there are legal starting points for SLS (Sect. 3), which constraints should be taken into account (Sect. 4) and what the next steps should be (Sect. 5).

2 What Is Soil and Land Stewardship?

A first obvious question that arises is how to interpret the concept of stewardship?

In Sect. 2.1 we examine the general—and etymological—meaning of stewardship, while in Sect. 2.2 we look for the specific sense of Soil and Land Stewardship in an environmental context.

2.1 Stewardship: A Boundary Object

The word steward derives from sty-wards, which means someone who looks after farm animals, and has links with the term warden.[9]

In the common sense,[10] a steward can be understood as:

- Someone who collects the rent for his lord;
- The manager of an estate;
- Man in his responsibility on earth.

[7]Ibid., 92. This type of contamination is characterised by long distance transport of low concentrations of contaminants that are deposited in soils as a sink. Diffuse contamination is more difficult to control than point pollution because it is linked to a multitude of sources spread over the land.

[8]Ibid., 92. Emerging pollutants (EPs) are defined as synthetic or naturally occurring chemicals or micro-organisms that are not commonly monitored in the environment, but which have the potential to enter the environment and cause known or suspected adverse ecological and (or) human health effects.

[9]Worrel and Appleby (2000), p. 264.

[10]We have derived this common meaning from Van Dale's Comprehensive Dictionary of the Dutch Language.

Although the third meaning of stewardship has a religious background—man as a steward of God on earth[11]—this interpretation already hints at the definition of Soil and Land Stewardship used for this contribution (see Sect. 2.2).

As a boundary object, stewardship is recognised as a mobilising principle that everyone can work with from their own background.[12] A *boundary object* is a term found in sociology and stands for a concept, framework or instrument that is both flexible enough to acquire different meanings and thus adapts to the local needs and contradictions of the various actors that make use of it, but on the other hand is sufficiently robust to maintain a common or recognisable identity.[13] As a result, the boundary object can be used in different fields of work[14] (e.g. art, human sciences, social sciences and natural sciences).

Stewardship arises on the border between *care* (the value system—the why?), *knowledge* (knowledge about the soil and the land—the what?) and *agency* (participating in stewardship actions—the how?).[15] These three important dimensions of stewardship help to facilitate communication between *all* the different meanings of stewardship. In this way, one gets a grip on this flexible boundary object.[16]

2.2 Soil and Land Stewardship

The fact that stewardship implies an important responsibility towards society was already expressed in the third meaning of stewardship.

In an environmental context, stewardship is often defined as "the responsible use and conservation of natural resources taking full and balanced account of the interests of society, future generations and other species, as well as of private needs, and accepts significant accountability to society[17]".

On the other hand, *The United States Environmental Protection Agency* (EPA) defines stewardship as "the responsibility for the quality of the environment, shared by all those whose actions may have an impact on the environment[18]".

---

[11] Worst (1915), http://www.gutenberg.org/files/24080/24080-h/24080-h.htm (consulted on September 27, 2021). WORST writes that the land on which the harvest depends is a favour from God to mankind. According to WORST, no stewardship brings with it such enormous responsibilities as the stewardship of the land. The stewardship represents the farmer's obligation to society.

[12] Bal et al. (2018), p. 13.

[13] Star (2010), pp. 601–617.

[14] Ibid.

[15] Enqvist et al. (2018), pp. 17–37.

[16] Ibid.

[17] Refer to the glossary of the INSPIRATION project: www.inspiration-h2020.eu/page/stewardship (consulted on September 27, 2021). The main aim of INSPIRATION was to develop a Strategic Research Agenda (SRA) to inform environmentally friendly, socially acceptable and economically affordable soil and land use management that meets societal needs and challenges.

[18] United States Environmental Protection Agency (2005), p. 2.

Another definition states that "the duty of stewardship requires that the owner uses and maintains the land in a manner that does not interfere with any significant natural resource that it may contain[19]".

In general terms, therefore, a steward is an individual who devotes a substantial percentage of his thoughts and efforts to maintaining or enhancing the condition of some thing(s) or person(s), not primarily for its own sake.[20] The environmental steward will act for the sake of the (preservation of) the natural world.[21]

Soil and Land Stewardship goes further than just sustainable land or soil management, in the sense that SLS emphasises responsibility and *action*—putting it into practice—on the basis of this responsibility.[22] Responsibility is one of the fundamental aspects of stewardship.

In more legal terms, SLS can be described as "an easement for the sake of the Globe, interfering with strict ownership[23]". An environmental easement (or environmental land charge) imposes certain obligations on the owner of the burdened land in the interest of the environment.[24]

3 Are There Any (Legal) Points of Departure Regarding Soil and Land Stewardship?

Now that it is clear what we need to understand under the SLS principle, we will, in this section, further investigate what already (legally) exists about SLS. Because we consider stewardship as an action perspective for (more) soil care, we also extend our research focus to soil care.

In Sect. 3.1 we discuss the legal points of departure. In Sect. 3.2 we will address other—more controversial—leads for SLS and more soil care.

3.1 Legal Points of Departure

In this general introduction to the Soil and Land Stewardship concept, it is of course impossible to discuss all the legal levers. Therefore, in this contribution we will focus on the international dimension (Sect. 3.1.1), the European dimension (Sect. 3.1.2) and the so-called environmental care obligations (Sect. 3.1.3). Finally, we

---

[19] Karp (1993), p. 748.

[20] Butti et al. (2019), https://www.filodiritto.com/impact-eu-legislation-principles-and-case-law-national-contaminated-land-regimes#_ftn85 (consulted on September 27, 2021).

[21] Welchman (1999), p. 415.

[22] Block (2013).

[23] Nicole and Common Forum (2018), p. 14; Van De Voorde (2018), pp. 1195–1313.

[24] Van De Voorde (2018), p. 1206.

make a short trip to the Catalonian region of Spain to examine its Land Stewardship agreement, which has been codified in the Catalan Civil Code (Sect. 3.1.4).

3.1.1 The International Dimension

Despite the increasing attention on the international stage as to the importance of soil and land in our world, there are few legally binding international soil regulations. As far as the international level is concerned, we note that there are definitely initiatives, but that they are very fragmented. There is also often an overlap of mandates and activities between the various international organisations. In addition, we note that the initiatives at the international level each have their own limitations.

3.1.1.1 The International Treaties on Soil Pollution

In this section we will review certain international treaties. The international treaties listed below all have in common that each one, within its own (limited) scope, deals with soil pollution, one of the (most important) soil and land threats identified by the JRC.

The Basel Convention

The Basel Convention refers to the Convention on the Control of Transboundary Movements of Hazardous Wastes and Their Disposal. The Basel Convention was adopted on 22 March 1989 by the Plenipotentiaries in Basel, Switzerland, in response to a public outcry following the discovery in the 1980s in Africa and other parts of the developing world of deposits of toxic wastes imported from abroad.[25]

The provisions of the Convention center around the following principal aims:

- The reduction of hazardous waste generation and the promotion of environmentally sound management of hazardous wastes, wherever the place of disposal;
- The restriction of transboundary movements of hazardous wastes except where it is perceived to be in accordance with the principles of environmentally sound management; and;
- A regulatory system applying to cases where transboundary movements are permissible.[26]

[25] See: http://www.basel.int/TheConvention/Overview/tabid/1271/Default.aspx (consulted on September 27, 2021).

[26] Ibid.

The Basel Convention, with the important aims of waste minimisation and state-of-the-art recycling in accordance with agreed standards, also leads to better protection of the air, soil, water and thus human health.[27]

The Rotterdam Convention

The Rotterdam Convention on the Prior Informed Consent Procedure for Certain Hazardous Chemicals and Pesticides in International Trade was adopted on 10 September 1998 by a Conference of Plenipotentiaries in Rotterdam, the Netherlands.

The objectives of the Rotterdam Convention are:

- To promote shared responsibility and cooperative efforts among Parties in the international trade of certain hazardous chemicals in order to protect human health and the environment from potential harm;
- To contribute to the environmentally sound use of those hazardous chemicals, by facilitating information exchange about their characteristics, by providing for a national decision-making process on their import and export and by disseminating these decisions to Parties.[28]

This Convention is considered an important step towards improving international regulation of hazardous chemicals. In order to protect human health and the environment from potential harm, Parties to the Convention are encouraged to share responsibility and cooperate in the international trade of such substances. The Convention also seeks to contribute to the environmentally sound use of these substances. The basic principle of the Convention is that the export of a banned or severely restricted chemical listed in Annex III to the Convention can only take place with the prior informed consent (PIC) of the importing Party.

The Convention prevents contamination of the soil by the substances concerned.

The Stockholm Convention

On 22 May 2001 in Stockholm, Sweden, the Conference of Plenipotentiaries adopted the Stockholm Convention on Persistent Organic Pollutants.

The Stockholm Convention is a global treaty to protect human health and the environment from chemicals that remain intact in the environment for long periods, become widely distributed geographically, accumulate in the fatty tissue of humans and wildlife, and have harmful impacts on human health or on the environment. Exposure to Persistent Organic Pollutants[29] (POPs) can lead to serious health effects,

---

[27] Ibid.

[28] See: www.pic.int/TheConvention/Overview (consulted on September 27, 2021).

[29] E.g. the pesticide Aldrin was recognised as a POP under the Stockholm Convention. It is a pesticide applied to soils to kill termites, grasshoppers, corn rootworm, and other insect pests. Aldrin can also kill birds, fish, and humans. In one incident, aldrin-treated rice is believed to have killed hundreds of shorebirds, waterfowl, and passerines along the Texas Gulf Coast when these birds either ate animals that had eaten the rice or ate the rice themselves. In humans, the fatal dose

including certain cancers, birth defects, dysfunctional immune and reproductive systems, greater susceptibility to disease and damage to the central and peripheral nervous systems. Given their long-range transport trough soil, water and air, no government acting alone can protect its citizens or its environment from POPs. In response to this global problem, the Stockholm Convention, which was adopted in 2001 and entered into force in 2004, requires its parties to take measures to eliminate or reduce the release of POPs into the environment.[30]

Contamination of soils with these POPs is of increasing concern.[31]

The Minamata Convention

The Minamata[32] Convention on Mercury was agreed at the fifth session of the Intergovernmental Negotiating Committee on mercury in Geneva, Switzerland, on 19 January 2013 and adopted later that year on 10 October 2013 at a Diplomatic Conference (Conference of Plenipotentiaries) held in Kumamoto, Japan.[33]

The Convention draws attention to a global and ubiquitous metal that, while naturally occurring, has broad uses in everyday objects and is released into the atmosphere, soil and water from a variety of sources. Controlling the anthropogenic releases of mercury throughout its lifecycle has been a key factor in shaping the obligations under the Convention.

Major highlights of the Minamata Convention include a ban on new mercury mines, the phase-out of existing ones, the phase out and phase down of mercury use in a number of products and processes, control measures on emissions to air and on releases to land and water, and the regulation of the informal sector of artisanal and small-scale gold mining. The Convention also addresses interim storage of mercury and its disposal once it becomes waste, sites contaminated by mercury as well as health issues.[34]

This Convention is believed to have a strong focus on the prevention and remediation of soil contamination, but with a focus on mercury.

---

for an adult male is estimated to be about five grams. Humans are mostly exposed to aldrin through dairy products and animal meats. Studies in India indicate that the average daily intake of aldrin and its by-product dieldrin is about 19 micrograms per person. See: www.pops.int/TheConvention/ThePOPs/The12InitialPOPs/tabid/296/Default.aspx (consulted on September 27, 2021).

[30] See: www.pops.int/TheConvention/Overview/tabid/3351/Default.aspx (consulted on September 27, 2021).

[31] Ren et al. (2018), pp. 1154–1163.

[32] The Minamata Convention is named after a city in Japan where serious health damage occurred as a result of mercury pollution in the mid-twentieth century. See: http://www.mercuryconvention.org/ (consulted on September 27, 2021).

[33] Ibid.

[34] Ibid.

3.1.1.2 The Convention to Combat Desertification

An important international legal instrument is the United Nations Convention to Combat Desertification (the UNCCD), adopted on 17 June 1994.[35]

The UNCCD is a legally binding agreement aimed at combating desertification and aridity. The Convention is even considered to be "the only legally binding international agreement that brings environment, development and sustainable soil use closer together".[36]

The Convention claims leadership for the implementation of *Land Degradation Neutrality* (LDN), but was for a long time limited to the dry and desertified areas, mainly in Africa. In 2015, it was decided to extend the scope of the UNCCD to include the prevention of land degradation in all areas, not just the dry areas. The international community has decided to achieve a balance, or LDN, at which land is degraded and restored through concrete intervention.

Land Degradation Neutrality has been defined by the Parties to the Convention as: *"A state whereby the amount and quality of land resources, necessary to support ecosystem functions and services and enhance food security, remains stable or increases within specified temporal and spatial scales and ecosystems*[37]*"*.

3.1.1.3 The International Organisations

For this contribution we will select one international organisation: The Food and Agriculture Organisation of the United Nations (FAO).

The FAO would be the leading organisation of the United Nations in the area of soil and land. However, for many years, the FAO focused mainly on agricultural land and food production. But here, too, we are noticing a shift towards increasing attention for the general soil problem: a few years ago, the FAO organised a Symposium on Soil Pollution.[38] The Global Symposium on Soil Biodiversity was scheduled for April 2021.[39] The FAO also prepares the *Status of the World's Soil Resources* (SWSR) report. The SWSR is a reference document on the status of global soil resources that provides regional assessments of soil change. The information is based on peer-reviewed scientific literature, complemented with expert

---

[35] See: https://www.unccd.int/convention/about-convention/status-ratification (consulted on September 27, 2021).

[36] See: https://www.canada.ca/en/environment-climate-change/corporate/international-affairs/partnerships-organizations/united-nations-convention-combat-desertification.html (consulted on September 27, 2021).

[37] See: https://www.unccd.int/actions/achieving-land-degradation-neutrality (consulted on September 27, 2021).

[38] See: http://www.fao.org/about/meetings/global-symposium-on-soil-pollution/en/ (consulted on September 27, 2021).

[39] See: http://www.fao.org/about/meetings/soil-biodiversity-symposium/en/ (consulted on September 27, 2021).

knowledge and project outputs. It provides a description and a ranking of ten major soil threats that endanger ecosystem functions, goods and services globally and in each region separately. Additionally, it describes direct and indirect pressures on soils and ways and means to combat soil degradation. The report contains a synthesis report for policy makers that summarises its findings, conclusions and recommendations.[40]

3.1.1.4 The SDGs

As already explained in Sect. 1, the SDGs contain a specific objective relating to "life on land". SDG 15 reads as follows: "Protect, restore and promote sustainable use of terrestrial ecosystems, sustainably manage forests, combat desertification, and halt and reverse land degradation and halt biodiversity loss[41]".

The principle of Soil and Land Stewardship fits nicely into sub target 15.3 (see: Sect. 1), which specifically calls for "striving for a world that is neutral in terms of land degradation". However, although the SDGs are based on human rights and are intended to be universally applicable, they were not negotiated as a treaty, making them function at best as soft law.[42]

In order to achieve the SDGs by 2030, there is an urgent need for a transition towards sustainable soil and land management, aiming at multifunctional land use and restoration of degraded soils.[43]

However, there is a growing demand for international environmental regulation for the protection and sustainable use of soil and land.

The IUCN WCEL Soil, Desertification, and Sustainable Agriculture Specialist Group,[44] for instance, has in recent years intensively explored and discussed the possibilities for an international soil instrument.[45]

3.1.2 The European Dimension

In this section we focus, without wanting to be extensive, on the European dimension. The key instruments that must be explored are the withdrawn proposal of the

[40] See: www.fao.org/documents/card/en/c/c6814873-efc3-41db-b7d3-2081a10ede50/ (consulted on September 27, 2021).

[41] See: https://sdgs.un.org/goals/goal15 (consulted on September 27, 2021).

[42] Singh (2015), p. 58.

[43] Visser et al. (2019), 10.3390.

[44] The International Union for Conservation of Nature and Natural Resources: an international partnership consisting of more than 1300 members (countries, governmental and non-governmental organisations, indigenous peoples, etc.). See: https://www.iucn.org/ (consulted on September 27, 2021).

[45] See: https://www.iucn.org/sites/dev/files/content/documents/soil_specialist_group_mid_year_report_june_2019.pdf (consulted on September 27, 2021).

Soil Framework Directive (Sect. 3.1.2.1) and the Environmental Liability Directive (Sect. 3.1.2.2).

3.1.2.1 The Soil Framework Directive

Back in 2002, in its Communication *Towards a Thematic Strategy for Soil Protection*,[46] the European Commission identified the main eight threats to which soils in the European Union are confronted. These eight soil threats are consistent with those in the 2015 *Soil Threats in Europe* report (see: Sect. 1).

To implement this *Thematic Strategy on Soil Protection*, in 2006 the European Commission presented a proposal[47] for a Directive of the European Parliament and of the Council establishing a framework for the protection of soil and amending Directive 2004/35/EC.

As soil is under increasing environmental pressure across the EU, driven or exacerbated by human activity (such as inappropriate agricultural and forestry practices, industrial activities, tourism or urban development), the Commission established that a specific EU legislation on soil protection was missing. Some soil protection aspects are scattered in the EU legislation (e.g. provisions in the existing environmental EU legislation on water, waste, chemicals, industrial pollution prevention, nature protection and pesticides). However, due to their different objectives and scope, and due to the fact that they often aim to safeguard other environmental media, existing provisions, even if fully implemented, provide fragmented and incomplete protection of the soil as they do not cover all soils and all soil threats identified. Soil is a resource of common interest to the EU and failure to protect it will undermine sustainability and long-term competitiveness in Europe. Moreover, the European Commission also stressed that soil degradation has strong impacts on other areas of common interest to the EU, such as water, human health, climate change, nature and biodiversity protection and food safety.

With the proposal the Commission wished to fill this gap in specific EU legislation on soil protection and to establish a common strategy for the protection and sustainable use of soil, based on the following principles:

- The integration of soil care into other policies;
- The preservation of soil functions within the framework of sustainable use;
- Prevention of threats to soil and reduction of their adverse effects;
- The restoration of degraded soils to a level of functionality consistent at least with the current and approved future use.[48]

---

[46] COM 2002/0179.

[47] COM 2006/0232.

[48] Ibid., 2.

The proposal aimed to establish a framework for the protection of soil and the preservation of the capacity of soil to perform its environmental, economic, social and cultural functions.

Article 1(1) of the Proposal listed the following vital soil functions:

- Biomass production, including in agriculture and forestry;
- Storing, filtering and transforming nutrients, substances and water;
- Biodiversity pool, such as habitats, species and genes;
- Physical and cultural environment for humans and human activities;
- Source of raw materials;
- Acting as carbon pool;
- Archive of geological and archaeological heritage.

The Proposal also included a provision on precautionary measures for soil owners in Article 4 of the Proposal, which stated: "Member States shall ensure that any land user whose actions affect the soil in a way that can reasonably be expected to hamper significantly the soil functions referred to in Article 1(1) is obliged to take precautions to prevent or minimise such adverse effects."

Furthermore, the draft Directive addressed the problem of soil sealing by limiting or mitigating its effects through the use of construction techniques and products which will allow as many of the listed soil functions as possible to be maintained.[49]

Other important components of the proposal were:

- The identification of risk areas of erosion, organic matter decline, compaction, salinisation and landslides;[50]
- The establishment of an inventory of contaminated sites.[51]

According to the *Umweltbundesamt*, the German Environment Agency, the Soil Framework Directive could have made a useful contribution to the objective of Land Degradation Neutrality.[52]

Several Member States were opposed to such a Soil Framework Directive, however, in part because of the administrative burden it would entail. Other Member States also argued that the proposed framework did not comply with the principles of subsidiarity and proportionality. Some Member States even denied the cross-border dimension of soil problems and soil protection.

This protest ultimately, in 2014, led to the withdrawal of the proposal for a European Soil Framework Directive.[53]

[49] Article 5.

[50] Articles 6–8.

[51] Articles 10–11.

[52] Umweltbundesamt (2015), p. 39.

[53] *Official Journal* C 153, 21 May 2014, 3.

3.1.2.2 The Environmental Liability Directive

The Environmental Liability Directive of 21 April 2004 is also worth discussing.

The *rationale legis* of this Directive is set out in the preamble. There are currently many contaminated sites in the EU Community, posing significant health risks, and the loss of biodiversity has dramatically accelerated over the last decades. Failure to act could result in increased site contamination and greater loss of biodiversity in the future. Preventing and remedying environmental damage as much as possible is necessary.[54] The prevention and remediation of environmental damage should be implemented through the furtherance of the "polluter pays" principle. The fundamental principle of the Directive should therefore be that an operator whose activity has caused environmental damage, or the imminent threat of such damage, is to be held financially liable in order to induce operators to adopt measures and develop practices to minimise the risks of environmental damage so that their exposure to financial liabilities is reduced.[55]

In addition to damage to protected species and natural habitats and water damage, the Environmental Liability Directive also defines land (or soil) damage as environmental damage. Article 1(c) of the Directive defines land damage as "any land contamination that creates a significant risk of human health being adversely affected as a result of the direct or indirect introduction, in, on or under land, of substances, preparations, organisms or micro-organisms". The Environmental Liability Directive thus covers only one of the identified soil and land threats within the European Union, i.e. soil contamination.

Article 3(1) of the Directive further limits its scope to:

- Environmental damage caused by any of the occupational activities listed in Annex III, and to any imminent threat of such damage occurring by reason of any of those activities;
- Damage to protected species and natural habitats caused by any occupational activities other than those listed in Annex III, and to any imminent threat of such damage occurring by reason of any of those activities, whenever the operator has been at fault or negligent.

The Directive requires that, where environmental damage has not yet occurred but there is an imminent threat of such damage occurring, the operator shall, without delay, take the necessary preventive measures.[56] When environmental damage has occurred the operator must take prompt remedial action.[57] In accordance with the polluter pays principle, article 8(1) of the Directive stipulates that the operator shall bear the costs for the preventive and remedial actions taken pursuant to this Directive.

---

[54] Recital (1).

[55] Recital (2).

[56] Article 5.

[57] Articles 6–7.

Although the Court of Justice of the European Union, in older case law[58] on the Environmental Liability Directive, has favoured a strict interpretation of the polluter pays principle, certain legal doctrine notes that in more recent judgments, the European Court of Justice has attached greater importance to the Stewardship principle.[59]

The FIPA case, for instance, concerned a heavily contaminated industrial site located in the province of Massa Carrara in Tuscany, Italy. The soil contamination with various chemicals (including dichloroethylene and ammonia) was probably caused by the production of insecticides and herbicides from the 1960s through the 1980s. After this, two private companies—that were not directly responsible for the contamination found on the site—became the owners of various plots of land. The competent Italian authorities ordered these companies to take "emergency safety" measures in their capacity as "guardians of the land". The question arose whether these companies could be held liable for the soil contamination. According to the owners, this was in violation of the polluter pays principle. Because there was disagreement among the Italian administrative courts, the Italian *Consiglio di Stato* decided to stay proceedings and to refer the matter to the European Court of Justice for a preliminary ruling.[60]

In its judgement[61] the Court considered that "the Environmental Liability Directive must be interpreted as not precluding national legislation such as that at issue in the main proceedings, which, in cases where it is impossible to identify the polluter of a plot of land or to have that person adopt remedial measures, does not permit the competent authority to require the owner of the land (who is not responsible for the pollution) to adopt preventive and remedial measures, that person being required merely to reimburse the costs relating to the measures undertaken by the competent authority within the limit of the market value of the site, determined after those measures have been carried out". In this judgement, the European Court of Justice

---

[58] E.g. ECJ 9 March 2010, C-378/08, RAFFINERIE MEDITERRANEE et al.

[59] Butti et al. (2019), https://www.filodiritto.com/impact-eu-legislation-principles-and-case-law-national-contaminated-land-regimes#_ftn85 (consulted on September 27, 2021).

[60] A preliminary question is used by national courts to request the Court of Justice to interpret European rules. The question referred for a preliminary ruling was as follows:

> Do the European Union principles relating to the environment, laid down in Article 191(2) TFEU and in Articles 1 and 8(3) of Directive 2004/35 and recitals 13 and 24 thereto — specifically, the "polluter pays" principle, the precautionary principle and the principles that preventive action should be taken and that environmental damage should be rectified at source as a matter of priority — preclude national legislation, such as the rules set out in Articles 244, 245 and 253 of the Environmental Code, which, in circumstances in which it is established that a site is contaminated and in which it is impossible to identify the polluter or to have that person adopt remedial measures, do not permit the administrative authority to require the owner (who is not responsible for the pollution) to implement emergency safety and rehabilitation measures, merely attributing to that person financial liability limited to the value of the site once the rehabilitation measures have been carried out?

[61] ECJ 4 March 2015, C-534/13, FIPA et al.

abandoned strict application of the polluter pays principle and imposed far-reaching obligations on landowners, who did not cause the pollution themselves.

In the TTK case concerning the application of Hungarian legislation, the Court of Justice went even further. In a judgement[62] of 13 July 2017, the European Court of Justice ruled that "the Environmental Liability Directive does not preclude national legislation, such as that at issue in the main proceedings, which identifies another category of persons who, in addition to those using the land on which unlawful pollution was produced, share joint liability for the environmental damage, namely the owners of that land, without it being necessary to establish a causal link between the conduct of the owners and the damage established".

According to BUTTI et al., the burden that the Court of Justice imposes on landowners with this case law is severe, but on the other hand, this far-reaching responsibility is precisely what the application of the Stewardship principle requires.[63]

In this way, the Environmental Liability Directive, albeit within its limited scope, has also a role to play in the translation of the Soil and Land Stewardship principle.

3.1.3 The Environmental Care Obligations

A typical instrument that regulators use to impose more environmental care on legal subjects is the so-called "environmental care obligations".

A duty of care is a general obligation of the legal subject, if he knows or can reasonably suspect that his acts or omissions may endanger the legal good that is protected by the law in question, to take all measures that can reasonably be expected of him in order to prevent the danger, to limit it as much as possible, or to limit and undo its consequences.[64] A duty of care concerns a vaguely defined obligation that marks the responsibility of the legal subject with respect to a certain legal commodity. This responsibility is legally enforced under administrative law, criminal law, private law or through a combination of these methods.[65]

Duty of care obligations are becoming more and more common in environmental law. In this case, the protected legal assets are "ecological legal assets" (e.g. land and soil). A duty of care is a method to establish that everyone—not only the government—is responsible for a clean environment.[66]

But now, why is there a shift to duties of care? There are several reasons:

- The generally formulated duties of care have a safety net function: they make it possible to take legal action against serious forms of environmental pollution that

[62] ECJ 13 July 2017, C-129/16, TÚRKEVEI TEJTERMELŐ KFT.

[63] Butti et al. (2019), https://www.filodiritto.com/impact-eu-legislation-principles-and-case-law-national-contaminated-land-regimes#_ftn85 (consulted on September 27, 2021).

[64] Faure and Visser (1999), p. 322.

[65] Meeus (2008), p. 454.

[66] Ibid.

are not covered by concrete, specific environmental norms.[67] Given the speed and complexity of environmental developments, it is impossible for the regulator to include all factual situations in a "watertight" regulation.[68] In order to fulfil this safety net function, duty of care obligations contain vague phrases such as "take the necessary (appropriate) measures". In addition, the provisions are composed of often undefined concepts,[69] such as "harm, damage and inconvenience".

- There is a need to transfer more responsibility from the government to the citizen through self-regulation. For the legislator it is impossible to issue conclusive regulations given speed and complexity of environmental developments.[70] Duty of care obligations are regarded as basic standards that determine the citizen's own responsibility. Within the contours of duty of care obligations, citizens are free to determine how they shape their behaviour (e.g. with regard to environmental issues).

There is, however, a downside to the vague and general nature of duties of care. Because of these characteristics, duties of care can come into conflict with the principle of legal certainty and the principle of criminal legality. Legal certainty presupposes that regulations for citizens must be as specific and clear as possible. The principle of criminal legality—also known as the *lex certa* requirement—means that it must be clear in advance to those seeking justice whether a particular act or omission may have criminal consequences and, if so, what such consequences may be.[71]

Below we list a few examples of environmental duty of care obligations that are relevant to achieve more soil care:

- The duty of care for nature laid down in the Flemish regulations.[72] This environmental care obligation prescribes (freely translated):

> Any natural or legal person who, manually, by mechanical means or by using pesticides or fixed or mobile noise sources, intervenes in or in the immediate vicinity of natural and partly natural habitats or ecosystems, of wetlands, of natural and semi-natural vegetation, of wild native fauna or flora or migratory species of wild animals or their respective habitats or habitats, or on small elements of the landscape, and who knows or has reasonable grounds to suspect that such habitats, ecosystems, wetlands, vegetation, fauna, flora or small elements of the landscape may be destroyed or seriously damaged thereby, shall be required to take all reasonable steps to prevent, reduce or, if that is not possible, restore the destruction or damage.
>
> The provision of the previous paragraph shall also apply to those who order the interventions referred to in that paragraph.

[67] Houweling (1997), p. 145.
[68] Meeus (2008), p. 455.
[69] Meeus (2008), p. 456.
[70] Houweling (1997), p. 145.
[71] Meeus (2008), p. 456.
[72] Article 14, §1 Decree of 21 October 1997 on Nature Conservation and the Natural Environment.

Insofar as interventions in the soil are accompanied by a possible impact on the natural elements listed in the Article, these may fall under the application of the duty of care for nature and can (even) be enforced under criminal law.

- The specific soil care obligation in Dutch legislation.[73] Section 13 of the Soil Protection Act, that has been in force in the Netherlands since 1987, stipulates:

> Any person performing acts on or in the soil as referred to in sections 6 to 11[74] inclusive and who is aware or can reasonably suspect that such acts are likely to contaminate or impair the soil shall be obliged to take any measure that can be reasonably required of him in order to prevent the soil being so contaminated or impaired or, in the event of such contamination or impairment occurring, to take remedial action and to limit and to eliminate as much as possible the contamination or the impairment or the direct consequences thereof. If the contamination or impairment is the result of an unusual event, the measures shall be taken forthwith.

This duty of care can likewise be enforced under both administrative and criminal law.

- The waste duty of care in the UK Environment Protection Act.[75] In short, this waste duty of care means that anyone who produces, imports, keeps, stores, transports, treats or disposes of waste must take all reasonable steps to ensure that waste is managed properly.[76] A breach of the duty of care could lead to an unlimited fine if convicted in the Magistrates Court or in the Crown Court.

3.1.4 A Short Trip to the Spanish Region of Catalonia

In Catalonia, one of the 17 autonomous regions of Spain, the Land Stewardship agreement has, since 2017, been incorporated into the Catalan Civil Code. This is the first legal regulation of Land Stewardship agreements in Europe.

Thus, we think it is worth taking a moment to reflect on this.[77]

This codification is primarily the accomplishment of the Xarxa de Custòdia del Territori (XCT), a non-profit organisation that wishes to promote the principle of

[73] Section 13 of the Soil Protection Act.

[74] The specific soil care obligation applies to certain acts performed in or on the ground, that are listed in Sections 6 to 11 of the the Soil Protection Act. For instance:

- Disposal of wastes on or in the soil (Section 6);
- Depositing fertilisers on or in the soil (Section 7);
- Earth and foundation works (Section 8);
- Transport of such substances with the aid of pipelines or other conduits (Section 9);
- The use of anti-slip agents (Section 10);
- The performance of acts not coming under sections 6 to 10 inclusive which may cause erosion, compaction or salination of the soil (Section 11).

[75] Section 34 of the Environmental Protection Act.

[76] See: www.environmentlaw.org.uk/rte.asp?id=83 (consulted on September 27, 2021).

[77] The information for this section was drawn from: Xarxa de Custòdia del Territori (2017) and Xarxa de Custòdia del Territori (2018).

Soil and Land Stewardship to protect the natural, cultural and landscape resources and values of Catalonia and its environment.[78]

Article 149.1.8 of the Spanish Constitution of 1978 provides that the regional parliaments of the Spanish autonomous communities may develop their own private regulations. In 2017, the Catalan Parliament adopted the sixth book of the Civil Code. In this sixth book of the Codi civil de Catalunya, temporary[79] Land Stewardship agreements are regulated by law under Article 623–34, that states:

1. In the Land Stewardship contract, which is temporary and whose purpose is land property, the grantor partially or totally allows the use or management of the land to the holder, which must be a legal entity having Land Stewardship among its purposes, in exchange for carrying out advisory, outreach, planning or management and improvement activities, in order to conserve biodiversity, natural and cultural heritage and landscape or to make sustainable management of its natural resources.
2. The rights constituted in favour of the holder may be in personam or in rem, whenever they satisfy, in this case, the requirements stated by law.
3. The legal framework of the Land Stewardship contract, as to the determination of the obligations of the parties and their failure to comply, the duration or guarantees, is freely determined by the contractual parties.

Through these voluntary agreements between landowners and/or land users and Land Stewardship organisations, Catalonia hopes to preserve and (where necessary) restore its nature, biodiversity, ecological integrity and landscape values.

By the end of 2017, 765 Land Stewardship agreements had been created and 1.4% of Catalonia's territory (40,072 ha) was covered by such agreements.[80] On the website *Custodia Territori* (*Land Stewardship*), the different Land Stewardship agreements are shown on a map.[81]

Another important initiative of the XCT was the creation of a Land Stewardship Toolkit. In this manual XCT provides basic tools for Land Stewardship organisations in Europe, e.g. for drawing up Land Stewardship agreements, funding Land Stewardship initiatives and selecting Land Stewardship projects.[82]

3.1.5 Conclusion

The above exposition shows very clear that the legal points of departure regarding SLS—and hence for a global soil protection—are at the moment very scattered.

This points precisely to the need for a higher-level principle from which the responsibility can be derived to cover all soil threats.

---

[78] Nicole and Common Forum (2018), p. 15.

[79] The maximum duration of an agreement is 99 years.

[80] Xarxa de Custòdia del Territori (2017) and Xarxa de Custòdia del Territori (2018).

[81] See: http://custodiaterritori.org/VIII-inventari-acords-xct-2017/#9/41.9758/2.0023 (consulted on September 27, 2021).

[82] Rodríguez and Sabater (2018).

All of these haphazard legal principles could be brought together under the duty of care of the Soil and Land Steward, who must use and maintain soil in a manner that does not interfere with any significant natural resource that it may contain.

3.2 Other Points of Departure

The concept Soil and Land Stewardship as a lever for more soil care is gaining more and more recognition among specialists (e.g. lawyers, soil remediation experts, competent public authorities). In addition to SLS, however, there are also other—less conventional—concepts that are receiving increasing attention as a lever for soil care. Below we discuss soil as a common (Sect. 3.2.1), ecosystems or ecosystem services (Sect. 3.2.2), granting rights to nature (Sect. 3.2.3) and circular economy (Sect. 3.2.4).

3.2.1 Soil as a "Common"

In Sect. 4 we discuss the constraints on the application of the Soil and Land Stewardship principle. More than once, the tension between SLS and the classic concepts of property rights and private land ownership has surfaced.[83] In order to respond to this, attention has been paid to the concept of commons or common goods.

In order to ensure health and access to natural resources of present and future generations, the European Union must, according to some, recognise soil and land as a common asset essential to life on our planet and make the sustainable management of these resources a primary commitment.[84]

Without wanting to be exhaustive, we give a brief explanation of what is meant by the commons principle. We will structure this discussion as follows: the demise of the commons (Sect. 3.2.1.1), the tragedy of the commons (Sect. 3.2.1.2) and the revival of the commons (Sect. 3.2.1.3).

3.2.1.1 The Demise of the Commons

Roman Law already considered private property as an absolute right.[85]

Previously, in the Middle Ages, many lands and services were jointly managed[86] (e.g. communal rights of use).

---

[83] Bal et al. (2018), p. 12.

[84] Messa (2015), p. 17.

[85] Gutwirth and Stengers (2017), pp. 7–8.

[86] Akkermans (2018), p. 1448.

Nevertheless, the (complete) extermination of the commons is mainly situated at the end of the eighteenth century and the beginning of the nineteenth century:[87]

- In England, this was caused by the so-called enclosures, in which communal land was fenced off and taken into private ownership by the landowners. The farming communities were driven away from the farmland on which they had a traditional right of use. They were also forbidden to extract the natural resources they needed from the forests.
- With the Code Napoléon (21 March 1804), the double sovereignty of state and owner was implemented in France. This Code Napoléon, later renamed to the Code Civil, departed from the Roman principles of absolute ownership. The only remaining choice is between more or less power to the state or to private property.[88]
- The appropriation of the terra nullius during the European colonisations and the expropriation of the native inhabitants of the United States. This private appropriation was inspired by the John Locke doctrine, which linked exclusive and individual ownership to productivity and general prosperity.

3.2.1.2 The Tragedy of the Commons

After the demise of the commons at the end of the eighteenth century and the beginning of the nineteenth century it was for a long time believed, that the demise of the commons was the only right way in man's quest for progress.

Hardin explained this thinking in his *The Tragedy of the Commons*[89] (1968). According to Hardin, the true tragedy of the commons is that joint exploitation of land, which has not been appropriated by either the government or a private owner, is not viable in the long run. As an example, Hardin takes a meadow on which a group of people let their cows graze freely. Each person in the group obviously wants to maximise his return from the meadow and will find it difficult and therefore not be able to resist the temptation to put an extra cow on it, if only, because he knows that the others will do the same. This continues until the meadow is tragically exhausted and destroyed. The moral of the story is that the meadow will only survive if its use is determined and regulated by an authority concerned with its preservation. An authority which, of course, can only be an owner (a private owner) or the state.[90]

[87] Gutwirth and Stengers (2017), pp. 6–8.

[88] Ibid., 6.

[89] Hardin (2018), pp. 1234–1248.

[90] Gutwirth and Stengers (2017), p. 8.

3.2.1.3 The Revival of the Commons

Today, however, there is a growing focus on the importance of common goods.

The famous work *Governing the Commons* (1990) by Nobel Prize winner OSTROM,[91] pioneer in the field of the commons, plays an important role. OSTROM demonstrated through empirical research of many common pool regimes (e.g. fisheries, irrigation systems, water resources, forests, pastures and meadows) that under certain circumstances, the common management of natural resources is more efficient and sustainable than regimes such as private or public ownership. OSTROM identified a series of generic elements that promote the success of stable management of shared resources and embody their basic principles: the so-called design principles. The first design principle, for instance, involves defining the rights and responsibilities of members with regard to natural resources and the limits to their extraction.

Common goods are characterised by scattered ownership: they belong to everyone and no one person at the same time. All people have access to these goods, but nobody should have exclusive rights to them. The commons must be managed on the basis of the principles of equality and solidarity, considering the interests of future generations.[92]

Examples of such commons are communally-owned windmills, cooperative housing communities and large companies whose employees are co-owners. But also digital commons such as Youtube and Wikipedia—where everyone can make their creative contribution—are examples of the development of commons.[93]

It is necessary to consider certain scarce resources, such as (drinking) water, air and soil, as a common good, not only to ensure the protection of the environment, but also the survival of mankind.[94] The survival of these resources is essential to ensure our health and world peace. Such common ownership is no longer the common ownership of production resources as in the communist system,[95] but a cooperation between citizens or between citizens and public authorities in order to achieve sustainable objectives.[96]

---

[91] Ostrom (1990), p. 270.

[92] Rodota (2011), p. 2.

[93] Akkermans (2018), p. 1450.

[94] Rodota (2011), p. 3.

[95] Indeed, in Marx's conceptualization as asserted in *Capital: A Critique of Political Economy* with Friedrich Engels, soil is considered a common good that is property of the State and, hence, of the people. In this situation, the State is the owner of lands and goods coming from it and has the monopoly over its exploitation, thereby serving the public interest. However, the notion we use as resources as a common good differs from this interpretation, as we use it as a cooperation between citizens or between citizens and public authorities to realize certain environmental goals, making it conceptually distinct from how common goods were perceived in the Marxist tradition. See: Marx and Engels (1967), 654 p.

[96] Akkermans (2018), p. 1449.

In order for soil and land to be managed in a sustainable way by a Soil and Land Steward—keeping in mind the interests of future generations—soil will also have to be regarded as a common property and must be dissociated from traditional private land ownership.

3.2.2 Ecosystems or Ecosystem Services

Another lever for more soil care can be found in the concept of ecosystems or ecosystem services. Ecosystem services are the services that ecosystems provide to society, the so-called benefits of nature for human well-being.[97]

As ecosystems, soil and land also provide important services, such as:[98]

- Biomass production (e.g. agriculture and forestry);
- The storage, filtration and conversion of nutrients, substances and water.

In this way, the components of ecosystems form natural capital. For instance, the natural capital of soil and land plays an important role in climate control through soil moisture, temperature changes and carbon storage.

Only when the citizens, companies and public authorities realise the important benefits they obtain from these ecosystems will they decide to preserve the natural capital, thereby better protecting the environment.[99]

A tool to convince everyone of the value of ecosystem services is by linking these ecosystem services to their economic value. In 1997, COSTANZA et al. identified 17 categories of ecosystem services and estimated their value at 33 USD trillion a year.[100] This estimate was considered to be a minimal estimate and was later updated to 125 USD trillion per year.[101] This monetary valuation links the environment and the economy. On the other hand, there is the fear that the financial valuation of ecosystem services will corrupt their inherent value.[102]

However, there is strong support that governments should use concepts such as natural capital and ecosystem services to shape their soil and land policies.[103]

[97] Robinson et al. (2018), p. 2.

[98] Ibid., 1.

[99] Ibid., 3.

[100] Costanza et al. (1997), pp. 253–260.

[101] Costanza et al. (2014), pp. 152–158.

[102] Robinson et al. (2018), p. 10.

[103] Nicole and Common Forum (2018), p. 21.

3.2.3 Granting Rights to Nature

Some scientists argue that a sixth mass extinction wave is currently taking place.[104] This is caused by the devastating impact of humans on ecosystems since the industrial revolution in Western Europe and the export of this model through colonisation throughout the world.[105]

The introduction of nature and environmental protection laws in the twentieth century sought to reduce these effects (with varying degrees of success).

It is against this background that a revolutionary concept is emerging: giving rights to nature. The aim of this legal revolution is to see nature's rights legally recognised in a more direct way and to grant rights to nature, ecosystems or a specific animal or plant species through this rights-based approach.[106] Over the last decade, certain elements of nature have for the first time been considered legal subjects, carriers of rights and obligations:

- In 2017, at the request of the Maori,[107] the New Zealand Whanganui River was granted legal status as a legal entity.[108] Two guardians, one of the Crown,[109] and one of the indigenous people, were given the role of guardians of the river;[110]
- In Bolivia, Mother Earth's Law granted natural ecosystems rights equivalent to those of human beings. The law laid down that development in Bolivia can only be done in connection with nature and with ancestral knowledge and wisdom. The law rejects the commercialisation of natural resources and obliges the Bolivians to prevent damage to the environment, biodiversity, human health and cultural values.[111]

The supporters of assigning legal personality to nature see this as a logical next step in the fight for better environmental protection.[112]

SCHOUKENS sees few legal objections. At present, only natural persons and legal entities, such as companies, are recognised in our law. But even a company is only a fictitious thing. In itself, therefore, there is no legal distinction as to why nature should be excluded from being granted legal personality, or a specific part of it, such as a river. Nature may not have a specific voice, but neither do companies.[113]

[104] Barnosky et al. (2011), pp. 51–57.

[105] Schoukens (2017), p. 59.

[106] Ibid.

[107] The Maori are the indigenous people who first inhabited New Zealand.

[108] Knack (2017).

[109] Queen Elizabeth II of the United Kingdom is the head of state of New Zealand.

[110] See: https://www.visitwhanganui.nz/whanganui-river-will-become-legal-entity/ (consulted on September 27, 2021).

[111] Mo (2014).

[112] Schoukens (2017), p. 65.

[113] Schoukens (2017), p. 65.

According to SCHOUKENS, there are several arguments in favour of granting rights to nature:

- Existing environmental protection laws are ineffective and (often) only inadequately enforced. Despite some small success stories, the loss of biodiversity continues unabated. SCHOUKENS gives the example of illegal deforestation.
- Giving direct rights to nature makes it easier to defend nature's interests. Guardians or representatives of nature could, just as in the case of incapacitated adults, take legal action for nature at any time, without having to hold lengthy debates about their procedural interests and capabilities.[114]

On the other hand, there is criticism that it is not easy to determine nature's rights. A nature reserve has no voice, no personal desires and no personal experiences: in the end, it is man—the natural person—who assigns value to nature.[115] How far should this protection go? Is it practically possible to protect entire ecosystems or should this protection focus on specific ecosystems, such as the case of the river Whanganui?

Although SCHOUKENS acknowledges that the argument for granting rights to nature can seem far-fetched, we agree with him insofar as he writes that the trend towards the granting of legal personality to nature forces us to look at today's environmental crisis from a new legal perspective.[116] In this way, granting rights to nature can also be a steppingstone to put more effort into the protection of our soil and land.

3.2.4 Circular Economy

The last lever we discuss is the circular economy. The circular economy is an economic system of closed loops in which raw materials, components and products lose their value as little as possible (e.g. trough waste recycling) and renewable energy sources are used.[117] This system aims to prevent waste and pollution, keep products and materials in use and allow natural systems to recover and thrive.

A special application is the Doughnut Economics of RAWORTH.[118] With her Doughnut Economics, she tries to offer an answer to the exhaustion of natural resources, which is partly due to the classic interpretation of private property as the most absolute right. For her model, she does not take economic growth as a

---

[114] In Flemish environmental law, a person may only act before the administrative courts (e.g. the Council of State) insofar as he has a personal interest in this procedure. This makes it difficult to act in the interest of nature.

[115] Schoukens (2017), p. 69.

[116] Schoukens (2017), p. 71.

[117] HET GROENE BREIN, https://kenniskaarten.hetgroenebrein.nl/en/knowledge-map-circular-economy/what-is-the-definition-a-circular-economy/ (consulted on September 27, 2021).

[118] Raworth (2017), pp. 51–60.

starting point, but sustainability. She starts from a bandwidth (the doughnut) within which we conduct our economy and transactions in a sustainable manner. The lower limit, the doughnut's hole, is a basis of fundamental social rights; the upper limit is an ecological ceiling. RAWORTH's most important principle is that growth and profit are fine, but excessive growth and excessive profit bring more problems than they solve.[119]

3.2.5 Conclusion

These other levers we discussed under Sect. 3.2 are—some more than others—controversial. We are therefore realistic that a paradigm shift to these new concepts is not for the immediate future. On the other hand, we believe that these concepts can contribute to a greater environmental awareness, including a greater awareness of the importance of soil and land within our world. We therefore see these concepts as the key reasoning and foundation for future scenarios for more soil care.

4 What Are the Constraints We Must Take in Account?

Under Sect. 3 we have examined Soil and land Stewardship and other levers for more soil care. In order to work out practical, workable (legal) translations of SLS, we must of course also pay attention to the most important constraints. After all, soil and land remain exposed to the constant, often conflicting demands of society. Within the scope of this contribution, we identify the following main constraints: private property rights (Sect. 4.1), legal-technical constraints (Sect. 4.2) and the lack of (political) support (Sect. 4.3).

4.1 Private Property Rights

One of the main bottlenecks in the introduction of a principle such as Soil and Land Stewardship are classic private property rights.

4.1.1 The Right to Ownership Is a Fundamental Human Right

It is a common claim that private owners do not like interference from the government or outside in the way they use their soil and land. Nor will private owners be

[119] Ibid.

willing to remediate soil pollution when the cause of the pollution cannot be clearly identified[120] or when the private owners did not cause the pollution.

In Belgium, this self-centred way of thinking stems from a centuries-old provision of the Civil Code. Article 544 of the Belgian Civil Code indeed defines (private) property rights as: "the right to enjoy and dispose of a property in the most complete manner possible, provided that one does not make use of it in a manner contrary to the laws or regulations[121]".

This provision was taken unchanged from the French Civil Code and dates to 3 September 1807. The jurists who drafted the French Civil Code relied on the work of DI SASSOFERRATO, who used the Roman concept of property law: the right to dispose of a property at will, i.e. a power that is as extensive as possible.[122]

In this classic—most common—private law-inspired conception, property is considered as an individual right, in which the acquisition of value, increase in value and individual prosperity are the starting points.[123] Ownership is a fundamental human right.[124] It should therefore come as no surprise that ownership has been elevated to a fundamental human right in several important legal instruments; for instance, in the European Convention on Human Rights.[125]

4.1.2 The European Convention on Human Rights (ECHR) as a Classic Lever for Interference

Article 1 of the First Additional Protocol to the ECHR guarantees the right to property as a human right. At the same time, however, this provision opens the door to restrictions on ownership. This article stipulates:

> Every natural or legal person is entitled to the peaceful enjoyment of his possessions. No one shall be deprived of his possessions except in the public interest and subject to the conditions provided for by law and by the general principles of international law.
>
> The preceding provisions shall not, however, in any way impair the right of a State to enforce such laws as it deems necessary to control the use of property in accordance with the general interest or to secure the payment of taxes or other contributions or penalties.

---

[120] E.g. diffuse soil pollution.

[121] Vuye et al. (2001), p. 888. According to Vuye et al., the fantasy of many readers of this article 544 of the Belgian Civil Code may run wild when reading the words "*in the most complete manner*". However, the literal terms of Article 544 of the Belgian Civil Code already determine the primary limits of the right of ownership: no use may take place that is contrary to the laws or regulations. One of these boundaries is, for instance, the duty of care for nature (see Sect. 3.1.3).

[122] Akkermans (2018), p. 1444.

[123] Struycken (2007), p. 236.

[124] Vandenberghe (2006), p. 83.

[125] Convention for the Protection of Human Rights and Fundamental Freedoms of 4 November 1950.

In its case law on this provision, the European Court of Human Rights (ECtHR) divides Article 1 ECHR into three distinctive rules:[126]

- The first rule can be found in the first sentence of the first paragraph: this is the general rule guaranteeing the undisturbed enjoyment of property;
- The second rule in the second sentence of the second paragraph is a specific rule which subjects expropriation to certain conditions;
- The third rule in the second paragraph recognises the right of States to regulate property without expropriation.

Environmental protection measures are also covered by this third rule, as are property restrictions imposed with a view to urbanisation. For instance, where spatial planning instruments impose restrictions on the degree of sealing or an obligation to provide a water-permeable pavement, this is a restriction of ownership within the meaning of the third rule.

Simply because Article 1 ECHR opens the door to a restriction of ownership or interference in ownership, this cannot be done in an arbitrary manner. The right of ownership is still a fundamental right, which presupposes that the restrictions of ownership must always meet cumulative conditions. These conditions are the legality requirement, the finality requirement and the proportionality requirement.[127] We briefly discuss these conditions below:

- The legality requirement means that the interference must be based on a legal rule. This national rule of law must be sufficiently accessible and foreseeable.
- According to the finality requirement, interference is only possible to the extent that it pursues an aim in the public interest.
- The third condition, known as the proportionality test, is the one most often challenged. It must be examined whether the public authorities have pursued a reasonable balance between the requirements of the general interest and those of the protection of the fundamental right to property. This condition requires a specific evaluation.

We must always bear these principles in mind when working out future scenarios for the (legal) translations of Soil and Land Stewardship and soil care. Because the (centuries-old) traditional property rights and their limitations often involve fierce discussions and years of proceedings—even before the ECtHR—it is not surprising that more attention has been paid to the new concepts discussed above such as SLS, soil as a common good, ecosystems and ecosystem services, etc.

[126] ECtHR 23 September 1982, nr. 7151/75, SPORRONG AND LÖNROTH V. SWEDEN and ECtHR 24 June 1993, nr. 14556/89, PAPAMICHALOPOULOS AND OTHERS V. GREECE.
[127] Ibid.

4.2 Legal-Technical Constraints

In the discussion of duties of care (see Sect. 3.1.3), the legal-technical bottlenecks were already briefly discussed.

These duties of care are deliberately kept vague in order to function as a safety net. These rules make it possible to take legal action against serious forms of environmental pollution that are not covered by concrete, specific environmental standards.

However, this general formulation of the duty of care provisions risks conflicting with other important principles, such as the principles of criminal legality and legal certainty. The principle of criminal legality, which is also expressed in the Latin legal adage *nullum crime, nulla poene sine lege*,[128] is also guaranteed as a fundamental human right in European and international treaties, for instance:

- Article 7(1) ECHR:

  No one shall be held guilty of any criminal offence on account of any act or omission which did not constitute a criminal offence under national or international law at the time when it was committed. Nor shall a heavier penalty be imposed than the one that was applicable at the time the criminal offence was committed.

- Article 15(1) of the International Covenant on Civil and Political Rights:

  No one shall be held guilty of any criminal offence on account of any act or omission which did not constitute a criminal offence, under national or international law, at the time when it was committed. Nor shall a heavier penalty be imposed than the one that was applicable at the time when the criminal offence was committed. If, subsequent to the commission of the offence, provision is made by law for the imposition of the lighter penalty, the offender shall benefit thereby.

The principle of criminal legality is one of the most fundamental principles of criminal law. Duties of care which would impose criminal liability but which are too vague cannot be enforced under criminal law.

It is a difficult exercise to find the right balance between:

- avoiding overly vague phrases and concepts (breach of the principle of legality); and
- formulating a duty of care broadly enough to fulfil its safety net function.

4.3 The Lack of (Political) Support

Finally, the implementation of effective measures for more soil care also requires political and community support, which recognises the vital role of soil and land in our society.

---

[128] Literally translated: "no crime, nor punishment without (prior) law".

5 What Are the Next Steps?

Although the concept of Soil and Land Stewardship may at first glance appear to be somewhat unreachable and far-fetched, our analysis in Sect. 3 shows that the Stewardship principle, as an action perspective for more soil care, already has some (legal) translations at different levels (see Sect. 3.1).

Since the SLS concept as a lever for more soil care is attracting increasing attention from specialists from a wide range of disciplines, there are not only purely legal concepts but also other, less conventional concepts that can lead to more soil care (see Sect. 3.2).

An important conclusion we can draw from this contribution is that around the concept of SLS as an action perspective for (more) soil care, there is still much food for thought and further research. This is only normal. We feel that getting started with this concept is primarily an "organic" process: a way of creating support, generating insight and generating energy for active involvement.[129]

We think the following next steps would be very valuable in further implementing the Soil and Land Stewardship principle:

- Seeking more political and community support from the bottom up. We have to examine how to raise soil and land awareness among citizens, authorities and businesses. People must be encouraged to take care of the soil.
- Further investigate the added value of the Land Stewardship agreements and their actual effects in Catalonia. If they show good results, other (non-)governmental organisations could also make use of such agreements to protect the natural, cultural and landscape resources and values of their territory.
- The existing soil regulations still focus very intensively on one single soil threat: soil pollution. The international, European and national levels must create more attention for the other soil threats through binding legislation (e.g. an international binding soil instrument) or (at least) through awareness raising.
- This organic process must be handled in a multidisciplinary way: not only involving the administration competent for soil, but also other departments, e.g. the departments competent for agriculture, forest, environmental licensing and environmental enforcement.
- Research whether we can consider ecosystems legal generalities in order to be able to protect them via special rules.

References

Akkermans B (2018) Duurzaam goederenrecht: naar een herijking van ons goederenrechtelijk stelsel? TPR 4:1437–1470

[129] Bal et al. (2018), p. 13.

Bal N, Luyten E, Vandenbroeck P, Vanheusden B (2018) Soil + Land Stewardship. Een handelingsperspectief voor bodemzorg. Bodem 12:11–13

Barnosky AD, Matzke N, Tomiya S, Wogan GOU, Swartz B, Quental TB, Marshall C, McGuire JL, Lindsey EL, Maguire KC, Mersey B, Ferrer EA (2011) Has the Earth's sixth mass extinction already arrived? Nature 417:51–57

Block P (2013) Stewardship: choosing service over self-interest. Berrett-Koehler, Oakland, p 312

Butti L, Toniolo B, Campigotto S (2019) The impact of EU legislation, principles and case law on the national contaminated land regimes, https://www.filodiritto.com/impact-eu-legislation-principles-and-case-law-national-contaminated-land-regimes-0 (consulted on September 27, 2021)

Communication from the Commission to the Council, the European Parliament, the Economic and Social Committee and the Committee of the Regions, *Towards a Thematic Strategy for Soil Protection*, COM/2002/0179, https://eur-lex.europa.eu/legal-content/EN/TXT/?uri=CELEX%3A52002DC0179 (consulted on September 27, 2021)

Convention for the Protection of Human Rights and Fundamental Freedoms of 4 November 1950

Convention of 10 October 2013 on Mercury

Convention of 10 September 1998 on the Prior Informed Consent Procedure for Certain Hazardous Chemicals and Pesticides in International Trade

Convention of 17 Juni 1994 to Combat Desertification

Convention of 22 March 1989 on the Control of Transboundary Movements of Hazardous Wastes and Their Disposal

Convention of 22 May 2001 on Persistent Organic Pollutants

Costanza R, D'Arge R, De Groot R, Farber S, Grasso M, Hannon B, Limburg K, Naeem S, Oneil RV, Paruelo J, Raskin RG, Sutton P, Van Den Bel M (1997) The value of the world's ecosystem services and natural capital. Nature 387:253–260

Costanza R, De Groot R, Sutton P, Van Der Ploeg S, Anderson SJ, Kubiszewski I, Farber S, Turner RK (2014) Changes in the global value of ecosystem services. Global Environ Change 26:152–158

Directive 2004/35/CE of the European Parliament and of the Council of 21 April 2004 on environmental liability with regard to the prevention and remedying of environmental damage, *Official Journal* L 143, 30 April 2004, 0056 – 0075

ECJ 13 July 2017, C-129/16, TÚRKEVEI TEJTERMELŐ KFT

ECJ 4 March 2015, C-534/13, FIPA et al.

ECJ 9 March 2010, C-378/08, RAFFINERIE MEDITERRANEE et al.

EctHR 23 September 1982, nr. 7151/75, SPORRONG AND LÖNROTH V. SWEDEN

EctHR 24 June 1993, nr. 14556/89, PAPAMICHALOPOULOS AND OTHERS V. GREECE

Enqvist JP, West S, Masterson VA, Haider LJ, Svedin U, Tengö M (2018) Stewardship as a boundary object for sustainability research: linking care, knowledge and agency. Landsc Urban Plan 179:17–37

Faure M, Visser M (1999) De strafrechtelijke bescherming van het leefmilieu in België, Duitsland en Nederland. Modellen van strafbaarstelling en hun bewijsrechtelijke implicaties. Intersentia, Antwerpen – Groningen, p 494

Gutwirth S, Stengers I (2017) Het recht en de heropleving van de commons. TMR 1:4–24

Hardin G (2018) The tragedy of the commons. Science 162:1234–1248

Houweling P (1997) Zorgplichtbepalingen in de nieuwe art. 8.40 Wm-besluiten. Milieu en recht 7(8):142–148

International Covenant on Civil and Political Rights of 19 December 1966

Karp JP (1993) A private property duty of stewardship: changing our land ethic. Environ Law 23:735–762

Knack, "Nieuw-Zeelandse rivier wordt als eerste ter wereld rechtspersoon", 15 March 2017., https://www.knack.be/nieuws/planet-earth/nieuw-zeelandse-rivier-wordt-als-eerste-ter-wereldrechtspersoon/article-normal-827223.html (consulted on September 27, 2021)

Marx K, Engels F (1967) Capital: a critique of political economy. International Publishers, New York, p 645

Meeus R (2008) De arresten nr. 36/2008 en nr. 82/2008 van het Grondwettelijk Hof: het moeilijke onderscheid tussen de grondwettige en ongrondwettige open structuur van strafrechtelijk gesanctioneerde milieuzorgplichtbepalingen. TMR 4:454–472

Messa M (2015) Slow Food Position Paper on Soil, https://www.slowfood.com/wp-content/uploads/2020/12/ING_suolo-4.pdf (consulted on September 27, 2021), p 20

MO, "Boliviaanse natuur heeft voorlopig alleen rechten op papier", 19 mei 2014., https://www.mo.be/nieuws/natuur-heeft-rechten-bolivie-maar-alleen-op-papier (consulted on September 27, 2021)

Nicole and Common Forum, "Land Stewardship: Investing in The Natural, Social and Economic Capital of Industrial Land", 2018., https://rwsenvironment.eu/subjects/soil/publications/land-stewardship/ (consulted on September 27, 2021), p 44

Ostrom E (1990) Governing the commons: the evolution of institutions for collective action. University Press, New York – Cambridge, p 270

Proposal for a Directive of the European Parliament and of the Council establishing a framework for the protection of soil and amending Directive 2004/35/EC, COM/2006/0232, https://eur-lex.europa.eu/legal-content/EN/TXT/?uri=CELEX%3A52006PC0232, (consulted on September 27, 2021)

Raworth K (2017) Doughnut economics: seven ways to think like a 21st century economist. White River Junction, Vermont, p 384

Ren X, Zeng G, Tang L, Wang J, Wan J, Liu Y, Yu J, Yi H, Ye S, Deng R (2018) Sorption, transport and biodegradation – an insight into bioavailability of persistent organic pollutants in soil. Sci Total Environ 610–611:1154–1163

Robinson DA, Seaton F, Sharps K, Thomas A, Roberts FP, Van Der Ploeg M, Jones L, Stolte J, De La Bellacasa MP, Harrison P, Emmett B (2018) Soil resources, the delivery of ecosystem services and value. Oxford Res Encycl Environ Sci:1–44

Rodota S (2011) The role of common goods in fostering shared responsibility, linking sustainable social cohesion to the reduction of inequalities, https://ideas.repec.org/p/iuc/rpaper/2-11.html (consulted on September 27, 2021), p 10

Rodríguez P, Sabater X (2018) Land Stewardship toolkit. Basic tools for land stewardship organisations in Europe, https://elcn.eu/sites/default/files/2018-01/XCT%202014%20LandLife_TOOLKIT.pdf (consulted on September 27, 2021), p 125

Schoukens H (2017) De natuur als rechtspersoon: een noodzakelijke paradigmashift in tijden van massa-uitsterven? TMR 2017, 25 jaar TMR, 57–71

Singh J (2015) The sustainable development goals: the role of ethics. Sight Life 29(2):58–63

Star SL (2010) This is not a boundary object: reflections on the origin of a concept. Sci Technol Hum Values 35:601–617

Stolte J, Tesfai M, Øygarden L, Kværnø S, Keizer J, Verheijen F, Panagos P, Ballabio C, Hessel R (eds) (2016) Soil threats in Europe. JRC Technical Reports, https://publications.jrc.ec.europa.eu/repository/handle/JRC98673 (consulted on September 27, 2021), p 206

Struycken TH (2007) De numerus clausus in het goederenrecht. Wolters Kluwer Nederland, Deventer, p 870

Umweltbundesamt, "Legal Instruments to implement the objective "Land Degradation Neutral World" in International Law", 2015., http://www.umweltbundesamt.de/publikationen/legal-instruments-to-implement-the-objective-land (consulted on September 27, 2021), p 126

United States Environmental Protection Agency, Everyday choices: opportunities for environmental stewardship, Washington, 2005, p 19

Van De Voorde J, Carette N, Hubeau B, Van Den Broeck P (2018) De milieu-erfdienstbaarheid of milieugrondlast: een nuttig zakenrechtelijk instrument voor de milieubescherming. TPR 3: 1195–1313

Vandenberghe H (2006) Zakenrecht. Acco, Leuven, p 379

Visser SM, Keesstra S, Maas G, De Cleen M, Molenaar C (2019) Soil as a basis to create enabling conditions for transitions towards sustainable land management as a key to achieve the SDGs by 2030. Sustainability 11(6792):19

Vuye H, Kokelenberg J, Van Sinay T (2001) Overzicht van rechtspraak. Zakenrecht 1994 – 2000. TPR 2:837–1199

Welchman J (1999) The virtues of stewardship. Environ Ethics 21(4):411–423

Worrel R, Appleby MC (2000) Stewardship of natural resources: definition, ethical and practical aspects. J Agric Environ Ethics 12:263–277

Worst JH (1915) The Stewardship of the soil, http://www.gutenberg.org/files/24080/24080-h/24080-h.htm (consulted on September 27, 2021)

Xarxa de Custòdia del Territori, "Land stewardship: an effective conservation tool", 2017., http://www.xct.cat/mm/file/2017/Fulleto%20internacional%202017.pdf (consulted on September 27, 2021), p 2

Xarxa de Custòdia del Territori, "Land stewardship: an effective conservation tool", 2018., https://elcn.eu/sites/default/files/2018-06/0104_rodrigo_land_stewardship_-_repaired.pdf (consulted on September 27, 2021), p 34

Innovative Phosphorus Governance: How to Address Recurring Regulatory Shortfalls—The Example of Germany, Costa Rica and Nicaragua

Jessica Stubenrauch

Abstract Phosphorus is a vital element for all life on earth. At the same time, phosphate rock is a finite resource. The scarcity of the essential nutrient on the one hand and ineffective agricultural phosphorus use and over-fertilisation on the other are global concerns that are not limited to several countries. This chapter provides effective governance approaches that promote sustainable phosphorus use in agriculture in line with the environmental targets set in the Paris Agreement and the Convention on Biological Diversity. The results build on a legal comparison between the industrial country Germany, the emerging economy Costa Rica and the developing country Nicaragua. It will therefore be shown that the phasing out of fossil fuels and the minimisation of livestock are directly connected with phosphorus governance and thus need to be addressed. In this respect, the application of economic policy instruments on a preferably broad geographical scale has proven to be advantageous compared to command-and-control approaches that have since maintained focus on single legal fields. Therefore, effective phosphorus governance requires a mix of both- economic policy instruments regulating major polluters and complementary command-and-control regulations for fine-tuning, e.g., to avoid hotspot problems or foster phosphorus recycling. These core findings are also applicable to other world regions.

J. Stubenrauch (✉)
Department of Environmental and Planning Law, Helmholtz-Centre for Environmental Research, Leipzig, Germany
e-mail: jessica.stubenrauch@ufz.de

H. Ginzky et al. (eds.), *International Yearbook of Soil Law and Policy 2020/2021*, International Yearbook of Soil Law and Policy 2020,
https://doi.org/10.1007/978-3-030-96347-7_17

1 Introduction

Phosphorus (P) is an essential and non-substitutable nutrient for plants, animals and humans and has a life-supporting function.[1] Global food security is therefore directly connected to the availability of the nutrient.[2] However, the P supply of soils varies considerably across the world. While in the global south soils tend to be undersupplied with P,[3] agricultural soils in Europe are frequently well-supplied or even oversupplied with P due to high livestock densities or longstanding stock fertilisation in the past.[4] Although a peak P is not foreseeable in the near future, phosphate rock is a finite resource located in only a few regions of the world.[5] Hence, most of the states in the world are highly dependent on the import of mineral P gained from phosphate rock, which increases the risk of supply shortages. Apart from that, the remaining predominantly sedimentary deposits of phosphate rock tend to be increasingly contaminated with heavy metals like cadmium (Cd) and uranium (U).[6] In spite of that, P is inefficiently used in agriculture worldwide and due to this, P cycles are globally disturbed.[7]

At the same time, global environmental targets set out by the Paris Agreement (PA) and the Convention on Biological Diversity (CBD) call for a major transformation of the agricultural sector. Article 2 para. 1 PA aims to keep global warming to well below 2 °C or even better 1.5 °C above pre-industrial levels. To limit global warming to 1.5 °C, worldwide net-zero emissions across all economic sectors are required no later than in approximately two decades.[8] That means that inevitably remaining emissions will have to be compensated by sinks, which would exhibit far-reaching consequences for the agricultural sector. However, so far, instead of being a net-sink for greenhouse gases (GHG), agriculture is a main driver of global climate change next to biodiversity loss.[9] Therefore, in the future, the contribution of the agricultural sector has to be reversed, so that climate and biodiversity standards are preserved. Besides oceans and forests, soils function as major carbon sinks and their capacity to store carbon can be increased (or reduced) depending on the agricultural management.[10] Beneficial agricultural practices which increase the carbon storage capacity of soils and at the same time foster a sustainable P use on

---

[1] The article was drafted in 2020, which represents the state of the legislation. Leinweber et al. (2018) and Schoumans et al. (2015).

[2] Cordell and White (2015).

[3] Mogollón et al. (2018). For a comparison between Germany and Costa Rica and Nicaragua Stubenrauch et al. (2018) and Stubenrauch (2019).

[4] Tóth et al. (2014).

[5] USGS (2020), p. 119.

[6] Kratz et al. (2016).

[7] Schoumans et al. (2015) and Sharpley et al. (2015).

[8] IPCC (2019a) and Ekardt et al. (2018a, b).

[9] IPCC (2019b) and IPBES (2019).

[10] Paustian et al. (2019), IPCC (2019b) and Batjes (1996).

farm level are compiled in Sect. 2. In addition, the dependence on fossil fuels in the agricultural sector has to be overcome to fulfil the PA. Thus, fossil-fuel based heavy machinery on the field and long transportation routes for agrochemicals, food and feedstuff have to be questioned as well. This also includes highly energy intensive fertiliser production and P recycling techniques. So far, the mining of phosphate rock, but even more so the processing of mineral nitrogen in the Haber-Bosch process depends to a large extend on fossil fuels.[11]

The global challenge that can be derived from the above is to close (P) nutrient cycles without continuing to be dependent on fossil fuels. Apart from that, local nutrient hotspots leading to P discharges into waterbodies and local supply bottlenecks with P need to be addressed. The required policy implications have hardly been fully investigated so far.[12] Thus, building on latest natural scientific findings, the chapter also aims to provide the most effective policy instruments for P use in agriculture as one cornerstone within an overall package of legal instruments so as to achieve the global climate and biodiversity targets. The governance analysis builds on the results of a legal comparison between the industrial nation Germany, the emerging economy Costa Rica and the developing country Nicaragua.[13] In doing so, the universal character of the United Nations Sustainable Development Goals (SDGs), which apply equally to all countries, is reflected.[14]

2 Sustainable Phosphorus Management in a Post-Fossil World

2.1 *Phosphorus Management on Farm Level*

Nutrient management is one crucial part in a comprehensive sustainable agricultural management concept. Key issues of an efficient P management on farm level are shown in Fig. 1.

The preservation and stimulation of natural soil fertility, e.g. by well-designed intercropping techniques combined with organic or green fertilisation strategies is of central importance to reduce P fertiliser input at source and above that also the off-farm inputs in general.[15] Organic fertilisation (e.g., compost, digestates, green manure or animal-derived fertilisers) can contribute to enhance soil organic matter and thus the natural soil fertility and the carbon reservoir of the soil.[16] Apart from

---

[11] Kitano (2012) and Kyriakou et al. (2020).

[12] Regarding a global P governance gap see Rosemarin and Ekane (2016) and Leinweber et al. (2018).

[13] Stubenrauch et al. (2018) and Stubenrauch (2019).

[14] United Nations (2015), p. 1.

[15] Glaze-Corcoran et al. (2020), Eichler-Löbermann et al. (2016) and Ohm et al. (2017).

[16] Vargas et al. (2019) and Paustian et al. (2019).

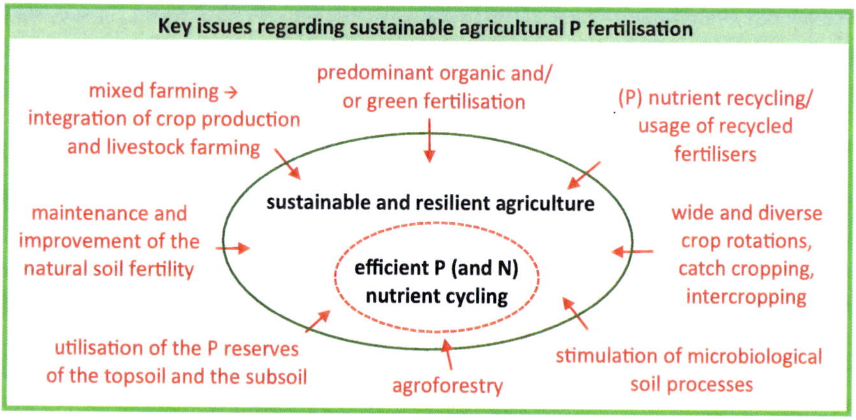

Fig. 1 Key issues regarding sustainable agricultural P fertilisation (own figure)

that, improved microbial activity by mycorrhiza fungi and bacteria are able to increase the plant availability of phosphate bound in the soil.[17] The cultivation of deep-rooted next to shallow-rooted plants can help to enhance the plant availability of the phosphate reservoirs in the subsoil.[18] At the same time, soil characteristics like soil texture, aggregate stability and water retention capacity improves.[19] An improved water retention capacity as well as increasing soil coverage through catch crops prevent nutrient losses due to erosion.[20] Thus, site-specific mixed and intercropping in wide crop rotations combined with crop-specific (preferably) organic fertilisation minimises both—the necessary P fertiliser inputs and the risk of P losses to adjacent ecosystems.[21] This is also true for the integration of trees into agricultural crop production in agroforestry systems.[22]

Furthermore, it is imperative to strengthen the combination of livestock farming and crop production in site-adapted mixed-farming systems,[23] and to end intensive livestock farming which is also a major reason for globally disturbed P cycles (see Sect. 2.2). This would therefore ease the local availability of organic fertilisers on the one hand and prevent detrimental nutrient hotspots on the other. These scientific findings are important for the required definition of good agricultural practice in agricultural legislation as well as for the design of the circular economy and fertiliser

[17] Saxena et al. (2014); Ceci et al. (2018) as well as Mącik et al. (2020) also point towards the importance of biofertilisers that contain microorganisms, and which can be injected to the soil as non-toxic alternative to synthetic agrochemicals.

[18] Sattari et al. (2012).

[19] Buczko et al. (2019), Koch et al. (2019) and Peine et al. (2019).

[20] Garske et al. (2020) and Sharpley (2015).

[21] Withers et al. (2017) and Macintosh et al. (2019). Especially in highly P-supplied soils a P fertilisation below plant removal rates regulary shows no negative influence on yields and severly minimises the risk of P discharges into adjacent ecoystems. Buczko et al. (2019).

[22] Techen et al. (2020), p. 187.

[23] Lemaire et al. (2014) and Poffenbarger et al. (2017).

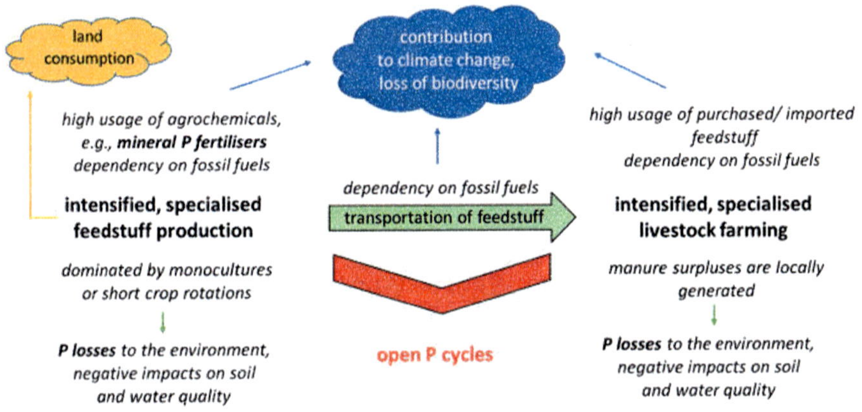

Fig. 2 Intensive livestock farming systems and open phosphorus cycles (own figure)

legislation (see Sect. 3.4). Notwithstanding this, efforts in the development of (non-energy intensive) P recycling techniques beyond farm level could minimise the dependency on P fertilisers gained from phosphate rock.[24]

2.2 *Intensive Livestock Farming as a Global Driver of Open Nutrient Cycles*

Industrialised and intensified livestock farming systems not only contribute significantly to global climate change and the loss of biodiversity but are also a major cause of globally disturbed P and nitrogen (N) cycles (Fig. 2).[25]

On the one hand, fodder production which is dominated by monocultures or short crop rotations requires a high input of agrochemicals including mineral P. This places an additional burden on the soil and as a result of this, its natural functions start to degrade. As a consequence of accelerated soil degradation, natural soil fertility declines and the risk of P discharges into waterbodies, which causes eutrophication, increases.[26]

On the other hand, high densities of animals lead to an accumulation of manure, causing nutrient hotspots. The risk of P discharges into water bodies increases and poses a major threat to the biodiversity of highly sensitive aquatic ecosystems.[27] Thus, a minimisation of livestock and overcoming the spatial decoupling of land and

[24] Promising alternative and innovative P fertilisers are for example struvite recovered from wastewater streams or sulfur-enriched bone char. See Hallas et al. (2019), Kataki et al. (2016) and Zimmer et al. (2019).

[25] Weishaupt et al. (2020) and Machovina et al. (2015).

[26] Leip et al. (2015); Stubenrauch (2019), pp. 101–103.

[27] Schindler et al. (2016) and Bloem et al. (2020).

livestock is imperative to close disturbed P cycles worldwide, as already pointed out in Sect. 2.1. Figure 2 furthermore illustrates the large dependency on fossil fuels of the intensified and globalised animal-derived food production system as well as its high land use. In summation, intensive livestock farming and fossil fuels are a major cause for various intertwined environmental problems and thus, next to P itself, have to be targeted thoroughly by effective policy instruments (see Sect. 3.2).[28]

3 Closing the Gap of Phosphorus Governance

Although the risk of potential P supply shortages has led to an increased public and scientific debate around P during the last ten years, comprehensive P governance approaches on an international and on most national levels are still missing.[29] Instead, even if there is awareness of the problem at the policy level, the political interventions regularly do not adequately address the various, closely intertwined environmental problems surrounding P effectively or fail due to typical governance problems.[30]

In the following, a theoretical discussion on the necessary design of governance instruments to avoid typical governance problems and achieve a large steering effect (Sect. 3.1) will serve as basis to develop and propose effective legal approaches that are able to govern P comprehensively (Sects. 3.2 and 3.4).

3.1 *Targets, Strategies and Design of Policy Instruments to Avoid Typical Governance Problems*

Before any discussion on the concrete design of policy instruments, it is essential to identify the targets against which their effectiveness should be measured. A specific international policy target for P use in agriculture does not exist and it is questionable as to whether such a target can indeed be defined explicitly.[31] A total phasing out of mineral P (like for fossil fuels) might partially counteract food security and at the same time could not cover all the environmental concerns connected to P use in agriculture. However, the efficacy of policy instruments can hardly be measured

[28] Machovina et al. (2015); Metson et al. (2016); Ekardt et al. (2018a, b); EU Commission (2020), p. 35.

[29] Rosemarin and Ekane (2016) and Leinweber et al. (2018).

[30] Garske (2020), Stubenrauch et al. (2018) and Ekardt (2019).

[31] The fulfilment of the SDGs is (partly indirectly) associated with sustainable P use in agriculture. In particular SDG 2 to end hunger as well as the SDGs 6, 12, 13, 14, 15 and 17. However, the SDGs represent soft law, from which no legal obligations for the states can be derived. See United Nations (2015) and regarding P in detail Stubenrauch (2019), pp. 192–193.

Table 1 Efficiency, consistency and frugality strategies in phosphorus governance (own table)

Efficiency	Consistency	Frugality
Increases in efficiency through:	*Achievement of consistency through*:	*Minimisation of/reduced consumption of:*
– **efficient soil P management**	– **recycling of waste**	– **animal-derived foodstuff**
(1) crop- and soil-specific fertilisation	(1) organic fertilisation at farm level	– **non-regional, non-seasonal products**
• precise determination of the quantity of fertiliser use (e.g. via precision farming), time and place of fertilisation	• application of solid dung/(farmyard) manure	– **long transport routes for fertilisers, animal feed and foodstuff by**
	• utilisation of green manure or compost	• promotion of mixed farming
(2) maintenance of the natural soil fertility and the humus content of the soil	(2) P recycling	• promotion of direct marketing of local agricultural production
• site-adapted crop cultivation	• recovery of P from relevant source streams	• establishment of regional economic cycles
• mixed cropping and intercropping		
• diverse crop rotation		
– **efficient animal husbandry**, e.g. phase feeding	– **substitution of fossil fuels by renewable energies**	– **energy-intensive technologies**
– **reduction of food losses** (spoilage and disposal rate)	• with regard to P recycling, machinery in the field, transportation	

against a diffuse set of criteria.[32] Instead, the PA and the CBD set a clear framework within which future agriculture systems and P fertilisation need to operate within (see Sect. 2). Thus, the fulfilment of these legally binding targets can also serve as a benchmark for the effectivity of P governance.

The necessity to develop policy instruments which aim for a more efficient P use and which enhance P recycling is undisputed.[33] However, in light of the of the PA and the CBD, it becomes clear that efficiency and consistency strategies alone will probably not be sufficient to reach the ambitious targets. Instead, frugality strategies are an indispensable regulatory component to effectively respond to the challenges of natural resource, climate and biodiversity protection (within the remaining time span).[34] As shown in Sect. 2.2, to this end, the minimisation of livestock farming and the phasing out of fossil fuels are key elements. The relevant efficiency, consistency and frugality strategies to be included into P governance are compiled in Table 1.

[32] Ekardt (2019), pp. 5–7.

[33] Schoumans et al. (2015), Sharpley et al. (2015) and Rosemarin and Ekane (2016).

[34] Stubenrauch (2019), pp. 90–95; Leinweber et al. (2018).

Finally, policy instruments have to be designed in a way that typically occurring governance problems are avoided. Within transformational governance research policy instruments that have not yet existed need to be developed. Based on the knowledge of behavioural motivations of humans,[35] and on the assessment of the impacts of current policy instruments, insights about potential governance problems can be derived. Typical governance problems include rebound effects, spatial or sectoral shifting effects, lack of target stringency or enforcement problems as well as problems of depicting.[36] In this respect, the effectiveness of certain groups of policy instruments vary, also depending on their level of implementation (regionally, nation state, transnational or international). While command-and-control measures tend to be rather prone to various governance problems, in particular in the agricultural sector,[37] economic policy instruments entail certain advantages:

- *Shifting effects:* Stricter legislation on, e.g., agricultural fertilisation in a particular country or world region could lead a relocation of production to other world regions. This is why policy instruments should ideally be implemented on a broad geographical scale or include border adjustments to minimise shifting effects.[38]
- *Rebound effects:* Rebound effects occur if, e.g., efficiency gains of the minimised P need of one certain plant are reversed by an overall increase in crop production and P fertilisation. Rebound effects also arise when halted overexploitation of resources in one sector causes (possibly even more severe) detrimental effects in other sectors. Where absolute reductions are required (quantity control) economic policy instruments have proven to be advantageous compared to command-and-control regulations.[39]
- *Enforcement and lack of target stringency*: Besides a lack of stringency towards the targets of the PA and the CBD, command-and-control regulations in particular tend to suffer from weak enforceability. In the agricultural sector, a stringent enforcement of the latter would imply consistent monitoring of an almost infinite number of operations on a large number of farms, which is simply not feasible. This is why command-and-control instruments should only be one component of effective agricultural P governance.[40]
- *Problems of depicting*: Among others, the exact determination of GHG fluxes between the atmosphere and soils (including organic soils such as peatlands),[41] and the exact measurement of carbon contents of soils remains challenging,

[35] Regarding the behavioural motivation of humans, knowledge and cost-benefit considerations of single actors are typically overrated, while conceptions of normality, trained habits, values, emotions, personal search for recognition or cognitive dissonances are underestimated. Ekardt (2019).

[36] Ekardt (2019), pp. 243–246; Paul et al. (2019); Ekardt et al. (2020).

[37] Paul et al. (2019).

[38] Ekardt (2019), pp. 243–246; Will (2019).

[39] Ekardt (2019), pp. 243–246.

[40] Ekardt (2019), p. 271; Paul et al. (2019); Garske and Ekardt (2021).

[41] Ekardt et al. (2020).

especially with regard to larger, partly diverse areas. In fact, there is "an inverse relationship between the cost and the uncertainty of the measurements."[42] This must also be taken into account when designing economic policy instruments, which should preferably be based on easy-to-grasp parameters.[43]

3.2 Economic Policy Instruments

Considering the aforementioned theoretical discussion on the design of policy instruments economic policy instruments should primarily focus on major, easy-to-grasp polluters to address various environmental problems simultaneously and be implemented on a broad geographical scale. Next to fossil fuels and livestock farming also, P use in agriculture might be directly steered by economic policy instruments. Apart from that, targeted subsidies may contribute to enhanced agricultural and fertilisation practices.

3.2.1 Fossil Fuels

Cap and trade schemes, like emission trading schemes (ETS) are considered to be one of the most cost-effective ways to effectively limit the amount of fossil fuels and phase them out in accordance with the PA.[44] Taking into account the aforementioned governance problems, policy intervention at the first level of trade seems to be most promising. Thus, fossil fuels could be taken out of the market most efficiently by an upstream pricing of emissions.[45] Cap and trade systems already exist in several countries around the world, including in the European Union (EU ETS).[46] Emissions caused by the production of mineral fertilisers, waste-water treatment plants and the processing, cooling, and packaging of food and feedstuff are already covered by the EU ETS. However, a clear alignment of the EU ETS with the targets of the PA is missing and various loopholes cause low carbon prices that reduce the effectiveness of the cap-and-trade scheme.[47] The EU ETS therefore requires a thorough reform to exploit its full steering potential.[48] With regard to the comparably small Central American countries Costa Rica and Nicaragua an implementation of a cap-and-trade system at transnational level would be most effective. According to Art. 3 (h), (i) of

[42] Paustian (2019), p. 569.

[43] Ekardt (2019), p. 5; Garske (2020), pp. 293–295.

[44] Narassimhan et al. (2018), Zhang and Zhang (2019) and Bocklet et al. (2019).

[45] Ekardt (2019), Coria and Jaraitė (2019) and Weishaupt et al. (2020).

[46] Worldbank (2019); Herold et al. (2019), pp. 29–33.

[47] Hepburn and Teytelborn (2017). On discussions on actual reforms Bocklet et al. (2019) and Perino (2019).

[48] With regard to the current developments on EU level see Herold et al. (2019), pp. 29–33.

the Tegucigalpa Protocol,[49] the Central American Integration System[50] aims to promote the sustained economic, social, cultural and political development of the member states,[51] and to carry out concerted action to protect the environment. Thus, an implementation of an ETS would theoretically be conceivable. However, against the background of current political developments in some of the member states, such a political progress is unlikely.

3.2.2 Livestock Farming

The most effective approach to minimise the absolute number of animals in line with the PA is a separate cap-and-trade scheme for animal products combined with a livestock-to-land ratio (Flächenbindung) on farm level.[52] While the cap for fossil fuels needs be lowered towards zero, the cap for livestock farming—being an important component to close nutrient cycles (and for food security in some world regions)—cannot be that strict. Therefore, the cap for livestock farming cannot be integrated into a joint ETS scheme that also covers fossil fuels.[53] To minimise the enterprises involved in the proposed cap and trade system, the processing and supplying industry (slaughterhouses, dairies or egg producers) could be the regulatory points. Output-based emission values[54] per kilogram of product (slaughter weight) based on regional or national average values is a feasible control variable.[55] This approach minimises problems of depicting—compared to the challenging determination of GHG emissions at farm level. The additional consideration of production methods (manure management, pasture or stable farming) for the calculation of emission values favours grazing, small-scale and mixed farming systems, which promote the closing of nutrient cycles. At the same time, intensive livestock farming systems are impeded.[56] In contrast to these economic instruments, command-and-control instruments which define a maximum number of permitted animals per farm could easily be circumvented by splitting up livestock holdings.

[49] Tegucigalpa Protocol to the Charter of the Organization of Central American States (ODECA), https://www.sica.int/busqueda/busqueda_archivo.aspx?Archivo=prot_320_1_04052005.htm (last assessed 26.08.2021).

[50] Sistema de la Integración Centroamericana (SICA).

[51] Member States are Costa Rica, Nicaragua, Honduras, El Salvador, Guatemala, Panama, Belize, Dominican Republic.

[52] Weishaupt et al. (2020). Again, just like for a cap-and-trade system on fossil fuels, a precise design of border adjustment in line with the world trade law would avoid shifting effects to other countries that do not participate in the scheme.

[53] Weishaupt et al. (2020).

[54] Meat taxes are regularly based on that as well. Wirsenius (2011) and Bähr (2015).

[55] Weishaupt et al. (2020), p. 12.

[56] Weishaupt et al. (2020).

Furthermore, also compared to a tax on meat or feed imports, there would be no quantity control of the total number of animals which is required by the PA.[57]

However, to additionally adapt livestock farming into local conditions, a livestock-to-land ratio to avoid nutrient surpluses should be implemented on farm level.[58] As a consequence, P (and N) nutrient hotspots are avoided and feedstuff production at farm level is promoted. Additional planning law instruments or an import restriction for animal feed would probably then not be necessary to avoid over-fertilisation and to close nutrient cycles.[59] However, this also depends on the fact as to whether the aforementioned governance approaches to restrict livestock farming are implemented effectively.

3.2.3 Phosphorus

The cap-and-trade schemes for fossil fuels and livestock products, implemented on a transnational level like the EU or Central America, would substantially diminish various P (and N) induced environmental problems. The fossil-fuel based production of mineral fertilisers would be limited by a cap and prices for mineral fertilisers would be expected to rise.[60] However, additional economic instruments focussing directly on P might also contribute to close disrupted nutrient cycles and foster P recycling.[61] Because the implementation of a tax or an absolute cap on phosphate rock to limit extraction in the mining countries is not realistic, a *tax or cap on raw phosphate-containing mineral P* (or on the imports of them) in the importing countries seem to be more promising. This policy approach would also restrict the usage of potentially harmful mineral P fertilisers.[62] Farms that do not use or minimise the use of mineral fertilisers (e.g., organic farming), would not be affected by rising prices and thus benefit. Others would be encouraged to change their agricultural practices. However, demand for mineral P from the phosphate rock is rather inelastic. Therefore, the price increase would have to be relatively high to provide a significant steering effect.[63] Besides that, hot spot problems would not be addressed so that a strict regulation of livestock farming according to Sect. 3.2.2 remains indispensable.

---

[57] Regarding respective regulatory approaches see SRU (2015), p. 65; Möckel et al. (2014), pp. 551–552.

[58] Weishaupt et al. (2020), p. 27.

[59] On respective proposed instruments see SRU (2015), p. 356; Möckel et al. (2014), pp. 397–426.

[60] In particular regarding N, but also regarding P, that is often sold together as NP or NPK fertiliser together with potassium (K). Ekardt et al. (2018a, b) and Pfromm (2017).

[61] Grames et al. (2019), p. 324.

[62] Garske (2020), pp. 295, 297.

[63] Garske (2020), p. 298; Heckenmüller et al. (2014); Ecotec et al. (2001), p. 273.

The introduction of a producer-oriented *surplus levy* is often proposed to limit nutrient surpluses.[64] Based on a nutrient balance, the levy would tax the difference between P input and output on farm level if the difference exceeds allowed maximum values. Due to the detrimental effects of P surpluses, it would be reasonable to impose the levy from a P surplus above zero or to establish regionally differentiated limit values—depending on the environmental situation.[65] To this end, the elaboration of a nutrient balance on farm level would prove to be imperative. In Germany, according to § 1 (2) of the German Regulation on Nutrient Balances (Stoffstrombilanzverordnung, StoffBilV),[66] larger farms over 30 hectare or with more than 50 livestock units are obliged to elaborate a nutrient balance. However, for emerging or developing countries which so far do not even have an applied fertiliser legislation, as it is the case for Costa Rica and Nicaragua (see Sect. 3.4.2), this would be an extremely high and probably unfeasible administrative burden. Apart from that, the legal analysis of the adopted nutrient balance in Germany shows that the regulation remains partly very imprecise and leaves various loopholes.[67] This once again highlights that a stringent management of the livestock-farming sector in line with the PA (see Sect. 3.2.2) would avoid manure surpluses and thus nutrient hotspots far more effectively. In contrast, if a regulation of the livestock sector is not implemented, legal loopholes might continue to be found or the law might even be violated, as there is still a large amount of manure that has to be disposed of. Thus, a P (or also N) surplus levy alone could most probably not achieve the desired governance effect.

Last but not least, the implementation of a *nutrient trading scheme* that includes the market exchange of pollution allocations between different (mostly point) sources is not likely to be a suitable legal instrument to steer sustainable P use in agriculture. This is because diffuse, nonpoint pollution is hard to capture within the appropriated scheme.[68] Apart from that, regional characteristics in soil and water quality could hardly be considered. Furthermore, nutrient trading schemes cannot

[64] Regarding N see SRU (2015), pp. 349–251; Möckel (2017a).

[65] Garske (2020), p. 316.

[66] Ordinance on the management of nutrients on the farm and operational nutrient balances (Verordnung über den Umgang mit Nährstoffen im Betrieb und betriebliche Stoffstrombilanzen) of 14.12.2017, Federal Law Gazette I, p. 3942. From 2023 on this applies also to farms larger than 20 hectare or, in the case of a farm receiving manure (§ 3 (3) StoffBilV).

[67] For example, nutrient inputs or losses are calculated largely with average values and maximum P balances are not specified. See in detail Douhaire (2019), pp. 151–153.

[68] https://www.cbf.org/document-library/cbf-guides-fact-sheets/nutrient_trading_fact_sheet_-_pa_and_va374c.pdf (last accessed on 26.08.2021), with regard to P see also Withers et al. (2017), p. 14; Garske (2020), pp. 322–327.

protect highly vulnerable areas because buying a nutrient certificate would still allow P surpluses in these areas.[69]

3.3 Subsidies

Changing agricultural practices might be cost-intensive at first, yet in the long-term the need for purchased inputs is expected to be lower.[70] This is why subsidies can be an important element to accelerate the transformation towards sustainable, decarbonised agriculture with closed nutrient cycles.

However, first of all, the elimination of harmful subsidies that head in the opposite direction and potentially distort the global agricultural market is indispensable.[71] Therefore, the Common Agricultural Policy (CAP) of the EU that provides direct income support for farmers needs to be clearly aligned with the overarching targets of climate and biodiversity protection. However, the draft for the CAP,[72] from 2023 to 2027 does not show such a clear alignment.[73] Above all, Member States are required to draw up strategic plans for the implementation of the future CAP. The Commission proposed to keep the two-pillar structure, divided into mainly hectare-based payments under the first pillar (market support) and support for rural management under the second pillar. Yet, the second pillar remains underfunded. The first pillar includes statutory management requirements (SMRs) and standards for good agricultural and environmental conditions (GAECs)—forming enhanced conditionality. They are minimum requirements to receive direct payments.[74] Going beyond this, eco-schemes are to be established in the strategic plans of the Member States. These are sustainable agricultural practices that can be taken up by farmers on a voluntary basis and shall contribute as "key tools,"[75] to meet the Green Deal targets. The funding share for eco-schemes, however, probably

---

[69] Garske (2020), pp. 322–327.

[70] Paustian et al. (2019), p. 568.

[71] Ekardt (2019), p. 235; TEEB (2010), p. 27.

[72] Proposal for a Regulation of the European Parliament and of the Council establishing rules on support for strategic plans to be drawn up by Member States under the Common agricultural policy (CAP Strategic Plans) and financed by the European Agricultural Guarantee Fund (EAGF) and by the European Agricultural Fund for Rural Development (EAFRD) and repealing Regulation (EU) No 1305/2013 of the European Parliament and of the Council and Regulation (EU) No 1307/2013 of the European Parliament and of the Council. COM(2018) 392 final.

[73] Heyl et al. (2021), Pe'er et al. (2020) and Fortuna (2020).

[74] Here, GAEC 7 that shall provide for securing no bare soil in most sensitive periods and GAEC 8 concerning crop rotation are to be highlighted as they are able to foster a higher diversification of agricultural production. Also, the Farm Sustainability Tool for Nutrients (GAEC 5) could contribute to minimise external inputs enhance site-adapted fertilisation with innovative fertilisers. However, it remains unclear thus far whether particularly the latter will indeed remain as a mandatory prerequisite to receive full pillar 1 payments (Annex III CAP-Proposal).

[75] European Commission (2021), p. 1.

only accounts for 20% or a maximum of 30% of total first pillar funding.[76] Apart from that, only the interventions for rural development of the second pillar go beyond minimum requirements and include inter alia environmental, climate and other management commitments and thus also foster organic farming approaches and closed nutrient cycles (Art. 64 CAP-Proposal). However, because the overall targets of the CAP are not specific enough to be easily translatable into quantified targets and Member States are attributed a high sovereignty in shaping the CAP the accomplishment of sustainable fertilisation practises primarily based on organic and innovative fertilisers cannot be guaranteed. Rather, a 'race to the bottom' concerning environmental standards in general in the single Member States might take place.[77] Further research into alternative legal approaches of the CAP funding system in order to exclusively reward farmers for good agricultural and fertilisation practices also incorporating the targets of the EU's Farm to Fork Strategy,[78] is therefore highly desirable. Up until the new CAP enters into force in 2023, there might be a window of opportunity to correct these omissions and fundamentally reform the CAP. This would also be a step forward for third countries, which for their part, often have only punctual agricultural subsidy schemes, if at all.[79]

In Costa Rica, for example, only individual support through a kind of investment grant (Reconocimiento por Beneficios Ambientales, RBA) is provided for with regards to specific good agricultural practices on small and medium farms.[80] This is dependent on the short- or long-term environmental or economic impact of the measure which is divided into positive individual or societal effects, whereby 20%, 25% or 30% of the investment costs are covered. The purchase and use of organic fertilisers as well as the production of organic fertilisers and the necessary infrastructure are eligible for funding.[81] This can contribute to minimising the dependence on mineral fertilisers and to the of closing nutrient cycles at the regional level. In addition, expenditures for intercropping or for the creation of terraces and hedges to reduce erosion and increase natural soil fertility are promoted by the RBA.[82]

[76] Röder and Matthews (2021).

[77] European Court of Auditors (2019).

[78] European Commission (2020). The F2F Strategy aims, inter alia, at increasing agricultural climate resilience, fostering organic farming and reducing and optimising the use of external inputs by 2030. Clear, however, legally non-binding targets are set in this respect: the share of arable land dedicated to organic farming should be enhanced to 25%. The overall use and risks of chemical and hazardous pesticides shall be reduced by 50%. Nutrient losses shall be reduced by 50% and the use of fertilisers by 20%. According to an earlier statement of the EU Commission 30% of mineral fertiliser shall additionally be replaced by recycled nutrients (EU Commission 2018).

[79] Anderson (2016) and Cantore (2012).

[80] Stubenrauch (2019), pp. 355–358. Programa de Reconocimiento por Beneficios Ambientales (Environmental Benefits Recognition Program) based on law No. 8835 Apoyo y Fortalecimiento del Sector Agrícola of 14.06.2010, Law Gazette 114 of 14.06.2010, last revised at 14.02.2019.

[81] Normativa para la aplicación y asignacion de Reconocimiento de los Beneficios Ambientales (Implementation rules for the application and allocation of Environmental Benefit Recognition), Ministry of Agriculture and Livestock, Costa Rica, 2010.

[82] Ibid.

Compared to the subsidy-scheme of the EU's CAP, however, the provided funding for the individual farmer remains at a very low financial level.

In Nicaragua support is primarily provided through development aid projects which aim to implement the right to food—some of which have an agroecological focus.[83] Besides that, based on the strong agroecological movement in Nicaragua, support is provided through free ecological farm consultations by trained supervisors. Basically, they identify potential improvements within the specific farm and help to consolidate the utilisation of sustainable agricultural practices that also cover sustainable fertilisation techniques, based on renewable and organic resources.[84] Free farm advisory services could also promote change beyond this specific Nicaraguan example by actively supporting the individual farmer in the conversion of agricultural practices.[85] It is noteworthy, that Nicaragua is the only country that has established a specific law to promote agroecological movement and practices.[86] The legal concept of good agricultural production practice (Buenas Prácticas Productivas, BPP) is introduced, which combines agricultural practices that;- (a) guarantee compliance with international and national legal standards, (b) contribute to food security and environmental protection, and (c) produce high-quality agricultural products that do not endanger human health or biodiversity.[87] The successful implementation of the agroecological legislation in Nicaragua is also supported by the FAO.[88] The legal anchoring of sustainable agricultural practices and the corresponding funding opportunities make Nicaragua a role model—even if the agroecological movement as a whole lacks sufficient support from the Ortega regime, from which it could highly benefit.[89]

3.4 *Command-and-Control Regulations: Assessing the Examples of Germany, the European Union, Costa Rica and Nicaragua*

Command-and-control regulations can complement economic policy instruments, e.g., to avoid soil contamination and to hot-spot problems and to foster P recycling.

[83] Stubenrauch (2019), pp. 358–363.

[84] http://maonic.org/ (last accessed on 26.08.2021).

[85] With regard to Germany see also Möckel (2020), online: https://www.ufz.de/index.php?de=36336&webc_pm=17/2020 (last accessed on 26.08.2021).

[86] Ley de Fomento a la Producción Agroecológia u Orgánica No. 765 (Law for the Promotion of Agroecological or Organic Production) of 14.04.2011, Law Gazette 124 of 5.07.2011.

[87] Technical Implementing Regulation No. 11 037-12 (NTON No. 11 037-12 Caracterización, Regulación y Certificación de Unidades de Producción Agro Ecológica) of 30.04.2013, Law Gazette No. 123 of 03.07.2013.

[88] http://www.fao.org/agroecology/database/detail/en/c/1242091/ (last accessed on 26.08.2021).

[89] Stubenrauch (2019), pp. 335–340, 374–376.

Table 2 Limit values for cadmium concentrations in mineral fertilisers (own table)

Region	Legal norm	Limit value
Europe	Regulation (EC) No 2003/2003	–
	EU Fertilising Products Regulation	60 mg Cd/kg P_2O_5[a]
Germany	DüMV	50 mg Cd/kg P_2O_5[b]
Central America	RTCA 65.05.54:09	–
Costa Rica	RTCR 485:16	80 mg Cd/kg P_2O_5
Nicaragua	Law No 291	–

[a] Valid from 16 July 2022
[b] Only valid for fertilisers not labelled as EC-fertilisers

Besides that, the protection of highly sensitive ecosystems requires supplementary command-and-control regulations.

In the following, compared to its theoretical steering potential, the effectivity of the agricultural, environmental and planning legislation of Germany/the EU, Costa Rica and Nicaragua, for sustainable P use will be assessed and conclusions for enhanced policy approaches will therefore be drawn.

3.4.1 Product-Related Fertiliser Legislation

The product-related fertiliser legislation regulates the marketing of fertilisers and should ensure that only high-quality P fertilisers are permitted for agricultural use and that organic or recycled P fertilisers are promoted over mineral P fertilisers.

In the EU, since 2003, Regulation (EC) No 2003/2003,[90] lays down the rules for the marketing of mineral fertilisers (EC fertilisers). Organic or recycled P fertilisers are not included in the regulation (Art. 3 Regulation (EC) No 2003/2003). Thus, compared to alternative fertilisers, for years mineral fertilisers from phosphate rock have benefited from easier transnational market access in Europe. The new EU Fertilising Products Regulation,[91] addresses this issue. The regulation covers for the first time organic and recycled fertilisers from secondary raw materials and other fertilising products and aims to support their promotion. The regulation entered into force on 25 July, 2019. However, parts of the regulation, for example, limit values for potential pollutants, which only apply from 16 July 2022 (see Table 2). The

[90] Regulation (EC) No 2003/2003 of the European Parliament and of the Council of 13.10.2003 relating to fertilisers, OJ L 304 of 21.11.2003, p. 1, last revised by Regulation (EC) No 162/2007 of the Commission of 19.02.2007, OJ 2007 L 51, p. 7.

[91] Regulation (EU) 2019/1009 of the European Parliament and of the Council of 5 June 2019 laying down rules on the making available on the market of EU fertilising products and amending Regulations (EC) No 1069/2009 and (EC) No 1107/2009 and repealing Regulation (EC) No 2003/2003 [2019] OJ L 170/1.

German Fertiliser Ordinance[92] (Düngemittelverordnung, DüMV) regulates the marketing of fertilisers not labelled as "EC fertilisers" (§ 2 (1) DüMV).

In Costa Rica, Regulation RTCR 485:2016,[93] based on the Plant Protection Act No 7664[94] lays down the rules for the marketing of fertilisers. In Nicaragua, based on the Basic Law on Animal and Plant Health No 291,[95] the Institute for the Protection and Health Monitoring of the Agricultural Sector,[96] is responsible for the registration and quality control of fertilisers. The Central American regulation regarding fertilisers and agricultural amendments (RTCA 65.05.54:09[97]) applies. A more specific national regulation, as that that is in Costa Rica, does not exist in Nicaragua.

One critical point of mineral P fertilisers gained from phosphate rock is their accelerating contamination with heavy metals. Cd average values of over 60 mg Cd/kg P_2O_5, with maximum values over 80 mg Cd/kg P_2O_5 have been measured[98] and U concentrations between 61 and 70 mg U/kg P_2O_5 with maximum concentrations of up to 200 mg U/kg P_2O_5.[99] While there are no limit values for U, limit values for Cd are established for mineral P fertilisers:

For human health and soil protection, it is concerning that the respective regulations establish no or no strict limit values for Cd (and for U).[100] By setting strict limit values for heavy metals in mineral P fertilisers, recycled and organic fertilisers could be promoted substantially due to the increasing price of mineral fertilisers. However, even beyond this unused steering potential, recycled and other innovative P

---

[92] Fertiliser Ordinance (Verordnung über das Inverkehrbringen von Düngemitteln, Bodenhilfsstoffen, Kultursubstraten und Pflanzenhilfsmitteln, DüMV) of 05.12.2012, Federal Law Gazette I, p. 2482, last revised at 02.10.2019, Federal Law Gazette I, p. 1414.

[93] Costa Rican Technical Regulation on Chemicals, Fertilisers and Agricultural Amendments No 485:2016 (Decreto N° 39994-MAG, Reglamento Técnico Costarricense (RTCR) 485:2016 Sustancias Químicas, Fertilizantes y Enmiendas Para uso Agrícola, Tolerancias y Límites Permitidos para la Concentración de los Elementos y Contaminantes) of 24.10.2016, Law Gazette 241 of 15.12.2016.

[94] Plant Protection Act No 7664 (Ley de Protección Fitosanitaria) of 13.04.1997, Law Gazette 83 of 02.05.1997, last revised at 15.05.2013.

[95] Basic Law on Animal and Plant Health No 291 (Ley Básica de Salud Animal y Sanidad Vegetal) of 16.04.1998, Law Gazette 136 of 22.07.1998, last revised at 04.03.2015.

[96] Instituto de Protección y Sanidad Agropecuaria, IPSA.

[97] Central American Technical Regulation on Chemicals, Fertilisers and Amendments for Agricultural Use No 65.05.54:09 (Reglamento Técnico Centroamericana (RTCA) 65.05.54:09 Fertilizantes y Enmiendas de uso Agrícola. Requisitos para el Registro) of 05.01.2014, online: https://members.wto.org/crnattachments/2009/sps/CRI/09_1825_00_s.pdf (last accessed on 26.08.2021).

[98] Kördel et al. (2007), Dittrich and Klose (2008) and Kratz et al. (2016).

[99] Smidt et al. (2011). However, the number of samples was relatively small with 75 fertilisers examined in Germany and 39 in Brazil.

[100] For the constant weakening of the limit values for cadmium over time in the EU Fertilising Products Regulation see Stubenrauch (2019), p. 277; Ulrich (2019).

fertilisers[101] are not particularly promoted by the aforementioned legislation. Instead, the existence of parallel markets continues to be emphasised.[102]

3.4.2 Good Agricultural Practice as Part of the Environmental and Applied Fertiliser Legislation

The environmental and/or applied fertiliser legislation has the potential to define the good agricultural practice and thus to regulate an efficient and resource-saving P fertilisation on farm level which is optimally adapted to the location, the existing soil properties and the crops grown in the daily agricultural practice. In Costa Rica and Nicaragua, an applied fertiliser legislation that is legally binding does not exist.[103] It is therefore questionable to what extend sustainable agricultural practices are anchored in the environmental legislation of the countries.

In Germany, the legal requirements for good agricultural practice are generally laid down in the environmental legislation. According to § 5 (2) No 1, 2 of the Federal Nature Conservation Act (Bundesnaturschutzgesetz, BNatschG),[104] agricultural use should be site-adapted and should maintain the natural soil fertility. Land use should not go beyond what is necessary to achieve a sustainable yield. In addition, biotope connectivity should be ensured, livestock farming should be in a "balanced relationship" with crop production[105] and erosion control should be implemented (§ 5 (2) No 3 BNatSchG). However, according to the Federal Administrative Court, these requirements remain vague and are not sufficiently specified so that legal enforcement is impossible.[106]

According to 17 (2) No 1-7 of the Federal Soil Conservation Act (Bundesbodenschutzgesetz, BBodSchG[107]), site-adapted soil cultivation that considers local climatic conditions should be performed, soil structure maintained and improved, natural structural elements preserved, soil compaction and soil erosion avoided and the soil's biological activity as well as the humus content maintained by a sufficient supply of organic matter or a reduced cultivation intensity. However,

[101] Concerning the advantages of sulfur-enriched bone char as an innovative P fertiliser see Zimmer et al. (2019).

[102] In detail Stubenrauch (2019), pp. 282–286; Douhaire (2019), p. 171.

[103] Regarding respective voluntary (albeit rather weak) international certification standards GLOBALG.A.P. and Rainforest Alliance see Stubenrauch (2019), pp. 308–315.

[104] Federal Nature Conservation Act (Gesetz über Naturschutz und Landschaftspflege, BNatSchG) of 27.09.2009, Federal Law Gazette I, p. 2542, last revised at 04.03.2020, Federal Law Gazette I, p. 440.

[105] This could be interpreted as the mandatory livestock-to-farm ratio proposed in Sect. 3.2.2.

[106] Möckel et al. (2014), p. 147. Federal Administrative Court, judgment of 01.09.2016 – 4 C 4.15.

[107] Federal Soil Conservation Act (Gesetz zum Schutz vor schädlichen Bodenveränderungen und zur Sanierung von Altlasten, BBodSchG) of 17.03.1998, Federal Law Gazette I, p. 502, last revised at 27.09.2017, Federal Law Gazette I, p. 3465.

once again these requirements for good agricultural practice suffer from a lack of precision and are thus not enforceable.[108]

The Fertiliser Act (Düng[109] in combination with the recently amended Applied Fertiliser Ordinance (DüV)[110] (*lex specialis*) establishes further provisions for agricultural fertilisation. According to the current amendment, heavily polluted "red" areas are subject to more stringent fertilisation restrictions (§§ 13, 13a DüV), every fertilisation act has to be documented within two days (§ 10 (2) DüV), fertilisation is forbidden on flooded, water saturated, frozen or snow-covered soils (§ 5 (1) DüV) and stricter distance regulations to water bodies were introduced (§ 5 (3) DüV). The fertilisation requirement for P has to be determined (only) every six years or once during crop rotation by means of representative soil samples (§ 4 (4) No 2 DüV).[111] However, the amount of fertilisation is primarily based on the nutrient requirements of the plant without taking into account the environmental conditions of the site in a sufficient manner. In particular, a mandatory P fertilisation below the nutrient requirement of plants in soils that are highly oversupplied with P (Abreicherungsgebot) has—despite scientific evidence[112]—still not been integrated into the DüV. In contrast, nutrient surpluses are—apart from derogations in the red areas—still permitted for P within the nutrient balance (§ 11a DüngG, StoffBilV). Furthermore, livestock farming is not restricted and organic fertilisers are not particularly promoted over mineral fertilisers. All in all, the further tightened requirements for fertilisation practices still mainly focuess on nitrogen (N) which is due to the threat of a repetitive infringement of the EU Nitrates Directive. However, a total reduction of P and N nutrient emissions and an adequate protection of adjacent ecosystems cannot be achieved by the amended DüV.[113]

These shortcomings can also not be remedied by further soil protection related provisions in the BBodSchG and the BNatSchG. The BBodSchG (§ 3 (1) No 1–11) as well the BNatSchG (§ 5 (2) No 6 BNatSchG) are subsidiary to the *lex specialis* on agricultural land use and agricultural fertilisation. Apart from that, according to § 14 (2) BNatSchG, agricultural and forestry use does not constitute to being an intervention into nature if good agricultural practice required by the BBodSchG (that is however insufficiently specified) is maintained.[114] This applies to fertilisation, classified as day-to-day farming practice. Thus, agricultural P

[108] Möckel et al. (2014), p. 118.

[109] Fertiliser Act (Düngegesetz, DüngG) of 09.01.2009, Federal Law Gazette I, p. 54, 136, last revised at 05.05.2017, Federal Law Gazette I, p. 1068.

[110] Applied Fertiliser Ordinance (Verordnung über die Anwendung von Düngemitteln, Bodenhilfsstoffen, Kultursubstraten und Pflanzenhilfsmitteln nach den Grundsätzen der guten fachlichen Praxis beim Düngen, DüV) of 26.05.2017, Federal Law Gazette I, p. 1305, last revised at 28.04.2020.

[111] For N fertilisation requirements have to be determined every three years.

[112] VDLUFA (2018), Buczko et al. (2019) and Garske et al. (2020).

[113] Möckel et al. (2020), online: https://www.ufz.de/index.php?de=36336&webc_pm=17/2020 (last accessed on 26.08.2021).

[114] Critcial on this Möckel (2017b).

fertilisation is altogether not sufficiently regulated. Binding land use plans and other spatial planning instruments that aim to adapt agricultural land use to the local conditions and regulate P fertilisation accordingly do not exist.

Costa Rica According to Art. 2 of the Soil Conservation Act No 7779,[115] a regular inventory of soils to balance the natural soil potential and their economic production capacity shall take place. To this end, the Ministry of Agriculture and Livestock is required to establish a National Land Management and Soil Protection Plan with binding guidelines for agricultural land use (Art. 11 Law 7779). Management practices that increase the vegetation cover and water infiltration rate of soils, reduce soil contamination and preserve natural soil fertility by maintaining the organic matter of the soil shall be promoted (Art. 12 Law 7779). According to Art. 14 Law 7779, every two years the plan shall be revised and adapted to current circumstances. Additionally, Regional Soil Protection and Restoration Plans are to be drawn up (Art. 15 Law 7779). Sensitive soils in water catchment areas shall be defined and proposals for the best agricultural use developed so that adequate soil protection is guaranteed (Art. 16 Law 7779). The agricultural practices shall also include mandatory regulations for fertiliser and pesticide use, that can be restricted according to the site-specific conditions (Art. 19 Law 7779). Strategies for organic fertilisation and erosion control shall be included in the regulations as well. The Ministry of Agriculture and Livestock and the Ministry of the Environment and Energy shall finally control the usage of agricultural inputs and equipment (Art. 30 Law 7779). These legal requirements would be equivalent to binding land use planning that aims to ensure sustainable fertilisation based on the natural conditions. However, neither the national nor the regional land use plans have yet to be completely drawn up and it is unlikely this will ever happen.[116] Thus, due to lacking implementation, good agricultural practice and fertilisation is also not implemented by soil protection legislation.

An environmental impact assessment for agricultural projects in accordance with the Basic Act on the Environment No 7554,[117] which aims to maintain a healthy and ecologically balanced environment, is generally only required for changes in land use (Art. 3 No 3a, 17 in conjunction with Art. 4 Implementing Regulation No. 31849[118]).[119] Exceptions are listed in Annex 2 of Implementing Regulation

[115] Soil Conservation Act No 7779 (Ley de Uso, Manejo y Conservación de Suelos) of 30.04.1998, Law Gazette No 97 of 21.05.1998, last revised at 25.06.2012.

[116] Stubenrauch (2019), p. 231. Corresponding binding land use plans are, as already mentioned in Sect. 3.2.2, partly demanded with regard to Germany. SRU (2015), pp. 397–423.

[117] Basic Act on the Environment No 7554 (Ley Orgánica del Ambiente) of 13.11.1995, Law Gazette 215 of 13.11.1995, last revised at 13.05.2019.

[118] Implementing Regulation on the procedure for an environmental impact assessment No 31849 (Reglamento General sobre los Procedimientos de Evaluación de Impacto Ambiental (EIA)) of 24.05.2004, Law Gazette 125 of 28.06.2004, last revised at 25.06.2019.

[119] Since there are fewer and fewer small-scale farms in Costa Rica and, formerly small areas are being bought up and merged e.g. for the rapidly expanding monocultural pineapple or oilpalm cultivation, this exception carries serious consequences. See Stubenrauch (2019), p. 209.

No. 31849. For such agricultural projects, the environmental impact has to be assessed in an attenuated procedure. For example, technical studies on soil properties or the hydrology of the site are frequently not required (Art. 17 Law 7554, Art. 4 No 2 in conjunction with Annex 2 Regulation 31849). It is only necessary to prove compliance with the (as shown hardly implemented) environmental legislation by a documented good environmental practice.[120] Thus, despite existing regulations in environmental law, agricultural practices, including fertilisation strategies, remain largely legally uncontrolled in Costa Rica.

Nicaragua The Nicaraguan Constitution obliges the citizens to maintain a clean environment and to protect "Mother Earth" and its natural resources and all life-sustaining natural processes (Art. 60 of the Nicaraguan Constitution). The Basic Act on the Environment and Natural Resources No 217[121] pursues the overall objective of maintaining a clean environment (atmosphere, biodiversity, soil and water), protecting natural resources and ensuring their rational and sustainable use is in accordance with the Constitution (Art. 1 Law 217). Deterioration of the environmental status is to be avoided, pollution of ecosystems is to be regulated, controlled and prevented, and water bodies are to be used sustainably. In addition, unsustainable production and consumption patterns are to be eliminated in order to protect, preserve and improve natural resources and the environment (Art. 4 No. 2 Law 217). A balance is to be established between the preservation of vital ecosystem functions and the productive capacity of the country (Art. 12 in conjunction with Art. 13 No. 1 Law 217). Consequently, soils shall be used according to their natural conditions, such as physical and chemical properties. The production capacity should be in line with their natural balance (Art. 105 Law 217). Harmful soil management practices that cause soil erosion or soil degradation shall be avoided. In case of severe land degradation, soil protection areas with restricted usability can be determined by local administrations (Art. 107 Law 217). However, these optional provisions only apply if land degradation has already occurred. Precise requirements for good agricultural practices are not included in Law 217. Besides that, sector-specific soil protection legislation does not exist. Thus, the general principles of nature and soil protection laid down in the Constitution and in Law 217 are not sufficiently specified to be translated into concrete measures.

Furthermore, according to Decree 20-2017,[122] agricultural projects of any kind are in principle no longer covered by an environmental impact assessment,[123] which became in general, severely restricted by the Nicaraguan president. It can therefore

[120]In Spanish: Buenas Prácticas Ambientales.

[121]Basic Act on the Environment and Natural Resources No 217 (Ley General del Medio Ambiente y los Recursos Naturales) of 27.03.1996, Law Gazette No 105 of 06.06.196, last revised at 17.01.2014.

[122]System of environmental impact assessment Decree No 20-2017 (Decreto Ejecutivo Sistema de Evaluación Ambiental) of 29.11.2017, Law Gazette No 248 of 29.11.2017.

[123]This therefore also applies to the large-scale and regularly input intensive cultivation of palm oil, peanuts or sugar cane.

be stated that in Nicaragua agricultural land use and fertilisation practices are not subject to any regulations.

3.4.3 Circular Economy Legislation

The recovery of P from wastewater streams can be specifically promoted by circular economy law- Germany is a pioneer in this respect. The German Sewage Sludge Ordinance (Klärschlammverordnung, AbfKlärV[124]) was amended in 2017 as part of the Federal Government's Resource Efficiency Programme II.[125] It establishes a mandatory recovery of P from sewage sludge which has to contain at least 20 grams P per kilogram dry matter and prohibits direct agricultural application of sewage sludge. From 01 January 2029, these provisions will apply for wastewater treatment plants that cover more than 100,000 inhabitants and from 01 January 2032, for wastewater treatment plants which cover more than 50,000 inhabitants (Art. 5, 6 Ordinance on the reorganisation of sewage sludge recycling[126]). Thus, consistency strategies will be strengthened. However, relatively long transitional periods inhibit rapid progress. Besides that, decentral and therefore potentially more comprehensive and possibly less energy-intensive P recycling practices are not legally covered and fostered. However, especially against the background that less animal-derived fertilisers will be available in the future and mineral P fertilisation based on rock phosphates is to be limited, all sources of potential P recycling should be considered.

Fostering decentralised P recycling would stand as being an opportunity for countries which so far do not have a large-scale wastewater system (and thus also no respective legislation that focuses on the recycling of nutrients) as well. This also applies to Costa Rica and Nicaragua where rural areas in particular lack adequate wastewater treatment. There, the installation of separation or composting toilets,[127] as well as the direct integration of P recovery technology into newly planned wastewater plants should be considered.[128] Besides that, future development aid projects should incorporate the recovery of nutrients from wastewater. This is because a successful decentralised nutrient recovery could permanently increase the security of P supply these areas.

---

[124] Sewage Sludge Ordinance (Verordnung über die Verwertung von Klärschlamm, Klärschlammgemisch und Klärschlammkompost, AbfKlärV) of 27.09.2017, Federal Law Gazette I, p. 3465, last revised by Art. 6 of the Regulation of 27.09.2017, Federal Law Gazette I, p. 3465.

[125] BMU (2016).

[126] Ordinance on the Reorganisation of Sewage Sludge Recycling (Verordnung zur Neuordnung der Klärschlammverwertung) of 27.09.2017, Federal Law Gazette I, p. 3465.

[127] Regarding the successful recycling of P from urine from separation toilets see https://www.eawag.ch/en/department/eng/projects/aurin-fertilisers-from-urine/ (last accessed on 26.08.2021) and further ecological-based sanitation Porterfield et al. (2018) and Sharma et al. (2019).

[128] In detail about the so far rather unused potential of decentralised P recycling in Germany, Costa Rica and Nicaragua Stubenrauch (2019), pp. 179–183.

4 Conclusions

This chapter has shown that the major environmental challenges including natural resource, climate and biodiversity protection and the lack of P governance require more than a simple adjustment of existing policy instruments. Instead, motivational problems and systematic weaknesses of existing policy instruments call for an absolute reduction of quantities of resources and pollutants. The key to implement-efficiency, consistency and frugality strategies in P governance thus lies in an optimised policy mix of economic policy instruments and command-and-control regulations. While economic policy instruments are considered to be most effective to govern a phasing out of fossil fuels and reduce the total number of livestock, command-and-control regulations can complement these overarching regulations to prevent hot spot problems, foster P recycling and ensure the efficient usage of high-quality fertilisers. To avoid negative governance effects, efforts should be undertaken to implement policy instruments on a broad geographical scale, preferably an international or transnational level like the EU or the Central American Integration System. In contrast, introducing instruments (mainly) at local or nation state level is likely to be insufficient due to impending shifting effects.

However, the assessment of the status quo of the legislation in Germany/the EU, Costa Rica and Nicaragua shows, that thus far, neither major polluters such as fossil fuels and intensive animal husbandry are consistently controlled nor are good agricultural practices with alternative fertilisers legally mandatory in detail. As far as environmental legislation is concerned, soils and waters are in each case insufficiently protected from the negative impacts resulting from agricultural P (and N) fertilisation practices. Thus, the environmental legislation remains toothless, while product related fertiliser legislation ought to be tightened and mandatory requirements for the good agricultural practice have to be established.

The necessary transformation process can be supported by a corresponding subsidy scheme. However, the CAP in the EU has so far been insufficiently oriented towards the future challenges in the agricultural sector and is rather opposed to accelerated change. In contrast, the individual support programs in Costa Rica and in Nicaragua based on the agroecology approach are worth highlighting. At the same time, however, the support provided to individual farmers in these two countries remains comparatively low.

It can be assumed that the findings on the inadequate regulation and promotion of the specific agricultural fertilisation practice from one industrialised, emerging and developing country are to a certain extend exemplary for the situation in many other countries. This once again shows the difficulty of regulating individual fertilisation processes in detail, and that instead, the framework conditions for the agricultural sector must be changed, preferably by implementing economic policy instruments. Thus, it is still a long way to go to close the P governance gap consistently and at the same time reconcile the needs of developed and developing countries.

Acknowledgements The author gratefully thanks the German Federal Ministry of Education and Research (BMBF) for funding the BonaRes project InnoSoilPhos (No. 031B0509) and Katharine Heyl for proofreading. This research was also performed within the scope of the Leibniz ScienceCampus Phosphorus Research Rostock.

References

Anderson K (2016) Market-distorting policies: long-run trends and short-run insulation. In: Agricultural trade, policy reforms, and global food security. Palgrave studies in agricultural economics and food policy. Palgrave Macmillan, New York

Bähr CC (2015) Greenhouse gas taxes on meat products: a legal perspective. Transnatl Environ Law 4:153–179

Batjes NH (1996) Total carbon and nitrogen in the soils of the world. Eur J Soil Sci 47:151–163

Bloem E, Haneklaus S, Schnug E (2020) Regulatory scenarios to counteract high phosphorus inputs into the Baltic Sea. In: Negm AM, Omran EE, Kubiak-Wojcicka K, Zelenakova M (eds) Water-resources quality and management in Baltic Sea countries. Springer Nature, Switzerland, pp 127–140

BMU (German Federal Ministry for the Environment, Nature Conservation and Nuclear Safety) (2016) Deutsches Ressourceneffizienzprogramm II. Programm zur nachhaltigen Nutzung und zum Schutz der natürlichen Ressourcen. BMU, Berlin, Germany

Bocklet J, Hintermayer M, Schmidt L, Wildgrube T (2019) The reformed EU ETS - intertemporal emission trading with restricted banking. Energy Econ 84:104486

Buczko U, Steinfurth K, Laak M (2019) Meta-analysis of the yield response to phosphorus fertilization based on long-term field experiments. Agric For 65(4):7–14

Cantore N (2012) Impact of the Common Agricultural Policy on food price volatility for developing countries. Overseas Development Institute, London, United Kingdom

Ceci A, Pinzari F, Russo F, Maggi O, Persiani AM (2018) Saprotrophic soil fungi to improve phosphorus solubilisation and release: in vitro abilities of several species. Ambio 47(1):30–40

Cordell D, White S (2015) Tracking phosphorus security: indicators of phosphorus vulnerability in the global food system. Food Secur 7:337–350

Coria J, Jaraitė J (2019) Transaction costs of upstream versus downstream pricing of CO_2 emissions. Environ Resour Econ 72:965–1001

Dittrich B, Klose R (2008) Schwermetalle in Düngemitteln. Bestimmung und Bewertung von Schwermetallen in Düngemitteln, Bodenhilfsstoffen und Kultursubstraten. Sächsische Landesanstalt für Landwirtschaft, Dresden, Germany

Douhaire C (2019) Rechtsfragen der Düngung. Eine steuerungs- und rechtswissenschaftliche Analyse vor dem Hintergrund unions- und völkerrechtlicher Verpflichtungen und politischer Zielsetzungen zum Umwelt- und Ressourcenschutz; Schriften zum Umweltrecht. Duncker & Humblot, Berlin, Germany

Ecotec, CESAM, CLM, University of Gothenburg, UCD, IEEP (2001) Study on the economic and environmental implications of the use of environmental taxes and charges in the European Union and its Member States. Final Report. Brussels, Belgium

Eichler-Löbermann B, Bachmann S, Busch S, Schiemenz K, Krey T, Pfahler V, Uptmoor R (2016) Management options for an efficient utilization of phosphorus in agroecosystems. In: Schnug E, De Kok LJ (eds) Phosphorus in agriculture: 100% zero. Springer, Dordrecht, Netherlands

Ekardt F (2019) Sustainability. Transformation, governance, ethics, law; environmental humanities: transformation, governance, ethics, law. Springer International Publishing, Basel, Switzerland

Ekardt F, Wieding J, Garske B, Stubenrauch J (2018a) Agriculture-related climate policies – law and governance issues on the European and global level. Carbon Clim Law Rev 12(4):316–331

Ekardt F, Wieding J, Zorn A (2018b) Paris Agreement, precautionary principle and human rights: zero emissions in two decades? Sustainability 10(8):2812

Ekardt F, Jacobs B, Stubenrauch J, Garske B (2020) Peatland governance: the problem of depicting in sustainability governance, regulatory law, and economic instruments. Land 9(3):83

European Commission (2018) Circular economy: agreement on Commission proposal to boost the use of organic and waste-based fertilisers. EU Commission, Brussels, Belgium

European Commission (2020) Communication from the Commission to the European Parliament the Council, the European Economic and Social Committee and the Committee of the Regions. A Farm to Fork Strategy for a fair, healthy and environmentally-friendly food system, COM (2020) 381 final, Brussels, Belgium

European Commission (2021) List of potential agricultural practices that eco-schemes could support, Brussels, Belgium. Available via https://ec.europa.eu/info/sites/info/files/food-farming-fisheries/key_policies/documents/factsheet-agri-practices-under-ecoscheme_en.pdf

European Court of Auditors (2019) Opinion No 7/2018 pursuant to Article 322(1)(a) TFEU) concerning Commission proposals for regulations relating to the Common Agricultural Policy for the post-2020 period. COM(2018) 392, 393 and 394 final, OJ C41/1, Luxembourg, Luxembourg

Fortuna G (2020) MEPs agree in two-year bridging CAP divide. Available via https://www.euractiv.com/section/agriculture-food/news/meps-agree-on-two-year-bridging-cap-divide/

Garske B (2020) Ordnungsrechtliche und ökonomische Instrumente der Phosphor-Governance. Unter Berücksichtigung der Wirkungen auf Böden, Gewässer, Biodiversität und Klima. Metropolis, Marburg, Germany

Garske B, Ekardt F (2021) Economic policy instruments for sustainable phosphorus management: taking into account climate and biodiversity targets. Environ Sci Eur 33:56

Garske B, Stubenrauch J, Ekardt F (2020) Sustainable phosphorus management in European agricultural and environmental law. Rev Eur Comp Int Environ Law 29(1):107–117

Glaze-Corcoran S, Hashemi M, Sadeghpour A, Jahanzad E, Keshavarz Afshar R, Liu X, Herbert SJ (2020) Understanding intercropping to improve agricultural resiliency and environmental sustainability. Adv Agron 162:199–256

Grames J, Zoboli O, Laner D, Rechberger H, Zessner M, Sánchez-Romero M, Rskawetz A (2019) Understanding feedbacks between economic decisions and the phosphorus cycle: a general equilibrium model including material flows. Resour Policy 61:311–347

Hallas FJ, Mackowiak LC, Wilkie CA, Harris GW (2019) Struvite phosphorus recovery from aerobically digested municipal wastewater. Sustainability 11(2):376

Heckenmüller M, Narita D, Klepper G (2014) Global availability of phosphorus and its implication for global food supply: an economic overview. Kiel Working Paper No. 1897. Kiel, Germany

Hepburn C, Teytelborn A (2017) Reforming the EU ETS: where are we now?. In: Pittel K, Ian P, Vollebergh H (eds) Energy tax and regulatory policy in Europe. Reform priorities. MIT Press, Cambridge, Massachusetts, London, England, pp 1–29

Herold A, Cook V, Baron Y, Cames M, Gores S, Graichen J, Kasten P, Mehlhart G, Siemons A, Urrutia C, Wolff F (2019) EU environment and climate change policies. State of play, current and future challenges. European Parliament, Brussels, Belgium

Heyl K, Döring T, Garske B, Stubenrauch J, Ekardt F (2021) The Common Agricultural Policy beyond 2020: a critical review. Rev Eur Comp Int Environ Law 30:95–106

IPBES (Intergovernmental Platform on Biodiversity and Ecosystem Services) (2019) The global assessment report on biodiversity and ecosystem services. Summary for Policymakers. IPBS, Bonn, Germany

IPCC (Intergovernmental Panel on Climate Change) (2019a) Global warming of 1.5°C. An IPCC Special Report on the impacts of global warming of 1.5°C above pre-industrial levels and related global greenhouse gas emission pathways, in the context of strengthening the global response to the threat of climate change, sustainable development, and efforts to eradicate poverty. IPCC, Geneva, Switzerland

IPCC (2019b) Climate Change and Land. An IPCC Special Report on Climate Change, Desertification, Land Degradation, Sustainable Land Management, Food Security, and Greenhouse Gas Fluxes in Terrestrial Ecosystems. Summary for Policymakers. Approved Draft; IPCC: Geneva Switzerland

Kataki S, West H, Clarke M, Baruah DC (2016) Phosphorus recovery as struvite: recent concerns for use of seed, alternative Mg source, nitrogen conservation and fertilizer potential. Resour Conserv Recycl 107:142–156

Kitano M, Yasunori I, Yamazaki Y, Hayashi F, Kanbara S, Matsuishi S, Yokomama T, Kim S, Hara M, Hideo H (2012) Ammonia synthesis using a stable electride as an electron donor and reversible hydrogen store. Nat Chem 4:934–940

Koch S, Kahle P, Lennartz B (2019) Biogas digestate application modifies solute transport conditions in soils and increases the release of phosphorus. Vadose Zone J 18:190031

Ko̎rdel W, Herrchen M, Müller J, Kratz S, Fleckenstein J, Schnug E, Saring T, Haamann H, Reinhold J (2007) Begrenzung von Schadstoffeinträgen bei Bewirtschaftungsmaßnahmen in der Landwirtschaft bei Düngung und Abfallverwertung. UBA, Dessau-Roßlau, Germany

Kratz S, Schick J, Schnug E (2016) Trace elements in rock phosphates and P containing mineral and organo-mineral fertilizers sold in Germany. Sci Total Environ 542:1013–1019

Kyriakou V, Garagounis I, Vourros A, Vasileiou E, Stoukides M (2020) An electrochemical Haber-Bosch process. Joule 4:142–158

Leinweber P, Bathmann U, Buczko U, Dohaire C, Eichler-Löbermann B, Frossard E, Ekardt F, Jarvie H, Krämer I, Kabbe C, Lennartz B, Mellander PE, Nausch G, Ohtake H, Tränckner J (2018) Handling the phosphorus paradox in agriculture and natural ecosystem: scarcity, necessity, and burden of P. Ambio 47(1):3–19

Leip A, Billen G, Garnier J, Grizetti B, Lassaletta L, Reis S, Simpson D, Sutton MA, de Vries W, Weiss F, Westhoek H (2015) Impacts of European livestock production: nitrogen, sulphur, phosphorus and greenhouse gas emissions, land-use, water eutrophication and biodiversity. Environ Res Lett 10(11)

Lemaire G, Franzluebbers A, de Faccio Carvalho PC, Dedieu B (2014) Integrated crop-livestock systems: strategies to achieve synergy between agricultural production and environmental quality. Agric Ecosyst Environ 190:4–8

Machovina B, Feeley KJ, Ripple WJ (2015) Biodiversity conservation: the key is reducing meat consumption. Sci Total Environ 536(1):419–431

Mącik M, Gryta A, Frąc M (2020) Biofertilizers in agriculture: an overview on concepts, strategies and effects on soil microorganisms. Advances in agronomy. Academic Press, 2020. ISBN 0065-2113

Macintosh KA, Donnacha GD, Withers PJA, McDowell RM, Douglas RS, Laura TJ, Bruuslema TW, O'Flaherty V, McGrath JW (2019) Transforming soil phosphorus fertility management strategies to support the delivery of multiple ecosystem services from agricultural systems. Sci Total Environ 649:90–98

Metson GS, Cordell D, Ridoutt B (2016) Potential impact of dietary choices on phosphorus recycling and global phosphorus footprints: the case of the average Australian city. Front Nutr 3(35). https://doi.org/10.3389/fnut.2016.00035

Möckel S (2017a) Rechtsgutachten zur Klärung von Rechtsfragen zur Erhebung einer Abgabe auf Stickstoffüberschuss und einer Abgabe auf stickstoffhaltigen Mineraldünger durch den Landesgesetzgeber. Halle, Germany

Möckel S (2017b) Novellierungsbedarf beim BNatSchG aus ökologischer und europarechtlicher Sicht. ZUR 2017:195–205

Möckel S, Köck W, Schramek J, Rutz C (2014) Rechtliche und andere Instrumente für vermehrten Umweltschutz in der Landwirtschaft. UBA, Dessau-Roßlau, Germany

Möckel SM et al (2020) Düngeverordnung: zu kurz gesprungen. Helmholtz Center for Environmental Research, 1. https://www.ufz.de/index.php?de=36336&webc_pm=17/2020

Mogollón JM, Beusen AHW, van Grinsven HJM, Westhoek H, Bouwman AF (2018) Future agricultural phosphorus demand according to the shared socioeconomic pathways. Glob Environ Change 50:149–163
Narassimhan E, Gallagher KS, Koester S, Alejo JR (2018) Carbon pricing in practice: a review of existing emissions trading systems. Clim Pol 18(8):967–991. https://doi.org/10.1080/14693062.2018.1467827
Ohm M, Paulsen HM, Moos JH, Eicherl-Löbermann B (2017) Long-term negative phosphorus budgets in organic crop rotations deplete plant-available phosphorus from soil agronomy for sustainable development 37(17)
Paul C, Techen AK, Robinson JS, Helming K (2019) Rebound effects in agricultural land and soil management: review and analytical framework. J Clean Prod 227:1054–1067
Paustian K, Collier S, Baldock J, Burgess R, Creque J, DeLonge M, Dungait J, Ellert B, Frank S, Goddard T et al (2019) Quantifying carbon for agricultural soil management: from the current status toward a global soil information system. Carbon Manag 10:567–587
Pe'er G, Bonn A, Bruelheide H, Dieker P, Eisenhauer N, Feindt PH, Hagedorn G, Hansjürgens B, Herzon I, Lomba Â, Marquard E, Moreira F, Nitsch H, Oppermann R, Perino A, Röder N, Schleyer C, Schindler S, Wolf C, Zinngrebe Y, Lakner S (2020) Action needed for the EU Common Agricultural Policy to address sustainability challenges. People Nat 2(2):305–316
Peine M, Vitow N, Grafe M, Baum C, Zicker T, Eichler-Löbermann B, Schulz S, Schloter M, Leinweber P (2019) Effect of triple superphosphate and biowaste compost on mycorrhizal colonization and enzymatic P mobilization under maize in a long-term field experiment. J Plant Nutr Soil Sci 182(2):167–174
Perino G (2019) Reply: EU ETS and the waterbed effect. Nat Clim Chang 9(736)
Pfromm PH (2017) Towards sustainable agriculture: fossil-free ammonia. J Renew Sustain Energy 9:034702
Poffenbarger H, Artz G, Dahlke G, Edwards W, Hanna M, Russell J, Sellers H, Liebman M (2017) An economic analysis of integrated crop-livestock systems in Iowa. U.S.A. Agric Syst 157:51–69
Porterfield KK, Heisey S, McNicol L, Kramer SB, Ryals R (2018) Recycling organic waste streams to improve soil and crop health using ecological-based sanitation. American Physical Union 2018(12)
Röder N, Matthews A (2021) Eco-schemes a work in progress. Available online: http://capreform.eu/eco-schemes-a-work-in-progress/ (last accessed on 26.08.2021)
Rosemarin A, Ekane N (2016) The governance gap surrounding phosphorus. Nutr Cycl Agroecosyst 104:265–279
Sattari SZ, Bouwman AF, Giller KE, van Ittersum MK (2012) Residual soil phosphorus as the missing piece in the global phosphorus crisis puzzle. PNAS 109(16):6348–6353
Saxena J, Minaxi, Jha A (2014) CLEAN - Soil Air Water 42(9):1248–1252. https://doi.org/10.1002/clen.201300492
Schindler DW, Carpenter SR, Chapra SC, Hecky RE, Orihel DM (2016) Reducing phosphorus to curb lake eutrophication is a success. Environ Sci Technol 50:8923–8929
Schoumans OF, Bouraoui F, Kabbe C, Oenema O, van Dijk KC (2015) Phosphorus management in Europe in a changing world. AMBIO 44(2):180–192
Sharma B, Vaish B, Monika M, Singh UK, Singh P, Singh RP (2019) Recycling of organic wastes in agriculture: an environmental perspective. Int J Environ Res 13:409–429
Sharpley AN, Bergström L, Aronsson H, Bechmann M, Bolster CH, Börling K, Djodjic F, Jarvie HP, Schoumans OF, Stamm C et al (2015) Future agriculture with minimized phosphorus losses to waters: research needs and direction. AMBIO 44:163–179
Smidt G, Landes F, Machado de Carvalho L, Koschinsky A, Schnug E (2011) Cadmium and Uranium in German and Brazilian Phosphorous Fertilizers. In: Merkel B, Schipek M (eds) The New Uranium Mining Boom. Springer Geology, Berlin, Heidelberg, Germany, pp 167–175
SRU (German Advisory Council on the Environment) (2015) Stickstoff, Lösungsstrategien für ein drängendes Umweltproblem. SRU, Berlin, Germany

Stubenrauch J (2019) Phosphor-Governance in ländervergleichender Perspektive - Deutschland, Costa Rica, Nicaragua. Ein Beitrag zur Nachhaltigkeits- und Bodenschutzpolitik. Metropolis, Marburg, Germany

Stubenrauch J, Garske B, Ekardt F (2018) Sustainable land use, soil protection and phosphorus management from a cross-national perspective. Sustainability 10:1–23

Techen AK, Helming K, Brüggemann N, Veldekamp E, Reinhold-Hurek B, Lorenz M, Bartke S, Heinrich U, Amelung W, Augustin K, Boy J et al (2020) Section four: soil research challenges in response to emerging agricultural soil management practices. Adv Agron 161:179–240

TEEB (The Economics of Ecosystems and Biodiversity) (2010) Mainstreaming the economics of nature. A synthesis of the approach, conclusions and recommendations of TEEB. TEEB, Malta

Tóth G, Guicharnaud RA, Tóth B, Hermann T (2014) Phosphorus levels in croplands of the European Union with implications for P fertilizer use. Eur J Agron 55:42–52

Ulrich A (2019) Cadmium governance in Europe's phosphate fertilizers: not so fast? Sci Total Environ 650(1):541–545

United Nations (2015) Transforming our world: the 2030 Agenda for Sustainable Development. A/RES/70/1, United Nations General Assembly, New York, USA

USGS (United States Geological Survey) (2020) Mineral commodity summaries. USGS, Reston, Virginia, United States

Vargas R, Cuevas-Corona R, Yigini Y, Tong Y, Bazza Z, Wiese L (2019) Unlocking the potential of soil organic carbon: a feasible way forward. In: Ginzky H, Dooley E, Heuser IL, Kasimbazi E, Markus T, Qin T (eds) International yearbook of soil law and policy 2018, pp 373–395

VDLUFA (Association of German Agricultural Investigation and Research Institutes) (2018) Standpunkt. Phosphordüngung nach Bodenuntersuchung und Pflanzenbedarf. VDLUFA, Speyer, Germany

Weishaupt A, Ekardt F, Garske B, Stubenrauch J, Wieding J (2020) Land use, livestock, quantity governance, and economic instruments—sustainability beyond big livestock herds and fossil fuels. Sustainability 12(5):2053

Will U (2019) Climate border adjustments and WTO law extending the EU emissions trading system to imported goods and services. Leiden, Netherlands

Wirsenius S, Hedenus F, Mohlin K (2011) Greenhouse gas taxes on animal food products: rationale, tax scheme and climate mitigation effects. Clim Chang 108:159–184

Withers PJA, Hodgkinson RA, Rollett A, Dyer C, Dils R, Collins AL, Bilsborrow PE, Bailey G, Sylvester-Bradley R (2017) Reducing soil phosphorus fertility brings potential long-term environmental gains: a UK analysis. Environ Res Lett 12:1–20

Worldbank, Ecofys (2019) State and trends of carbon pricing 2019. World Bank, Washington, DC, USA

Zhang Y, Zhang J (2019) Estimating the impacts of emissions trading scheme on low-carbon development. J Clean Prod 238:117913

Zimmer D, Panten K, Frank M, Springer A, Leinweber P (2019) Sulfur-enriched bone char as alternative P fertilizer: spectroscopic, wet chemical, and yield response evaluation. Agriculture 9(1):21

SoiLEX, The New Tool of the Global Soil Partnership to Strengthen Soil Governance

Hugo Bourhis, Natalia Rodriguez Eugenio, Rosa M. Poch, Clara Lefèvre, and Ronald Vargas

Abstract In its efforts to strengthen soil governance at national and global levels and fill the gap that has formed around soil legislation, the Global Soil Partnership has launched SoiLEX, a global database of national legislation on soil protection, sustainable management and restoration. This platform, which is the result of a collaborative approach with FAO members and national experts, aims to facilitate access to existing national instruments on soil protection and management and will serve to improve our knowledge on how soil is legally covered.

1 Introduction

Soil constitutes one of the core components of a productive agricultural sector, responsible for the provision of about 95 percent of our food. Soil is also a key provider of other equally important ecosystem services, essential for sustaining life on terrestrial ecosystems, and key player in the achievement of many of the Sustainable Development Goals (SDGs).[1] However, global soil resources are under increasing pressures. The Food and Agriculture Organization of the United Nations (FAO) and the Intergovernmental Technical Panel on Soils (ITPS) estimated that about one third of world's soils are in poor or very poor status due to ten major threats or degradation processes.[2]

Despite the importance of soil and all the different stakeholders that benefit from it, soil governance is in many cases weak or inexistent. Based on the FAO definition

[1] Keesstra et al. (2016).

[2] FAO and ITPS (2015).

H. Bourhis · N. Rodriguez Eugenio (✉) · C. Lefèvre · R. Vargas
Land and Water Division, Global Soil Partnership Secretariat, Food and Agriculture Organization of the United Nations, Rome, Italy
e-mail: natalia.rodriguezeugenio@fao.org

R. M. Poch
University of Lleida, Department of Environment and Soil Sciences, Lleida, Spain

H. Ginzky et al. (eds.), *International Yearbook of Soil Law and Policy 2020/2021*,
International Yearbook of Soil Law and Policy 2020,
https://doi.org/10.1007/978-3-030-96347-7_18

of governance,[3] soil governance will be understood in this paper as the formal and informal rules, organisations and processes through which public and private actors articulate their interests; define and prioritise issues; and make, implement, monitor and enforce soil-related decisions. This definition encompasses not only legal regulations, but also market incentives, norms, habits, all of which have an impact on decision-making by influencing the way priorities are defined and implemented.

From a legal perspective, soil management is addressed in a scattered manner in current legislation, ranging from agricultural legislation to regulations on mining, urban development, or environmental protection. Soil threats are not only affecting agricultural soils, but all soils, regardless of their cover and use, as evidenced in the Voluntary Guidelines for Sustainable Soil Management.[4] It is therefore important that legal instruments address all of these uses and the impacts they have on soil functions and ecosystem services in order to ensure that sustainable soil management (SSM) is properly implemented.

Achieving SSM is a thriving area of concern and has been addressed in various international fora. Soils are addressed in the three Rio Conventions: the Convention on Biological Diversity (CBD), the United Nations Framework Convention on Climate Change (UNFCCC) and the United Nations Convention to Combat Desertification (UNCCD), although the role given to soils in specific actions and targets and the recognition of their importance as an ally in achieving the objectives are not fully developed in these three conventions.[5] The UNCCD promotes sustainable land management, understood as the integrated management of natural resources by which socio-economical and biophysical values and ecosystem services are enhanced and conserved as a solution to achieve its target of land degradation neutrality. SSM is also considered a cross-cutting issue in 4 of the SDGs of the 2030 Agenda for Sustainable Development, for which soil health represents a necessity to achieve food security and nutrition (SDG 2), human health and well-being (SDG 3), and the preservation of terrestrial ecosystems (SDG 15). The Paris Agreement within the UNFCCC emphasises the need for Parties to commit to "building resilience in socio-economic and ecological systems [. . .] through sustainable management of natural resources".[6] The Intergovernmental Panel on Climate Change (IPCC) has recently emphasized the role of soil organic carbon sequestration through SSM as a cost-effective solution to climate change mitigation and adaptation,[7] and thus contributing to the achievement of the SDG 13.

Many voices have been raised to put soils in the spotlight of the international community and to try to facilitate the dialogue between the scientific community and

---

[3] FAO (2021) Governance for Sustainable Development of Food and Agriculture. In: Food and Agriculture Organization of the United Nations. http://www.fao.org/policy-support/governance/en/ . Accessed 22 September 2021.

[4] FAO (2017) Voluntary Guidelines for Sustainable Soil Management.

[5] Montanarella and Alva (2015).

[6] UNFCCC (2016), art. 7.

[7] IPCC, *Special Report on Climate Change and Land*, 2019.

legislators. Several governmental and non-governmental organizations and academia have also sought to join efforts to ensure that soils are duly recognized as providers of key ecosystem services for development and human well-being. Through its Soil, Desertification, and Sustainable Agriculture Specialist Group, the International Union for Conservation of Nature (IUCN) has greatly contributed to the advancement of national legal soil-related frameworks.[8] Another example of a collaborative action is the adoption of the biennial Global Soil Week, which started in 2012 in the framework of the FAO's Global Soil Partnership (GSP), and represents a multi-stakeholder platform for sharing experiences on SSM and discussing soil governance needs.[9] The GSP, since its inception in 2012, has continued investing in strengthening synergies with other initiatives, organizations and governments to advocate for SSM and soil governance.

However, comprehensive soil governance has not yet been fully achieved and national legal frameworks do not yet reflect the cross-cutting role of soils that has been sought at the international level. Not every country has a complete range of soil-related legislation, covering all possible uses of soils, and less than twenty countries have a single comprehensive soil act that addresses any possible use and impact on soil resources. In order to better understand how soils are legally protected and to facilitate access to existing national legal instruments, the GSP has developed SoiLEX, an online global database of soil-related legal instruments, which gathers in a systematic way legal instruments with an impact on soil protection, conservation and restoration in the large majority of FAO Members and its regions, where applicable.

2 The Framework of the GSP to Strengthen Soil Governance

Effective soil governance requires the involvement of all actors and land users, many of whom are not always aware of the medium or long-term impact of their activities on soils nor of the existing legal instruments that regulate their activities and interventions. Soil awareness is therefore a key component of strengthening soil governance. This should include raising awareness of the importance of soil for ecosystems and human well-being, as well as the pressures anthropogenic activities cause to soils, and the measures in place to protect, sustainably manage and restore soils when they have been degraded.

Indeed, in the absence of an international convention on soils advocating for sustainable soil management, the GSP was established in 2012 with the aim of

[8] IUCN (2016) Soil, Desertification, and Sustainable Agriculture. In: International Union for Conservation of Nature. https://www.iucn.org/commissions/world-commission-environmental-law/our-work/soil-desertification-and-sustainable-agriculture. Accessed 22 September 2021.

[9] Shahid (2013).

positioning soil in the global agenda and ensuring its sustainable management by all stakeholders. The GSP covers many aspects of soil governance by raising awareness on soil and promoting sustainable soil management, facilitating the implementation of the legal framework with reliable scientific data and information, promoting capacity building, knowledge sharing and skills acquisition, supporting multi-stakeholder platforms and partnerships to build consensus, and strengthening the soil policy and legal frameworks in which the SoiLEX initiative is embedded.

As part of its efforts to improve soil governance, the GSP has produced a number of normative documents such as the Revised World Soil Charter,[10] the Voluntary Guidelines for Sustainable Soil Management,[11] and the International Code of Conduct for the Sustainable Use and Management of Fertilizers.[12] These documents provide guidance to all stakeholders on how to translate the principles of SSM into concrete actions. These non-binding documents can serve as a basis for effective actions such as the development of regulations, awareness programmes and other initiatives to address the main soil threats in each country.

Together with the formulation of normative tools, the GSP also raises awareness of the importance of soils and their specific threats, through the observance of the UN World Soil Day, which is held every year on 5 December, and the organisation of annual symposia on a specific soil threat. Each year, one of these threats is highlighted: 2017 has put forward soil organic carbon loss, 2018 soil pollution, 2019 soil erosion, 2020[13] soil biodiversity loss and 2021 covered soil salinization. These annual themes generate a growing interest from the general public, which can subsequently be translated into political will, regulations or programmes. Indeed, these annual science-policy meetings are a platform for exchanges between all the stakeholders concerned, presenting the state-of-the-art research on these subjects and providing guidance on the way forward through the creation of outcome documents setting recommendations.

While the provision of guidelines, recommendations and the promotion of SSM play a fundamental role in strengthening soil governance, the improvement of the quality and availability of soil data and information is also crucial for the adoption of informed-decisions. Improved soil knowledge and data provide de basis to evaluate the impact of interventions and regulations on soil and to assess progress over time. As such, the GSP supports the generation and compilation of reliable and harmonized data and information to better understand the state of soils and monitor trends of their evolution through the Global Soil Information System (GloSIS) and the Global Soil Laboratory Network (GLOSOLAN). GloSIS, based on the support of its International Network of Soil Information Systems (INSII), connects national soil information systems through common standards following a decentralized federated

[10]FAO (2015).

[11]FAO (2017) Voluntary Guidelines for Sustainable Soil Management.

[12]FAO (2019) The International Code of Conduct for the Sustainable Use and Management of Fertilizers.

[13]Postponed to 2021.

approach. GLOSOLAN works on the harmonization and standardization of protocols and methods, making the comparison of soil data possible among countries and regions. Furthermore, GLOSOLAN supports laboratories in improving their analytical performance by providing laboratories with tools and guidelines for (internal and external) quality control procedures, equipment purchasing, and wet and dry chemistry (spectroscopy).

Finally, to facilitate access to existing legal instruments and to better understand how soils are legally protected, the GSP has put in place the SoiLEX initiative. SoiLEX, by focusing more on the legal branch dealing with soil protection and soil degradation prevention, responds to a need to better understand the status of the legal framework associated with soil protection in each country and to obtain information on the coverage of the different soil threats.

3 SoiLEX: A Wiki on Soil Legislation

3.1 Objectives

While some tools, such as FAOLEX[14] and ECOLEX[15] already compile national legislation and policies on food, agriculture, forestry, and natural resources management, and therefore include some legislation related to soil, the soil-related instruments are currently difficult to find because of the wide variety of topics they cover. Specifically, FAOLEX has three keywords to filter the search on soil legislation: "soil conservation/improvement", "soil pollution/quality" and "soil rehabilitation". However, these three categories are insufficient for specific searches related to the different threats to soil health and other soil-related topics. Indeed, soil-related instruments may also fall under other topics, such as fertilisers, desertification, erosion and others, and are not always present under the three soil-related keywords. As a consequence, by using the three existing soil keywords one cannot easily understand how soil is legally considered in each country.

The aim of SoiLEX is to respond to this need and fill the knowledge gap that exists around soil legislation by serving as a research tool for soil stakeholders at every level. SoiLEX presents the legal instruments hierarchized and organized according to keywords relevant to the soil stakeholder community.

[14] FAO (2021). FAOLEX. http://www.fao.org/faolex/en/. Accessed 22 September 2021.

[15] FAO, IUCN and UNEP (2021). ECOLEX. https://www.ecolex.org/. Accessed 22 September 2021.

3.2 Development of the SoiLEX System

3.2.1 Initial Database

To create SoiLEX, the GSP built its database on the FAOLEX database, which is to date one of the largest global electronic collection of legal frameworks and instruments related to food, agriculture and natural resources management. The platform, which contains approximately 180,000 legal and policy texts from 185 countries, is constantly updated with some 8000 new documents added every year. FAOLEX database includes over 24,000 documents that contain the word "soil", although only about 2500 documents are classified with the three soil-related keywords.

As an initial step in building SoiLEX database, all documents tagged with the three soil keywords were extracted from FAOLEX and filtered to keep only the documents that explicitly mention soil. Thus, all documents relating to land tenure, land management and sustainable agricultural practices, which can have positive effects on soils, were removed from the initial selection when the soil was not explicitly mentioned in the document. This decision was taken to avoid having too many documents in the SoiLEX database to be able to put forward all the documents covering soil explicitly.

For European Member States, the FAOLEX database has been complemented by the "EU Soil Wiki",[16] an online inventory of national and regional soil-related policy instruments. This project, carried out from January to October 2016, was a collaborative approach with all European Member States to identify the main gaps in soil protection at European level by examining the coverage of the different soil threats identified.[17] This approach ultimately aimed to provide a baseline on which to build new policy actions on soil in the EU.

3.2.2 Structure of the SoiLEX Database

Fourteen keywords were selected in order to better group the existing legislations through concepts more familiar to the soil science community. These keywords include the ten soil threats identified in the Voluntary Guidelines for Sustainable Soil Management,[18] plus soil conservation/protection, soil restoration/remediation, soil quality and soil monitoring. These last four keywords have been added to allow SoiLEX to include legal instruments relating to soil data collection and monitoring, soil standards and threshold values as well as instruments defining measures and guidance on how to restore degraded soils.

---

[16] EC (2021). SoilWiki. https://webgate.ec.europa.eu/fpfis/wikis/pages/viewpage.action?spaceKey=SOIL&title=Home. Accessed 22 September 2021.

[17] Frelih-Larsen AC et al., *Updated Inventory and Assessment of Soil Protection Policy Instruments in EU Member States | Ecologic Institute: Science and Policy for a Sustainable World*, 2017.

[18] FAO (2017) Voluntary Guidelines for Sustainable Soil Management.

About 2500 documents tagged with the three soil-related keywords in FAOLEX were selected for a first screening. Through this screening, about 1600 soil-related legal instruments were selected and classified according to the SoiLEX keywords. The legislation was therefore categorised by the following keywords:

1. **Soil fertility/nutrient imbalance**: any legislation to foster soil nutrient balance and cycles; regulations of the use of synthetic and organic fertilizers, including composition, limits and testing; regulations on nutrient reuse and recycling, etc.
2. **Soil salinization/sodification**: any legislation preventing the accumulation of salts or sodium in the soil, including those addressing irrigation water quality and use of wastewater in irrigation; use of soil amendments and industrial waste disposal; any legislation promoting alternatives use of salt-affected soils such as in biofuels and halophyte carbon sequestration; and legislation on groundwater exploitation especially with regard to coastline aquifer protection.
3. **Soil acidification**: any legislation regulating the inefficient use of nitrogen and ammonium based fertilisers, the emission of acidifying substances, etc.
4. **Soil sealing and land take**: any legislation focusing on urbanisation, land use change, etc.
5. **Soil biodiversity loss**: any legislation preventing the loss of soil biodiversity, regulating the use of integrated pest management, regulating the introduction and control of invasive or alien species, etc.
6. **Soil waterlogging/water management**: any legislation addressing flooding, irrigation management (not quality), drought management, etc.
7. **Soil compaction**: any legislation regulating the use of heavy machinery, tilling, etc.
8. **Soil pollution/contamination**: any legislation to prevent or control soil pollution, both diffuse and point source pollution, including waste management, mining, sewage sludge application, pesticides application, fertilizers use, distance to/of emitting industries, etc.
9. **Soil erosion**: any legislation to prevent or control loss of soil by wind, water and tillage erosion, such as agricultural practices, forest management, landslide control measures, revegetation of slopes, control of road drainage, and stabilization of riverbanks and their riparian areas.
10. **Soil organic carbon loss**: any legislation to enhance soil organic matter content, reuse of crop residues, agricultural practices, especially recarbonization of (degraded) farmland, and measures for payment for ecosystem services.
11. **Soil restoration/remediation**: any legislation that define measures and guidance on when and how to restore degraded soil, including remediation conditions of polluted soils and soils affected by mining activities.
12. **Soil conservation/protection**: any legislation protecting soil by addressing in a comprehensive and holistic way all the ten treats mentioned above.
13. **Soil quality**: any legislation specifying soil standards and values, including those specifying methods and protocols to assess soil properties.
14. **Soil monitoring**: any legislation establishing a monitoring system of soil resources and providing guidance on how, when and for what (for example,

compliance with international commitments such as Conventions) to monitor soil properties and status. Legislations requesting soil surveys must also be included in this category.

In addition to it, a classification system was integrated to better understand the relevance of each document in relation to the keywords. Thus, each document is classified according to an algorithm taking into account (1) its relevance of the document for the selected keyword, (2) the scope of the document (national/regional/local), (3) the nature of the document (binding/non-binding[19]), and (4) the date of the document. Binding national instruments with a high relevance for a specific keyword can thus be found as the first documents available on the platform. The ranking system thus helps to better understand which documents are the most important for each country according to each selected topic at present.

3.2.3 Verification of the Database at the National Level

In order to verify that the information for each country contained in the initial database was correct and up to date, a survey was circulated in the first semester of 2020 with the GSP National Focal Points,[20] and with all relevant stakeholders within each country (other ministries, universities, research institutes, institutions and others). The questionnaire was also intended to gather information on the legal instruments with relevance to soil that were missing from the SoiLEX database. Respondents to the survey assessed and provided the most accurate representation possible of the current national legal framework for soil protection and prevention of soil degradation.

It is worth pointing out the great diversity of ministries in which the National Focal Points are based. Indeed, despite a strong dominance of the Ministry of Agriculture (76 percent) and the Ministry of Environment (10 percent), some GSP Focal Points are based in other ministries such as the Ministry of Industry, Development, Economic Affairs and Foreign Affairs, to name but a few. This wide variety of ministries working on SSM reflects the crosscutting nature of soil, but could also result in a certain bias in the responses collected at the national level.

The survey contained questions relating to all the platform's classification filters. Thus, each respondent to the survey had to classify each legal instrument present in the SoiLEX database with the corresponding keywords. A maximum of three keywords could be selected for each instrument, and these keywords had to be

[19]Legally binding instruments are instruments that impose legal obligations or duties. On the contrary, non-binding instruments do not have legal or binding force.

[20]National Focal Points are individuals appointed by the ministry of agriculture of each FAO member, selected for their expertise on soil management and conservation issues and are aimed at improving communication and collaboration between the GSP and member countries. Focal points are mostly senior officials in ministries (64 percent), researchers in research institutes (33 percent) or in universities (3 percent).

classified according to their level of importance in the document. Respondents were then asked to confirm the scope of each instrument and its nature. If any documents were missing, they were asked to complete all the classification filters presented above and share the missing document. Finally, a question on whether the country (or a State in Federal States) had a Soil Act was raised.

82 responses from 59 different countries were received. This collaborative approach allowed to validate more than 1100 soil-related documents. North America, Latin America and the Caribbean, Europe, and Eurasia were the regions that responded the most to the survey. Conversely, the Near East and North Africa, Asia, sub-Saharan Africa and the Pacific are the regions where the project received the least feedback.

The dissemination of the survey at the national level also improved extensively the database as 285 new documents from 40 countries were added to SoiLEX, and shared with FAOLEX to improve the completeness of both databases. More than three quarters of the new documents provided were binding instruments from Europe and Eurasia, providing a valuable source of new information for users of both platforms.

In parallel to the validation carried out by the national experts, the GSP Secretariat reviewed the remaining documents in the database when no feedback was received from the country. Thus, the GSP Secretariat examined 640 documents in order to complete the platform. The review of the documents consisted of removing the repealed documents, ensuring that soil was explicitly mentioned in each document and classifying the document on the basis of the keywords present on SoiLEX.

A total of 1745 documents from 175 countries were therefore processed and validated, either by the GSP National Focal Points and other national experts or by the GSP Secretariat, and made available for the launch of SoiLEX in January 2021.

3.3 *SoiLEX Applications*

SoiLEX facilitates the search for national soil legal instruments, the understanding of the different legal areas relevant to soil management and protection, as well as the exchange of experiences in soil governance between countries and regions. The development of legal studies on soils, based on the information available in SoiLEX, will also make it possible to identify the main gaps in soil protection legal frameworks and explore options for strengthening soil governance.

As such, SoiLEX will serve as a tool to:

- Support the development of adequate soil legislation as a contribution to sustainable soil management adapted to the socio-economical and political particularities of each region;
- Enhance the capacity of national, regional, local authorities and other stakeholders in soil legislation;

- Develop new knowledge, documenting existing good practices, identifying and adapting legal tools in relevance to specific soil threats;
- Empower civil society by giving them access to legislative materials.

4 Preliminary Analysis of the SoiLEX Database

4.1 Overview

Of the 1745 documents present on SoiLEX in January 2021, Europe is the region with the most documents available (564 documents), followed by Latin America and the Caribbean (439 documents) and Eurasia (217 documents). Figure 1 presents the number of documents for each of the GSP regions present on SoiLEX as of January 2021.

These data is only reflecting number of documents analysed and cannot be used to highlight the performance of specific regions on soil governance, as quantity does not in any way mean better soil protection or governance. Indeed, a federal state, by virtue of its system, systematically provides more documents. Canada and the Russian Federation, for example, are the two countries with the most documents available on the SoiLEX platform due to their subregional organization in which each region has autonomy to develop new legal instruments or to transpose national documents into regional ones.

Of the 1745 documents currently available online, 1503 are binding instruments and 242 are non-binding instruments. Most of the non-binding instruments catalogued on SoiLEX come from sub-Saharan Africa, where soil is covered by national poverty reduction and rural development programmes or strategies. The

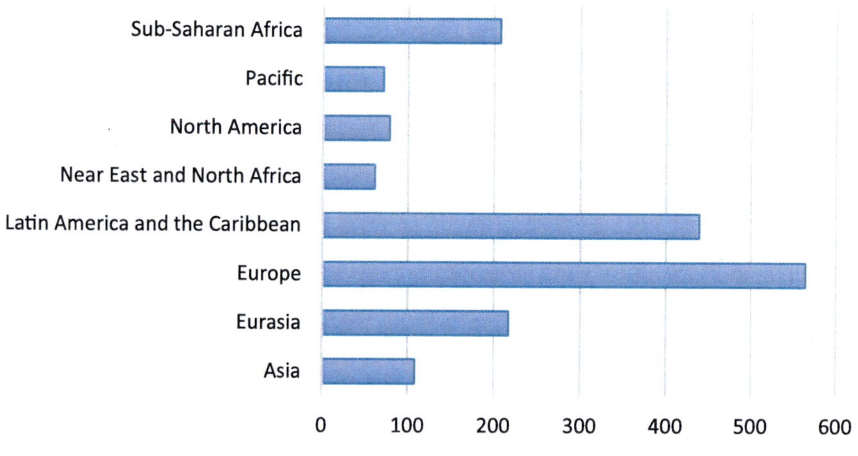

Fig. 1 Number of documents present on SoiLEX by GSP region (as of January 2021)

proportion between binding and non-binding instruments could result from the legal scope of SoiLEX, which thus has fewer policy instruments.

4.2 *Thematic Coverage of Soil-Related Legal Instruments*

Based on the documents currently present in the SoiLEX database and the keywords selected for each of these documents, **soil pollution** is the most covered threat among the 10 threats presented in the Status of the World's Soil Resources report.[21]

A number of reasons can explain this result. Firstly, environmental pollution in general has become a global problem known and faced worldwide. As a result, environmental pollution is addressed from a legal standpoint in all countries. However, although soil pollution is less well known than water pollution or air pollution, soils benefit from broader environmental protection frameworks aimed at covering environmental pollution as a whole. The Lithuanian *Law on Environmental Protection*, for example, consolidated in 2016, mentions soil only in the definition of the environment (Article 1) and thus deals with environmental pollution as a whole throughout the document.

Secondly, soil pollution is caused by a wide variety of anthropogenic activities and thus is covered by different legal instruments. Documents related to environmental protection, mining, waste management, management of hazardous substances, or use of wastewater and sewage sludge often point out the need to prevent soil pollution. As a result, the legal framework associated with the prevention and remediation of soil pollution often goes beyond the agricultural sector, and has a more broad scope than some threats still associated solely with agriculture. The scope of China's *Soil Pollution Prevention and Control Law* (2018), attests to this observation. Soil pollution monitoring is particularly addressed and covers agriculture, construction, mining and other activities that can lead to soil pollution. This broad spectrum of how and to what extent soil pollution is covered may explain the large number of documents present under this threat.

After soil pollution, **soil erosion** and **soil fertility loss** (nutrient imbalance) are the two main threats most covered. Soil erosion coverage can be explained by the large number of countries that have a National Adaptation Programme of Action (NAPA). NAPAs provide a framework for countries to address the challenge of climate change by identifying priority adaptation activities. Established in the framework of the United Nations Framework Convention on Climate Change (UNFCCC) in 2001, these programmes of action often cover desertification and soil erosion, particularly in sub-Saharan Africa.

Another factor to be taken into account is the creation of the United Nations Convention to Combat Desertification (UNCCD) in 1994, which advanced the issue of desertification and soil erosion at the global and national level. Under the

---

[21] FAO and ITPS (2015).

UNCCD, more than 120 countries have committed themselves to achieving land degradation neutrality, which has been defined as "a state whereby the amount of quality of land resources, necessary to support ecosystem functions and services and enhance food security, remains stable or increases within specified temporal and spatial scales and ecosystems".

Soil fertility loss is also widely covered, mainly through an agricultural prism, with regulations on fertilisers and their management, and regulations to ensure the fertility of agricultural land. It should be noted that soil fertility loss is also largely covered under the NAPAs mentioned above.

Soil biodiversity is often part of broader environmental frameworks aimed at preserving and conserving the environment. For example, many Forest Laws cover soil biodiversity with the aim of preserving habitat characteristics to ensure the biological diversity of forests. Soils are then mentioned as particularly important natural habitats for forest diversity, as is the case in Finland's *Forest Act* (Section 10). In addition, regulations on fertilisers and soil improvers may also cover soil biodiversity in order to prevent accumulation of substances in soil organisms. However, there is a need to further integrate soil biodiversity into countries' National Biodiversity Strategies and Action Plans (NBSAPs) in the future because soil biodiversity is often neglected or referenced only indirectly.[22]

Soil organic carbon (SOC) is linked to a number of binding and non-binding instruments, covering topics such as use of compost, environmental protection, sustainable development, fertilisers and climate change. Soil is the major terrestrial reservoir of carbon and has a great potential to store atmospheric carbon, thus contributing to mitigate climate change. Indeed, Mexico's *General Law on Climate Change* emphasises the importance of soils in climate change adaptation and mitigation, as well as the need to adopt sustainable soil management practices in order to maintain and increase their role as carbon sinks. As such, the law aims to halt deforestation, expand vegetation cover and restore degraded land in response to climate change (Article 34). In addition, it also establishes a special fund for projects that contribute to climate change mitigation and adaptation through soil conservation and restoration to improve carbon sequestration (Article 82). It should be noted that in recent years, SOC has gained great interest with respect to climate change mitigation and adaptation, as evidenced by its inclusion in the Paris Agreement in 2015 and the creation of the 4per1000 initiative[23], but still few countries include SOC in legal frameworks. It is also important to note that the SoiLEX database is still in its early stages and that not all the thematic areas with a possible influence on soil management have been analysed in depth, so it is possible that in the coming years, as SoiLEX grows, this and other topics will gain greater relevance and representation within the database.

Soil sealing is closely associated with urban planning and construction. As such, spatial planning covers this threat by providing building standards and guidelines on

[22] FAO et al. (2020).

[23] Rumpel et al. (2018).

which soils can be sealed to prevent the conversion of productive agricultural soils. In addition to urban planning regulations, flood regulations also cover soil sealing, as soil sealing prevents the influx of water into the soil and can therefore lead to more severe flooding. For European countries, this applies to the transposition of the EU *Floods Directive* adopted in 2007, which explicitly sets out requirements to reduce flood risks, including urban floods. As part of the EU's new Green Deal to achieve the ambitious goal of a climate-neutral EU by 2050, soil sealing is included as a priority. Soil sealing is increasingly being recognised by policy makers as a major degradation problem in urban areas and un-sealing measures are being considered.[24]

The legal coverage of **soil salinization** is closely associated with irrigation and water management regulations, which also contribute to prevent **waterlogging** in agricultural areas. For example, under section 8 of the South African *Conservation of Agricultural Resources Act*, the Ministry of Agriculture has established an irrigation improvement scheme to prevent waterlogging and salinization of irrigated land. Subsidies can thus be paid to participants in the scheme for hydraulic works to prevent soil salinization. A similar approach has been adopted in Ukraine, which, through its *Water Code*, obliges measures to be taken with regard to water used for irrigation of agricultural land, to prevent flooding, waterlogging, salinization and to ensure that water complies with established quality standards (Art. 65).

Soil compaction and **soil acidification** are the least covered threats according to the SoiLEX database. As presented in the Voluntary Guidelines for Sustainable Soil Management, soil compaction "is related to the degradation of soil structure due to imposed stresses by machinery and livestock trampling".[25] The increasing mechanisation of agriculture and the significant increase in the weight and size of agricultural machinery in recent decades have had a strong impact on the soil system. It has been estimated that the weight of fully loaded combine harvesters increased about six fold in half a century, from 4 Mg in 1958 to 25 Mg in 2009.[26] This increase in the weight of agricultural machinery has a negative effect on the soil functions. Soil compaction reduces soil aeration, water drainage and infiltration, which limits root growth and ultimately crop production.[27] This threat should therefore be better covered in national regulations if soils are to be protected and sustainably managed.

Between 10 and 20 percent of German cultivated land has been reported as compacted due to anthropogenic activities.[28,29] This report led to bringing this threat to soil and agriculture to the forefront of the political agenda. The German Environment Agency is currently investigating the introduction of a system to classify soils

[24] Tobias et al. (2018).

[25] FAO (2017), p. 12.

[26] Schjønning et al. (2015).

[27] FAO (2017) Voluntary Guidelines for Sustainable Soil Management.

[28] Schneider and Don (2019).

[29] Meunier (2018).

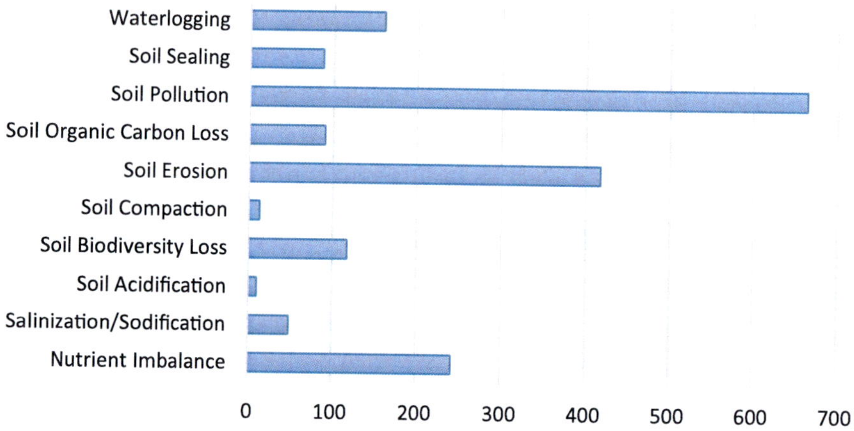

Fig. 2 Number of documents present on SoiLEX by soil threats (as of January 2021)

on the basis of physical properties in order to better preserve their functions.[30] This system would then be closely associated with the prevention of soil compaction by requiring manufacturers of agricultural machinery to offer options for the purchase and deployment of machines better adapted to each soil type.[31]

While **soil acidification** can result from acid precipitation and the deposition of acidifying gases or particles from the atmosphere,[32] soil acidification in agricultural land is mainly associated with the use of nitrogen-based fertilizers.[33] It should be stressed that, despite the number of legal instruments relating to the use and management of fertilisers present on SoiLEX, soil acidification is not covered exhaustively in the vast majority of cases. This tendency could also be explained because soil acidification is not explicitly mentioned as a threat to be prevented or remediated but rather implicit in a number of documents associated with sustainable fertilizer management.

Figure 2 presents in more details the number of documents for each of the 10 soil threats present on SoiLEX as of January 2021.

While most countries cover the various soil threats presented above with separate instruments, very few countries have comprehensive laws dealing with soil protection and prevention of soil degradation. Indeed, only 19 countries currently have Soil Acts. The Bulgarian *Soils Act*, for example, specifically identifies erosion, acidification, salinization, compaction, organic matter decline, contamination, sealing and landslides as processes leading to soil degradation (Article 12). In parallel, the law also mentions other threats present in the Voluntary Guidelines for Sustainable Soil

[30] Horn R (2019) Don't take soil compaction lightly. In: The Furrow | The John Deere Magazine. https://thefurrow.co.uk/dont-take-soil-compaction-lightly/. Accessed 22 September 2021.

[31] Ibid.

[32] Goulding (2016).

[33] FAO (2017) Voluntary Guidelines for Sustainable Soil Management.

Management, such as soil biodiversity loss and nutrient imbalance. The law also prohibits 12 identified practices that can lead to soil degradation (Article 13) and emphasises the need to apply sustainable soil management practices (Article 3).

However, it should be stressed that not all Soil Acts cover all threats and that some Acts only address soil conservation as a whole. For example, the Namibian *Soil Act* refers more to soil conservation than to specific threats. Soil erosion and soil pollution are the only two explicit threats mentioned in the Namibian Act. Therefore, the existence of a Soil Act should not be seen as synonymous with coverage of all soil threats, or as a better protection of the soil compared to other, less specific instruments.

5 Way Forward

The SoiLEX database is a living tool that will be constantly updated to ensure that the platform is as accurate and up-to-date as possible to reflect each country's legal framework on soil. The platform will continue growing as a collaborative initiative and relaying on national responses received through the SoiLEX website.[34] Other FAOLEX keywords will also be reviewed in the near future to identify other potential legal areas having an impact on soil protection, sustainable management and restoration.

The GSP has also set up a multidisciplinary Working Group on Soil Legislation. Fourteen countries are currently represented in the working group, from all regions of the world except the Near East and North Africa, and North America, providing legal expertise in their research areas and regions of interest. New inputs will be reviewed and processed by the Working Group on Soil Legislation to assess the keywords relevance.

While compiling national laws and maintaining the database up-to-date is of paramount importance, it is only the first step to strengthen soil governance and the development of adequate soil legislation. In order to make SoiLEX more useful to governments, policymakers, land users, academia, civil society and other stakeholders, the working group will also contribute to conducting regional and national studies on soil legislation. As such, the working group will document existing good practices, identify gaps and needs for improvement and assist national governments to create an enabling legal environment to adopt sustainable soil management. These efforts will build on the analytical work already done on the subject by other GSP partners and initiatives (such as IUCN, the FAO Legal Division, or the International

[34] FAO (2021). Soils Portal. http://www.fao.org/soils-portal/soilex/en/. Accessed 22 September 2021.

Development Law Organization (IDLO)[35] among others) and will be combined with other GSP initiatives to improve knowledge of the state of national resources, to see how countries with similar soil types and problems could learn from each other based on the results of their legal actions to achieve sustainable soil management. Under this framework, SoiLEX aims to support countries in achieving their international commitments, such as the 2030 Agenda on Sustainable Development, the Paris Agreement on Climate Change or the post-2020 Global Biodiversity Framework.[36]

A central element of these targeted studies will be the evaluation of the level of implementation of the various legal instruments listed in SoiLEX. Indeed, despite many documents covering the different soil threats, soil degradation persists.[37] On this basis, it is then valuable to better understand the level of implementation of these instruments and what are the feasible and practical solutions for a better application on the ground. Coming back to the FAO's definition of governance, it is then essential to better understand the implementation, monitoring and enforcement of these instruments.

6 Conclusion

As part of the Global Soil Partnership's efforts to improve soil governance, SoiLEX aims to fill the knowledge gap on soil legislation around the globe, to facilitate access to existing legal instruments, and to understand how soils are legally protected in each country. This living platform will grow and evolve over time based on the responses and feedback gathered at the national level. Following its bottom-up approach, the GSP will continue to strengthen its networks and establish new connections with other partners and initiatives with expertise in soil legislation and governance to enhance and grow the SoiLEX database. On the basis of the information available on SoiLEX, it will thus be possible to identify the priority areas to be improved in soil-related legal frameworks, identify and replicate successful experiences, and explore options for strengthening soil regulatory frameworks.

[35] IDLO (2021). International Development Law Organization, Sustainability. https://www.idlo.int/what-we-do/sustainability. Accessed 22 September 2021.

[36] CBD (2021). Preparations for the Post-2020 Biodiversity Framework. Third meeting of the Open-ended Working Group on the Post-2020 Global Biodiversity Framework. https://www.cbd.int/conferences/post2020. Accessed 22 September 2021.

[37] FAO and ITPS (2015).

References

FAO (2015) Revised World Soil Charter. FAO, Rome, Italy

FAO, ITPS (2015) Status of the world's soil resources (SWSR): Main Report. Food and Agriculture Organization of the United Nations and Intergovernmental Technical Panel on Soils, Rome

FAO, ITPS, GSBI et al (2020) State of knowledge of soil biodiversity - status, challenges and potentialities. Food and Agriculture Organization of the United Nations, Rome, Italy

Goulding KWT (2016) Soil acidification and the importance of liming agricultural soils with particular reference to the United Kingdom. Soil Use Manag 32:390–399. https://doi.org/10.1111/sum.12270

Keesstra SD, Bouma J, Wallinga J et al (2016) The significance of soils and soil science towards realization of the United Nations Sustainable Development Goals. SOIL 2:111–128. https://doi.org/10.5194/soil-2-111-2016

Meunier C (2018) Daten zur Umwelt 2018: Umwelt und Landwirtschaft. Umweltbundesamt

Montanarella L, Alva IL (2015) Putting soils on the agenda: the three Rio Conventions and the post-2015 development agenda. Curr Opin Environ Sustain 15:41–48. https://doi.org/10.1016/j.cosust.2015.07.008

Rumpel C, Lehmann J, Chabbi A (2018) '4 per 1,000' initiative will boost soil carbon for climate and food security. Nature 553:27–27. https://doi.org/10.1038/d41586-017-09010-w

Schjønning P, van den Akker JJH, Keller T et al (2015) Driver-Pressure-State-Impact-Response (DPSIR) analysis and risk assessment for soil compaction-a European perspective. Adv Agron:183–237. https://doi.org/10.1016/bs.agron.2015.06.001

Schneider F, Don A (2019) Root-restricting layers in German agricultural soils. Part I: extent and cause. Plant Soil 442:433–451. https://doi.org/10.1007/s11104-019-04185-9

Shahid S (2013) First Global Soil Week 18-22 November 2012 briefing about the event soils, climate change, challenges and agriculture. World Forum on Climate Change Agriculture and Food Security Newsletter 3:2013

Tobias S, Conen F, Duss A et al (2018) Soil sealing and unsealing: state of the art and examples. Land Degrad Dev 29:2015–2024. https://doi.org/10.1002/ldr.2919

Soil Protection, Food Security and the Nexus Between Climate Governance and Trade in Agriculture

Oliver C. Ruppel

Abstract This chapter examines—both from an international and national law perspective—the interface between soil protection and food security, climate governance and trade in agriculture. Although these different spheres are most often viewed in isolation, an attempt is made to analyse them more holistically with the aim of identifying their intrinsic connectedness. Soils also delivers valuable services in carbon sequestration making it crucial for fighting climate change. Despite this, the right to food does barely display in the soil, climate and trade discussion. So where is the connect between Soil, the UNFCCC, the Paris Agreement and the rules of the WTO? Soils are the world's second largest carbon sink after the oceans. And while the Paris Agreement fails to explicitly mention 'soil', it still falls under Article 5(1) that obliges parties to take action to conserve and enhance, as appropriate, sinks and reservoirs of greenhouse gases. Of course the WTO is not an environmental protection agency but existing WTO rules fail to make agricultural trade patterns per se sustainable or resilient in terms of climate change mitigation efforts. We are however today in a time of climate crisis making it more necessary than ever to reconcile conceptual differences of the past. Industrial transition to climate neutrality and decarbonisation should be guided by the *leitmotiv* to place soil protection for sustainable food security at the centre of any economic policy (This abstract and chapter is largely based on Ruppel, FAO global soil partnership webinar on soil governance on 13 January 2021, 2021a.).

Special thanks are given to Dr. Katharina Ruppel-Schlichting for her critical substantive input in the preparation of writing this chapter and to Ms. Larissa-Jane Houston for assistance in making it formally publishable.

O. C. Ruppel (✉)
Faculty of Law, Stellenbosch University, Stellenbosch, South Africa

Research Center for Climate Law, University of Graz, Graz, Austria
e-mail: ruppel@sun.ac.za

H. Ginzky et al. (eds.), *International Yearbook of Soil Law and Policy 2020/2021*, International Yearbook of Soil Law and Policy 2020,
https://doi.org/10.1007/978-3-030-96347-7_19

1 Introduction

Soils are essential ecosystems that deliver valuable services such as the provision of food and carbon sequestration, among others. Therefore, soil is crucial for fighting climate change, protecting human health, safeguarding biodiversity and ecosystems and ensuring food security.[1] While the demand for food must be satisfied under changing climatic (and other such as COVID-19) conditions, land degradation and soil depletion are of growing concern.[2] Conventional food and agricultural trade, as well as global value chains, may need to be reconsidered to ameliorate the concerns. While the impacts of climate change on global food production and food security are well known, the effects of climate change on food safety are much less so. Changes in global food systems and the increased globalisation of the food supply means that populations worldwide are at risk of exposure to various food safety hazards. This can, among other things affect food security, national economies and international trade.[3]

> Trade, as a mechanism, was theoretically intended to efficiently allocate immovable land resources so as to match global supply and demand, thus leading to improved global prosperity, while simultaneously distributing environmental pressures among the least sensitive ecosystems. In reality, however, trade has displaced considerable environmental pressures from developed to developing countries, which usually have weaker enforcement capabilities of environmental standards [...]. Over the past fifty years the value of the international agricultural and food trade has increased around fivefold, however the profits resulting from this appreciation have not been evenly distributed throughout the regions of the world.[4]

Global trade has turned arable land into a mobile resource. Especially developed and emerging economies are exporting their hunger for land to the developing world. They import land in the form of products grown abroad.[5] Having said this with regard to international trade and looking through the ecological crystal ball into the future, it becomes increasingly clearer that the interest in the element ‘soil’, which is both linked to and at the same time detached from the land, can be international and at the same time domestic in nature.

> The climate affects the soil – but the soil also influences the climate. Carbon dioxide and other greenhouse gases are especially important. The soil is a huge carbon sink; it contains more carbon than the atmosphere and all terrestrial vegetation combined. Relatively small changes in the amount of organic matter in the soil can have a major effect on the atmosphere – and on global warming [...]. With the right management, soil has the potential to absorb large amounts of carbon, and so combat global warming. Restoring the soil’s ability to lock away carbon is an important way to reduce the impact of climate change.[6]

---

[1] European Commission (2020).

[2] Larbodière et al. (2020), p. 78.

[3] FAO (2020a).

[4] Raschio (2017).

[5] Heinrich Böll Foundation and Institute for Advanced Sustainability Studies (2015).

[6] *Ibid.*

Therefore, international soil protection and (climate) governance is related to different mechanisms, including international organisations, treaties and conventions and other multilateral and bilateral agreements, such as trade agreements. Not only has the term 'soil' been relatively rarely mentioned in international agreements, but the central function of soil in the ecosystem and in society has also been undervalued. Both soil and land have received far too little protection. The onus to protect the soil is largely on national governments that are accountable to the international community through international agreements and play a pivotal role in coordinating, planning and determining policy priorities and distributing resources. Global climate mitigation efforts must, however, also be coordinated with national and regional policies in an interrelated system of multilevel governance. Soil and land must be seen as cross-cutting themes in policymaking in order to receive adequate protection. While outlining the various levels and cross-cutting themes hereafter, strategies for a better common future are sought.

2 International Developments in Law

Both at national and at international level in the realm of law, improving soil governance includes options for enhancing coordination and coherence. In terms of the existing state of affairs, the European Soil Charter of 1972 was held to have been the first international document relating to soil.[7] The World Charter on Nature[8] and Agenda 21[9] have been criticised as being inappropriate to aid in soil conservation, as their wording is too broad to establish clear norms,[10] and other international developments have proved to be more relevant.

The World Soil Charter[11] and the World Soils Policy[12] were negotiated by the United Nations Environment Programme (UNEP) in coordination with the United Nations Food and Agriculture Organization (FAO) and were adopted in 1982. Both instruments contain non-binding guidelines and principles relating to soil conservation[13] and were intended to aid states in formulating domestic policies. However, in light of modern environmental practices, these instruments are considered to be outdated.[14] In 2015, the World Soil Charter has been revised,[15] to reflect major policy developments and conceptual advances with relevance to soils. The revised

[7] Alori and Nwapi (2015).

[8] United Nations (1982).

[9] United Nations Conference on the Environment and Development (1992).

[10] Alori and Nwapi (2015), p. 106.

[11] FAO (1982).

[12] UNEP (1982).

[13] Alori and Nwapi (2015), p. 105.

[14] *Ibid.*

[15] FAO (2015a).

Charter comprises nine principles and furthermore, guidelines for action by individuals, the private sector, groups, the science community, governments and international organisations to ensure that soils are managed sustainably and that degraded soils are rehabilitated or restored.

2.1 Global Soil Partnership

The Global Soil Partnership (GSP) is a globally recognised mechanism established in 2012. Its mission is to position soils in the Global Agenda through collective action. Its key objectives are to promote sustainable soil management (SSM) and improve soil governance to guarantee healthy and productive soils and support the provision of essential ecosystem services towards food security and improved nutrition, climate change adaptation and mitigation, and sustainable development. The GSP was established prior to the 2015 International Year of Soils and aided the implementation and coordination of the roll-out of the year-long activities.[16] Recognising the critical role of soils for food security and ecosystem services, FAO was requested to establish a GSP. The GSP is striving to become an interactive, responsive and voluntary partnership, open to governments, regional organisations, institutions and other stakeholders at various levels. The GSP is hosted by FAO in view of its global mandate.[17]

The International Year of Soils then resulted in a wealth of awareness activities across the globe, in addition to putting soils back on the international policy agenda.[18] This has led to a new international dialogue concerning the protection and rehabilitation of soils and sustainable farming practices in general.[19] New scientific knowledge has been gained over the past decades, "especially with respect to new issues that emerged or were exacerbated, like soil pollution and its consequences for the environment, climate change adaptation and mitigation and urban sprawl impacts on soil availability and functions".[20]

Consequently, the World Soil Charter has been revised and was unanimously endorsed in June 2015, during the course of the International Year of Soils, by the member states of the FAO during the 39th Session of the FAO Conference.[21] The revised guidelines intended to ensure that "soils are managed sustainably and that

---

[16]FAO (2015b).

[17]*Ibid.*

[18]FAO (2015c).

[19]*Ibid.*

[20]Cf. http://www.fao.org/fileadmin/user_upload/GSP/docs/ITPS_Pillars/annexVII_WSC.pdf accessed 07 June 2021.

[21]The revised World Soil Charter is organised into a preamble, nine principles, and guidelines for action.

degraded soils are rehabilitated or restored".[22] The actions therein target individuals and the organised private sector, governments and international organisations, which triggered a growing international dialogue concerning the protection and rehabilitation of soils and sustainable farming practices.[23] While tools such as FAOLEX and ECOLEX compile national legislation and policies, and include some legislation on soil protection and soil degradation prevention, a newly established working group on soil legislation will in the time contribute to reviewing and updating the SoiLEX database containing all soil-related legal instruments adopted in each country. SoiLEX is envisaged to become a global database on national legislation on soil protection, conservation and restoration to facilitate access to information on the existing legal instruments in force and bridge the gap between the various soil stakeholders. The online platform will facilitate the search for national soil legal instruments, the understanding of the different legal areas relevant to soil management and protection, and the exchange of experiences in soil governance between countries and regions.[24]

2.2 *United Nations Convention to Combat Desertification*

The United Nations Convention to Combat Desertification (UNCCD) is the main international legal document to combat desertification and mitigate the effects of drought in affected countries through effective action at all levels supported by international cooperation. This instrument is the only international treaty specifically addressing land-related issues, while the definition of desertification therein clearly relates to soil conservation.[25] In 2015 the UNCCD introduced the new concept of land degradation neutrality (LDN), which was later adopted as a target of Goal 15 of the Sustainable Development Goals (SDGs). In doing so, the UNCCD

> laid the groundwork for developing and establishing the concept of LDN. After adoption of the SDGs, the CCD claimed leadership for implementation of target 15.3 on LDN. It decided to integrate LDN in its work and has engaged in various activities. Besides a target setting programme this includes elaborating guidance material. In particular, the CCD published a Scientific Conceptual Framework that is intended to apply to all land and guide all parties in implementing LDN. Although the legal and political constraints make the UNCCD's potential difficult to assess, it could continue to pursue a leading role in implementing the LDN target and serve as forum for discussing soil-related issues between developing and developed countries.[26]

[22] FAO (2015a) Section 3.

[23] FAO (2015c).

[24] FAO (2020b).

[25] Land degradation in arid, semi-arid and dry sub-humid areas resulting from various factors including climatic variations and human activities.

[26] Bodle et al. (2020).

So far, however, the tangible effect of the UNCCD remains limited, as the focus is primarily placed on capacity-building, as opposed to creating binding obligations per se.[27]

> According to the target and indicators of LDN, a first step towards rebuilding and restoring soil capital is to halt land degradation. An abundance of options is available to land managers in pursuit of this goal, but the appropriate solutions in each context are not always known and many hurdles still impede progress. Land health needs to be recognised as a precondition for resilient agricultural production and incentivised to trigger the desired shift and impact at scale.[28]

Because land is fixed in quantity, there is ever-increasing competition to control land resources and capitalise on the flows of goods and services from the land. This has the potential to cause social and political instability, fuelling poverty, conflict and migration. For that reason, the implementation of LDN requires multi-stakeholder engagement and planning across scales and sectors, supported by national-scale coordination that utilises existing local and regional governance structures.[29] In light of this, LDN has been defined by the Parties to the Convention as:

> A state whereby the amount and quality of land resources, necessary to support ecosystem functions and services and enhance food security, remains stable or increases within specified temporal and spatial scales and ecosystems.[30]

The objectives of LDN are to maintain or improve the sustainable delivery of ecosystem services; to maintain or improve productivity, in order to enhance food security; to increase resilience of the land and populations dependent on the land; to seek synergies with other social, economic and environmental objectives; and to reinforce responsible and inclusive governance of land.[31]

It has been stated in recent studies that there is an overlap and potential competition and conflict between the UNCCD and the FAO, which also claims leadership regarding international soil. Both regimes are major international actors with high participation and political legitimacy in this field. Moreover, there seems to be an overlap with the Convention on Biological Diversity (CBD) in terms of legal scope and mandate regarding soil biodiversity. Here the CBD is probably the more relevant international instrument, as the diversity within species and ecosystems is closely linked and reliant upon the conservation of soils and ecosystems. It aims at conserving biological diversity, promoting the sustainable use of its components, and encouraging equitable sharing of the benefits arising out of the utilisation of genetic resources.[32] The CBD's Strategic Plan for Biodiversity (2011–2020) takes a broad perspective on preconditions for the conservation and sustainable use of

[27] Alori and Nwapi (2015), p. 107.

[28] Larbodière et al. (2020), p. 78.

[29] Cf. https://www.unccd.int/actions/achieving-land-degradation-neutrality accessed 07 June 2021.

[30] *Ibid.*

[31] Global Environment Facility (2019).

[32] Bodle et al. (2020), p. 53.

biodiversity. It does so by, inter alia, focusing on food security. The Aichi Biodiversity Targets (which form part of the Strategic Plan) aim at achieving five strategic goals. Soil health and soil biodiversity are cross-cutting among these targets.[33]

In 2019, under the New Delhi Declaration, countries agreed to make 'land degradation neutrality by 2030' a national target for action. The governments which are party to the UNCCD met in the Indian capital of New Delhi over 10 days for COP14, adopting a series of breakthrough measures in the accord known as the New Delhi Declaration. Besides the LDN agreement—whereby countries pledged to halt the degradation of land to the point where ecosystems and land use can no longer be supported—there was a landmark decision to boost global efforts to mitigate and manage the risks of crippling drought. Countries will now also be expected to address insecurity of land tenure, including gender inequality; promote land restoration to reduce land-related carbon emissions; and mobilise innovative sources of finance from public and private sources to support the implementation of these decisions at a national level.[34]

The New Delhi Declaration recognises that desertification, land degradation and drought undermine health, development and prosperity in all regions and acknowledges that dryland ecosystems are areas of special focus. Striving to achieve land degradation neutrality has the potential to act as an accelerator for achieving a number of the Sustainable Development Goals and as a catalyst for attracting sustainable development finance to implement the Convention. Those practices which conserve and restore land and soil affected by desertification, land degradation, drought and floods contribute towards achieving land degradation neutrality and can also have multiple long-term benefits for the health, well-being and socio-economic development of entire societies, and especially for the livelihoods of the rural poor. It makes explicit reference to the findings of the Intergovernmental Science-Policy Platform on Biodiversity and Ecosystem Services Land Degradation and Restoration Assessment and its Global Assessment Report on Biodiversity and Ecosystem Services, as well as the Intergovernmental Panel on Climate Change Special Report on Climate Change, Desertification, Land Degradation, Sustainable Land Management, Food Security, and Greenhouse Gas Fluxes in Terrestrial Ecosystems, which recognise the role of the land-use sector. Furthermore, the New Delhi Declaration states that effective and responsible governance and stewardship of natural resources, especially land and water, for combating desertification, land degradation and drought and for the sustainable development of current and future generations is of the essence. It calls for diverse multi-stakeholder participation, as appropriate, at local, subnational, national and regional levels and for all sectors of society, including civil society organisations, local government and the private sector, which will be crucial to achieving the objectives of the UNCCD, the long-term goals of the Paris Agreement and the development of an ambitious post-2020 global biodiversity framework, taking into consideration land-based solutions for

[33] *Ibid* 59.

[34] United Nations (2019).

climate action and biodiversity conservation and the mutually supportive implementation of the three Rio conventions.[35]

2.3 *United Nations Framework Convention on Climate Change*

The 1992 the United Nations Framework Convention on Climate Change (UNFCCC) was adopted to regulate levels of greenhouse gas concentration in the atmosphere, so as to, inter alia, avoid the occurrence of climate change on a level that would compromise initiatives in food production. Article 2 of the UNFCCC defines the parties' ultimate objective as the stabilisation of greenhouse gas concentrations in the atmosphere at a level that would prevent dangerous anthropogenic interference with the climate system.

So far, international climate policy has mostly focused on emission sources and thus on the avoidance of greenhouse gas emissions, for example from the electricity sector, the production industry, transport, and land-use changes. This will, however, be increasingly complemented by the preservation and enhancement of emission sinks to remove carbon dioxide from the atmosphere. To achieve the global climate targets adopted by the UNFCCC, alternative mitigation methods, as for example through programmes for re- or afforestation and the restoration of ecosystems, become more and more relevant. Despite the fact that the combination of bioenergy and carbon capture and storage, increased carbon sequestration in soils,[36] and the direct capture of CO_2 from ambient air need to be further researched and are not yet at the stage of market maturity, the carbon removal approach has considerable potential, while soils are the world's second largest carbon sink after the oceans.[37]

> The most potentially devastating impacts of industrial modes of agricultural production stem from their contribution to increased greenhouse gas emissions. Together, field-level practices represent approximately 15 per cent of total human-made greenhouse gas emissions, inter alia from the loss of soil organic carbon in croplands.[38]

The 2015 Paris Agreement, as part of the UNFCCC regime, binds its parties regarding activities on their respective territories and under their control. As such, the Paris Agreement supplements the UNFCCC and the Kyoto Protocol of 1997 by incorporating existing elements of this regime. Both the UNFCCC and Kyoto Protocol adopted rules on reporting and accounting for emissions from land use,

---

[35] Cf. https://www.unccd.int/news-events/new-delhi-declaration-investing-land-and-unlocking-opportunities accessed 07 June 2021.

[36] Soil carbon sequestration is the process of capturing atmospheric CO_2 through changing land management practices to increase soil carbon content. Various land management practices promote soil carbon sequestration.

[37] Geden and Schenuit (2020), p. 5.

[38] De Schutter (2014).

land use change and forestry (LULUCF). These rules determine how parties have to report LULUCF in their regular emission inventories, which under the Kyoto Protocol is also relevant for accounting whether parties meet their emission reduction targets.[39]

According to Article 2, the Paris Agreement's overarching objective is to keep the increase in global temperature well below 2 °C, or even 1.5 °C. Parties are required to prepare and present individual climate plans (Nationally Determined Contributions—NDCs) every five years that set out how the party intends to contribute to the collective objectives. Under the Paris Agreement, the Principle of Common but Differentiated Responsibilities (CBDR) is an obligation for all parties when formulating their NDCs. This is the result of protracted negotiations about the role and impact of historic and present, and of relative and absolute, GHG producers. Although the Paris Agreement does not specify how to take the CBDR principle into account, principles of justice and equity help to improve the understanding of the normative implications of climate law under the Paris Agreement. While equity as a normative concept has a sense of fairness, justice plays an important role in legal-political decisions in relation to climate policy in particular and through differentiation in obligations.[40] GHG emissions have global, not merely national, effects, which on the basis of the need to contain the potential proliferation of trade distortions due to climate policies in terms of equity, may justify sanctioning the inaction by large GHG emitters, which can have a serious impact on local food production affected by global warming.[41]

Through the sustainable development mechanism in Article 6, the Paris Agreement allows the space to harness the lowest cost mitigation options worldwide. This may incentivise policymakers to enhance mitigation ambition by speeding up climate action.[42] This implies that global climate policy development and the future of the carbon market also relate to mechanisms which support and encourage sustainable climate policies in host countries as production-based accounting does not necessarily reflect a country's contribution to global emissions because globalisation and consumption can prompt emissions beyond borders.

By signing the Paris Agreement (and in particular Article 14 therein), parties agreed on long-term goals backed by national plans that are collectively reviewed in the global stocktake, which is key to increasing ambition. While the first planned stocktake is scheduled for 2023, it has already become apparent today that the improved accuracy of carbon stock estimates would allow for more targeted interventions and better monitoring of the NDCs—which has equal significance for the protection of soil in the context of agricultural production.[43] Whereas the UNFCCC

---

[39] Bodle et al. (2020), p. 17.

[40] Lawrence and Reder (2019), pp. 511–531.

[41] Häberli (2018), p. 34.

[42] Tänzler et al. (2019).

[43] Interestingly, in its NDCs, the Republic of South Africa, states that policy instruments under development include regulatory standards and controls for specifically identified GHG pollutants

does not explicitly provide for specific trade measures, the Kyoto Protocol contains more detailed obligations related to the reduction of greenhouse gases and provides for trade-affecting techniques such as tax impositions on carbon dioxide emissions and the elimination of subsidies adversely affecting the objective of the UNFCCC.[44] In addition, the parties to the Paris Agreement explicitly recognise

> [...] the fundamental priority of safeguarding food security and ending hunger, and the particular vulnerabilities of food production systems to the adverse impacts of climate change;

while Article 2(1)(b) of the Paris Agreement provides for

> [in]creasing the ability to adapt to the adverse impacts of climate change and foster climate resilience and low greenhouse gas emissions development, in a manner that does not threaten food production [...].

Notwithstanding the legally neutral wording of Article 2(1) when read in isolation, achieving its purpose

> [...] is mandatory not for any one state or group of states, of course; it is mandatory for the state parties collectively. This straightforward logical implication of the Paris Agreement does not seem to have been noticed before, despite its potentially profound consequences.[45]

The Paris Agreement further requires parties to engage in adaptation planning and implementation that takes into account "vulnerable people, places and ecosystems" and builds "the resilience of socio-economic and ecological systems, including through economic diversification and sustainable management of natural resources". Of course, soil as well as land use, land degradation and sustainable land management are closely linked to climate change in terms of carbon capture and storage and the emissions from deforestation and agriculture. This is underlined by Article 4 of the Paris Agreement, which explicitly includes the target "to achieve a balance between anthropogenic emissions by sources and removals by sinks of greenhouse gases in the second half of this century", although the Paris Agreement fails to explicitly mention 'soil', 'land' or 'agriculture'. As such, the Paris Agreement only indirectly addresses soil protection in the general context of climate change. And despite the importance of land use and soil management for climate change, the UNFCCC, the Kyoto Protocol and the Paris Agreement have not established a comprehensive regime with regard to land-related climate change measures.[46] Article 5(1) of the Paris Agreement obliges parties to take action to conserve and enhance, as appropriate, sinks and reservoirs of greenhouse gases as referred to in

---

and emitters; Cf. https://www4.unfccc.int/sites/ndcstaging/PublishedDocuments/South%20Africa%20First/South%20Africa.pdf accessed 07 June 2021.

[44] Ruppel (2018), p. 37.

[45] Zahar (2020), pp. 165–188.

[46] Bodle et al. (2020), p. 53.

Article 4(1)(d) of the Convention. Yet, in fact, agriculture first appeared in the ongoing climate negotiations under the Koronivia joint work on agriculture programme at COP 23 in 2017:

> The decision officially acknowledges the significance of the agriculture sectors in adapting to and mitigating climate change. Countries agreed to work together to make sure that agricultural development ensures both increased food security in the face of climate change and a reduction in emissions. The joint work is expected to address six topics related to soils, nutrient use, water, livestock, methods for assessing adaptation, and the socio-economic and food security dimensions of climate change across the agricultural sectors.[47]

In order to achieve the aforementioned, countries should take all appropriate measures according to their capabilities to progressively achieve the protection of the interests of all concerned. And when speaking of 'all concerned' in the context of global food security, this phrase is by no means an exaggeration. Much of the work to translate the Paris Agreement and the NDCs into concrete climate interventions in agriculture is in progress.

> Food systems are responsible for 21-37% of global greenhouse gas emissions and a major driver of deforestation and land degradation, yet there is still widespread food insecurity and malnutrition. Managing the land sector (agriculture, forestry, wetlands, bioenergy) sustainably and holistically could contribute up to 30% of the global climate mitigation effort [...].[48]

In 2018 the Paris Agreement adopted a transparency framework which, inter alia, included rules for reporting on and accounting for land use and land-use change, which is expected to eventually replace the existing UNFCCC framework. This may open opportunities also to shape new rules complementing the UNFCCC's Koronivia joint work on agriculture.

A wide range of policy instruments is needed to strengthen the mutually supportive role of the Paris Agreement and other international agreements when it comes to the protection of soil. The AFOLU (agriculture, forestry, and other land use) sector plays an important role in the 1.5 °C pathways and is, inter alia, responsible for food production. Changes in the AFOLU sector are driven by demand changes, efficiency of production, and policy assumptions. While demand for agricultural products and other land-based commodities is influenced by consumption patterns, including dietary preferences and food waste (affecting demand for food), policy assumptions relate to the level of land protection, the treatment of food waste, policy choices about the timing of mitigation action, the choice and preference of land-based mitigation options, and interactions with other sectors and trade.[49]

---

[47] Cf. http://www.fao.org/climate-change/our-work/what-we-do/koronivia/en/ accessed 07 June 2021.

[48] Palahí et al. (2020).

[49] Rogelj et al. (2018).

2.4 *Sustainable Development Goals and Land Degradation Neutrality*

In 2015 Sustainable Development Goals (SDGs) were formulated as a successor to the Millennium Development Goals (MDGs). The 2030 Agenda for Sustainable Development includes 17 SDG and 169 targets. The SDGs provide a universally accepted framework to foster global collaboration with a strong emphasis on the rule of law and human rights. While Agenda 2030 is aimed at fostering and renewing multilateralism and international cooperation on the global but common challenges, SDG Goal 2 is aimed at 'Zero Hunger'. Unfortunately, however,

> the world is not on track to achieve Zero Hunger by 2030. If recent trends continue, the number of people affected by hunger would surpass 840 million by 2030. According to the World Food Programme, 135 million suffer from acute hunger largely due to man-made conflicts, climate change and economic downturns. The COVID-19 pandemic could now double that number, putting an additional 130 million people at risk of suffering acute hunger by the end of 2020.[50]

In addition, SDG Goal 15 (Life on Land) stipulates the following:

> Protect, restore and promote sustainable use of terrestrial ecosystems, sustainably manage forests, combat desertification, and halt and reverse land degradation and halt biodiversity loss.

Target 15.3 of the same goal envisages that "[b]y 2030, combat desertification, restore degraded land and soil, including land affected by desertification, drought and floods, and strive to achieve a land degradation-neutral world." Although not legally binding, the SDGs, and in particular the LDN target in SDG 15.3, have at least established a political consensus for continued dialogue that guides national policies and governmental action for national land and soil policies.[51]

> The SDGs include economic and social development goals that potentially involve trade-offs with environmental sustainability. In particular, Zero Hunger (SDG 2) can be considered as in direct competition with Life on Land (SDG 15). One of society's most urgent challenges is to satisfy the rights of people to a 'good life', including adequate food and nutrition, while remaining within the planetary boundaries. In other words, we need to reconcile agriculture and the environment to "end hunger, achieve food security and improved nutrition and promote sustainable agriculture" (Zero Hunger, SDG 2) and also "protect, restore and promote sustainable use of terrestrial ecosystems, sustainably manage forests, combat desertification, and halt and reverse land degradation and halt biodiversity loss" (Life on Land, SDG 15).[52]

The inclusion of the LDN target in SDG 15.3 symbolises the international community's commitment towards a land degradation neutral world by 2030. The United Nations General Assembly recently identified that land-based solutions, as part of nature-based solutions, constitute promising options to evaluate and consider

[50] Cf. https://www.un.org/sustainabledevelopment/hunger/ accessed 07 June 2021.

[51] Bodle et al. (2020).

[52] Larbodière et al. (2020), p. 8.

in connection with sequestering carbon and enhancing the resilience of people and ecosystems affected as well as adverse effects of climate change. At present, 123 countries have set or committed to set their LDN targets. LDN means a paradigm shift in land management policies and practices. It is a unique approach that, inter alia, counterbalances the expected (unavoidable) loss of productive land with the recovery of degraded areas. It strategically places the measures to conserve, sustainably manage and restore land in the context of land-use planning. To achieve the target of land degradation neutrality by 2030, large amounts of financial resources must be mobilised. New financial instruments and intermediaries, as well as enabling conditions, are needed to catalyse private capital to attain LDN. The creation of the LDN Fund is a commendable step in this regard. In its resolution 73/233, the United Nations General Assembly acknowledged that the LDN Fund was an innovative scheme to finance the 2030 Agenda for Sustainable Development, including the achievement of a land degradation neutral world.[53]

> Sustainable development broadly supports and often enables the fundamental societal and systems transformations that would be required for limiting warming to 1.5°C above preindustrial levels.[54]

The SDGs call for more sustainable production and consumption patterns and agricultural and food systems that protect natural resources, using integrated and partnership-based approaches with the involvement of various actors. Possible supply chain approaches, for example, intervene at the point of end consumption of such products, the production of which in distant, politically sovereign states causes sustainability risks. Corresponding sets of rules that have already been used for supply chains differ with regard to the targeted individual sustainability goals, addressed sectors, incentive structure and degree of obligation. So far, voluntary regulations have dominated the focus on sustainability. These take various sustainability goals of the FAO Voluntary Guidelines for Responsible Land Use into account. Setting international standards has proved to be difficult, owing to the variety of circumstances that exist around the world. This is especially true for agricultural practices, which have to respond to differences in climate, soils and ecosystems, and are an integral part of cultural diversity.

National and international strategies in support of the progressive realisation of the right to adequate food seem to be very much in line with the recommendations of the Committee on Economic, Social and Cultural Rights in its general comment No. 12 on the right to adequate food (para. 21).[55] According to Article 25(1) of the Universal Declaration of Human Rights (UDHR), everyone has the right to a standard of living adequate for the health and well-being of himself and of his family, including food, while Article 11 of International Covenant on Economic, Social and Cultural Rights (ICESCR) recognises the right of everyone to an adequate

---

[53] Cf. https://www.un.org/pga/74/event/high-level-dialogue-on-desertification-land-degradation-and-drought/ accessed 07 June 2021.

[54] Rogelj et al. (2018).

[55] De Schutter (2014).

standard of living for himself and his family, including adequate food; as a fundamental right of everyone to be free from hunger. Article 11(2) ICESCR in recognising the fundamental right of everyone to be free from hunger, compels Parties to take measures to (a) improve methods of production, conservation and distribution of food by making full use of technical and scientific knowledge, by disseminating knowledge of the principles of nutrition and by developing or reforming agrarian systems in such a way as to achieve the most efficient development and utilization of natural resources; (b) taking into account the problems of both food-importing and food-exporting countries, to ensure an equitable distribution of world food supplies in relation to need. Similarly, Article 24(2)(c) of the Convention on the Rights of the Child (CRC) obliges Parties to take appropriate measures to combat disease and malnutrition, including within the framework of primary health care, through, inter alia, the provision of adequate nutritious foods, taking into consideration the dangers and risks of environmental pollution.

At the national level, this often hinges on the holistic integration of the right to food being intrinsically interconnected to land and climate issues, sectors, levels of government, and policy instruments that address the increasing volatility. To counter increasing food insecurity, the LDN goal can only be achieved by means of a system-level holistic action that can reduce, avoid and reverse land degradation, through sustainable land management. Successful implementation of LDN interventions require, inter alia, proper policy and regulatory mechanisms and science-policy interaction.

3 Land Degradation Neutrality-Relevant Policies

In order to achieve LDN, cross-sectoral policy coherence is critical—between institutions, sectors, levels of governance—to resolve fragmentation, incoherence, sectorisation, lack of connectivity, and conflicting interests. Besides agriculture and forestry, other land-use policies need to be included in LDN planning, such as water, mining, settlements, and infrastructure.[56]

Policies responding to degradation and desertification, inter alia, include improving market access, enhancing gender empowerment, expanding access to rural advisory services, strengthening land tenure security, realising payments for environmental services, improving decentralised natural resource management, investing in research and development, and in modern renewable energy sources, monitoring of desertification and desert storms, developing modern renewable energy sources, and developing and strengthening climate services. Policies supporting economic diversification include investing in irrigation, expanding agricultural commercialisation, and facilitating structural transformation in rural economies.

---

[56] Cf. http://earthobservations.org/documents/gwp20_22/GEO-LDN.pdf accessed 07 June 2021.

Policies and actions also include promoting indigenous and local knowledge, soil conservation, agroforestry, crop-livestock interactions as an approach to managing land degradation, and forest-based activities such as afforestation, reforestation, and changing forest management. Further measures identified for the achievement of LDN include effective financial mechanisms (for implementation of land restoration measures and the long-term monitoring of progress), parameters for assessing land degradation, and detailed plans with quantified objectives and timelines.[57]

In the land–climate interface, effective policy responses must include carbon pricing, emissions trading schemes (including net CO_2 emissions from agriculture), carbon taxes,[58] regulations limiting GHG emissions and air pollution, forest conservation (mix of land-sharing and land-sparing) through participation, incentives for ecosystem services and secure tenure, protecting the environment, microfinance, crop and livelihood insurance, agriculture extension services, agricultural production subsidies, low export tax and import tariff rates on agricultural goods, dietary awareness campaigns, taxes on and regulations to reduce food waste, improved shelf life, sugar/fat taxes, and instruments supporting sustainable land management (including payment for ecosystem services, land-use zoning, REDD+, standards and certification for sustainable biomass production practices, legal reforms on land ownership and access, legal aid, and legal education), as well as reframing these policies as entitlements for women and small agricultural producers.[59] Similarly, border carbon adjustments can help level the playing field and prevent emissions leakage,[60] which occurs when climate action in one region merely shifts emissions elsewhere.[61]

Such is, for instance, European Union (EU) soil policy action, based on Article 191 of the Treaty on the Functioning of the EU (TFEU), which requires Union policy to aim at preserving, protecting and improving the quality of the environment, protecting human health, a prudent and rational utilisation of natural resources, promoting measures at international level to deal with regional or worldwide environmental problems, and in particular combating climate change.

[57] Hurlbert et al. (2019).

[58] Such was for example the Carbon Tax Act 15 of 2019, a relatively new addition to South Africa's legislative record, aiming to provide for the imposition of a tax on the carbon dioxide (CO_2) equivalent of greenhouse gas emissions; and to provide for matters connected therewith. This aim is expected to be achieved by the deployment of a range of measures to support the system of desired emissions reduction outcomes, including the appropriate pricing of carbon, the use of emissions offsets and economic incentives for rewarding the efficient use of energy to provide appropriate price signals to help nudge the economy towards a more sustainable growth path. Such tax phased in over time allows for learning, while the tax revenue can for example finance additional climate change mitigation efforts. Whether a carbon tax yields a better result, for global food security, than carbon sequestration, depends on many different factors. Taxation for climate change mitigation could be included under any broad (NDC) commitment to reduce emissions or in the promotion of green technologies.

[59] Hurlbert et al. (2019).

[60] Peters et al. (2011), pp. 8903–8908.

[61] Kasturi et al. (2019), pp. 25–30.

The EU considers improved soil protection fundamentally important due to (1) the transboundary impacts of soil degradation, such as CO2 emissions from soil organic carbon and loss of biodiversity, hampering EU food security through reduced production of food commodities traded in the internal market, hampering water quality across borders through contaminants and sediments in river basins, food safety concerns from soil contaminants; (2) the absence of a level playing field for economic operators subject to very different national soil protection regimes, leading to a distortion of the internal market; and (3) the risk that the EU and its Member States will fail to fulfil international and European commitments in the field of environment, sustainable development and climate.[62]

EU Member States have to ensure that greenhouse gas emissions from land use, land use change or forestry are offset by at least an equivalent removal of CO_2 from the atmosphere in the period 2021 to 2030. The Regulation on the inclusion of greenhouse gas emissions and removals from land use, land use change and forestry (LULUCF) into the 2030 climate and energy framework was adopted by the Council on 14 May 2018, following the European Parliament vote on 17 April 2018. It is also in line with the Paris Agreement, which points to the critical role of the land use sector in reaching long-term climate mitigation objectives.[63]

Moreover, the EU Biodiversity Strategy for 2030 announced the update of the 2006 EU Soil Thematic Strategy to address soil and land degradation in a comprehensive way and to help achieve land degradation neutrality by 2030. The Biodiversity Strategy for 2030 highlights that it is essential to step up efforts to protect soil fertility, reduce erosion and increase soil organic matter. The European Court of Auditors recommended to the European Commission to aim at a better understanding of land degradation and desertification in the EU, to assess the need to enhance the EU legal framework for soil, and to step up actions towards delivering EU and international commitments, and particularly land degradation neutrality by 2030. The European Parliament called on the EU and the Member States to make strong commitments towards sustainable food systems, agriculture and forestry, including requirements and strategies for the protection of soils. The European Environment Agency called for a comprehensive and coherent policy framework to protect land and soil. The strong political will and momentum guides the new EU soil strategy, which is already underway to consolidate, complement and steer action in the different policy areas that affect and depend on soil and guide the implementation of sustainable soil and land management practices. It is expected to cover key horizontal aspects from funding instruments to developing knowledge, research, communication and international cooperation. This will be done in close coordination and complementarity with other European Green Deal initiatives, including the upcoming Zero Pollution Action Plan, and other initiatives resulting from the EU Biodiversity Strategy for 2030 and the Farm to Fork Strategy. This also includes the legally binding EU nature restoration targets that the Commission will propose in

[62] European Commission (2020).

[63] Cf. https://ec.europa.eu/clima/policies/forests/lulucf/en accessed 07 June 2021.

2021 and which should contribute to the achievement of the objectives of the new Soil Strategy and the restoration of degraded soils.[64]

In addition, the European Commission aims to enshrine the so-called 'no-debit rule' in EU law by incorporating land use into the EU's emission-reduction efforts. The actions of farmers to secure carbon stored in soils will thus contribute to achieving the EU's commitment under the Paris Agreement on climate change to reduce greenhouse gas emissions by at least 40% by 2030 compared to 1990 levels.[65]

4 Land Degradation Neutrality-Relevant National Legal Imperatives

Soil management should be an integral part of land management and many jurisdictions address soil within land management instruments or within soil-specific legislation. Soil management can be found in sectoral legislation on agriculture, land, environment or even water. While some legislative approaches distinguish between different soil types and characteristics in order to determine the specific interventions warranted to achieve the appropriate quality for the land use selected, others focus on issues such as soil quality, contamination and pollution, soil conservation and soil rehabilitation.[66] The fact is that the maintenance and protection of soil are vital to allow a continued reliance upon it.[67] Soil maintenance and protection need to be governed by legislation and policy, along with prescribed and practised enforcement measures. Incentives and deterrents for sustainable land use ensure that private land use is in line with social and other policy objectives and promote certain practices, fertilizers, subsidies, etc. No doubt, sustainable agriculture, rural development and the upliftment of poor and marginalised communities require a cross-sectoral involvement of environment, climate change, land rights, gender equality, traditional and indigenous affairs, health, economy and trade, among others.

It is national rules and norms that must first and foremost guide, constrain, and shape human interaction in terms of soil and law. Both formal laws, policies, and structured processes and informal norms, conventions, and consultation must be taken into full consideration. Sub-national efforts, especially at the level of communities, are also significant in the context of the land.

Thus, at the national level, any comprehensive legislative soil assessment requires an analysis of the Constitution (in respect of social, economic and environmental rights, and responsible branches of government for legislating and enforcing those rights); the examination of the country's international (monist or dualist), regional

[64] European Commission (2020).

[65] Cf. https://ec.europa.eu/clima/policies/forests/lulucf/en accessed 07 June 2021.

[66] FAO (2020c), p. 132.

[67] Ruppel and Ginzky (2021).

and bilateral agreements and legally binding obligations reflected or not reflected in national legislation (including conflicting and ambiguous legislation); an analysis of existing regulatory (sectoral and cross-sectoral) policy, and legal and institutional frameworks; the identification of existing regulatory gaps, bad practices and responsibility overlaps; the identification of institutional and governance frameworks; and the scrutiny of implementation, monitoring, standardisation and enforcement challenges and pinpointing of the reasons (cost, lack of capacity, political will, human resources, etc.) thereof.[68]

Moreover, effective national soil legislation should take into consideration aspects related to dispute resolution (courts and tribunals, administrative environmental tribunals, alternative dispute resolution, and customary courts); and access to justice (standing, costs, geographic accessibility, timeliness, availability of counsel, and non-governmental advocacy organisations).[69]

Apart from the specific national legislative approach, contextual factors also play a role in the assessment of optimal soil protection. These include demography (distribution of wealth, population density, age structure, urban/rural split, education and literacy, gender equity); economy (contribution of natural resource/extractive sector to the state economy, per capita income, evenness of development); politics (fragility, corruption perception, rule of law generally); corruption (in the control of natural resources/concessions, in management of natural resource revenues, in the enforcement process); and civic engagement.[70]

Integrated land-use planning must balance competing land uses and land tenure conflicts, property rights, conservation aspects, settlements, and agricultural areas. The interactions between land tenure and land policy and how they relate to sustainable development have largely been neglected by policymakers and planners. The same applies to past, present and future large-scale land acquisitions, today often referred to as land-grabbing. Despite these developments, land ownership has always been exclusive and to some extent also unfair. The question therefore remains: what social contract in the future can form a more inclusive and equitable land scheme, especially including land tenure? More research seems to be required to better understand the complex interconnections of land, climate, water, society and food, including new models that allow incorporation of considerations of justice, inequality and human agency in socio-environmental systems.[71]

Progression in terms of land tenure rights is also necessary to enable soil protection to obviate the effect that land tenure can have on soil. Human influence on the land and natural resources, such as soil, is accelerating as a result of the growth in population and the associated increase of food requirements.[72] While scarce, land is more than a source of food security, income and shelter. Land is also

[68] FAO (2020c), p. 63.

[69] UNEP (2019).

[70] *Ibid.*

[71] Hurlbert et al. (2019).

[72] Kanianska (2016).

related to cultural identity and is thus often a source of economic tension or political competition.[73] Therefore, reliable land tenure arrangements serve to reduce poverty, support sustainable livelihoods, enable social stability and housing, and foster environmental protection. Moreover, food security and food availability are highly dependent on secure access to and the productivity of land.[74] A sound legal framework is, for instance, necessary for any registration programme to function and bring about the desired outcomes, such as the administration of land usage and protection of natural resources.[75]

So far, LDN-relevant sustainable soil management and soil protection lack adequate legal instruments because of the lack of perception and understanding of the importance of soil for sustainable development. Relying on the work done so far by the FAO Legal Office—which maintains a comprehensive, up-to-date legislative and policy database (FAOLEX), one of the world's largest electronic collection of national laws, regulations and policies on food, agriculture and natural resources management—the GSP Secretariat has extracted and analysed current soil-related legal instruments from all member countries. These soil-specific legal instruments are presented in the SoiLEX database to be used as a facilitator of national analysis, by providing easily accessible models of effective soil policy that can guide country efforts to develop new legislation.[76]

5 Soil Agronomy, Circular Bioeconomy and Supply Chain Management

Driven by a global population projected to rise to over 10 billion people by 2050 (from 7.6 billion today) and an increase in the 'consuming class' with the purchasing power to demand more food per capita (including food with a higher environmental footprint), the world could require a doubling in agricultural production from 2005 levels in order to meet demand. Such a trajectory is unsustainable.[77]

A more circular bioeconomy should be based on healthy, biodiverse and resilient ecosystems that provide the basis of sustainable well-being for society at large. This can be achieved by means of functioning ecosystem services and sustainable management of biological resources, leading to the circular transformation in food, feed, energy and biomaterials within the ecological boundaries of the ecosystem that it relies on.[78]

[73] FAO (2020c), p. 93.

[74] Sandrey (2019/2020).

[75] Tlale (2018), p. 266.

[76] FAO (2019).

[77] World Economic Forum (2020).

[78] Palahí et al. (2020), p. 15.

An alignment of the food and agricultural sector with the SDGs and the Paris Climate Agreement can enhance the interplay between environmental, nutritional, social and governance-related factors that need to be tackled while examining how business indicators might be developed in support of greenhouse gas (GHGs) emissions reductions.[79]

The food and land-use system could significantly benefit from a fundamental shift towards productive and regenerative agriculture. Transforming agricultural landscapes and farming practices for both food and non-food agriculture through a combination of traditional farming techniques, advanced precision technologies, and bio-based inputs can increase biodiversity, enrich soils, improve water management and enhance ecosystem services while improving yields. This transition, however, requires greater understanding and adoption of the appropriate agronomic solutions, including re-alignment of agricultural subsidies, while navigating trade-offs between improving yields and strengthening biodiversity outcomes.[80]

> Long-term scenarios of adoption of agroecological approaches and provision of ecosystems services will help evaluate the benefits and potential trade-offs for society and provide desirable pathways for policy makers. Agriculture and conservation actors should seek consensus over indicators of sustainability and farmers should be supported to improve their performance against those indicators developing scientifically established agroecological approaches.[81]

In the context of end-use and market integration, this, inter alia, involves exploiting economies of scope across products, supply chains, land-use, transportation, city planning, and climate mitigation.[82] It is, for instance, possible to design cities in a way that reduces the amount of soil lost. Designing greener cities with commercial areas scattered throughout the city avoids paving over large areas, shortens transport distances, and leaves room for open landscapes and gardens.[83]

Agricultural policy is influenced by more than just political forces. An understanding of food supply chains is important in the context of soil management practices. Improving soil management is important to the food and farming supply chain, increasing the performance of soil to produce higher yields that are more resilient to physical and financial shocks as well as the effects of climate change. The increased number of regulatory policies can be seen as the result of higher standards of living, which have boosted consumers' demand for safe and high-quality products, and of growing problems with water, air and soil pollution. Economic growth is a driver of poverty reduction but is also responsible for environmental destruction. In a world already confronted with climate change, diminishing exhaustible natural resources, and other threats to the environment, international trade contributes further to these harmful developments through the production of agricultural

---

[79] Sachs (2020).

[80] Ruppel and Ginzky (2021).

[81] Larbodière et al. (2020), p. 79.

[82] Hurlbert et al. (2019).

[83] Heinrich Böll Foundation and Institute for Advanced Sustainability Studies (2015).

goods and global transportation as part of the supply chain. Moreover, "the modernisation of food supply chains, together with the implementation of agricultural policies focused more on the production of commodities than on food, has led to the marginalisation of local food systems, which is a trend that must be reversed".[84]

The value and protection of soil is interrelated with production and supply chain management. The environmental effects of production differ widely across countries owing to differences in climate, land availability, soil fertility, use of technology, energy sources, laws and institutions, and other factors.[85] This indicates that it can also be preferable to produce agricultural goods where this is most environmentally efficient. By promoting specialisation, competition, economies of scale, innovation and technology transfer at a global level, trade can both help to lower the production costs and help to achieve better environmental outcomes.

Supply chain management regulation is a relatively new attempt to address the economic and ethics problem of negative externalities in international trade by legal rule. It has become necessary because of a failure of the first-best solution of global rules and the failure of states to address the issue.

> Subsidies, norms and standards in particular can block as well as encourage the desired transformation. Industry concentration or business strategies can be incompatible with the necessary re-diversification of crop rotations and agricultural landscapes. Supply chain arrangements often lead to the concentration of value capture and decision-making power in the downstream part of the supply chain, preventing farmers from transitioning to more sustainable practices.[86]

Global financial flows should be increasingly redirected towards sustainable value chains and healthy landscapes, which would also benefit soil protection and LDN. Science can help to develop indicators and standards that can guide those investments towards more sustainable practices, products and processes along the food supply chain.[87] An early identification of emerging risks lies at the heart of protecting public health and the environment and, by identifying such risks, for example in the food supply chain, can assure effective and timely measures to protect consumers.

As supply chain regulation also has inherent effects on international trade, it is crucial for its legality to also meet the requirement of world trade law. There is more leeway for individually agreed tariff preferences in compliance with regulations than at the World Trade Organisation (WTO) level. For this reason, various actors, including those from the supplier regions, should urgently be involved in the formulation of the details. Only in this way can the special difficulties be identified and eliminated, for example through development policy measures, so that sustainability can actually be achieved. After all, an overall strategy made up of classic and expanded approaches to regulating supply chains would be flexible enough to react

[84] De Schutter (2014).

[85] UNEP & WTO (2018), p. 31.

[86] Larbodière et al. (2020), p. 68.

[87] *Ibid* 81.

to changes in supply chains. In principle, the current direction of supply chains can be reversed in the future. Today's suppliers could become processors at the end of supply chains with greater added value—which could have positive development impacts for currently supplying developing countries.[88]

So, for example, in a letter, India and South Africa requested the WTO for a waiver from certain provisions of the Agreement on Trade-Related Aspects of Intellectual Property Rights (TRIPS Agreement) for the prevention, containment and treatment of COVID-19. The letter addressing the WTO Council for Trade-Related Aspects of Intellectual Property Rights on 2 October 2020 states:

> The WTO has cautioned that the "Pandemic represents an unprecedented disruption to the global economy and world trade, as production and consumption are scaled back across the globe". We have witnessed a break down in global supply chains coupled with growing supply-demand gaps.[89]

6 Global Agricultural Commodities Trade

Agricultural commodities trade is subject to changes, reflecting the uneven and disproportionate impact of climate change on agricultural sectors across the globe. International trade can play an important role in adaptation efforts, and thus contributing to soil protection and food security in many countries.

The fact that the WTO Agreement on the Application of Sanitary and Phytosanitary Measures (SPS Agreement) makes reference to Codex food safety standards, means that the Codex Alimentarius for international food standards also has far reaching implications for international food trade. WTO members that wish to apply stricter food safety measures than those set by Codex may be required to justify these measures scientifically. The Codex standards are based on the assumptions and decisions of the Codex Alimentarius Commission, a joint body of the Food and Agriculture Organisation (FAO) and the World Health Organisation (WHO) of the United Nations and serve in many cases as a basis for national legislation.[90]

International food trade has existed for thousands of years but traditionally food was mainly produced, sold and consumed locally. Over the last century the amount of food traded internationally has grown exponentially, and a quantity and variety of food never before possible, with billions of tonnes of food produced, marketed and transported globally.[91] By for instance moving food from surplus to deficit areas, trade can not only provide a mechanism to address production shortfalls due to extreme weather but can also contribute towards reducing greenhouse gas (GHG)

---

[88] Rudloff and Wieck (2020).

[89] Cf. https://docs.wto.org/dol2fe/Pages/SS/directdoc.aspx?filename=q:/IP/C/W669.pdf&Open=True accessed 07 June 2021.

[90] Cf. http://www.fao.org/fao-who-codexalimentarius/en/ accessed 07 June 2021.

[91] *Ibid.*

emissions and adjusting agricultural production in a more efficient and local manner. This, in turn, could also be in the interest of soil protection and LDN.

While the work to translate the Paris Agreement and the NDCs into concrete climate interventions in agriculture is in progress, a wide range of policy instruments is available, from investments in innovative technologies to subsidies that provide incentives to farmers to adopt climate-smart agricultural practices, and regulations to reduce emissions of agricultural activities to carbon taxes.[92] Most of these policy instruments are covered by agreements under the WTO, which is the forum for governments where international trade agreements are negotiated.

6.1 The World Trade Organisation

The WTO provides a system of trade rules covering goods, services and intellectual property, as well as a legal and institutional framework for the implementation and monitoring of these agreements, and a venue for settling disputes arising from the interpretation and application of WTO agreements. Administering WTO trade agreements, monitoring national trade policies, providing technical assistance and training for developing countries, and cooperating with other international organisations are further functions of the WTO.[93]

The WTO's founding and guiding principles remain the pursuit of open borders, the guarantee of the most-favoured-nation principle and non-discriminatory treatment by and among members, and a commitment to transparency in the conduct of its activities. The opening of national markets to international trade, with justifiable exceptions or with adequate flexibilities, can encourage and contribute to sustainable development, raise people's welfare, reduce poverty, and foster peace and stability. At the same time, the liberalisation of markets must be accompanied by sound domestic and international policies which contribute to economic growth and development according to each member's needs and aspirations.[94]

Although the WTO is primarily concerned with reducing trade barriers and eliminating discriminatory treatment in international trade, world trade law is increasingly framed by the concept of sustainable development. The agreement establishing the WTO (unlike the General Agreement on Tariffs and Trade (GATT)) has anchored the objective of sustainable development and the need to protect and preserve the environment within its Preamble:

> Recognizing that their relations in the field of trade and economic endeavour should be conducted with a view to raising standards of living, ensuring full employment and a large and steadily growing volume of real income and effective demand, and expanding the production of and trade in goods and services, while allowing for the optimal use of the

[92] FAO (2018).
[93] WTO (1994).
[94] *Ibid.*

> world's resources in accordance with the objective of sustainable development, seeking both to protect and preserve the environment and to enhance the means for doing so in a manner consistent with their respective needs and concerns at different levels of economic development.

Although this statement in the Preamble is more of a policy goal than a binding principle, it should have significant weight in decision-making and dispute resolution and can make an important difference to the agreement's operation in practice. Yet, the WTO is, of course, not an environmental protection agency. So far, its competence in the field of trade and environment is limited to trade policies and to the trade-related aspects of environmental policies that have a significant effect on trade. In addressing the link between trade and environment, the two fields should increasingly complement each other.[95] How can the WTO trading system help with the implementation of the Paris Climate Agreement, mitigate climate change, enhance soil protection and contribute to food security?[96] The WTO has tremendous potential to contribute to decarbonisation and, relatedly, has significant potential to help mitigate climate change.[97] This hypothesis raises the question: How can progressive trade liberalisation be reconciled with the protection of non-economic interests where the trading system can contribute to mitigating climate change, shifting from trade as a major cause of environmental harm to trade as a tool for environmental protection?[98] And what does this mean for the protection of soil and the promotion of food security in the context of climate change, which is likely to affect agricultural production more and more across various sectors?

6.2 Trade in Agriculture Under the WTO

In the decades following the Second World War, both the United States and nations of Western Europe provided generous subsidies to their agricultural producers and imposed both tariff and non-tariff import barriers to protect these producers from foreign competition. The 1947 General Agreement on Tariffs and Trade (GATT) generally exempted agriculture from the GATT's trade liberalisation obligations:[99]

> Trade in agriculture has been distorted by subsidies and protectionism to the detriment of both producers and consumers. Trade in agricultural products at the same time contributes to global food security by helping countries to obtain food supplies from world markets. Agricultural imports can be risky if they crowd out more expensive local production. This can have negative income effects for producers and thereby continuously weaken local agriculture. In the case of acute supply bottlenecks, such effects can often only be quickly

[95] Ruppel (2018).

[96] Leal-Arcas and Morelli (2018).

[97] *Ibid* 6.

[98] Leal-Arcas and Morelli (2018), p. 29.

[99] Gonzalez (2014), p. 105.

remedied by imports, provided that enough food is available on the world market at affordable prices.[100]

Since the start of the *Brexit* negotiations, the agri-food sector suffered under a lack of certainty regarding the future relationship between the European Union (EU) and the United Kingdom (UK). Existing supply chains and trade flows for agricultural goods and food products, within the EU but also with respect to imports from and export to third countries, suggest a significant challenge for farmers and food businesses in the UK, in Ireland, across the EU and around the world. Issues of relevance range from market access to plant protection, food safety, and food and quality labelling.[101]

International trade of food commodities induces a virtual transfer of embodied land, carbon, and other land based resources, while most of the environmental impacts of agricultural production remain in the producing countries the role of trade in food security is expected to increase due to climate change, population growth and changing diets.[102] The causes of, and contributing factors to, global food insecurity are numerous and complex. It is clear, however, that the WTO and international trading rules play an important role in the pursuit of global food security.[103]

GATT Article XXI forms a controversial WTO provision recognising certain flexibilities for states in the international trading system, permitting ordinarily trade-restrictive measures for the purpose of national security. Article XXI(b)(iii) on "security exceptions" states that nothing in the GATT must be construed to prevent any WTO Member "from taking any action which it considers necessary for the protection of its essential security interests" in times of "emergency in international relations". This provision can justify certain trade restrictions introduced in pursuit of certain political objectives.

In 2019 in the case of *Russia — Measures Concerning Traffic in Transit* the WTO Dispute Settlement Panel found that "essential security interests" could be generally understood as referring to those interests relating to the quintessential functions of the state. The Panel observed that the specific interests at issue will depend on the particular situation and perceptions of the state in question and can be expected to vary with changing circumstances. For these reasons, the Panel held that it is left in general to every Member to define what it considers to be its essential security interests.[104]

According to Article XI of the GATT, supply risks explicitly justify otherwise prohibited trade restrictions and bans for food. Since trigger criteria and deadlines are not regulated, export bans are implemented rapidly, which in principle drives prices up and results in supply risks for other import-dependent countries. On the import side, protective tariffs can seal off sectors in particularly threatening

[100] Rudloff and Wieck (2020).

[101] Fratini Vergano European Lawyers (2018).

[102] Zhou et al. (2020).

[103] Stewart and Manaker Bell (2015).

[104] WTO (2019).

situations, as is often the case for reasons of supply to stimulate production. In bilateral agreements, the weaker partners often condemn this protection option as too restrictive. At the same time, caution should be given against premature isolation, as it often makes sense to secure supplies through less expensive imports. In principle, the WTO complicates such incentives for specifications on production processes that do not lead to physical product differences, as is usually the case when considering sustainability.[105]

Article XI GATT has been violated in the context of a number of environmental disputes in which countries have imposed bans on the importation of certain products; it therefore has relevance for trade and environment discussions. Article XX grants general exceptions from the aforementioned GATT rules. Article XX (b) lists measures necessary to protect human, animal or plant life and health; Article XX(g) lists measures relating to the conservation of exhaustible natural resources. WTO members may be exempted from GATT rules in specific instances. However, measures must be necessary (necessity-test). If the conditions set by Article XX are fulfilled, they must still pass the test of the introductory clause (Chapeau) of Article XX. According to the Chapeau, measures may not be pronounced as arbitrary and unjustifiable discrimination between countries where the same conditions prevail, and they may not constitute a disguised restriction on international trade. GATT rules provide significant scope for members to adopt national environmental protection policies.[106] GATT rules impose only one requirement in this respect—that of non-discrimination. WTO members are free to adopt national environmental protection policies provided that they do not discriminate between imported and domestically produced like products (NT principle), or between like products imported from different trading partners (MFN clause). Non-discrimination is one of the main principles on which the multilateral trading system is founded. It secures predictable access to markets, protects the economically weak from the more powerful, and guarantees consumer choice.[107]

Certification and appropriate, non-deceptive labelling in line with WTO rules, in particular the Agreement on Technical Barriers to Trade (TBT), can enable consumers to make sustainable food choices avoiding unjustified barriers to trade. International food-safety as well as plant and animal health standards, based on the SPS Agreement, are essential for reaping the benefits of agricultural trade and for avoiding potential risks to human, animal and plant health, while unjustified sanitary and phytosanitary restrictions on food trade can exacerbate food insecurity.[108]

In 2011, the World Health Organization (WHO) reported that several European countries have experienced outbreaks caused by a dangerous strain of Escherichia

[105] Rudloff and Wieck (2020).

[106] Ruppel (2018).

[107] On the trade and environment negotiations: Cf. https://www.wto.org/english/tratope/envire/envirnegotiationse.htm accessed 07 June 2021.

[108] Cf. https://www.gffa-berlin.de/wp-content/uploads/2020/02/GFFA-Communique-2020-EN.pdf accessed 07 June 2021.

coli (E. coli) bacteria, while some countries have also reported deaths related to the outbreaks. These E. coli are strains of the bacterium Escherichia coli that produce either Shiga or Shiga-like toxin while only a minority of these cause illness in humans. The ones that do are collectively known as enterohemorrhagic E. coli (EHEC) and are major causes of foodborne illness. Responses then have involved actions that implement and affect international legal regimes on public health and international trade imposing restrictions on international trade in terms of the Agreement on the Application of Sanitary and Phytosanitary Measures (SPS Agreement).[109] Following the detection of EHEC bacteria on Spanish cucumbers on 26 May 2011 (which, however, later proved to be of a different type than the E. coli associated with the outbreak), the competent ministries in various German federal states adopted regulations requiring traders to provide proof of safety for Spanish cucumbers, which was tantamount to a de facto import ban, leading the Spanish government to assert that it would seek compensation for its farmers, who lost sales costing millions of euros. On 1 June 2011, the EU Commission removed the warning label on Spanish cucumbers from the European rapid alert system, citing the latest investigation results. German courts then had to decide on the compensation for the damage suffered by Spanish cucumber suppliers. In turn, outside the EU, Russia imposed an import ban on 2 June 2011, on raw vegetables from all EU countries, prompting complaints from EU members that such a ban was not justified and was disproportionate. The EHEC outbreaks have triggered various trade responses internationally, under EU law and the SPS Agreement. Ultimately they also underscored a problem with international legal obligations under the WHO and WTO regimes that seek to regulate trade-related responses to disease outbreaks.[110]

Article XX of the GATT 1994 states that measures "necessary to protect human, animal or plant life or health" (b) and those "relating to the conservation of exhaustible natural resources [...]" (g) can be interpreted as a legally accepted exception. While this could be most relevant for the protection of soil, a typical measure that can fall under this exception may be requiring export countries to comply with certain policies prescribed by the importing country.[111]

Further exceptions in accordance with Article XX of the GATT could possibly also be used to justify border carbon adjustment (BCA) measures as a tool for addressing carbon leakage. Such measures could, for instance, be the inclusion of certain imported goods in a Carbon Emissions Trading Scheme (ETS), a customs duty, or a border tax.

> To adhere to WTO principles of non-discrimination, countries cannot ask for more or different compliance from importers than they ask of their own firms producing comparable products. That means that only price-based climate policies can be associated with a price at the border. A domestic carbon tax can be complemented by a border tax.[112]

[109] Fidler (2011).

[110] Fidler (2011).

[111] Van den Bossche and Zdouc (2017).

[112] Droege and Fischer (2020).

The Paris Agreement does not explicitly state but implies counteracting like products and services with a higher footprint, which can take place in several different ways, and not necessarily through discrimination against only foreign goods (MFN/NT).[113] The Paris Agreement does not prescribe border carbon adjustment measures. Whether a carbon tax yields a better result for global food security than carbon sequestration remains open for discussion beyond the scope of this article. Yet, the rapidly increasing food import volumes or price decreases may legitimise safeguarding action by countries which have had to transform their non-tariff barriers (NTBs) into tariffs. Basically, however, rules and limits apply to four categories of protection and support policies. Border protection should be limited to tariffs. The maximum rates (bound/scheduled) should not be increased without compensation as per Article XXVIII of the GATT. Import quotas are prohibited under Article XI of the GATT. A time-limited border protection is available against imports threatening or jeopardising local production, which are generally available safeguards under Article XIX of the GATT.[114] In addition, in Article XXIV(5) of the GATT, WTO members may exclude customs unions and bilateral or regional free-trade areas from compliance with WTO disciplines in certain circumstances. These regional agreements are important, as they establish disciplines which might affect both the adoption of domestic and international carbon rules and measures to promote sustainable development and environmental cooperation.[115]

6.3 *The WTO Agreements on Agriculture, Subsidies and Countervailing Measures*

The Agreement on Agriculture (AoA) entered into force at the time of the inception of the WTO on 1 January 1995. In principle, all WTO rules on trade in goods apply to agriculture. These rules include, inter alia, the GATT and pacts such as those dealing with sanitary and phytosanitary measures, customs valuation, import licensing, pre-shipment inspection, safeguarding measures, subsidies in general, and various standards, regulations and labelling requirements that imports have to meet (known as "technical barriers to trade"). The AoA was negotiated in the Uruguay Round (1986–1994) and was a significant step towards fairer competition and a less distorted sector. WTO member governments agreed to improve market access and reduce trade-distorting subsidies in agriculture. The AoA seeks to reform trade in agricultural products and provides the basis for market-oriented policies. In its Preamble, the Agreement reiterates the commitment of members to reform agriculture in a manner which protects the environment. Under the Agreement, domestic

[113] Häberli (2018), p. 20.

[114] *Ibid* 8.

[115] Gehring and Hepburn (2013).

support measures with minimal impact on trade (known as green box policies) are excluded from reduction commitments (contained in Annex 2 of the Agreement).[116]

The AoA primarily covers three aspects which need to be adapted to improve international agricultural trade, namely market access, export competition and domestic support. Market access is set out in Articles 4 and 5 of the AoA, and requires member states to convert their non-tariff barriers into tariffs and then to reduce those tariffs to improve agricultural trade market transparency and to strengthen the connection between domestic and international agricultural markets. The AoA also highlights the need for stricter regulation of domestic support measures under Articles 3, 6 and 7 to avoid their use for protectionist strategies which promote unfair competition, and categorises domestic agricultural support measures into 3 boxes according to the level of their trade-distorting effect, namely amber box, blue box and green box measures. Exemptions for reductions in support measures include green box subsidies which are considered minimal or non-trade distorting and include support for public stockholding for food security purposes and domestic food aid, as well as development measures which assist support of agricultural and rural development objectives. Export competition, as set out in articles 9 and 10 of the Agreement on Agriculture, required member states to make reduction commitments on their export subsidies. Article 20 recognises the importance of taking into account non-trade concerns and special and differential treatment for developing country members, resulting in many developing countries, through negotiating groups bringing forward proposed amendments to the AoA on the elimination of export subsidies, the use of public stockholding in the context of food security purposes and trade remedies such as special safeguard mechanisms.

The AoA in Article 21(1) stipulates that the GATT and all other WTO agreements on trade in goods (officially Annex 1A of the Marrakesh Agreement establishing the WTO) apply but if there is a conflict, then the rules in the Agriculture Agreement prevail.[117] While the “AoA professed to ameliorate the double standards in global agricultural trade”, it has been said that it

> was riddled with ambiguities that enabled wealthy countries to continue to subsidize their agricultural producers while requiring market openness in developing countries. Since most developing countries had already liberalised their markets pursuant to structural adjustment programs, the impact of the AoA was to preclude these countries from adopting these subsidies in the future beyond de minimis levels. Agricultural subsidies in the United States and European Union, however, actually increased in the aftermath of the AoA.[118]

In terms of agricultural product subsidies there is no outright prohibition, but because they are considered to distort trade, they are limited for all WTO members. The conditions for unlimited governmental programmes are narrowly defined. The Developing Country Green Box (Article 6(2) AoA) allows, for instance, certain

[116] Cf. http://www.fao.org/climate-change/our-work/what-we-do/koronivia/en/ accessed 07 June 2021.

[117] WTO (2015).

[118] Gonzalez (2014), p. 106.

credit schemes and subsidies, for example for irrigation construction, and even for the running costs of low-income and resource-poor producers. Article 6(2) provides in relevant parts that measures which are "an integral part of the development programmes of developing countries [. . .] shall be exempt from domestic support reduction commitments that would otherwise be applicable to such measures". These are "*investment subsidies* which are generally available to agriculture in developing country Members and *agricultural input subsidies* generally available to low-income or resource-poor producers in developing country Members".[119]

Nevertheless, there is still a need to update global trade rules to reflect market and policy shifts that have occurred in recent years and to address contemporary agricultural and food challenges in reducing trade-distorting agricultural support of the past.[120] This does not come as a surprise, as the AoA has given rise to a relatively large number of disputes reflecting the fact that agriculture is a sensitive sector in many member countries. In its 2019 recent panel report DS511 on China – Domestic Support for Agricultural Producers the DSB found that China was not in compliance with its domestic support commitments pursuant to Articles 3(2) and 6(3) of the Agreement on Agriculture after the United States contended that China has provided market price support to its agricultural producers of wheat and rice in excess of its commitments under the AoA.[121]

The agriculture, forestry, and other land use (AFOLU) sector is an important sector that services national food requirements and export earnings for many developing countries around the world. It is unique in the sense that it is the only sector within which both sources and sinks for greenhouse gases can be found. AFOLU plays a central role in food security, sustainable development and climate change mitigation and adaptation and could also be considered as a valid motive under Article 6(2) AoA through measures that do not distort trade. Effective climate-smart support to farmers can also improve the comparative advantage of agriculture in countries that will be negatively affected by changing climate, allowing them to become competitive and achieve a better balance in export and import performance.[122]

Further relevant provisions for trade in agricultural products are found in the WTO Agreement on Subsidies and Countervailing Measures (SCM). SCM exerts discipline over the use of subsidies and regulates the actions that countries can take to counter the effects of subsidies. Under the agreement, a country may use the WTO's dispute-settlement procedure to seek the withdrawal of the subsidy or the removal of its adverse effects. Alternatively, a country can launch its own investigation and ultimately charge extra duty (countervailing duty) on subsidised imports

---

[119] Häberli (2018), p. 9.

[120] Cf. https://www.gffa-berlin.de/wp-content/uploads/2020/02/GFFA-Communique-2020-EN.pdf accessed 07 June 2021.

[121] Cf. https://docs.wto.org/dol2fe/Pages/SS/directdoc.aspx?filename=q:/WT/DS/511R.pdf&Open=True accessed 07 June 2021.

[122] FAO (2018).

found to be detrimental to domestic producers.[123] In line with Article 13 AoA, the SCM agreement now also applies to agricultural export (and import displacement) measures. Although export subsidies—a long-term concern of many competitive agricultural product exporters—were finally prohibited in 2015, there still is no agreement on the implementation details (e.g. schedule changes) nor on the rules tightening mandated for all export competition measures under the Doha Development Agenda (DDA). This failure is also reflected in the stalling reform process under Article 20 of the AoA to agree on additional disciplines making trade patterns more sustainable, more resilient under a climate change perspective.

> The key for an economic impact assessment of agricultural subsidies in a climate perspective would probably be the contribution of a differentiating subsidy under the Paris Agreement. Here again, not all countries are equal. Some temperate climate countries may actually benefit from global warming, with little or no justification for a subsidy. For countries located closer to the Equator, adaptation subsidies and Official Development Assistance (ODA) might find economic justification especially for farmers without meaningful support from their governments.[124]

This could contribute to global efforts to control atmospheric greenhouse gas concentrations, foster AFOLU-related mitigation pathways and at the same time lead to improved soil conditions. In this regard, trade could become more central in climate change mitigation efforts, and this would also benefit soil protection and LDN. If trade could provide the necessary signals to farmers to produce low carbon footprint products, emissions could be reduced globally. In practice, this would necessitate the imposition of a carbon tax (or an equivalent mitigation measure) on agricultural products domestically, combined with a corresponding tariff adjustment at the border to discriminate against high carbon footprint imports.

WTO provisions offer flexibility for waivers or exemptions from complying with the non-discrimination principle. While sufficient space for policy discussions needs to be pursued at the intersection of the WTO and the Paris Agreement, the principle of differentiated responsibilities, respective capabilities, and the special and differential treatment of developing countries remain ever relevant when discussing and implementing transformative policies for climate change adaptation and mitigation to make agriculture meet contemporary challenges.[125]

7 Conclusion

COVID-19 has propelled the discourse on the resilience of societies, also with regard to more long-term crisis phenomena such as climate change and environmental degradation. It further

---

[123] Cf. http://www.wto.org/english/tratope/scme/scme.htm, accessed 07 June 2021.

[124] Häberli (2018), p. 25.

[125] FAO (2018).

> revealed the global interdependence between society, nature and economic activity. Their containment requires drastic measures and, as a consequence, new fundamental questions arise about the values of economic activity, the mutual responsibility between individuals but also between societies, and the relationship between humans and nature. The corona pandemic underlines the vulnerability of economic and social (sub)systems as a result of globalisation.[126]

The WHO emphasises that social and environmental processes are stressing the world's arable lands and pastures essential for the provision of food, water and quality air. Land degradation and desertification can affect human health through complex pathways.[127] The fact, that the Norwegian Nobel Committee has decided to award the 2020 Nobel Peace Prize to the World Food Programme (WFP) for its efforts to combat hunger is a clear reflection of the growing significance of food security in our time. The WFP is the world's largest humanitarian organisation addressing hunger and promoting food security and the primary instrument of the United Nations for realising this goal. In the face of the COVID-19 pandemic, the WFP itself has stated, that "until the day we have a medical vaccine, food is the best vaccine against chaos". The Norwegian Nobel Committee emphasised that providing assistance to increase food security not only prevents hunger but can also help to improve prospects for stability and peace in the world.[128]

Ending hunger and achieving food security is at the heart of the SDGs. At the same time, climate change is affecting agriculture and food security and will make the challenge even more difficult. In fact, food insecurity and climate change already undermine basic human rights of entire populations. While these rights may eventually be enforced through litigation, they have yet to be explicitly recognised in the context of increasing climate-related litigation efforts[129]—it is merely a question of time.[130]

Politically, COVID-19 exposed a new "wealth of social conflicts". They concern all the major issues of the day, from the digitalisation of education, to supply chains in the age of globalisation and the false promises of the "technological paradigm" (Pope Francis: Laudato si').[131] Food security as a primary justice concern connected

---

[126] Ginzky et al. (2020).

[127] Cf. https://www.who.int/globalchange/ecosystems/desert/en/ accessed 07 June 2021.

[128] Cf. https://www.nobelprize.org/prizes/peace/2020/press-release/ accessed 07 June 2021.

[129] On 17 February 2020 a climate relevant judgment was handed down by the Western Cape High Court in *Philippi Horticultural Area Food and Farming Campaign and Another v MEC for Local Government, Environmental Affairs and Development Planning: Western Cape and 12 Others* (*Philippi case*) (16779/17) [2020] ZAWCHC 8; 2020 (3) SA 486 (WCC). In this case, climate change was recognised as a relevant factor to be considered when decision-makers were called upon to approve developments that might detrimentally impact upon the environment. The decision confirms the South African courts' recognition of climate change, as a relevant factor to be considered when approving developments. Relevant considerations, according to the court, include the potential impacts of the development on inter alia food security, climate change and land reform.

[130] Ruppel (Forthcoming 2021c).

[131] Mertes (2020), pp. 721–722.

with climate change must therefore also be viewed in the context of food production and distribution, where producers and consumers are located on different continents.

> COVID-19 has the potential to significantly disrupt both food supply and demand while agricultural production, food prices and food availability will also be negatively impacted.[132]

In fact, a better understanding of food security must go beyond a developmental or humanitarian understanding thereof, it must even include linkages with geopolitics.[133]

As the historical impact of global supply chains on nature has been largely negative, characterised by unsustainable practices in agriculture and other sectors. What must, however, be prevented is an increasing securitisation of trade where countries put up trade barriers on just about everything under the pretext of security.[134] Instead of disrupting global value chains, a shift toward more responsible supply chain management practices offers an opportunity to avoid harm and even positively impact nature.[135]

> The design of the mitigation portfolios and policy instruments to limit warming to 1.5°C will largely determine the overall synergies and trade-offs between mitigation and sustainable development.[136]

New strategies for a better future need to be implemented as it becomes clear that the international soil governance framework remains highly fragmented, while the displayed international law instruments cover different aspects of soil protection in a relatively unconcerted manner. While a variety of institutions and instruments exist, more effective global governance mechanisms and coordination efforts are needed to integrate also the circular bioeconomy concept into the existing international conventions for climate, biodiversity and desertification.[137] Whereas misguided aid, trade and development policies, as well as large-scale land acquisitions that benefit wealthy nations and transnational corporations are often conducted at the expense of the poor,[138] they are also most vulnerable and threatened by climate change, which depresses food production and increases food prices.[139]

International trade, while essential for food security, also creates vulnerabilities through supply disruptions, growing unilateralism and competition over agricultural resources that can be both a cause and a consequence of geopolitical rivalry.[140] The WTO is a crucial institution for the governance of international trade, yet it has been

---

[132] United Nations (2020).

[133] Zhou et al. (2020).

[134] Narlikar (2020).

[135] Deutz et al. (2020).

[136] IPCC (2018).

[137] Palahí et al. (2020), p. 43.

[138] Gonzalez (2014), p. 104.

[139] Niang et al. (2014).

[140] Zhou et al. (2020).

characterised by frequent deadlocks in the past and has suffered from credibility loss due to the persistent failure of the Doha Development Agenda. The WTO is now subject to ongoing trade war dynamics and a dysfunctional appellate body, all of which further exacerbates the need for reforms.[141]

Moreover, some countries have been hoarding food items to ensure supplies for their population in light of the COVID-19 pandemic. Such grain exporters, including Russia, the Ukraine and Kazakhstan, have been said to take a nationalist turn by restricting or planning to restrict exports to ensure enough supplies for their own populations. The is an alarming example on how food availability can easily be threatened in a trade system that encourages import dependence and export-oriented agriculture, but cannot require countries to export food, which could be detrimental to countries that depend on imported food.[142] Trade wars are a huge threat to food security.[143] Further examples are Russia's ban on Western food imports since 2014 and the ongoing trade war between China and the USA that began in 2018 disrupting normal agricultural flows. These geopolitical frictions hamper reform consensus to revive multilateral institutions, including the WTO,[144] which should be empowered beyond the trade effects of trade.[145]

Countries should increase efforts through the international architecture, specifically the WTO, to develop green trade agreements that facilitate and incentivise increased trade in commodities produced without conversion of natural habitats. While subsidies are, for the most part, deployed within the country granting the subsidies and can only be reformed through the actions of domestic governments, reforming harmful subsidies still requires an international effort. International organisations can facilitate changing the status quo on subsidies reform and encourage governments to cooperate on ways to implement change.[146]

On 18 January 2020 at the 12th Berlin Agriculture Ministers' Conference on the occasion of the Global Forum for Food and Agriculture (GFFA) a Communiqué was developed with some very useful considerations complementing in summary the conclusion of this article: The Communiqué inter alia recommends that trade policies should form part of a larger integrated and coherent strategy towards improved sustainability of agriculture and food systems. Fostering trade for global food security requires strengthening trade rules and transparent and inclusive local, regional and global value chains. Spill-over effects from national policies on other countries are necessary. To harness the benefits of trade for sustainable agricultural development and to create new economic opportunities, sustainable farming practices are important, just as promoting risk management tools and strengthening land tenure rights among others. While every country must have the right to develop its

[141] Narlikar (2020).
[142] Pandey (2020).
[143] Cf. https://unctad.org/news/trade-wars-are-huge-threats-food-security accessed 07 June 2021.
[144] Zhou et al. (2020).
[145] Messenger (2017), p. 86.
[146] Deutz et al. (2020), p. 66.

own agricultural model to feed its population, respect for the needs of other countries and international obligations remains key. Policies must therefore assure that trade can meet global challenges, facilitates the sustainable and efficient use of land, protects biodiversity and prevents overexploitation and degradation of land and natural resources. In particular, nationally appropriate measures to conserve natural resources and combat climate change that are respectful of international commitments related to sustainable development, e.g. the Paris Agreement on Climate Change, the Rio Declaration on Environment and Development and the Convention on Biological Diversity.[147]

WTO reform to better accommodate climate change measures is an increasingly urgent issue. Such reform could entail legal changes, namely amending the WTO agreements to accommodate climate change measures; introducing a waiver that temporarily relieves WTO members from their legal obligations under the WTO agreements when pursuing climate action; adopting an authoritative interpretation clarifying the scope of WTO rules in relation to climate policies; and introducing a time-limited peace clause pursuant to which WTO members will not challenge the climate policies of other members. Such changes would, however, involve complex political processes that—for a variety of reasons—would be difficult to implement in practice.[148]

In the meantime, existing flexibilities under current WTO law should be utilised to advance climate action, while it is not unlikely that conflicts between the trade and climate regimes will sooner or later surface in the WTO's dispute settlement system.[149] It has been rightfully stated that international courts and tribunals must become the new environmental sentinels in international law. In the interest of global soil protection and for the sake of sustainable food security, the challenge will be to bridge the gap where measures claiming to implement the Paris mitigation commitments collide with present trade rules. This will require commitment to overcome substantial barriers at various institutional (and conceptual) levels as well as adequate and corresponding regulatory frameworks.[150] With more ambitious NDCs expected in the future, countries can take trade-related climate measures that are likely to assume increasing importance.[151]

The fact is that the climate protection goals of the Paris Agreement can only be reached if, in addition to the decarbonisation of the global economy, more areas of land are used to extract carbon dioxide (CO_2) from the atmosphere.[152]

---

[147] Cf. https://www.gffa-berlin.de/wp-content/uploads/2020/02/GFFA-Communique-2020-EN.pdf accessed 07 June 2021.

[148] Häberli (2018).

[149] Kasturi et al. (2019).

[150] For more information on the relevance of investment, see Ruppel and Shifotoka (2017) and Ruppel and Borgmeyer (2018).

[151] Kasturi et al. (2018), p. 6.

[152] WBGU – German Advisory Council on Global Change (2020), p. 8.

> Land is a global commons: humankind must accept and assume its responsibility for land in order to mitigate climate change, conserve biodiversity and safeguard food security; it must discharge this responsibility nationally and enforce it internationally.[153]

And soil as a natural system is essentially a capital stock (similar to financial, built, and human capital) that provides a flow of services to people. These 'ecosystem services', which include fertile soil must be considered in the context of regenerative or conservation agriculture, yielding a public good.[154] In this light, greater efforts are needed to combine economic and environmental performance in "determining soil as a natural capital and an asset that needs a price tag even though it is actually priceless".[155]

References

Alori E, Nwapi C (2015) The international legal regime for sustainable soil. In: Ako R, Olawuyi D (eds) Food and agricultural law: readings on sustainable agriculture and the law in Nigeria. Afe Babalola University Press, Ado-Ekiti

Bodle R, Stockhaus H, Wolff F, Scherf CS, Oberthür S (2020) Improving international soil governance – analysis and recommendations. German Environment Agency, Dessau-Roßlau

De Schutter O (2014) Report of the Special Rapporteur on the Right to Food: United Nations General Assembly, A/HRC/25/57. http://www.ohchr.org/EN/HRBodies/HRC/RegularSessions/Session25/Documents/A_HRC_25_57_ENG.DOC. Accessed 07 June 2021

Deutz A, Heal GM, Niu R, Swanson E, Townshend T, Zhu L, Delmar A, Meghji A, Sethi SA, Tobin- de la Puente J (2020) Financing nature: closing the global biodiversity financing gap. The Paulson Institute, The Nature Conservancy, and the Cornell Atkinson Center for Sustainability, Chicago

Droege S, Fischer C (2020) Pricing carbon at the border: key questions for the EU. ifo DICE Report 18, 30–34

European Commission (2020) Roadmap: new soil strategy - healthy soil for a healthy life. Ref. Ares (2020) 6391319 - 05/11/2020. https://ec.europa.eu/info/law/better-regulation/have-your-say/initiatives/12634-New-EU-Soil-Strategy-healthy-soil-for-a-healthy-life. Accessed 12 Nov 2020

FAO / Food and Agriculture Organization of the United Nations (1982) World Soil Charter. http://www.fao.org/fileadmin/user_upload/GSP/docs/WSCharter/World_Soil_Charter.pdf. Accessed 07 June 2021

FAO / Food and Agriculture Organization of the United Nations (2015a) Revised Soil Charter. http://www.fao.org/3/a-i4965e.pdf. Accessed 05 June 2021

FAO / Food and Agriculture Organization of the United Nations (2015b) Global Soil Partnership. http://www.fao.org/globalsoilpartnership/iys-2015/en/. Accessed 07 June 2021

FAO / Food and Agriculture Organization of the United Nations (2015c) After the International Year of Soils, we won't take soils for granted anymore. http://www.fao.org/soils-2015/news/news-detail/en/c/353737/. Accessed 07 June 2021

FAO / Food and Agriculture Organization of the United Nations (2018) The state of agricultural commodity markets: agricultural trade, climate change and food security. FAO, Rome

[153] *Ibid.*

[154] Deutz et al. (2020), p. 192.

[155] Ruppel (2021b).

FAO / Food and Agriculture Organization of the United Nations (2019) Global Soil Partnership Plenary Assembly. GSPPA.VII/2019/4. http://www.fao.org/3/ca4769en/ca4769en.pdf. Accessed 07 June 2021
FAO / Food and Agriculture Organization of the United Nations (2020a) Climate change: unpacking the burden on food safety. Food safety and quality series, No. 8. http://www.fao.org/3/ca8185en/CA8185EN.pdf. Accessed 12 Nov 2020
FAO / Food and Agriculture Organization of the United Nations (2020b) Contribute to the GSP newest tool: SoiLEX. http://www.fao.org/global-soil-partnership/resources/highlights/detail/en/c/1274929/. Accessed 07 June 2021
FAO / Food and Agriculture Organization of the United Nations (2020c) Legislative approaches to sustainable agriculture and natural resources governance. FAO Legislative Study No. 114. https://www.unenvironment.org/resources/publication/legislative-approaches-sustainable-agriculture-and-natural-resources. Accessed 07 June 2021
Fidler DP (2011) International law and the E. coli outbreaks in Europe. Insights: American Society of International Law 15(14). https://www.asil.org/insights/volume/15/issue/14/international-law-and-e-coli-outbreaks-europe. Accessed 07 June 2021
Fratini Vergano European Lawyers (2018) From trade perspectives 18 of 5 October 2018. http://www.fratinivergano.eu/static/upload/1/1/18.10_.05_TP_Issue_18-2018_.pdf. Accessed 07 June 2021
Geden O, Schenuit F (2020) Unconventional mitigation carbon dioxide removal as a new approach in EU climate policy. SWP Research Paper No. 8. German Institute for International and Security Affairs, Berlin
Gehring MW, Hepburn J (2013) Climate, trade and investment law in the global green economy. In: Ruppel OC, Roschmann C, Ruppel-Schlichting K (eds) Climate change: international law and global governance volume I: legal responses and global responsibility. Nomos, Baden-Baden
Ginzky H, Löwe C, Neßhöver C (2020) Lessons from the corona crisis: new guiding principles required for environmental and sustainability policy? – A discussion paper. German Environment Agency. https://www.umweltbundesamt.de/sites/default/files/medien/1410/publikationen/pp_neueleitmotive_final_bf.pdf. Accessed 30 Oct 2020
Global Environment Facility (2019) As per the draft of the 2019 Guidelines for Land Degradation Neutrality by the Global Environmental Facility. https://www.thegef.org/sites/default/files/council-meeting-documents/EN_GEF.STAP_.C.56.Inf_.02_Guidelines%20for%20Land%20Degradation%20Neutrality.pdf. Accessed 07 June 2021
Gonzalez CG (2014) World poverty and food insecurity. Pa State J Law Int Aff 3(2):100–126
Häberli C (2018) Potential conflicts between agricultural trade rules and climate change treaty commitments. The State of Agricultural Commodity Markets (SOCO) 2018. Background Paper. FAO, Rome
Heinrich Böll Foundation and Institute for Advanced Sustainability Studies (2015) Soil Atlas 2015, Facts and Figures about Earth, Land and Fields. Heinrich Böll Foundation and the Institute for Advanced Sustainability Studies, Berlin
Hurlbert M, Krishnaswamy J, Davin E, Johnson FX, Mena CF, Morton J, Myeong S, Viner D, Warner K, Wreford A, Zakieldeen S, Zommers Z (2019) Risk management and decision making in relation to sustainable development. In: IPCC climate change and land: an IPCC special report on climate change, desertification, land degradation, sustainable land management, food security, and greenhouse gas fluxes in terrestrial ecosystems. Cambridge University Press, Cambridge
IPCC / Intergovernmental Panel on Climate Change (2018) Summary for policymakers. In: Global warming of 1.5°C. An IPCC special report on the impacts of global warming of 1.5°C above pre-industrial levels and related global greenhouse gas emission pathways, in the context of strengthening the global response to the threat of climate change, sustainable development, and efforts to eradicate poverty. Cambridge University Press, Cambridge
Kanianska R (2016) Agriculture and its impact on land-use, environmental, and ecosystem services. Matej Bel University, Slovakia

Kasturi D, van Asselt H, Droege S, Mehling M (2018) Making the international trade system work for climate change: assessing the options. Climate Strategies, London

Kasturi D, van Asselt H, Droege S, Mehling M (2019) Towards a trade regime that works for the Paris Agreement. Economic and Political Weekly, December 21 2019, 25–30

Larbodière L, Davies J, Schmidt R, Magero C, Vidal A, Arroyo Schnell A, Bucher P, Maginnis S, Cox N, Hasinger O, Abhilash PC, Conner N, Westerburg V, Costa L (2020) Common ground: restoring land health for sustainable agriculture. IUCN, Gland, Switzerland

Lawrence P, Reder M (2019) Equity and the Paris Agreement: legal and philosophical perspectives. J Environ Law 31(3):511–531

Leal-Arcas R, Morelli A (2018) The resilience of the Paris Agreement: negotiating and implementing the climate regime. Georgetown Environ Law Rev 31(1):1–64

Mertes K (2020) Jenseits von richtig und falsch. Stimmen der Zeit 145:721–722

Messenger G (2017) Substantial development and the commodities challenge: the eventual greening of the World Trade Organisation. Trade Law Dev 54:54–88

Narlikar A (2020) Reforming the World Trade Organization. Heinrich Böll Stiftung, 17 January 2020. https://www.boell.de/en/2020/01/17/reforming-world-trade-organization. Accessed 07 June 2021

Niang I, Ruppel OC, Abdrabo MA, Essel A, Lennard C, Padgham J, Urquhart P (2014) Africa. In: Field CB, Barros VR, Dokken DJ, Mach KJ, Mastrandrea MD, Bilir TE, Chatterjee M, Ebi KL, Estrada YO, Genova RC, Girma B, Kissel ES, Levy AN, MacCracken S, Mastrandrea PR, White LL (eds) Climate change 2014: impacts, adaptation, and vulnerability. Part B: regional aspects. Contribution of Working Group II to the Fifth Assessment Report of the Intergovernmental Panel on Climate Change. Cambridge University Press, Cambridge

Palahí M, Pantsar M, Costanza R, Kubiszewski I, Potočnik J, Stuchtey M, Nasi R, Lovins H, Giovannini E, Fioramonti L, Dixson-Declève S, McGlade J, Pickett K, Wilkinson R, Holmgren J, Trebeck K, Wallis S, Ramage M, Berndes G, Akinnifesi FK, Ragnarsdóttir KV, Muys B, Safonov G, Nobre AD, Nobre C, Ibañez D, Wijkman A, Snape J, Bas L (2020) Investing in nature as the true engine of our economy: a 10-point action plan for a circular bioeconomy of wellbeing. Knowledge to Action 02. European Forest Institute, Joensuu

Pandey A (2020) Will coronavirus spark a wave of food nationalism?. https://www.dw.com/en/will-coronavirus-spark-a-wave-of-food-nationalism/a-52952081. Accessed 07 June 2021

Peters GP, Minx JC, Weber CL, Edenhofer O (2011) Growth in emissions transfers via international trade from 1990 to 2008. Proc Natl Acad Sci 108(21):8903–8908

Raschio G (2017) Land value chains global land outlook. UNCCD Working Paper. https://knowledge.unccd.int/sites/default/files/2018-06/7.%20Land%2BValue%2BChains__G_Raschio.pdf. Accessed 07 June 2021

Rogelj J, Shindell D, Jiang K, Fifita S, Forster P, Ginzburg V, Handa C, Kheshgi H, Kobayashi S, Kriegler E, Mundaca L, Séférian R, Vilariño MV (2018) Mitigation pathways compatible with 1.5°C in the context of sustainable development. In: IPCC global warming of 1.5°C. An IPCC special report on the impacts of global warming of 1.5°C above pre-industrial levels and related global greenhouse gas emission pathways, in the context of strengthening the global response to the threat of climate change, sustainable development, and efforts to eradicate poverty. Cambridge University Press, Cambridge

Rudloff B, Wieck C (2020) Nachhaltige Lieferketten im Agrarsektor: Wert schöpfen statt Zuliefern. Unternehmerpflichten politikfeldübergreifend in eine EU-Strategie einbinden. Stifung Wissenschaft und Politik-Aktuell 2020/A 70, September 2020. https://www.swp-berlin.org/10.18449/2020A70/. Accessed 07 June 2021

Ruppel OC (2018) International trade, environment and sustainable development. In: Ruppel OC, Kam Yogo ED (eds) Environmental law and policy in Cameroon - towards making Africa the tree of life. Nomos, Baden-Baden

Ruppel OC (2021a) FAO global soil partnership webinar on soil governance on 13 January 2021. http://www.fao.org/global-soil-partnership/resources/events/detail/en/c/1364883/. Accessed 07 June 2021

Ruppel OC (2021b) Soil protection and the right to food for a better common future. Int J Environ Policy Law 51(1–2):57–73

Ruppel OC (Forthcoming 2021c) Climate change, responsibility and liability in South Africa: the legal system, public and private law considerations. In: Hinteregger M, Kirchengast G, Meyer L, Schnedl G, Schulev-Steindl E, Steininger K (eds) Climate change, responsibility and liability. Nomos, Baden-Baden

Ruppel OC, Borgmeyer T (2018) The BRICS partnership from a South African perspective: sustainable development space in a new global governance. In: Ndulo M, Kayizzi-Mugerwa S (eds) Financing innovation and sustainable development in Africa. Cambridge Scholars Publishing, Newcastle

Ruppel OC, Ginzky H (eds) (2021) African soil protection law: law and constitution in Africa. Law and constitution in Africa series, vol 41. Nomos, Baden-Baden

Ruppel OC, Shifotoka F (2017) Foreign direct investment protection in Africa – contemporary legal aspects between BITS and BRICS. In: Yusf AA (ed) African yearbook of international law, vol 21, issue 1. Leiden, Brill

Sachs J, Schmidt-Traub G, Sofra E, Tozzi C, Maher AK, Antonelli M, Cordes KY, Cresti S, Espinosa G, Ocampo-Maya C, Riccaboni A, Rossi A, Sachs LE (2020) Fixing the business of food. How to align the agrifood sector with the SDGs. Barilla Foundation; UN Sustainable Development Solutions Network; Columbia Center on Sustainable Investment; Santa Chiara Lab University of Siena, Parma

Sandrey R (2019/2020) African agricultural land: past, present and future. In: Chidede T, Mbakhwa F, Moobi M, Phaleng L, Sandrey R, Wagner N, Viljoen W, Zosuliwe K (eds) Agricultural and food security in Africa. Trade Law Centre, Stellenbosch, South Africa

Stewart TP, Manaker Bell S (2015) Global hunger and the World Trade Organization: how the international trade rules address food security. Pa State J Law Int Aff 3(2):113–155

Tänzler D, Groß J, Li L, Warnecke C, Kurdziel MJ, Tewari R, Cames M, Healy S (2019) Analysing the interactions between new market mechanisms and emissions trading schemes: opportunities and prospects for countries to use Article 6 of the Paris Agreement. German Environment Agency, Dessau-Roßlau

Tlale MT (2018) A critical evaluation of the South African Land Tenure Policy: a comparison with selected aspects of the Kenyan and Tanzanian Law, PhD Thesis. North-West University, South Africa

UNEP / United Nations Environment Programme (1982) The World Soils Policy was adopted by UNEP. https://library.wur.nl/isric/fulltext/isricu_i34280_001.pdf. Accessed 07 June 2021

UNEP / United Nations Environment Programme (2019) Environmental Rule of Law: First Global Report. https://www.unenvironment.org/resources/assessment/environmental-rule-law-first-global-report. Accessed 07 June 2021

UNEP / United Nations Environment Programme & WTO / World Trade Organization (2018) Making trade work for the environment, prosperity and resilience. UNEP, Nairobi

United Nations (1982) World Charter for Nature. A/RES/37/7. 48th Plenary Meeting, 28 October 1982. https://undocs.org/en/A/RES/37/7. Accessed 07 June 2021

United Nations (2019) Delhi Declaration: countries agree to make 'land degradation neutrality' by 2030, a national target for action. https://news.un.org/en/story/2019/09/1046332. Accessed 07 June 2021

United Nations (2020) United Nations Coordinated Appeal Global Humanitarian Response Plan Covid-19. https://www.unocha.org/sites/unocha/files/Global-Humanitarian-Response-Plan-COVID-19.pdf. Accessed 12 Nov 2020

United Nations Conference on the Environment and Development (1992) Agenda 21. UN Doc a/CONE151/4. https://sustainabledevelopment.un.org/content/documents/Agenda21.pdf. Accessed 07 June 2021

Van den Bossche P, Zdouc W (2017) The law and policy of the World Trade Organization, 3rd edn. Cambridge University Press, Cambridge

WBGU – German Advisory Council on Global Change (2020) Rethinking land in the anthropocene: from separation to integration. WBGU, Berlin

World Economic Forum (2020) The future of nature and business. World Economic Forum, Geneva. http://www3.weforum.org/docs/WEF_The_Future_Of_Nature_And_Business_2020.pdf. Accessed 07 June 2021

WTO / World Trade Organization (1994) Agreement Establishing the World Trade Organization. https://www.wto.org/english/docs_e/legal_e/04-wto.pdf. Accessed 07 June 2021

WTO / World Trade Organization (2015) The WTO Agreements Series: Agriculture. https://www.wto.org/english/res_e/booksp_e/agric_agreement_series_2.pdf. Accessed 07 June 2021

WTO / World Trade Organization (2019) Russia - Measures Concerning Traffic in Transit: Report Of the Panel, WT/DS512/R: https://docs.wto.org/dol2fe/Pages/SS/directdoc.aspx?filename=q:/WT/DS/512R.pdf&Open=True. Accessed 07 June 2021

Zahar A (2020) Collective obligation and individual ambition in the Paris Agreement. Transnatl Environ Law 9(1):165–188

Zhou J, Dellmuth LM, Adams KM, Neset TS, von Uexkull N (2020) The geopolitics of food security: barriers to the sustainable development goal of zero hunger. Insights on Peace and Security No. 11. Stockholm International Peace Research Institute (SIPRI), Stockholm